ANNALS OF GLACIOLOGY

Proceedings of the Second Symposium on Remote Sensing
in Glaciology held at the University of Cambridge,
United Kingdom, 8-9 and 11-12 September 1986

Published by the International Glaciological Society
Cambridge CB2 1ER, England

VOL 9, 1987

ISSN = 0260-3055
ISBN = 0 946417 00 8

+

DRD⁵

5

S

Cover illustration: Multi-spectral composite of TM bands 4, 5, and 6 of quadrant 1 of Landsat scene 5034407520, recorded on 8 February 1985, shows (on the left of the image) lakes at the eastern side of Jutulstraumen, as it flows into the Fimbul Ice Shelf in the northern part of Dronning Maud Land, East Antarctica. Warm areas at lower elevations show as red, high temperatures associated with nunataks show as orange-yellow, and high clouds show as yellow-green. The image measures 90 km across. (Reproduced by permission of EOSAT.) (See Orheim and Lucchitta, p. 113.)

Text set by Beverley Baker
Printed in the Netherlands by Lochem Druk BV

87   00029

ANNALS OF GLACIOLOGY VOL 9, 1987

CONTENTS

ABSTRACTS OF PAPERS PRESENTED AT THE SYMPOSIUM BUT NOT PUBLISHED IN THIS VOLUME

ABSTRACTS OF PAPERS ACCEPTED FOR BUT NOT PRESENTED AT THE SYMPOSIUM

NOTE

Our normal policy of arranging articles alphabetically by first author
has been modified in order to meet requests for some illustrations to
appear on facing pages and in order to create a special section for
articles with colour illustrations. These articles, on pages 97 − 144,
are printed on special paper in 16-page units with 4-colour
processing. (An article that fell outside these units has its colour
illustration inserted as an appendix.)

The cost of printing these articles has been met by the authors'
organizations. We are grateful to them, and to the many other authors
whose organizations have provided financial support for their articles.

We also thank British Petroleum Exploration Company Limited for
their grant to help with the publication of this volume.

PREFACE

The 1986 Symposium on Remote Sensing in Glaciology was the second one organized by the International Glaciological Society on this topic (the first one was held in September 1974); it was held in the University Chemical Laboratory, Cambridge, England on 8–9 and 11–12 September. Nine special lectures were presented on 10 September about the development of glaciological studies during the past half century as part of the separate 50th Anniversary celebrations of the Society. Organization of the symposium and local arrangements were taken care of by the Society's headquarters office. After the Symposium, a Golden Jubilee Tour to Switzerland took place on 13–20 September, organized by the Laboratory of Hydraulics, Hydrology and Glaciology of the Federal Institute of Technology (ETH), Zürich, and co-sponsored by the Glacier Commission of the Swiss Academy of Sciences.

The International Glaciological Society is grateful to the staff of King's College for the pleasant accommodation and memorable meals served to the symposium participants in a unique setting. Special thanks are also warranted for the hospitality provided during an evening reception hosted by David J. Drewry, Director, and his friendly staff at the Scott Polar Research Institute (SPRI). On another evening, Gordon de Q. Robin, former Director of SPRI, was honoured by the Society as recipient of the prestigious Seligman Crystal for his lifetime of significant geophysical achievements in the science of glaciology.

The symposium attracted 91 participants from 17 countries. Forty-five papers were presented in 8 plenary sessions; 10 papers were given in a poster session. Seventeen abstracts included in this volume were accepted by the papers committee but the papers were not presented. The papers committee for the symposium was chaired by Richard S. Williams, Jr., and included David J. Drewry, Preben Gudmandsen, René O. Ramseier, Konrad Steffen, Gerd Wendler, and H. Jay Zwally. All papers and abstracts included in this proceedings volume were technically reviewed and edited in accordance with the usual standards of the Society. With the cheerful and meticulous guidance of the Chief Scientific Editor, Richard S. Williams, Jr., the Associate Editors, listed at the front of this volume, worked hard on the scientific editing of the papers, which were then sent to I.G.S. headquarters office for copy editing, typesetting and preparation for printing. The Chief Scientific Editor undertook more than the normal workload because of the illness of one of the editorial team. He deserves our special thanks for his efforts.

Hans Röthlisberger
President 1984–87

Photograph by K. Steffen

The week-long symposium and fiftieth anniversary festivities were most successful and memorable. Immediately following the symposium, however, Eric L. Richardson, the competent and affable House Editor for the Annals of Glaciology, became seriously ill. Although it seemed at times that he would recover, sadly, he passed away on 14 January 1987. Eric's many friends mourn his loss and offer their deepest-felt condolences to his widow, Hilda Richardson, Secretary General of the Society. He will be missed; volume 9 of this Annals of Glaciology represents his last professional effort and is dedicated to him posthumously.

The Chief Scientific Editor is especially indebted to Hilda Richardson for taking over the reins as house editor, coordinating the work on the many papers still in technical review or edit during the difficult weeks following the symposium; and to Raymond J. Adie, House Editor, Journal of Glaciology, for exceptional service as volunteer surrogate copy editor for Annals volume 9. Their sense of professionalism during a trying time was without peer.

Richard S. Williams, Jr.
Chief Scientific Editor
Annals Volume 9

Annals of Glaciology 9 1987
© International Glaciological Society

TEXTURE OF POLAR FIRN FOR REMOTE SENSING

by

R.B. Alley

(Geophysical and Polar Research Center, University of Wisconsin-Madison,
1215 W. Dayton Street, Madison, WI 53706, U.S.A.)

ABSTRACT

Knowledge of the texture of polar firn is necessary for interpretation of remotely sensed data. We find that dry polar firn is an irregularly stratified, anisotropic medium. Grains in firn may be approximated as prolate spheroids with average axial ratios as high as 1.2 or greater and with a preferred orientation of long axes clustered around the vertical. Such elongate grains are preferentially bonded near their ends into vertical columns, so that grain bonds show a preferred horizontal orientation. The grain-size distribution is similar in most firn and the normalized distribution is stationary in time, but the distribution is somewhat different in depth hoar. Fluctuations of firn properties are large near any depth, but decrease with increasing depth. With increasing depth, anisotropy of surfaces decreases, bond size relative to grain size decreases slightly, and number of bonds per grain and fraction of total grain surface in bonds increase. Grain size increases linearly with age below 2 to 5 m, but increases more rapidly in shallower firn.

1. INTRODUCTION

Most remote-sensing techniques applied to polar ice sheets sample the firn to some depth. Interpretation of data collected then requires knowledge of firn properties (density, texture, and their variations) and how the radiation being sensed interacts with those properties. For example, Zwally 1977) has shown that microwave brightness temperatures depend on grain size in firn in a strong, nonlinear manner, suggesting that knowledge of grain size, its distribution, and its depth variation (and perhaps grain shape and other factors) is necessary to interpret microwave data. Here we present some data on the texture of firn that may be of interest in formulating models for remote sensing.

2. METHODS

We measure textural quantities on thin sections of firn using standard metallographic techniques. These techniques are detailed elsewhere (e.g. Gow 1969; Underwood 1970; Kry 1975; Gubler 1978; Narita and others 1978; Alley 1986; all of these sources except Underwood [1970] also include applications of the techniques to firn and snow, and present results similar to some of those reached here). We make all measurements on photographs of thin sections viewed in reflected light, and thus on truly two-dimensional surfaces, but use transmitted, cross-polarized light to aid in identification of grain boundaries.

Many quantities can be calculated without *a priori* assumptions about firn geometry (Table I). These are based on the counting measures ρ_c and $N_i^{\,j}$. The fraction of randomly placed points that falls within grains on a section plane is denoted ρ_c, and is an unbiased estimator of the volume fraction of ice (relative density) in the bulk sample. $N_i^{\,j}$ is the number of intersections per unit length between randomly placed test lines (which may be directed) and traces of surfaces on a section plane. N_g and N_f refer to intersections with traces of grain bonds and free (ice–air) surfaces, respectively, and N^h and N^v refer to horizontally directed and vertically directed test lines on a vertical section plane, respectively. Quantities that can be calculated from these measurements include mean free paths in

individual grains (L_g), pores (L_p) and solid regions of ice (L_s), and mean spacings along test lines between centers of grains (l_g), pores (l_p), and ice regions (l_s); on vertical sections these quantities can be determined in both vertical (L^v, l^v) and horizontal (L^h, l^h) directions. We can also calculate the total area per unit volume of ice–ice contact (S_g) and ice–air contact (S_f), the fraction of surface on the average grain involved in bonds (β), and the non-random fractions of total ice–ice surface oriented horizontally (ω_g) and ice–air surface oriented vertically (ω_f; the anisotropy parameters, ω, are the differences between the surfaces encountered by vertical and horizontal traverses on a section normalized by the total surface present, and $\omega_g=\omega_f=0$ if all surfaces are oriented randomly). Equations are given in Table I, and are derived in Underwood (1970).

Other textural quantities can be calculated only if certain reasonable but unsubstantiated geometric assumptions are made. Some such quantities are the grain size (A,

TABLE I. STEREOLOGIC QUANTITIES CALCULATED FROM SIMPLE COUNTING MEASUREMENTS FOR SAMPLE FROM SITE 4530, ICE STREAM A, WEST ANTARCTICA. (All quantities are defined in methods section of text.)

		Vert. Sect.		Horz. Sect.
Lengths (mm)		Horz.	Vert.	
L_g	$\dfrac{2\rho_c}{N_f+2N_g}$	0.57±.02	0.71±.04	0.58±.02
L_p	$\dfrac{2(1-\rho_c)}{N_f}$	0.75±.03	1.18±.06	0.85±.03
L_s	$\dfrac{2\rho_c}{N_f}$	0.78±.03	1.22±.06	0.75±.03
l_g	$\dfrac{2}{N_f+2N_g}$	1.13±.03	1.40±.06	1.23±.02
l_p	$\dfrac{2}{N_f}$	1.53±.04	2.40±.10	1.60±.03
l_s	$\dfrac{2}{N_f}$	1.53±.04	2.40±.10	1.60±.03
Densities (relative)				
ρ_c			0.51±.01	0.47±.02
Areas $(S_g, S_f$ in mm^2/mm^3)				
β	$\dfrac{2S_g}{2S_g+S_f}$		0.31±.01	0.23±.01
Isotropic firn				
S_g	$2N_g$			0.38±.02
S_f	$2N_f$			2.49±.04
Anisotropic firn				
S_g	$N_g^h+N_g^v$		0.53±.03	
S_f	$\dfrac{\pi}{2}N_f^h+\left(2-\dfrac{\pi}{2}\right)N_f^v$		2.41±.05	
ω_g	$\dfrac{N_g^v-N_g^h}{N_g^v+N_g^h}$		0.12±.06	
ω_f	$\dfrac{N_f^h-N_f^v}{N_f^h+\left(\dfrac{4}{\pi}-1\right)N_f^v}$		0.31±.03	

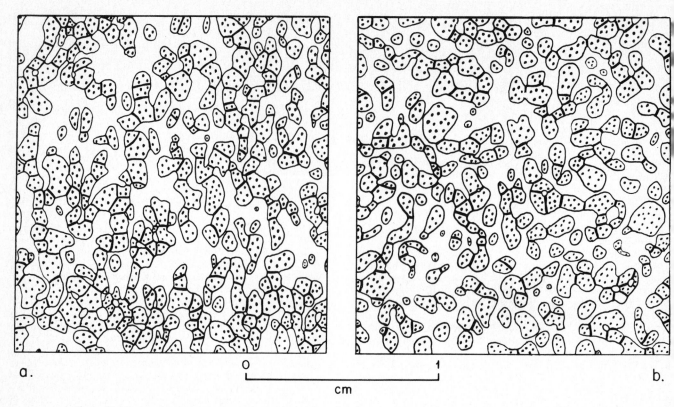

Fig.1. Tracings of portions of (a) vertical and (b) horizontal thin sections from 2.7 m deep at site 4530, ice stream A, West Antarctica.

defined here as 3/2 the average grain cross-sectional area on a plane of section; Alley 1986), the sphericity (ϕ, which measures how closely a given grain cross-section approaches a circle and is defined here as the ratio of the radii of the largest inscribed circle in a grain to the smallest circumscribed circle about that grain; Alley and others 1982), the relative bond size (α, defined here as the ratio of the radii of the average bond to the average grain; Alley 1986) and the coordination number or number of bonds per grain (n_3; Alley 1986). The assumptions involved in these calculations are discussed by Underwood (1970), Kry (1975), and Alley (1986), among others.

3. RESULTS

Consider Fig.1, which shows portions of vertical and horizontal thin sections from about 2.7 m depth at site 4530 on the downstream part of ice stream A, West Antarctica. This is a warm (≈ -25 to $-30\,^\circ$C), low accumulation (≈ 0.1 m/a water; Bull 1971) site. The samples were cut from a fine-grained layer of bulk density 472 kg m^{-3}; however, the layer did not appear perfectly homogeneous. The horizontal section was cut from just above the vertical.

It is immediately evident from inspection that the firn is isotropic in a horizontal plane but is anisotropic and inhomogeneous vertically; compared to other samples, the anisotropy here is strong but not atypical. The thin layer near the bottom of the vertical section was visible on the firn core as a resistant crust.

Data collected from these sections are shown in Tables I and II. All data from the vertical section were taken above the thin crust. If the vertical and horizontal sections sampled the same homogeneous firn, then the point-count densities, total specific surfaces, and intercept lengths in the horizontal direction would be the same for the two sections. However, both stratigraphic inspection of the firn core and careful examination of Fig.1 show that vertical inhomogeneity exists within the layer studied; thus, differences between the horizontal and vertical sections are expected. Errors arising from counting are given in the tables; errors arising from violation of assumptions cannot be evaluated accurately.

Orientation data for these sections are shown in Fig.2. It should be evident that the grains have no preferred

TABLE II. STEREOLOGIC QUANTITIES REQUIRING GEOMETRIC ASSUMPTIONS, FOR SAMPLE FROM SITE 4530, ICE STREAM A, WEST ANTARCTICA. (Variables are defined in methods section of text. All assumptions may be violated slightly, so error limits cannot be constructed exactly. We expect that all values are within ±20% and may be within ±10 or less.)

	Vert. Sect.	Horz. Sect.
A (mm^2)	0.85	0.70
ϕ	0.57	0.65
α	0.57	0.58
n_3	4.6	3.4

orientation in the horizontal plane but a strong vertical orientation, and that grains tend to be joined at their ends by horizontal bonds to form vertical columns. One possible model of a grain in firn is a prolate spheroid with axial ratio b. If we assume that all grains have the same value of b and have their long axes oriented vertically, then $b = L_g^v/L_g^h = 1.2$ (see Table I). Because orientations of long axes of grains show a distribution, $b > 1.2$ ($b \approx 1.4$ may be a good estimate, but we cannot demonstrate this rigorously).

The normalized grain-size distribution for the horizontal sample, calculated using the Saltykov area method (Underwood 1970: 123-126) is shown in Fig.3. The Saltykov method, which assumes that grains are spherical, may be reasonably accurate for large grains but can become erratic or nonphysical for the smallest grain sizes; the smoothed curve in Fig.3 was drawn up so as to be physically realistic. This distribution is quite similar to the steady distribution expected by Hillert (1965) for normal grain growth, with the maximum grain radius equal to about twice the average grain radius, and also is similar to distributions we see in other firn samples. Although further work is required, our data indicate that the grain-size distribution is similar to Fig.3 for most firn in the upper 10 m, with the exceptions that depth hoar tends to show a stronger peak at smaller grain sizes and that distributions in the top 0.5 m tend to be somewhat variable. (Results for depth hoar also may be less reliable than for other firn

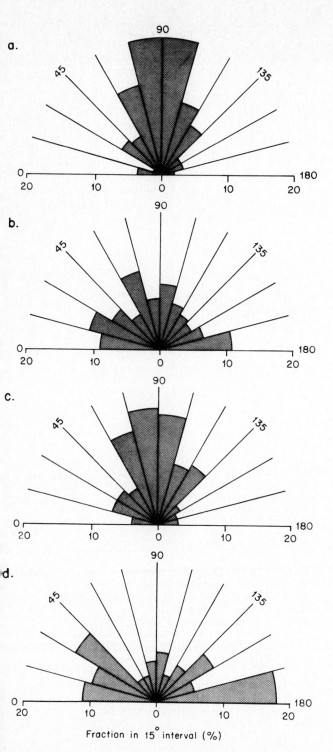

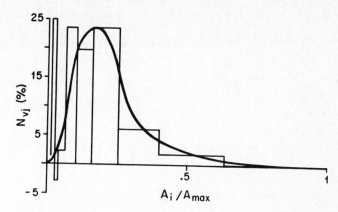

Fig.3. Normalized grain-size distribution for horizontal section shown in Fig.1, treating grains as spheres. N_{vj} is the fraction of grains per unit volume having the maximum cross-sectional area, A_i, divided by the maximum cross-sectional area of the largest grain in the sample, A_{max}.

collected from a core. Points to notice especially include: 1) grain size increases linearly with age below about 2.5 m (the regression line for data from 5–40 m is shown in the figure; see also Gow 1969; Alley and others 1986[a], [b]); but increase is more rapid in shallow firn where vapor transport down temperature gradients dominates grain growth (Colbeck 1983); 2) vertical anisotropy forms at shallow depth and then decreases steadily below about 2–3 m to essential isotropy below about 15 m; 3) density variations and other variations typically (but not always) decrease with increasing depth. Also, coordination number increases with depth whereas the relative size of grain bonds is large near the surface and decreases slightly with depth.

4. CONCLUSIONS

Polar firn is an irregularly stratified, inhomogeneous, anisotropic material. Measurements of density, specific surfaces and their anisotropy, and mean free paths in ice, individual grains, and air in specified directions can be made from thin sections with high accuracy and with no or few untestable assumptions. Grain size, grain-size distribution, relative bond size, coordination number, grain shape, and other parameters also can be calculated from measurements on thin sections, but all require assumptions regarding geometry that cannot be checked rigorously.

Grains in firn typically are elongated vertically and align vertically in columns. Grains may be modeled as prolate spheroids in which the axial ratio can vary from nearly 1 (for deeper firn or for firn in high-accumulation areas where little time is allowed for development of anisotropy in near-surface regions) to perhaps 1.4 or more in strongly anisotropic firn. Long axes of grains cluster about the vertical but show some distribution.

Grain size increases linearly with age (and almost linearly with depth) below about 2-5 m, but increases more rapidly in shallower firn. Anisotropy of free surfaces decreases with increasing depth to about zero below about 15 m. The fraction of grain surface involved in bonds and the coordination number increase rapidly with depth, but the bond size relative to grain size decreases slightly with increasing depth. Variations in these parameters near any depth are quite large in shallow firn, but tend to decrease slightly with increasing depth.

Collection of data such as these is a slow process, and other sorts of data could be collected. Optimization of data collection will require identification of those parameters that are most important for understanding remotely sensed data.

ACKNOWLEDGEMENTS

This work was supported in part by the U.S. National Science Foundation under grants DPP-8315777 and DPP-8412404. I thank the Polar Ice Coring Office, Lincoln, Nebraska for core recovery, S Shabtaie for access to samples, S C Colbeck and S T Rooney for helpful

Fig.2. Orientation data for sections shown in Fig.1; 90° is vertical on vertical sections, 0° and 180° are horizontal, and full scale is 20% of grains on a section in a 15° interval. a) Long axes of grains, vertical section. b) Long axes of grains, horizontal section. c) Angles between centers of grains in contact, vertical section. d) Orientations of grain bonds, vertical section.

because of the irregular shapes of depth-hoar grains.) To accord with other workers (Gow 1969; Duval and Lorius 1980) we also report grain size as an average cross-sectional area in Table II and Fig.4 (see Methods section, above).

We have been studying the variability and diagenesis of firn in some detail on the ridge between ice streams B and C on the Siple Coast of West Antarctica, and some of our data are shown in Fig.4. Ridge BC is similar in accumulation rate ($\approx$0.08 m/a water) and temperature ($\approx$−26.5°C) to site 4530. Data in Fig.4 from the upper 2 m are from a detailed pit study, and deeper data were

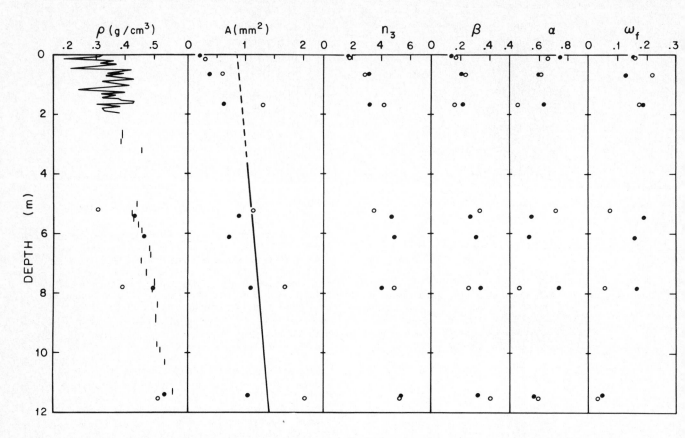

Fig.4. Textural data for firn from ridge BC, West Antarctica. Shown are density (ρ), grain size (A), coordination number (n_3), fraction of grain surface in bonds (β), relative bond size (α), and anistropy of ice–air surfaces (ω_f). Errors are similar to, or slightly larger than, those in Tables I and II. Open circles are depth hoar; solid circles are typical and fine-grained firn. Density data include a pit profile in the upper two meters (continuous line), measurements made on long core sections (vertical bars), and measurements made by point counting thin sections (open and solid circles). The regression line for A is fitted to data from 5–40 m deep, and is included to emphasize the rapid rate of increase of grain size in near-surface firn.

comments, and S H Smith for figure preparation. This is contribution number 463 of the Geophysical and Polar Research Center, University of Wisconsin-Madison.

REFERENCES

Alley R B 1986 Three-dimensional coordination number from two-dimensional measurements: a new method. *Journal of Glaciology* 32(112): 391–396

Alley R B, Bolzan J F, Whillans I M 1982 Polar firn densification and grain growth. *Annals of Glaciology* 3: 7–11

Alley R B, Perepezko J H, Bentley C R 1986[a] Grain growth in polar ice: I. Theory. *Journal of Glaciology* 32(112): 415–424

Alley R B, Perepezko J H, Bentley C R 1986[b] Grain growth in polar ice: II. Application. *Journal of Glaciology* 32(112): 425–433

Bull C 1971 Snow accumulation in Antarctica. *In* Quam L O (ed) *Research in the Antarctic. A symposium presented at the Dallas meeting of the American Association for the Advancement of Science – December 1968.* Washington, DC, American Association for the Advancement of Science: 367-421 (Publication 93)

Colbeck S C 1983 Theory of metamorphism of dry snow. *Journal of Geophysical Research* 88(C9): 5475-5482

Duval P, Lorius C 1980 Crystal size and climatic record down to the last ice age from Antarctic ice. *Earth and Planetary Science Letters* 48(1): 59-64

Gow A J 1969 On the rates of growth of grains and crystals in South Polar firn. *Journal of Glaciology* 8(53): 241-252

Gubler H 1978 Determination of the mean number of bonds per snow grain and of the dependence of the tensile strength of snow on stereological parameters. *Journal of Glaciology* 20(83): 329-341

Hillert M 1965 On the theory of normal and abnormal grain growth. *Acta Metallurgica* 13(3): 227-238

Kry P R 1975 Quantitative stereological analysis of grain bonds in snow. *Journal of Glaciology* 14(72): 467-477

Narita H, Maeno N, Nakawo M 1978 Structural characteristics of firn and ice cores drilled at Mizuho Station, East Antarctica. *Memoirs of National Institute of Polar Research*. Special Issue 10: 48-61

Underwood E E 1970 *Quantitative stereology.* Reading, MA, Addison-Wesley Publishing

Zwally H J 1977 Microwave emissivity and accumulation rate of polar firn. *Journal of Glaciology* 18(79): 195-215

Annals of Glaciology 9 1987
© International Glaciological Society

INTERNAL REFLECTING HORIZONS IN SPITSBERGEN GLACIERS

by

J. L. Bamber

(Scott Polar Research Institute, University of Cambridge, Cambridge CB2 1ER, U.K.)

ABSTRACT

A single pronounced internal reflecting horizon has been observed on radio echo-sounding from over 30 glaciers in Spitsbergen. They are often present along the entire length of the glacier, remaining at a fairly constant depth (100–200 m) below the ice surface. Echo-strength data from radio echo-sounding have been used to obtain reflection coefficients, for these horizons, of between −15 and −25 dB. Combined with results of ice-core studies, the possible causes of this internal layer are investigated. The presence of water is found to be the most likely explanation, indicating the existence, at depth, of a layer of temperate ice.

INTRODUCTION

A striking feature observed on radio echo-sounding (RES) records from many of the thicker (>200 m) glaciers in Spitsbergen is the existence of a single continuous internal reflecting horizon (IRH) ranging in depth from 100 to 200 m below the surface and persisting for up to 20 km (Fig.1). It is found in all parts of the glacier but has not been observed to reach the surface at the snout.

In 1983, the Scott Polar Research Institute (SPRI) undertook an extensive airborne RES programme in Svalbard (Drewry and Liestøl 1985). 60 MHz data were recorded digitally, affording precise measurement of echo strength and hence allowing the calculation of reflection coefficients (RCs). Thirty-four glaciers and six ice caps in Spitsbergen were sounded and a comprehensive network of flight lines flown over Nordaustlandet. Several Soviet RES programmes have also been undertaken using equipment operating at higher frequencies (620 or 440 MHz) (Macheret and others 1984). Previously in 1980, 33 glaciers had been sounded throughout Spitsbergen using analogue-recording techniques, the equipment being mounted in a helicopter (Dowdeswell and others 1984[a]).

Svalbard glaciers can be difficult to sound due to high dielectric absorption (much of the ice is close to the pressure melting-point), internal scattering, and thicknesses of up to 560 m, and has on occasion led to misinterpretation of reflections (Dowdeswell and others 1984[b]).

The glaciers of Svalbard are classed as sub-polar; they are not temperate throughout but may have zones close to or at the pressure melting-point. 10 m temperatures measured in the ablation zone on Austfonna (−8.6°C) are significantly lower than those recorded about 300 m higher in elevation (above the firn line), which range from −1°C (personal communication from A. Semb) to −6.4°C (Dowdeswell unpublished). This variability is due to the large amount of latent heat absorbed from melt water at depth in the firn in the accumulation zone. The thermal regime of these glaciers is thus complex and many have parts close to the pressure melting-point with water present at the bed. The size of the ice masses in Svalbard is also very varied, ranging from small valley glaciers less than 10 km long and 100 m thick to the ice caps on Nordaustlandet covering an area of 10 600 km².

DATA RECORDING IN 1983

The RES data were recorded as the geometric mean of eight individual wave forms, each comprising 256 samples taken at 100 ns intervals.

The ice-surface elevation changes were tracked automatically but for the weaker, noisier, and more intermittent bedrock and internal signals this was not possible. A differentiated continuous display of the digital data (known as a "Z"-scope display) was manually digitized. The errors of depth measurement incurred are negligible in comparison to the digital sampling length (equivalent to 8.4 m in ice, which defines the maximum resolution of an individual pulse).

Absolute elevations were calculated using terrain clearance from the radar sounder and a pressure altitude calibrated by overflying areas of open water. Interpolation between pulses enabled the *relative* surface elevation to be found to within 2 m (Drewry and Liestøl 1985). Ice

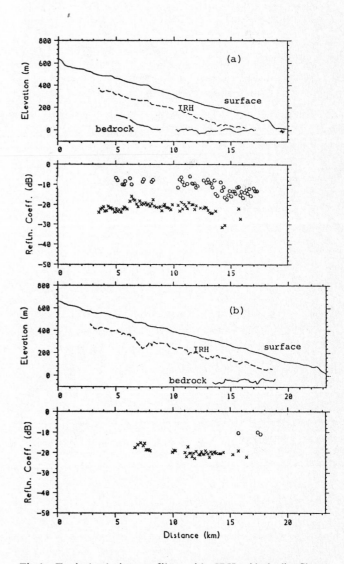

Fig.1. Typical glacier profiles with IRHs (dashed). Shown below are the respective RCs for the bedrock (circles) and IRH (crosses). (a) Uversbreen; (b) Kongsvegan; note loss of bedrock echo towards the accumulation zone.

thicknesses were calculated using a mean value for the permittivity of ice of 3.17 (equivalent to a velocity of 168 m/µs).

Navigation was carried out using a combination of a Tactical Airborne Navigation System (TANS) and visual landmarks recorded to ±1 s (equivalent to 60 m along track). Depending on the nature of the landmark, an error of, at maximum, 6 s (360 m) could also be introduced. Across-track errors are more variable but are nowhere more than 800 m and typically <400 m. Wherever possible, flight tracks followed the centre lines of the glaciers.

GLACIER PROFILES – GENERAL OBSERVATIONS

Two typical glacier profiles with very distinct IRHs are shown in Fig.1. They bear a stronger relationship to the surface than the bedrock, lying at a fairly uniform depth. They do not, however, represent a "smooth" surface and

Soviet data collected at 440 MHz indicate that they may not be continuous on a scale of centimetres.

This is in strong contrast to the sedimentary layer observed in Antarctic ice (e.g. Millar unpublished) which appear to follow the bottom topography more closely. In profile (b) the bedrock is only present in the ablation zone (due to a combination of signal obscuration by scatter echoes and higher total attenuation), another common feature. Bedrock RCs (plotted as circles) are usually between 5 and 15 dB stronger than the IRH values. The RC data relating to the IRHs are discussed in detail in the next section. On a qualitative basis, the radar records show markedly increased internal scattering below the IRH, which on several glaciers has led to the obscuration of the bottom echo and which is a common feature of RES data from temperate ice masses.

Of the 34 glaciers sounded in 1983, six had no resolvable returns (and showed signs of heavy scattering).

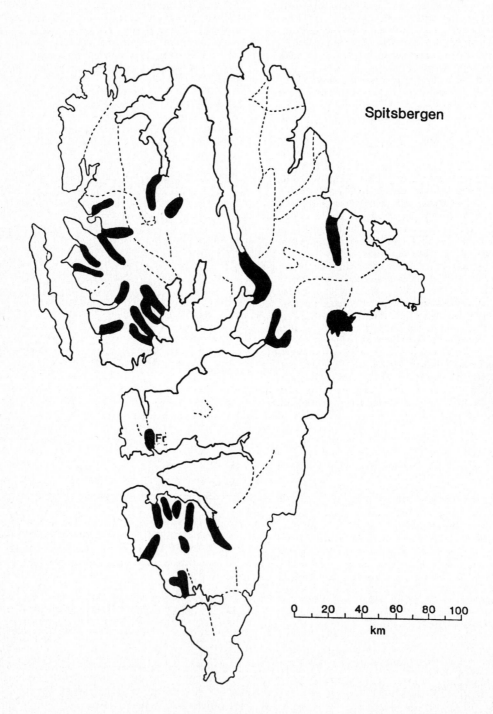

Fig.2. Geographical distribution of *all* IRHS observed (solid), including Soviet and SPRI 1980 data (Fr, Fridtjovbreen). The dashed lines show track sounded but absence of an IRH.

nd from the remaining 28 only three had no IRH present.
 1980, 38 glaciers were sounded, 20 of which had resol-
able echoes and of these ten had distinct IRHs. Although
his latter data set has a greater proportion of glaciers with
o IRH, they are generally thinner (<200 m), conforming to
he 1983 results. The geographical distribution of IRHs is
hown in Fig.2 and only indicates a limited pattern. One of
he coldest ice masses of the archipelago, Austfonna,
ossesses no IRH and fewer IRHs are observed on the
orth-eastern part of Spitsbergen, a region of relatively
igh-altitude mountain terrain covered by a thin ice layer
typically between 100 and 200 m thick). The dashed lines
epresent areas sounded but with an absence of IRHs. They
ndicate that the regional distribution of glaciers with IRHs
s not due solely to the areal coverage of flight lines,
hough the greater occurrence of IRHs on the western side
f the island may be caused by this.

REFLECTION COEFFICIENTS AND ICE-CORE ANALYSIS

The major consideration in obtaining experimental
alues of RCs from RES data is the estimation of the
dielectric absorption, B. This is a function of the complex
dielectric permittivity ϵ^*, where $\epsilon^* = \epsilon' - i\epsilon''$ and ϵ' is
he relative permittivity and ϵ'' is the dielectric loss factor.
* is in turn dependent primarily upon the ice temperature
hough the presence of impurities and/or water may also
ave a significant influence (Glen and Paren 1975). (Note
hat, for convenience, the asterisk will be dropped in future
eferences to the complex permittivity.)

For the majority of Spitsbergen glaciers the temperature
egime is poorly known. There is considerable uncertainty
n extrapolating results from one glacier to another
Zagorodnov and Zotikov 1981). Consequently, mean
absorption in the ice was calculated, where possible, from
he variation of attenuation with depth of the bedrock echo
Fig.3). This assumes that both the RC and ice properties
re invariant over the region covered. The justification for
his approximation can be assessed by the spread in points
bout the fitted regression line. Data from glaciers with a
orrelation coefficient of <0.8 were rejected. Values for B
vere typically between 4.0 and 4.5 dB/100 m. Attenuation
ften showed a poor correlation to ice thickness and for
hese glaciers a linear temperature profile, with surface
emperature $T_s = -4°C$ and the bed at the pressure
melting-point, was used. Data on the variation with
emperature of the absorption were taken from Jiracek
1965). The greatest errors in this approach are not due to
he uncertainty in temperature profile (changing T_s from
-4° to -1°C, for a 250 m column of ice, alters the total
bsorption by only 3.5 dB) but in the value of B assumed

for a given temperature. Errors in the total dielectric
absorption (and hence RCs) are estimated as no more than
±2 dB/100 m ice thickness.

RCs were calculated using the radar equation which
is, in log form

$$P_r = P_t + 20 \log \frac{\lambda G}{8\pi\left(H + z/n\right)} + RC - \frac{2BZ}{100} \quad (1)$$

where λ is free-space wavelength, G is forward antenna
gain, H is terrain clearance, Z is ice thickness, n is
refractive index of ice ($\equiv\sqrt{\epsilon'}$), B is the mean dielectric
absorption/100 m, P_r and P_t are receiver and transmitter
power in dB. IRH values of RCs lie in the range −15 to
−30 dB with an average value of approximately −20 dB;
errors depend on depth but are generally <5 dB.

INTERPRETATION

Reflection of E−M radiation from any surface implies
a change in dielectric properties at the point of reflection.

$$RC = \left| \frac{n_1 - n_2}{n_1 + n_2} \right|^2 \quad \dots \quad (2)$$

where

$$\eta = \left[\frac{\mu_0}{\epsilon_0\epsilon' + \frac{\sigma'}{i\omega}} \right]^{1/2} \quad \dots \quad (3)$$

μ_0 is permeability, σ' is real conductivity, ϵ_0 is permittivity
of free space, and ω is the angular frequency of radio
waves. In ice, a change in σ' or ϵ' can be caused by a
variation in density, impurities, temperature, crystal size and
orientation, or from a layer of another material such as
moraine. The uniformity in depth to the IRH virtually
precludes the possibility that it might be of a depositional
origin. If the most likely cause of the IRHs is due to a
change in water content, it is necessary to describe the
dielectric properties of water-laden ice. Such a two-phase
system, with components of differing σ' and ϵ', causes
interfacial polarization known as Maxwell–Wagner
polarization and the strength of this is dependent upon the
shape and size of the dispersoid (water). In the next section
it is suggested that the distribution of water is likely to be
highly anisotropic, being found in channels and conduits
rather than on an intragranular scale. Consequently, a
generalized mixture formula (known as the Looyenga
equation), which has been successfully applied to firn of
widely varying density (Glen and Paren 1975), was used:

$$\epsilon_m^{1/3} = \epsilon_i^{1/3} + V_w\left(\epsilon_w^{1/3} - \epsilon_i^{1/3}\right) \quad (4)$$

where V_w is fractional water content and $\epsilon_w' = 86$, $\epsilon_i' = 3.17$ (Smith unpublished) and $\epsilon_{air} = 1.0$.

Values for the RC of a boundary between air-filled
and water-filled ice have been calculated for water volumes
between 1 and 6% (Fig.4). The effect of water conductivity
is negligible. It can be seen that a water content of
approximately 5% is necessary to explain the observed RCs.
This does not represent the *mean* water content below the
IRH, but only that which is observed close to the
"wet"/"dry" boundary. It should be noted that there is a
two-fold increase predicted, in the water content, for an
increase of the RC from −25 to −20 dB. Given the
systematic errors inherent in calculating experimental values
of the RC, it is not possible to define a precise water
volume at the boundary, and even less so below it. What
can be established is that the average water content, at the
level of the IRH, is >3% and that this is considerably more
than the values typically found in temperate glaciers
(Raymond and Harrison 1975; Vallon and others 1976),
which range from 0.1% to 1% at maximum. In the next
section it is suggested that the primary influence on the
level of the IRH is the subglacial water pressure and, if the
ice above it is less permeable than below, then it is

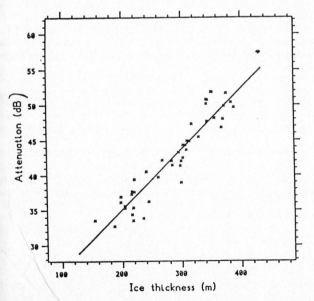

Fig.3. Graph of total absorption by the ice vs depth. Data
from the bedrock echoes from a glacier with an IRH
(Borebreen). B = 4.2 dB/100 m in this example.

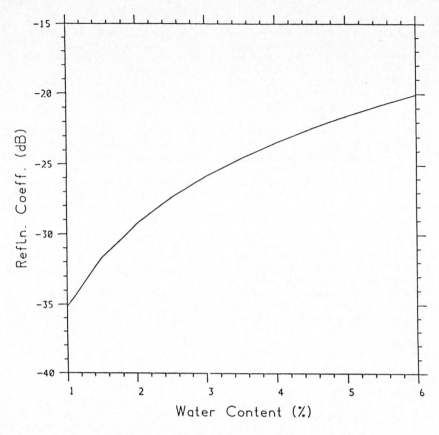

Fig.4. Theoretical reflection coefficients for a boundary between "dry" ice and ice with a given water content, calculated using the Looyenga mixture formula.

conceivable that the englacial water content will be a maximum near the interface.

A radio echo-logging experiment (Macheret and others 1984) on Fridtjovbreen (Fig.2) revealed a dielectric discontinuity at a depth coincident with an observed IRH. Above the IRH ϵ = 3.035 and below ϵ = 4.125. The authors calculated that such a change could be produced by a water content (f) of 3.8%. This is similar to the values predicted by the RCs but their measurements only went to a depth of about 30 m below the dielectric boundary and it is consequently difficult to deduce what the bulk water content might be. It is also difficult to envisage a mechanism by which such a high water content may be maintained throughout the lower layer.

Other causes for the IRHs have been considered but do not satisfactorily explain the observations from the IRHs. Paren and Robin (1975) included discussion of RC strengths arising from a density fluctuation in polar ice sheets and obtained a maximum RC of −51 dB for a discontinuity at a depth of 100 m. Even taking the largest variation possible between ice of density 830 kg/m^{-3} and 910 kg/m^{-3} gives an RC too low to explain the observed values. It also seems unlikely that such a layer would persist along the entire length of the glacier from accumulation to ablation zone. Dirt bands or other depositional discontinuities have even lower RCs (Millar unpublished). It is not only the RCs but the striking uniformity in depth and extent of the IRHs from Spitsbergen that reduces the possibility that they are sedimentary in origin.

A water *layer* of approximately 15 cm thickness or more could give similar values of RC but such a layer could not extend unbroken for 15 km or more and would still not explain the radio echo-logging results from Fridtjovbreen (Macheret and others 1984).

SIGNIFICANCE OF A WATER-SATURATED LAYER

A reasonable hypothesis for the IRHs is water saturation of the glacier below the IRH and it has several important implications. Interpretation of RES data possessing an IRH immediately indicates the general thermal regime of the glacier. The presence of water and hence the possibility of sliding at the bed is suggested and is hence useful in interpreting the dynamic properties. Duval (1977), for instance, has shown that there is an order of magnitude increase in the strain-rate for ice with V_w = 0.01 relative to that with a negligible concentration. Here lies a system with possible positive feed-back. Water production is directly proportional to the strain-rate, $\dot{e}$

$$q = \frac{2\dot{e}\tau}{L}, \qquad (5)$$

L is latent heat of fusion of ice, τ is shear stress, and q is water production due to strain heating.

This does not lead to an instability related to temperature as considered by Clarke and others (1977) where rapid warming and the attainment of pressure melting at the bed are the relevant factors, but is nonetheless a phenomenon regulated by heat production. The difficulty in developing this model is in understanding how drainage and permeability are related to the water content; at volume fractions as high as 1%, Nye and Frank's (1973) model of intragranular flow has been shown to be inapplicable (Lliboutry 1976).

The presence of water will affect the crystal fabric reducing strain hardening and crystal size (Lliboutry 1976). If impurities are concentrated at grain boundaries, as has been suggested by Wolff and Paren (1984), then flushing may have an important influence on the purity and hence dielectric properties of the ice.

FACTORS AFFECTING THE LEVEL OF THE IRH

Classical ground-water flow principles (namely, Darcy's law and Dupuit's assumptions which are that, for small slopes, the hydraulic head is equal to the surface gradient and that stream lines are parallel to the surface) can be used to find the piezometric surface of an unconfined aquifer based on basic well-known boundary conditions

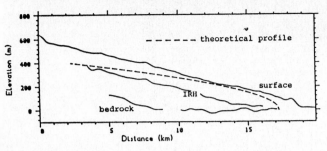

Fig.5. Comparison of the theoretical piezometric surface, as calculated from basic ground-water flow principles, with the IRH in Uversbreen.

(Fetter 1980). Such a system might represent ice possessing water at the grain boundaries as suggested by Nye and Frank (1973). This model gives rise to a parabolic relationship between hydraulic head and distance down-glacier, which is depressed by including a term for drainage at the bed. In Fig.5 the real profile is compared with one calculated from the aforementioned assumptions. It can be seen that poor agreement is obtained, the theoretical profile having a convex-upward profile.

RÖTHLISBERGER CHANNELS

Another model, considered by Röthlisberger (1972), is that the water flows in a small number of englacial or subglacial channels. The piezometric surface is now defined by the water pressure within these channels. The equation used was

$$\frac{dp}{dx} = \frac{D^{8/11}K^{-6/11}(nA)^{-8n/11}Q^{-2/11}(P - \frac{\rho_i}{\rho_w}p)^{8n/11}}{\left[1 - \left[\frac{1}{\rho_w g}D^{8/11}K^{-6/11}(nA)^{-8n/11}Q^{-2/11}(P - \frac{\rho_i}{\rho_w}p)^{8n/11}\right]^2\right]^{1/2}} - \rho_w g \tan \beta \dots \quad (6)$$

where $P = \rho_i gH$ (where H is ice thickness and ρ_i is ice density), $\tan \beta$ is bedrock gradient, $D = 3.63 \times 10^{10}$ (N m^{-2})$^{11/8}$ m$^{-3/8}$ (a physical constant), K is roughness coefficient of the channel, A is flow-law constant, and n is

flow-law index (assumed to be 3 in this case). It was integrated numerically to obtain a profile of the piezometric surface along the centre line of the glacier. This equation describes the water pressure, p, within a gradient conduit (i.e. one which follows the hydraulic grade line) and was used because it allows a continuous surface to be calculated, even for negative bedrock slopes, using the real surface and bedrock profiles. These were converted into an algebraic form using cubic spline curves, so that the integral could be evaluated at all points. The flow rate, Q, was estimated by calculating the water produced by strain heating in the bottom layer and bottom melting due to the geothermal heat flux. It was assumed to vary linearly, reaching a maximum at the snout.

Results of the modelling procedure, for two glaciers ((a) Tunabreen, (b) Borebreen) possessing IRHs are shown in Fig.6. For curve (1) A = 4 × 10^7 Pa s$^{1/3}$, K = 100 m$^{1/3}$ s^{-1}; (2) A = 6 × 10^7 Pa s$^{1/3}$, K = 100 m$^{1/3}$ s^{-1}, (3) A = 6 × 10^7 Pa s$^{1/3}$, K = 200 m$^{1/3}$ s^{-1}.

It can be seen that the general form of the IRHs is well described by this model. It is not suggested that the values of A or K represent those to be found within the glaciers or that flow will be in a single channel at the hydraulic grade line, but the analysis does indicate that the level of the IRH, along the glacier centre line, has a similar form to that described by a Röthlisberger channel. Physically, such a model implies that the ice has a large-scale permeability below the IRH, via conduits and channels anisotropically distributed within the glacier. These water bodies represent discrete radar scatterers (hence the heavy internal scattering observed below the IRH) as opposed to a uniform distribution of small-scale veins and lenses existing at grain boundaries (Nye and Frank 1973). The former model is supported by RES data collected by the Soviets (personal communication from Y.Y. Macheret) at

440 MHz, where on some glaciers the IRH appears to be composed of an array of individual scattering centres which is continuous on a kilometre scale but broken on the metre scale.

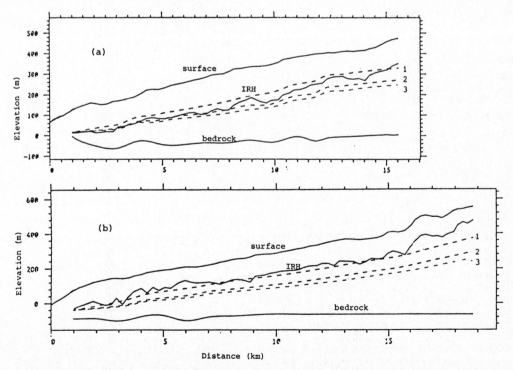

Fig.6. Comparison between the hydraulic grade line (dashed curves) for a Röthlisberger channel and the observed IRHs (solid line) on (a) Tunabreen and (b) Borebreen. For curve (1) A = 4 × 10^7 Pa s$^{1/3}$, K = 100 m$^{1/3}$ s^{-1}; (2) A = 6 × 10^7 Pa s$^{1/3}$, K = 100 m$^{1/3}$ s^{-1}; (3) A = 6 × 10^7 Pa s$^{1/3}$ K = 200 m$^{1/3}$ s^{-1}.

THERMAL REGULATION

One more factor that may have an important influence on the level of the IRH is heat conduction across the surface. Unlike tenperate glaciers, 10 m temperatures in Spitsbergen ice can be several degrees below $0\,°C$ (Dowdeswell unpublished) and it is possible that the IRH may reflect the pressure-melting isotherm. A simple calculation of the temperature gradient needed, to have a significant influence on the water level, suggests that this effect is of secondary importance.

The heat involved in freezing 3% of water is 9×10^6 J m^{-3}. Assuming a temperature gradient of $0.1\,°C$ m^{-1} at the interface, the heat flux across the boundary is 0.21 W m^{-2}. With these values, it would take about 496 d to freeze a layer 1 m thick. The water pressure, and hence water level, in temperate glaciers varies on a much larger scale than this on a seasonal basis. If the same is true for Spitsbergen glaciers, it seems unlikely that thermal regulation will be as important as water pressure. The small-scale roughness of the IRH supports this conclusion, as a thermally regulated boundary would be expected to be smooth on a 10 m scale. Although heat conduction across the boundary may not be important in defining its level, the fact that the overlying ice is cold, and possibly impermeable as a consequence, probably is important.

CONCLUSIONS

The most likely cause for the observed IRHs is the presence of a water-saturated zone. This hypothesis is supported by data from RES and bore-hole measurements. The concept of a two-layer glacier is thus invoked, an upper non-temperate layer and a lower temperate zone with potentially different properties which affect the dielectric and dynamic nature of the ice. Numerical modelling studies indicate that the most likely factor influencing the level of the IRH is the subglacial water pressure.

ACKNOWLEDGEMENTS

The author would like to thank Drs D J Drewry and J A Dowdeswell for their helpful advice and comments. Funding was through a UK NERC grant GR3/4463 and a NERC studentship to the author.

REFERENCES

Clarke G K C, Nitsan U, Paterson W S B 1977 Strain heating and creep instability in glaciers and ice sheets. *Reviews of Geophysics and Space Physics* 15(2): 235–247

Dowdeswell J A Unpublished Remote sensing studies of Svalbard glaciers. (PhD thesis, University of Cambridge, 1984)

Dowdeswell J A, Drewry D J, Liestøl O, Orheim O 1984[a] Airborne radio echo sounding of sub-polar glaciers in Spitsbergen. *Norsk Polarinstitutt. Skrifter* 182

Dowdeswell J A, Drewry D J, Liestøl O, Orheim O 1984[b] Radio-echo sounding of Spitsbergen glaciers: problems in the interpretation of layer and bottom returns. *Journal of Glaciology* 30(104): 16–21

Drewry D J, Liestøl O 1985 Glaciological investigations of surging ice caps in Nordaustlandet, Svalbard, 1983. *Polar Record* 22(139): 359–378

Duval P 1977 The role of the water content on the creep rate of polycrystalline ice. *International Association of Hydrological Sciences Publication* 118 (*General Assembly of Grenoble 1975 – Isotopes and Impurities in Snow and Ice*): 263–271

Fetter C W Jr 1980 *Applied hydrogeology*. Columbus, OH, Charles E. Merrill

Glen J W, Paren J G 1975 The electrical properties of snow and ice. *Journal of Glaciology* 15(73): 15–38

Jiracek G R Unpublished Radio echo sounding of Antarctic ice. (MSc thesis, University of Wisconsin, 1965)

Lliboutry L 1976 Physical processes in temperate glaciers. *Journal of Glaciology* 16(74): 151–158

Macheret Yu Ya, Vasilenko Ye V, Gromyko A N, Zhuravlev A B 1984 Radiolokatsionnyy karotazh skvazhiny na lednike Frit'of, Shpitsbergen [Radio echo logging of the bore hole on the Fridtjov glacier, Spitsbergen]. *Materialy Glyatsiologicheskikh Issledovaniy. Khronika. Obsuzhdeniya* 50: 198–203

Millar D H M Unpublished Radio echo layering in polar ice sheets. (PhD thesis, University of Cambridge, 1981)

Nye J F, Frank F C 1973 Hydrology of the intragranular veins in a temperate glacier. *International Association of Scientific Hydrology Publication* 95 (*Symposium on the Hydrology of Glaciers, Cambridge 7–13 September 1969*): 157–161

Paren J G, Robin G de Q 1975 Internal reflections in polar ice sheets. *Journal of Glaciology* 14(71): 251–259

Raymond C F, Harrison W D 1975 Some observations on the behavior of the liquid and gas phases in temperate glacier ice. *Journal of Glaciology* 14(71): 213–233

Röthlisberger H 1972 Water pressure in intra- and subglacial channels. *Journal of Glaciology* 11(62): 177–203

Smith B M E Unpublished Radio echo sounding studies of glaciers. (PhD thesis, University of Cambridge, 1971)

Vallon M, Petit J-R, Fabre B 1976 Study of an ice core to the bedrock in the accumulation zone of an Alpine glacier. *Journal of Glaciology* 17(75): 13–28

Wolff E W, Paren J G 1984 A two-phase model of electrical conduction in polar ice sheets. *Journal of Geophysical Research* 89(B11): 9433–9438

Zagorodnov V S, Zotikov I A 1981 Kernovoye bureniye na Shpitsbergene [Ice core drilling on Spitsbergen]. *Materialy Glyatsiologicheskikh Issledovaniy. Khronika. Obsuzhdeniya* 40: 157–163

Annals of Glaciology 9   1987
© International Glaciological Society

GLACIOLOGICAL INVESTIGATIONS USING THE
SYNTHETIC APERTURE RADAR IMAGING SYSTEM

by

R.A. Bindschadler

(NASA/Goddard Space Flight Center, Greenbelt, MD 20771, U.S.A.)

K.C. Jezek

(Thayer School, Dartmouth College and CRREL, Hanover, NH 03755, U.S.A.)

and

J. Crawford

(Jet Propulsion Laboratory, Pasadena, CA 91103, U.S.A.)

ABSTRACT

Numerous examples of synthetic aperture radar (SAR) imagery of ice sheets are shown and prominent features of glaciological importance which appear in the images are discussed. Features which can be identified include surface undulations, ice-flow lines, crevasses, icebergs, lakes, and streams (even lakes and streams which are inactive or covered by snow), and possibly, the extent of the ablation and wet snow zones. SAR images presented here include both L-band data from the Seasat satellite and X-band data from an airborne radar. These two data sets overlap at a part of eastern Greenland where a direct comparison can be made between two images. Comparison is also made between SAR and Landsat images in western Greenland. It is concluded that SAR and Landsat are highly complementary instruments; Landsat images contain minimal distortion while SAR's all-weather, day/night capability plus its ability to penetrate snow provide glaciologists with an additional and very powerful tool for research.

INTRODUCTION

The importance of ice sheets in the global climate system is becoming increasingly apparent, yet they remain the least explored and most poorly understood regions of the Earth. Vast size, remoteness, and inhospitable conditions have limited scientific investigations on the surface of the polar ice sheets, while frequent cloud cover and the extended polar night severely restrict visible observations from aircraft or from space. Before their role in global climate can be assessed, techniques must be developed which permit us to study repeatedly the entirety of these ice sheets on time-scales characteristic of the processes which drive them. Until we accomplish this, we will not be in a position to describe the present behavior of the ice sheets and, more importantly, what the likely future response will be to changing climate.

In this paper we discuss the utility of synthetic aperture radar (SAR) in studying ice sheets. SAR is an active microwave system with the commensurate benefits of high-resolution and all-weather/day-night capability. Although it has generally been held that the potential for successfully exploiting SAR data in glaciology is high (Rott and others 1985; Thomas and others 1985), there are very few data to support this contention. The data we present here significantly expand the types of ice-sheet surfaces imaged with SAR and illustrate that a SAR image of any region of the ice sheet is rich with glaciologically significant information. The images we present reveal surface relief on the glacier, the location of calving activity, and surface crevasses. Flow lines in outlet glaciers are easily detected and the data distinctly show surface lakes and streams (both active and relic). Some of our interpretations must remain conjectural; we present corroborative evidence wherever possible. We suggest that SAR imagery can be applied successfully to studies such as: providing measurements of the spatial extent of the ice sheet and ice shelves; identification of present and past flow features; location and configuration of ice rises; and the location of blue-ice zones for meteorite collection.

SYNTHETIC APERTURE RADAR OPERATION

Synthetic aperture radar (SAR) is a side-looking imaging sensor which transmits coherent electromagnetic pulses towards the Earth's surface and records the back-scattered energy. SAR instruments presently being used for remote-sensing applications are typically designed to operate in either the X-, C-, or L-bands of the microwave part of the spectrum. The major advantage of operating the radar at microwave frequencies is the inherent insensitivity of the signal to clouds or fog. Because SAR is an active device providing its own illumination of the surface, it can be operated at night as well as during the day.

The utility of SAR over real-aperture radars is the capability to achieve high resolution without a large antenna. Cross-track resolution is obtained by using a short pulse or by using a chirp and matched filters. However, unlike conventional radars, along-track resolution is obtained by using the shift in doppler frequency generated by the motion of the antenna relative to the target. The radar compiles amplitude and phase information over the length of the synthetic aperture by successively illuminating the Earth's surface. By processing the histories of amplitude and phase for a target, SAR can obtain a resolution much finer than achieved by conventional radar (Ulaby and others 1982). More importantly, when data are processed in this manner, resolution is independent of range, which makes it ideally suited for space applications.

Intensity variations observed in SAR images are caused by differing amounts of energy scattered back to the instrument. Factors which affect the back-scattered signal include the electrical properties of the material, the surface roughness, and volume inhomogeneities. All of these factors are frequency-dependent. Instrument parameters such as antenna depression angle, and polarization of the transmitted and received signals, also affect the received intensity.

SYNTHETIC APERTURE RADAR DATA OF ICE SHEETS

The Seasat satellite, launched in June 1978, provided a very limited amount of SAR data of ice sheets including one pass each in eastern and western Greenland, and two passes containing the ice cap Vatnajökull in Iceland (Fig.1). Since that mission, only a few images from the Shuttle

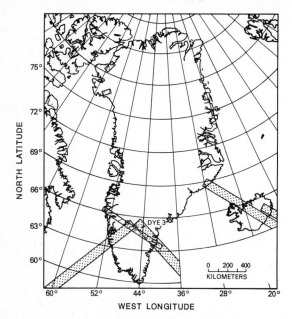

Fig.1. Location map of SAR data in Greenland. Airborne data (cross-hatched strips) are located in two short blocks in northern Greenland as well as a full transect in the south. Seasat data (stippled strips) consist of two swaths in southern Greenland. SAR data in the south overlap surface-strain networks represented in the figure by two spoked wheels and a stake line near station Dye-3.

Imaging Radar-A (SIR-A) containing isolated mountain glaciers have been added to the collection of SAR imagery from space. Seasat SAR operated at 1.275 GHz (23.5 cm), HH polarization, with a 25 m × 25 m ground resolution, a swath width of 100 km, and an antenna depression angle of 67° (Fu and Holt 1982). Thus, the nominal incidence angle in the center of the swath on horizontal ground was 23°.

Across the swath the incidence angle varied by ±3° with a lower incidence angle in the near range and larger incidence angle in the far range.

Airborne missions have added significantly to the data base of SAR images of ice. Most recently, a pair of missions dedicated to the collection of SAR data of ice was flown in Greenland including a full transect of the ice sheet (see Fig.1). The airborne data were collected with the STAR-1 system operated by Intera Corporation. This X-band radar operated at 9.4 GHz (3.2 cm), HH polarization, with a ground resolution of 12 m cross-track and 6 m along-track (Nichols and others 1986). The swath width was 45 km with a nominal antenna depression angle of 14° for an incidence angle of 76° on a horizontal surface. The range of incidence angles (again, referenced to a horizontal surface) varied by ±3°.

SAR IMAGES FROM SPACE

Fig.2 is one of the two Seasat SAR images of Greenland. These data were collected over western Greenland and optically processed to produce four film strips (see Fu and Holt 1982). Using maps of the Greenland coast published by the Greenland Geological Survey, it is easy to identify features of the coastline and confirm that the open ocean, coastal mountains, and ice sheet all have distinctive signatures. Concentrating in this paper on the ice sheet, there is a wide variety of glaciologically interesting features revealed in the SAR image. (The ice sheet in Fig.2 is shown in more detail in Fig.7a.) The lower outlet glaciers are marked by areas of intense crevassing as well as lineations running along the flow direction of the glaciers. These lineations are probably medial moraines or surface ridges and troughs. Up-stream, broader lineations mark the flow directions of the ice which feeds the glaciers. These features could be caused by a number of effects: longitudinal stretching of features as they accelerate down-flow, a less-pronounced ridge–trough system, drainage of water in the firn, or a combination of these effects. More obvious water-related features are the smaller, rounder bright spots and the serpentine lines which connect many of them. These features are quite likely surface lakes and stream channels between them. Large drainage streams many meters across have been observed in Greenland. Such a feature, even though narrower than the pixel resolution, would still be discernible due to their continuous nature, just as bridges and roads smaller than the resolution size can be seen on photographs from space. Their brightness

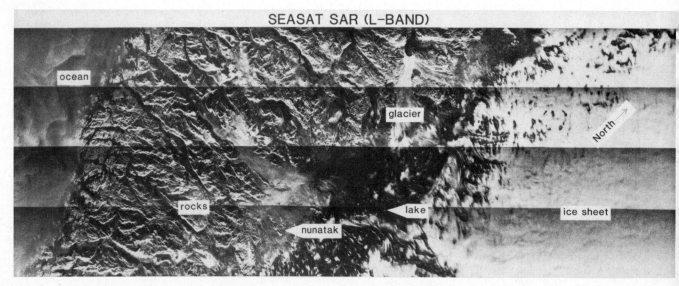

Fig.2 Optically processed Seasat SAR data of western Greenland collected on 9 October 1978. Width of swath is approximately 100 km. Illumination is from the bottom of the figure. Features are discussed in the text.

probably comes from the fact that the sides of the channels have facets which are oriented so as to provide a good reflection of the radar beam. This effect also causes the crevasse fields to show up as bright areas on the image. An observation which supports this explanation is that most of the crevasse fields and stream channels in Fig.2 are oriented normal to the illumination direction.

The ice sheet in Fig.2 also shows tonal variability on a broader scale. The darkest region lies just up-stream of the separate outlet glaciers and most likely corresponds to the ablation area of the ice sheet. Gudmandsen (personal communication) has confirmed that the equilibrium line in this region corresponds roughly to the boundary between the dark and light regions of the ice sheet. This interpretation is further supported by the presence of the surface lakes at the up-flow boundary of the darker region. We suggest that the reason this area is dark is that the bare ice or smooth ice surface of the ablation zone acts as a specular reflector for the incident energy. Unless a specularly reflecting surface is inclined normal to the incident beam, less energy from this surface is scattered back to the receiver as compared with the received energy from a more diffuse reflector such as the firn. This discrimination between specular and diffuse-reflecting surfaces is possible even if they are buried, as long as attenuation in the overlying medium is small. Smith and Evans (1972) have shown that L-band radar will have some capability of penetrating dry snow many tens of meters thick. Another alternative explanation of the dark area is that the dark area is wetter and, thus, more radar-absorptive. We feel this explanation is less likely because significant melting is known to occur at elevations above the equilibrium line (i.e. in the brighter area).

Gray patches within this dark region could be caused by a number of effects: residual snow (or deeper snow if the entire area has received recent snow); surfaces inclined more toward the receiver; rougher surfaces; or surfaces with more debris. It is clear that surface measurements simultaneous with SAR imagery are the only indisputable means of determining the proper interpretation of the tonal differences.

Further up-stream, the regional tone switches to a band of dark patches scattered on a light field. This band is approximately 25 km wide. The spatial scale of the gray patches is similar to the spatial scale of surface undulations as revealed on Landsat imagery, yet we feel it is unlikely that we are measuring topographic variations because the band does not extend to higher elevations while the undulations do continue. It is more likely that the tonal variations are affected by and possibly even controlled by topography. A wind crust might likely form on the windward side of undulations while the leeward side would contain less-consolidated drift. Another possibility is that enhanced melting occurs on south-facing slopes relative to

north-facing slopes due to the extra solar radiation. This extra solar input would cause more surface melt and form both a smoother surface and more sub-surface ice lenses – both of which would lead to a more specularly reflective target for the SAR. The observation that the gray patches diminish in both size and number in the higher elevations suggests that the cause of the gray patches is related to melt rather than wind. We feel this band corresponds to the slush zone characterized by Benson (personal communication) wherein the snow is saturated at the end of the melt season. The cause of this textural pattern could be due to drier snow with any resident water having been refrozen since the end of the melt season being the bright background, areas of wetter snow being gray, and areas of soaked snow or slush being darkest. Such a band has been observed at numerous places on the ice sheet. The dark streaks are interpreted as being drainage channels of excess water. Because the surface is snow or firn, these dark streaks would comprise slush and have diffuse edges rather than sharp boundaries incised into bare ice.

Furthest inland in Fig.2 the tonal variations are extremely slight. At these elevations, the effects of water and bare or smooth ice have probably diminished, leaving surface topography and volume inhomogeneities as the only causes of spatial variations of radar back-scatter.

The issue of sub-surface penetration is an important one which must be considered very carefully. Rott and others (1985) have shown that penetration depth of dry snow varies with frequency; the lower frequencies (e.g. L-band) penetrating further than higher frequencies (e.g. X-band). Also, Stiles and Ulaby (1982) have determined that penetration depth decreases with liquid water content. In these dependencies, we see a potential for being able to discriminate between surface (e.g. topographic) effects on back-scatter and sub-surface (e.g. ice lenses) effects in dry snow or between dry and wet snow by using multiple frequencies. This will allow the accurate determination of the boundary between the dry-snow zone and the percolation zone as well as the boundary between the percolation zone and the soaked (or wet-snow) zone. These boundaries are important in assessing the mass-balance situation on the ice sheet as well as mapping seasonal snow cover (Rott and others 1985). What is needed to evaluate the potential of this technique for ice-sheet studies is simultaneous SAR imagery of the same areas containing these different zones. Rott and Domik (1984) have used different SAR frequencies in a study area in the Ötztaler Alps in Austria but there is no percolation or dry-snow zone there. We have collected airborne SAR data at X-band which overlap the L-band Seasat SAR data in East Greenland but the airborne data were collected in April 1986 and the Seasat data in September 1978. The comparison of these data will be presented following the discussion of the airborne data alone.

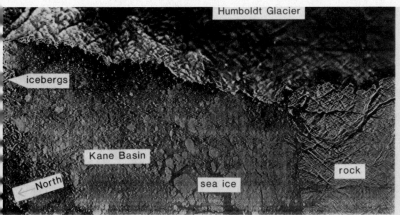
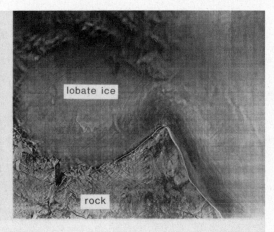

AIRBORNE SAR (X-BAND)

Fig.3. Airborne SAR data of the Humboldt glacier area in north-west Greenland collected on 14 March 1986. Width of swath is approximately 45 km. Illumination is from the bottom of the figure. White band represents interval when no data were collected.

AIRBORNE SAR IMAGE OF HUMBOLDT GLACIER

On 14 March 1986, SAR data of the Greenland ice sheet were collected by an aircraft using the STAR-1 SAR system described earlier (see Fig.1). Fig.3 is a part of these data showing the lower section of Humboldt glacier in north-western Greenland. The data are taken at X-band and have a nominal surface resolution of 10 m. The image reveals numerous tabular (longest about 1.6 km) and pinnacled icebergs frozen in between floes of sea ice in the Kane Basin. The concentration of icebergs adjacent to the regions of heavy crevassing in the glacier suggests that the calving process primarily occurs within the bright arcuate parts of the glacier. Heavy surface crevassing is observable in these areas and, based on the brighter surface there and the orientation of the SAR, the surface is probably steeper. Ice velocities in this steeper region are probably higher, a necessary condition to replenish the ice lost by calving. Most icebergs have apparently been swept northward along the coast away from the calving centers.

The interior of the glacier has a dark, mottled appearance. This is probably an indication of relief associated with glacier flow over complex basal topography. This assumption is based upon the similar texture between the surface morphology of the exposed bedrock on the flanks of the glacier and the light/dark patterns seen on the glacier. Moreover, melt-water ponds on the surface of the glacier are visible as are melt-water channels directed away from the ponds. We do not mean to suggest that we believe these ponds and channels are active, i.e. filled with water, in March; rather, we believe that they have a different topographic character than the surrounding ice and it is the SAR's sensitivity to this difference which provides evidence for their presence. Notice that a row of melt-water ponds is aligned along a dark linear feature that stretches across most of the image. We suggest that the lineament is a surface depression associated with basal topography.

The mottled appearance characteristic of the Humboldt glacier image changes markedly for the lobate ice adjacent to the glacier. The ice in this region terminates on land and the edge of the ice is very clearly marked by a bright line, probably caused by a very steep terminal face. The surface characteristic of this ice is more uniform. Grounded on rock, this ice is probably thicker and less active than Humboldt glacier and, thus, the roughness of the basal topography is not seen in the surface topography. This smooth surface is crossed by a large number of very thin lines which strike normal to the ice margin. This orientation strongly suggests these represent melt-water rivulets, much smaller than the large channels that drain the ponds seen on Humboldt glacier, but far more numerous. They collect only local surface melt-water as they form and run along the fall line of the ice surface. Ponds do not form on this smoother ice but, rather, an intricate, filamentary structure is revealed. In the southernmost ice (far right of Fig.3), the banding parallels the ice edge, while in the lobate ice the banding shows some additional convolution. This convolution resembles morainal features generated in zones of non-uniform ice flow. We suspect the formation mechanism for these features is similar and that the bands themselves are due to depositional or structural non-conformities within the ice.

AIRBORNE SAR IMAGERY OF CENTRAL GREENLAND

On 14 April 1986, the same instrument collected X-band SAR data on a single transect across the entire Greenland ice sheet. The flight path overlapped areas imaged by the Seasat SAR as well as areas studied by surface parties as part of the Southern Greenland Ice Sheet Program (Whillans and others 1984; Jezek and others 1985) (see Fig.1). A broad tonal patchwork pattern dominates the western margin of the ice sheet shown in Fig.4. We suggest that much of this pattern is correlated with basal topography given the general north-east strike of many of the linear features apparent in the image. This trend is seen in the exposed mountains in the left of Fig.4 and has been associated with regional, transcurrent faults (Watterson 1978).

Other features which appear in Fig.4 are the numerous lakes. Fig.5 compares a part of Fig.4 with an aerial photograph of surface lakes in Greenland taken in the summer of 1985 and kindly provided by H. Brecher. The lakes in the photograph are on the order of 500 m in diameter (personal communication from H. Brecher) and occur at the boundary between bare ice and firn. The photograph was taken further north than Fig.4: in the drainage basin of Jakobshavns glacier, where less surface melting occurs, so it is not surprising that the average size of the lakes is smaller than in the SAR image, but the pattern and shapes of the lakes in the two regions are strikingly similar.

The lakes just right of center in Fig.4 appear to have a smoother unmodified surface than the lakes further to the right. The texture of these two classes suggests that the smoother surface is a result of an undisturbed and possibly thicker ice cover, while the rougher surface may have been fractured and/or collapsed at the end of the previous melt season. Still other lakes are barely visible, perhaps having been inactive for more than 1 year. These would now be buried beneath accumulated snow but would still be evidenced by the smoother surface as compared with the surrounding undulating topography. The conjecture that buried lakes may be observable is not implausible given the capability of L-band radar to penetrate dry snow. In

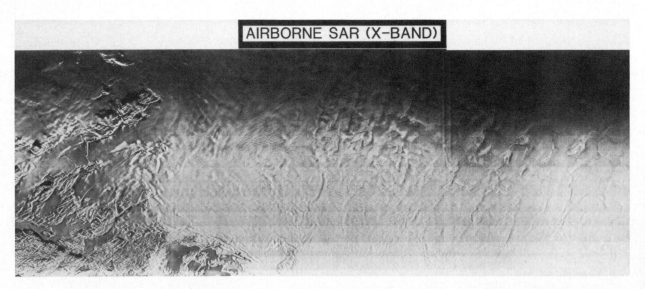

AIRBORNE SAR (X-BAND)

Fig.4. Western end of airborne SAR data in southern Greenland collected on 14 April 1986. Width of swath is approximately 45 km. Illumination is from the bottom. Antenna depression angle was 14° over the left two-thirds of the image but was increased to 15° over the remaining one-third of the image.

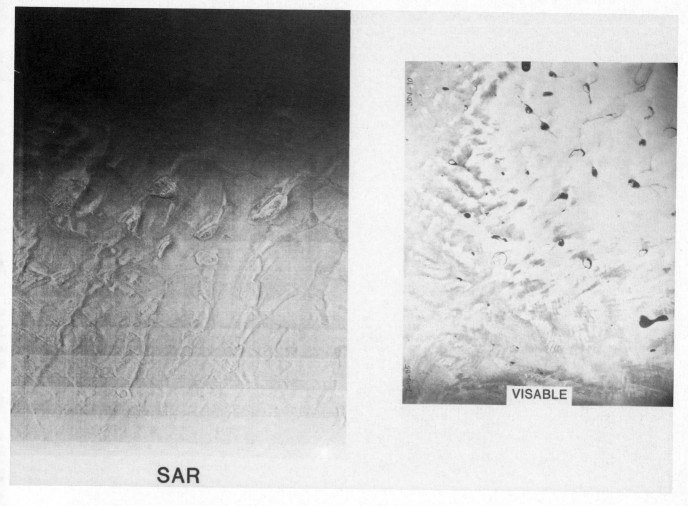

Fig.5. Comparison of part of Fig.4 with aerial photograph of surface lakes up-stream of Jakobshavns glacier, Greenland, taken on 10 July 1985. Scale of each image is approximately equal; height of SAR image is 45 km. (Photograph is presented by courtesy of H. Brecher.)

addition to sensing inactive and old lakes, the SAR appears to be detecting a branched drainage network through the winter snow cover.

In Fig.4 it can be seen that the maximum contrast occurs in a band about two-thirds up from the bottom of the image. It is in this band that the system of lakes and branched drainage network is most easily seen, probably due to the SAR's sensitivity to surface slope. The rapid degradation of contrast on either side of this region illustrates the sensitivity of SAR to incidence angle.

As the flight proceeded eastward, much of the contrast was lost due to changes in instrument settings. There was a tonal variation which continued to trend in the north-east direction. Surface measurements of ice-sheet elevation confirm this trend up to the central ice divide (Drew and Whillans 1984) and surface radar-sounding data strongly suggest that the topographic trends are associated with basal relief (Whillans and others in press). East of the ice divide (not shown), there is a weak indication from the SAR data that the structural trend shifts to the east-south-east, in agreement with the coastal geology (Watterson 1978).

COMPARISON OF AIRBORNE AND SEASAT SAR

The eastern end of the airborne data overlapped the Seasat SAR data of eastern Greenland (see Fig.1). Thus, the same area was imaged by the two instruments. Fig.6 presents the two images at the same scale. Contrary to the situation on the western margin, the ice rapidly descends from the divide to the coast without any evidence of a zone of surface lakes. There is a general agreement in the two images but a number of significant differences are also revealed. While some of the differences can be understood

in terms of the difference in instrument parameters, some may result from differences in the surface at the time of each image. This ambiguity highlights the need for simultaneous multi-frequency imaging; plans for such missions already exist.

The shadowing caused by the mountains is more extreme in the airborne data. This is due to the difference in antenna-depression angles for the two systems; the airborne data were taken with an antenna-depression angle of only 14° while the Seasat data had a much higher depression angle of 67°. Thus, the surface incidence angle was much closer to grazing for the airborne data. The higher resolution of the airborne data allowed the identification of crevasse fields which only appeared as bright patches on the Seasat image.

A consistent pattern of tonal variability is present in the left-hand side of both images but, again, the airborne data have a better resolution. It is difficult to assess the consistency of the broader-scale variation because the airborne data in particular suffer from both an increase in darkening from bottom to top across the swath and occasional changes in instrument settings which cause the bulk changes in contrast along the swath. Nevertheless, there appear to be some differences in tonal pattern between the two images which might be due to differences in the surface at the two times of the images. These include the darker areas on the Seasat image which occur at the lower elevations where the glaciers are discharging into the fiords and the upper right corner where the brighter airborne SAR return would indicate sea ice and icebergs which do not appear on the Seasat image. The darker Seasat region might well have the same explanation as in Fig.2: a smooth, bare-ice surface, possibly wet, and, therefore, either a specular or highly attenuating reflector. In April, when the

15

AIRBORNE SAR

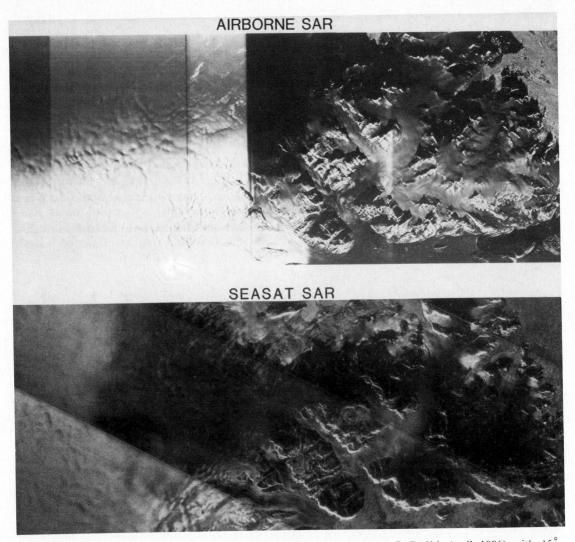

SEASAT SAR

Fig.6. Comparison of eastern Greenland imaged with airborne X-band SAR (14 April 1986) with 15° antenna depression angle and Seasat SAR (23 September 1978) with 70° antenna depression angle. In airborne image, instrument settings were changed numerous times across the image. In both cases, illumination is from the bottom of the figure. Width of each swath is 45 km.

airborne data were collected, it is likely that even the lower elevations are still covered with a blanket of snow and little melting, if any, has occurred. Similarly, icebergs and sea ice would be expected during April and not in September. Again, to firm up the interpretation, there is a need for coincident images at multiple frequencies.

COMPARISON OF SAR AND LANDSAT IMAGES

Hall and Ormsby (1983) have made a comparison of coincident SAR and Landsat MSS images for a few areas including the Malaspina glacier and Mount McKinley areas in Alaska. Their conclusion was that the SAR image contained no additional information which was not in the Landsat scene. Rott (1982) disagreed and demonstrated for Skeidarárjökull, Iceland, that Seasat SAR shows more detail than Landsat. Our position is that, because radar reflectivity varies much more strongly with the physical state of the ice or snow surface than does the optical reflectivity, SAR has a prominent role to play in glaciological research. Available comparisons between these two imagers show there can be information detected by either sensor which escapes the other. Consequently, it is most appropriate to view the two instruments as complementary rather than duplicative.

Another comparison was presented for the glaciers in the Karakoram of Pakistan using SIR-A and Landsat images (Ford and others 1983, p. 112–13). That comparison showed that, while the medial moraines exposed on the surface in the ablation (bare-ice) zone could be seen on both images, these moraines could be traced far up-stream into the accumulation area on the SAR image but not on the

Landsat image. This difference is probably due to the penetration capability of the SAR.

Ford and others (1980, p. 100–01) presented a comparison between Seasat SAR and Landsat in western Greenland but they did not discuss the glaciological significance of the features. Fig.7 repeats these two images for our discussion. Fig.7a is a part of the image already shown in Fig.2. The SAR image was collected by Seasat on 9 October 1978, while the Landsat scene (ID# 1448-14124) is from 14 October 1973. Many of the same features can be identified on both images even though the Landsat image has a much poorer resolution (80 m as compared with 25 m). Distortion of the SAR image prevents a direct overlaying of images without the aid of a sophisticated image-analysis computer, but the features in the SAR image such as crevassed outlet glaciers, a dark ablation zone, and mottled firn zone can be matched with regions of distinctive texture on the Landsat scene. The Landsat scene also helps identify the nunataks in the southern part of the image which are not particularly different from some of the smaller-scale ice features on the ice in the SAR image. Flow lines are far more distinct in the SAR image, particularly further up-stream. The scale of the patchy appearance of the SAR image is consistent with the scale of surface undulations observed with Landsat. One edge of the large V-shaped area in the SAR image shows up in the Landsat image as a ridge in the ice. Neither lakes nor streams are evident in the Landsat scene but show up dramatically with SAR.

A few other Landsat images were available for comparison with this West Greenland scene. Most were

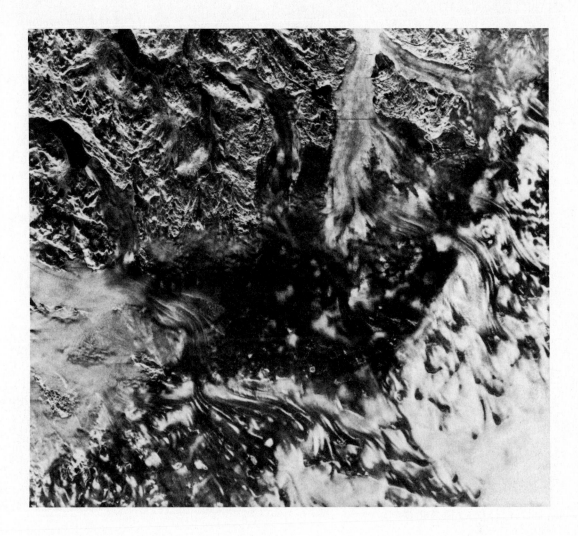

Fig.7a. Comparison of western Greenland images with Seasat SAR on 9 October 1978. Illumination is from the bottom of the image. Shape of Landsat scene is skewed to match SAR scene which has some distortion due to effects of surface relief.

cloud-covered but two are worth discussing: one taken in mid-winter (25 February 1981) and another taken on 27 September 1978, just 2 weeks before the SAR image. In the case of the mid-winter image, the low sun angle highlighted the surface topography slightly better than Fig.7b but, because the entire surface was snow-covered, it is difficult to see the margin of the ice sheet and impossible to see any difference between the firn and ablation areas. The image taken 2 weeks before the SAR image is important because it shows evidence of a fresh snowfall. This masked all but the most prominent surface features and rendered the image of little use. It also demonstrated that the SAR data were most likely collected over a surface covered with fresh snow, a situation which did not prevent the acquisition of valuable data (as was the case with Landsat).

We see the two sensors as highly complementary, their strengths meshing nicely. Landsat images contain minimal distortion because the instrument is nadir-pointing. Its multi-spectral characteristic has enabled enhanced images to highlight many important glaciologically significant features (Williams and Carter 1976; Swithinbank and Lucchitta 1986). These images can reveal subtle surface features when Sun elevations are low. SAR is sensitive to changes in the complex dielectric properties of the surface, surface roughness, and regional slope. Because the complex dielectric contrast between saturated snow and ice is greater than an order of magnitude, and because surface slopes on the interior of the ice sheet are small, much of the detail in SAR imagery is probably associated with variations in the surface composition. The all-weather capability of the SAR, as well as the fact that illumination angle and, to a lesser extent, radar frequency are operator-controlled, cannot be over-emphasized. We feel it is this characteristic, in particular, which will free the polar glaciologist from the frustrations of relying exclusively on visible imagery to study the polar ice sheets on a large scale.

SUMMARY

We have shown, by examples and comparisons, the power of SAR data in providing the glaciologist with a valuable source of data in image format. However, glaciological data are limited primarily because the SAR instruments developed so far have been designed exclusively for the oceanographic and the geologic communities. We think there is a need for the glaciological community to explore, in detail, the potential of SAR and the new instrument capabilities, such as multi-frequency and multi-polarization, presently being planned. To do this, complementary surface and laboratory measurements of the electrical properties of snow and firn are called for, particularly to determine the depth of penetration of SAR signals and to correlate precisely tonal variations in the SAR image with surface properties. In the coming years there

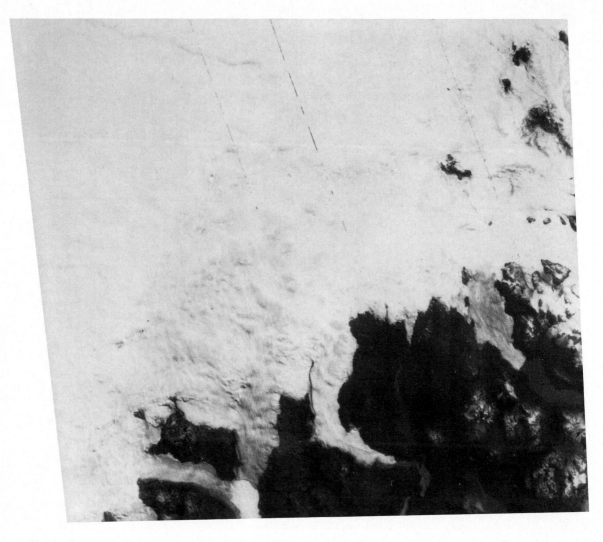

Fig.7b. Comparison of western Greenland imaged with unenhanced Landsat-1 (ID# 1448-14124), on 14 October 1973.

will be numerous SAR instruments flown in space providing opportunities to collect unique data for mapping and benchmark purposes. With these data, the glaciologist can begin to study the polar ice sheets as complete units and monitor their effect on, and response to, global climate.

REFERENCES

Drew A R, Whillans I M 1984 Measurement of surface deformation of the Greenland ice sheet by satellite tracking. *Annals of Glaciology* 5: 51–55

Ford J P *and 6 others* 1980 *Seasat views North America, the Caribbean, and western Europe with imaging radar.* Pasadena, CA, Jet Propulsion Laboratory (Publication 80-67)

Ford J P, Cimino J B, Elachi C 1983 *Space shuttle Columbia views the world with imaging radar: the SIR-A Experiment.* Pasadena, CA, Jet Propulsion Laboratory (Publication 82-95)

Fu L L, Holt B 1982 *Seasat views oceans and sea ice with synthetic aperture radar.* Pasadena, CA, Jet Propulsion Laboratory (Publication 81-120)

Hall D K, Ormsby J P 1983 Use of Seasat synthetic aperture radar and LANDSAT multispectral scanner subsystem data for Alaskan glaciology studies. *Journal of Geophysical Research* 88(C3): 1597–1607

Jezek K C, Roeloffs E A, Greischar L L 1985 A geophysical survey of subglacial geology around the deep-drilling site at Dye-3 Greenland. *In* Langway C C Jr, Oeschger H, Dansgaard W (*eds*) *Greenland ice core: geophysics. geochemistry and the environment.* Washington, DC, American Geophysical Union: 105–110 (Geophysical Monograph 33)

Nichols A D, Wilhelm J W, Gaffield T W, Inkster D R, Leung S K 1986 A SAR for real-time ice reconnaissance. *IEEE Transactions on Geoscience and Remote Sensing* (GE 24(3): 383–389)

Rott H 1982 Synthetic aperture radar capabilities for glacier monitoring demonstrated with Seasat SAR data. *Zeitschrift für Gletscherkunde und Glazialgeologie* 16(2), 1980: 255–266

Rott H, Domik G 1984 The SAR-580 Experiment on snow and glaciers at the Austrian test site. *In* Trevett J W (*ed*) *Proceedings of the SAR-580 Investigators Workshop.* Joint Research Centre, Ispra, Italy May 1984. Paris, European Space Agency

Rott H, Domik G, Mätzler C, Miller H, Lenhart K G 1985 *Study on use and characteristics of SAR for land snow and ice applications.* Innsbruck, Universität Innsbruck. Institut für Meteorologie und Geophysik (Mitteilung 1)

Smith B M E, Evans S 1972 Radio echo sounding: absorption and scattering by water inclusion and ice lenses. *Journal of Glaciology* 11(61): 133–146

Stiles W H, Ulaby F T 1982 Dielectric properties of snow. *CRREL Special Report* 82–18: 91–103

Swithinbank C, Lucchitta B K 1986 Multispectral digital image mapping of Antarctic ice features. *Annals of Glaciology* 8: 159–163

Thomas R H *and 8 others* 1985 *Satellite remote sensing for ice sheet research.* Washington, DC, National Aeronautics and Space Administration (Technical Memorandum 86233)

Ulaby F T, Moore R K, Fung A K 1982 *Microwave remote sensing active and passive. Vol 2. Radar remote sensing and surface scattering and emission theory.* Reading, MA, Addison-Wesley

Watterson J 1978 Proterozoic intraplate deformation in the light of south-east Asian neotectonics. *Nature* 273(5664): 636–640

Whillans I M, Jezek K C, Drew A R, Gundestrup N 1984 Ice flow leading to the deep core hole at Dye 3, Greenland. *Annals of Glaciology* 5: 185–190

Whillans I M *and 6 others* In press Glaciologic transect in southern Greenland; 1980–1981. GISP I data. *Byrd Polar Research Center. Report*

Williams R S Jr, Carter W D (*eds*) 1976 ERTS-1: a new window on our planet. *US Geological Survey. Professional Paper* 929

Annals of Glaciology 9 1987
© International Glaciological Society

REMOTE SENSING OF THE ROSS ICE STREAMS AND ADJACENT ROSS ICE SHELF, ANTARCTICA

by

C.R. Bentley, S. Shabtaie, D.D. Blankenship, S.T. Rooney, D.G. Schultz, S. Anandakrishnan and R.B. Alley

(Geophysical and Polar Research Center, University of Wisconsin-Madison, 1215 W. Dayton Street, Madison, WI 53706, U.S.A.)

ABSTRACT

In the first few seasons of the Antarctic Siple Coast project, the University of Wisconsin has concentrated on radar and seismic studies. Highlights of the results to date include the delineation of ice streams A, B, and C and the ridges in between, determination of the surface elevations over the area, discovery of a much more advanced grounding line than previously recognized and recognition of a broad, flat, barely grounded "ice plain" just inside the grounding line. Complex zones between and adjoining some of the ice streams, characterized by an interspersal of undisturbed ice and crevassed patches, give the impression of being transformed from sheet flow into stream flow in a process of ice stream expansion. An indicated negative mass balance for ice stream B could be the result of this "activation" process. Ice stream C, currently stagnant, exhibits terraces and reversals of surface slope, associated with zones of strong, steady basal radar reflections. These features suggest that subglacial water has been trapped by reversals in the hydraulic pressure gradient.

Low seismic P-wave and S-wave velocities in a meters-thick layer immediately below the ice strongly indicate a saturated sediment of such high porosity (~40%) and low effective (differential) pressure (~50 kPa, or 0.5% of the glaciostatic pressure) that it must be too weak not to be deforming. We presume this deforming layer to be a dilated till. Its base exhibits ridges and troughs parallel to the flow direction that resemble glacial megaflutes. We believe that at our site on the upper part of ice stream B the ice stream moves principally by deforming its bed. Analysis of seismographic recordings of micro-earthquakes that occur at the glacier bed shows that the micro-earthquakes are both small in energy and infrequent. This implies that virtually none of the energy of ice stream motion is dissipated by brittle fracture at the bed. If our models are correct, the subglacial deforming till becomes increasingly soft down-glacier, and/or the ice becomes decoupled from the till by intervening water, until on the "ice plain" basal drag is less important than longitudinal stresses in the dynamic balance. Our models also imply that the "ice plains" rest on "till deltas" that have been formed by the deposition of till carried along beneath the ice streams, and that the till deltas, and the grounding lines that bound them, are currently advancing in front of the active ice streams.

INTRODUCTION

During the past several years the Geophysical and Polar Research Center has been conducting a program of geophysical investigations of the Siple and Gould Coast regions of the West Antarctic ice sheet, and of the adjacent parts of the Ross Ice Shelf. Our research work has concentrated on radar and seismic studies of the glacier bed, but has also yielded a substantial amount of information concerning the glacier surface and internal characteristics. In this paper we give a review of the major findings of this program to date. This is not yet a comprehensive report on the results

of this work, since extensive analysis remains to be done on data already in hand; furthermore the field program will be continuing in the future. Elsewhere in this volume we report on two of the specific improvements in our remote-sensing capabilities that have been put to use effectively in recent seasons (Blankenship and others 1987; Schultz and others 1987).

The major focus of the research project being carried out by the University of Wisconsin in conjunction with The Ohio State University, NASA, and the University of Chicago is the "Ross ice streams", which are the dynamically dominant feature of the part of the West Antarctic inland ice sheet that feeds into the Ross Ice Shelf (Robin and others 1970[a]; Rose 1979). The work done to date has concentrated on ice stream B and Crary Ice Rise, but with extensive reconnaissance work over ice streams A and C, and ridges AB and BC (Fig.1).

MAPPING OF ICE STREAMS

A very important first result of the airborne radar program has been the mapping of ice streams A, B, and C and their marginal shear bands that is more detailed than the previous mapping by Rose (1979). As was shown previously by Robin and others (1970[a]) and Rose (1979), radargrams over the ice streams show strong scattered returns from inhomogeneities, especially crevasses, at and near the air–ice interface, producing a prolongation of the surface echo called "clutter". Clutter results from buried as well as surface crevasses, so it is an effective diagnostic feature of ice streams whose surfaces are characterized by crevasses, most of which are buried (Fig.2). The clutter from the marginal shear zones is often longer lasting than that from the other parts of the ice stream, presumably because the large open crevasses along the margins are particularly effective scatterers of energy. Using both the scatter from the bodies of the ice streams and the prolonged returns from their margins, which can be traced from cross-section to cross-section, Shabtaie and Bentley (1986; in press) produced the map of the ice streams that is shown in Fig.3.

It also has been possible to delineate the grounding lines that mark the boundary between the inland ice and the ice shelf (Shabtaie and Bentley 1986; in press). In places where the inland ice is moving relatively slowly, the grounding line is readily apparent from the surface elevation profile, and in fact can even be seen visually from the surface. However, that is not true down-stream of the active ice streams. The radar data have been used in two ways to delineate the grounding lines in those regions. In the first place, ice thicknesses can be compared with precise surface elevations obtained by satellite measurements at ground stations to determine whether the surface of the ice shelf is at or above buoyancy. Secondly, characteristics of the radar signals can be used directly. Two are particularly useful. (1) Hyperbolic returns occur from near the base of the ice; we interpret these as being due to bottom crevasses.

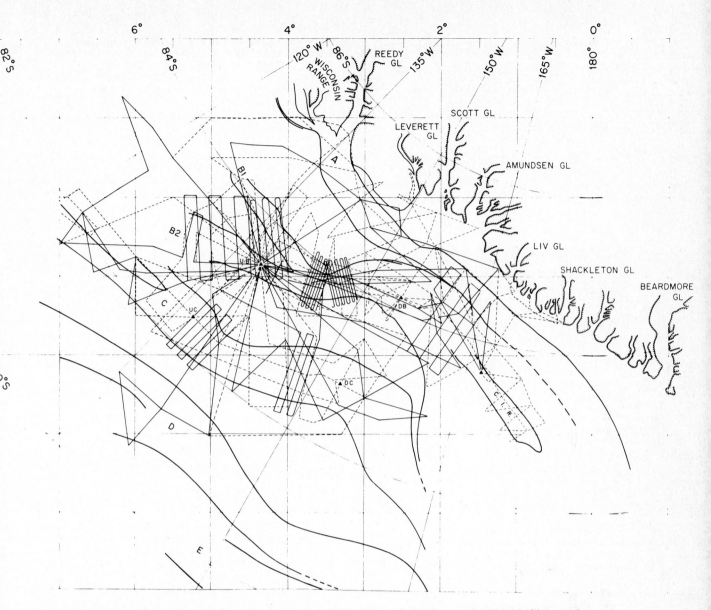

Fig.1. Map of Siple Coast region showing radar flight lines. Solid lines are 1985-86 flights using digital recording; dashed lines are 1984-85 flights with analog recording. Ice streams are marked by the letters A-E; C.I.R. is Crary Ice Rise. Base camps are denoted by triangles. In the rectangular grid coordinate system shown here and on other maps, the origin is at the South Pole and grid north is toward Greenwich. (Delineation of ice streams is not exact - Fig.3 should be used for greater accuracy.)

We have not observed bottom crevasses anywhere on the grounded ice, so we assume that the presence of bottom crevasses on a flight line means that the grounding line must be upstream. (2) The bottom echo from the grounded ice usually shows a fading pattern (i.e. rapid changes in echo strength) that is typical of inland ice sheets, whereas the bottom echo from the ice shelf is relatively smooth and strong. From this combination of data, Shabtaie and Bentley (in press) drew the grounding line in the position shown on the map in Fig.4, where it is contrasted with the position formerly mapped by Rose (1979), who did not have the benefit of precise satellite measurements of surface elevations.

The heads of the ice streams have not yet been mapped clearly. In fact, this may be difficult to do. Ice stream A is fed by Reedy Glacier and by the Shimizu/Horlick ice stream, both of which have their origins on the East Antarctic side of the Transantarctic Mountains. Ice streams B and C, on the other hand, clearly start within the West Antarctic inland ice, but with a complex boundary zone, especially in the case of ice stream B (Fig.3). Just what physiographic features (in addition to crevasses) or analytical characteristics of the ice mark the beginning of the ice streams, if indeed any clear demarka-

tion exists, remains to be seen from further analysis of the data on hand and concentrated study that is planned for the future.

The complexity of "ridge B1/B2" between ice streams B1 and B2 (Figs. 3 and 5) is particularly striking. Within this "ridge" several "islands" and "peninsulas" of clutter-free ice are separated by disturbed ice zones. We believe that the many irregular and discordant zones of crevasses are indicative of a region in transition; more specifically we believe that sections of the surrounding ice sheet are being transformed into stream flow, so that ice stream B is actually growing. We will return to this point later.

SURFACE ELEVATIONS

Another important product of airborne radar is the determination of the surface elevations of the ice sheet. Our radar-sounding flights are carried out as nearly as possible at a constant pressure elevation as determined by the barometric altimeter in the aircraft. Ice surface elevations relative to the aircraft are determined from the surface echo on the radargrams. As part of the Siple Coast project, a number of surface stations at which the elevation was measured by satellite observations were established.

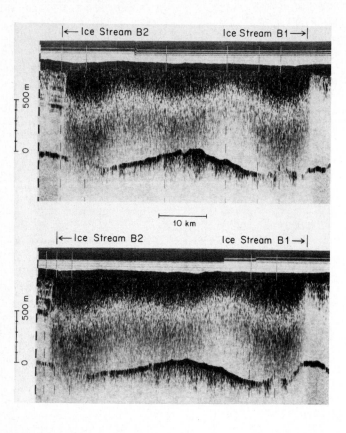

Fig.2. Radargrams showing two cross-sections of ice stream B around 3.3°W. Grid north is on the right. The lower section is about 7 km grid east (down-glacier) from the upper. Surface and bottom echoes and clutter, stronger near the ice-stream boundaries, are seen in each radargram. The boundary between ice streams B1 and B2 is not shown as its position is not certain.

Radar-sounding flights were designed to pass over as many of these ground control stations as possible. This provided excellent control on the absolute elevation of the surface beneath the flight lines. Furthermore, the flight lines crossed each other in many places, so that intertrack ties could be made. Analysis of the results has indicated that surface elevations obtained by radar are good to between ±10 and ±25 m, depending on the distance from the nearest tie point (Shabtaie and Bentley in press). Consequently, our new map (Fig.5) should be more accurate as well as more detailed than the regional map of Drewry (1983), who cited an error of 30-50 m.

ICE STREAM B

We now turn our attention to some of the specific characteristics of ice stream B that have been revealed by remote sensing. Ice stream B has two branches, called B1 and B2, which merge around grid 4°W; the boundary between the two persists downstream as a suture zone as far as grid 2°W (Fig.3). Even well below the junction at 3.3°W the bed continues to be marked by two distinct troughs (Fig.2). The trough beneath ice stream B1 is somewhat the more pronounced of the two, but neither is very deep compared to most ice streams (Bentley in press) – the total subglacial relief on the section of Fig.2 is less than 300 m. Farther upglacier, adjoining "ridge B1/B2", the trough beneath ice stream B1 is 300 m deeper than in Fig.2; downglacier it continues to shoal until it is 100 m less deep than the trough beneath ice stream B2, which has a less varying depth (Shabtaie and others in press[b]).

Seismic reflection studies conducted at Upstream B Camp (UB) revealed the existence of two reflectors near the base of the ice (Blankenship and others 1986; in press) Both compressional wave (P-wave) and shear wave (S-wave) reflections were recorded, each from both the upper and lower reflector. These reflections were seen not only on vertical but also on "wide angle" (oblique) reflection profiles, from which the P-wave and S-wave speeds in the layer were determined. The low P-wave speed, near to that in water, and the extremely low S-wave speed, less than 10% that for the P-waves, together mean that the layer must have a relatively high porosity (about 40%) and be saturated with water at high pressure – 99½% of the glacio-static overburden pressure must be supported by fluid pressure in the pores (Blankenship and others in press). The combination of high porosity and low effective pressure between sedimentary particles implies that the layer, which averages about 6 m thick along the sections that have been measured (Rooney and others in press), must be very weak. From this we have concluded that at least at Upstream B the ice stream moves primarily by deforming its bed rather than by sliding over it (Alley and others 1986; in press [a]).

The seismic reflections further showed that along the direction of ice movement the subglacial reflecting horizon is parallel to the bottom of the ice, but that transverse to the ice movement there is much less continuity in the subglacial reflectors (Rooney and others in press). The picture that emerges is one of ridges and valleys in the basal surface of the subglacial layer trending parallel to the axis of the ice stream. The alignment of these features along the direction of movement suggests that they are formed by erosion. The fact that there is no image of the ridges and troughs in the glacial-subglacial interface is further support for the model of a deforming bed that we have proposed. The largest of these flutes or grooves are over 10 m deep and more than a kilometer across, although more generally they are about 200-300 m across (Fig.6). The general appearance of these flutes is similar to glacial flutes observed on deglaciated terrain of the Laurentide ice sheet, such as in northern and central Alberta (Gravenor and Meneley 1958).

Other interesting features that show up on the seismic cross-sections are linear diffractors parallel to ice movement that suggest a chain of boulders entrained in the top of the till (other interpretations are also possible), isolated ridges parallel to ice flow where the till becomes very thin or pinches out entirely, and dipping reflectors beneath the deforming till that suggest an angular unconformity at the base of the deforming till (Rooney and others in press). Preliminary examination of other field records indicates that the till is probably continuous along a profile 12 km long parallel to the axis of the ice stream.

If our model of a basal deforming till is correct for the whole ice stream, it follows that the till is being transported downstream and deposited where the ice loses its ability to transport it (Alley and others 1986). A very flat region of barely grounded ice, which we propose to call an "ice plain", characterizes the downstream ends of ice streams A and B (Figs 4 and 5). We believe the bed beneath the "ice plain" to have been formed by the deposition, below formerly floating ice, of till carried along beneath the ice streams. The low to non-existent slope of the surface (first profiled, but not recognized as distinct from the ice shelf, by Robin and others (1970[b])), combined with the fast measured ice velocities in the "ice plain", suggest that the movement is primarily ice-shelf-like (dominated by longitudinal stresses) rather than ice-stream-like (dominated by basal shear stresses). It follows further that the transport of till will result in a growing "till delta" beneath the "ice plain", and an advancing grounding line. A similar process of transport was proposed long ago for the whole Ross Ice Shelf by Poulter (1950), although he believed tidal lifting would aid the forward motion of the ice. In fact, the "ice plain" lies several tens of meters above hydrostatic balance so tidal flotation does not occur. Strand cracks have been found at the "ice plain" grounding line near Crary Ice Rise (Bindschadler and others in press).

Supporting evidence for the existence of a soft bed directly beneath the "ice plain" comes from an examination of the seismic reflection records recorded as part of the Ross Ice Shelf Geophysical and Glaciological Survey

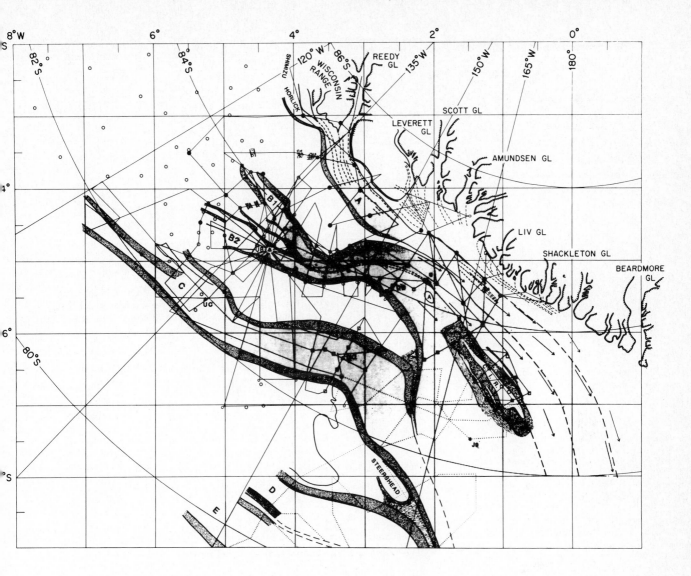

Fig.3. Map of Siple Coast region showing the boundaries of the ice streams and ice rises (a, A, and C.I.R.). Thin solid lines are 1984-85 flight lines; dotted lines are flights from the mid 1970s ("RIGGS"). 1985-86 flights are not shown. Solid circles, open circles, and open squares denote surface stations. From Shabtaie and Bentley (in press).

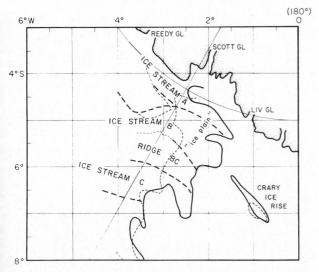

Fig.4. Simplified map of the Siple Coast showing new (solid) and old (light dashed) boundary lines and the region which we proposed to designate an "ice plain". The old boundary lines are from Rose (1979); the new from Shabtaie and Bentley (1986). Heavy dashed lines mark the lateral boundaries of the ice streams.

(RIGGS) in 1973-74 (Robertson unpublished; Robertson and Bentley in press). The best seismogram from each of six stations is reproduced in Fig.7. Stations E5, E6, and E7 are on the "ice plain". Stations E8 and F9 lie on the floating ice shelf over deep water down-stream from ice stream A. The reflection records at E5, E6, and E7 (Fig.7a) show poor or very poor single reflections that were originally interpreted by Robertson (unpublished) as being from the sea floor. However, the characteristics of the seismograms from E8 and F9 (Fig.7b), where both the ice-bottom and sea-bottom reflections are unequivocal, are clearly very different. We believe now that the reflections at E5, E6 and E7 are from within the sedimentary column, and that the ice/sediment reflection is not seen.

There is reason to believe that a reflection from the ice/sediment boundary might not be observable. If we assume that the sediments are the same as those beneath station UpB, except that they are no longer dilated, then we estimate the P-wave velocity at 1700 m s^{-1} and the density at 2.0 Mg m^{-3}; the acoustic impedance would then be 3400 Mg m^{-2} s^{-1}. For the base of the ice shelf, with a density of 0.91 Mg m^{-3} and a wave velocity of 3700 m s^{-1} (Robertson unpublished), the acoustic impedance is also 3400 Mg m^{-2} s^{-1}. Even if the presumed till were dilated, its acoustic imepdance, ~ 3000 Mg m^{-2} s^{-1}, would not be largely different from that of the ice. Thus, though conditions in the sea floor may vary considerably from place to place,

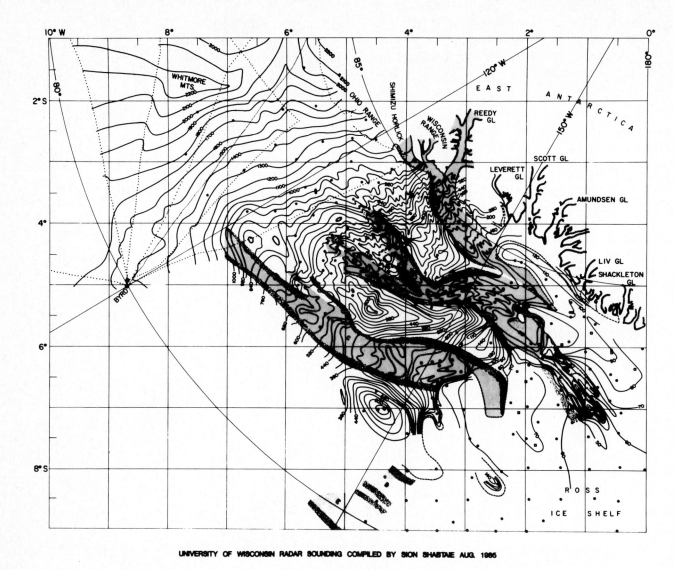

UNIVERSITY OF WISCONSIN RADAR SOUNDING COMPILED BY SION SHABTAIE AUG. 1985

Fig.5. Surface-elevation map of the Siple Coast region. Contour interval is 10 m up to 120 m, 20 m from 120 m to 200 m, 40 m from 200 m to 1000 m, and 100 m for higher elevations. The short-dashed lines denote the grounding line, long-dashed lines delineate some ice-stream boundaries, and dotted lines are oversnow traverse tracks. From Shabtaie and others (in press).

the acoustic impedance contrast could certainly be very small, so that the absence of a bottom reflection would not be surprising.

Also shown in Fig.7b is a reflection record from station F7 on the edge of the "ice plain" - this, the only case of a strong reflection on the "plain", suggests that there is a hard bottom at this spot (grid position 5.6°S, 1.8°W - see Fig.5).

Remote sensing also gives information about the internal characteristics of the ice. The same wide-angle seismic reflection profiles that yielded the seismic velocities in the subglacial layer also yield velocities in the ice itself. Wide-angle reflection measurements elsewhere have generally revealed that a substantial fraction of the ice sheet in most places is anisotropic (Bentley 1971; Blankenship unpublished). In contrast, the P-wave measurements at UpB show little or no anisotropy (Blankenship and others in press). On the other hand, the S-wave measurements, which are more sensitive to anisotropy, do indicate a difference between the speeds of shear waves with different polarizations.

Analysis of both S-wave reflections and waves that have been converted from P to S upon reflection from the bottom of the ice on a wide-angle profile along the flow direction indicates that transversely polarized S waves (SH waves) travel 1% or 2% faster than waves polarized in the plane of the profile. Furthermore, the very fact that P waves are converted into SH waves upon reflection is direct

evidence for the existence of some anisotropy at the base of the ice. Blankenship and others (1987) also observe a 2% difference in wave velocity for shear waves of different polarization generated by micro-earthquakes at the base of the ice stream (Fig.8).

Nevertheless, from the P-wave evidence, it appears that the ice in the ice stream is more nearly isotropic than sheet-flow ice elsewhere in West Antarctica and at Dome C. If the anisotropic structure at Byrd Station (Gow 1970; Bentley 1972; Kohnen and Gow 1979), in the catchment area for ice stream D, is typical also of ice in the catchment area for ice stream B, then a pronounced fabric must be destroyed somewhere along the flow path. Perhaps this occurs when the ice goes through a region of strong extensional stresses at the head of the ice stream where the ice is accelerated to the ice-stream speeds.

Geophysical remote sensing also provides us with information about the dynamics of ice stream B. Part of this information comes from the new passive seismic recording system that is described by Blankenship and others (1987). Twenty-six micro-earthquakes were detected and recorded during 85 hours of seismic monitoring. About half of the events were associated with crevassing near the surface of the ice stream, presumably near the margins. The most interesting phenomenon, however, was a group of 9 small micro-earthquakes occurring only minutes apart. Preliminary analysis of the location and focal

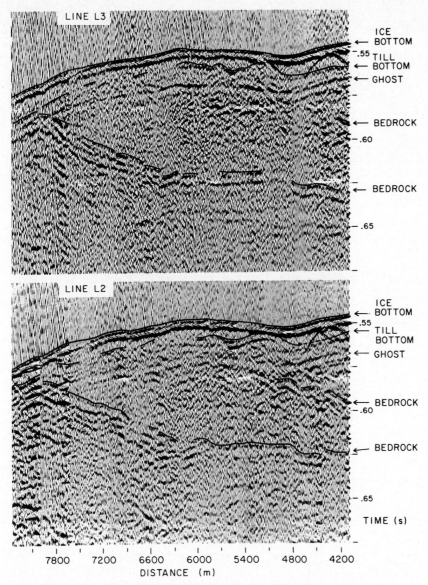

Fig.6. Two parallel 4 km long seismic reflection sections transverse to the ice-stream axis at camp UB. Ice-bottom and (mobile) till-bottom reflections are labeled; later marked events are reflections from sedimentary rock layers which appear truncated by the mobile till-bottom reflection. From Rooney and others (in press).

mechanism of these events (Blankenship and others 1987) shows that they were associated with slip on low-angle thrust planes at or near the bottom of the ice stream. The paucity of micro-earthquakes arising from the bottom of the ice stream leads Blankenship and others (1987) to conclude that virtually none of the energy of ice stream motion is dissipated by brittle fracture at the bed.

The complex "ridge B1/B2" zone between ice streams B1 and B2, which appears to be breaking up as the ice streams grow, has already been described. We find support for the hypothesis that ice stream B is growing headward from a consideration of its mass balance. The flux out of ice stream B into the Ross Ice Shelf is perhaps as much as twice as large as the input into the entire drainage basin of the ice stream — the apparent imbalance is 10-20 km^3 a^{-1} (Shabtaie and Bentley in press). Our co-operative work with Whillans (Shabtaie and others in press[a]) suggests that about half of that can be attributed to the zone in apparent transition, which has an area of about 10^4 km^2. That would imply a mean surface lowering of one or two meters per year. Surface lowerings several times this have been measured locally by Whillans and others (in press), as one might expect if the ice at a particular location is being transformed from sheet flow to ice stream flow.

The dynamics of the ice stream are bound to be fundamently affected if the deforming till is extensive underneath the ice stream. Alley and others (in press[b])

have developed a simple one-dimensional model for ice-stream flow on deforming till. The model assumes continuity of ice and till and a linear-viscous till rheology. The steady-state solution requires that the viscosity decrease downstream; we believe this probably results from a combination of decreasing effective pressure in the till and partial decoupling between the ice and the till. Numerical experiments with fixed ice thicknesses at the ends show that a perturbation at one end causes a wave of adjustment to travel the length of the ice stream in about 50 years, with a new steady-state attained in about 200 years. This rapid adjustment suggests that any non-steady effects observed in the ice stream probably represent continuing changes in the ice sheet above the ice stream or in the ice shelf downstream. As we discussed above, we believe that in ice stream B the deviation from steady-state reflects on-going changes at the head of the ice stream.

ICE STREAM C

Ice stream C is of particular interest because it has ceased its rapid motion. It now moves less than 10 m/yr both at its grounding line (Thomas and others 1984; Bindschadler and others in press) and at Upstream C camp (UC in Figs 1 and 3) (I.M. Whillans personal communication, 1985). Yet it clearly has been an active ice stream in the past — despite the absence of surface crevasses

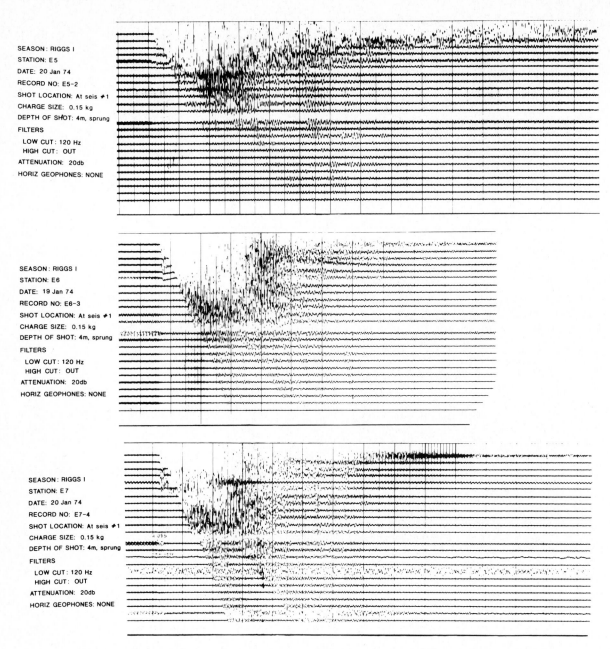

SEASON : RIGGS I
STATION: E5
DATE: 20 Jan 74
RECORD NO: E5-2
SHOT LOCATION: At seis #1
CHARGE SIZE: 0.15 kg
DEPTH OF SHOT: 4m, sprung
FILTERS
  LOW CUT: 120 Hz
  HIGH CUT: OUT
ATTENUATION: 20db
HORIZ GEOPHONES: NONE

SEASON : RIGGS I
STATION: E6
DATE: 19 Jan 74
RECORD NO: E6-3
SHOT LOCATION: At seis #1
CHARGE SIZE: 0.15 kg
DEPTH OF SHOT: 4m, sprung
FILTERS
  LOW CUT: 120 Hz
  HIGH CUT: OUT
ATTENUATION: 20db
HORIZ GEOPHONES: NONE

SEASON : RIGGS I
STATION: E7
DATE: 20 Jan 74
RECORD NO: E7-4
SHOT LOCATION: At seis #1
CHARGE SIZE: 0.15 kg
DEPTH OF SHOT: 4m, sprung
FILTERS
  LOW CUT: 120 Hz
  HIGH CUT: OUT
ATTENUATION: 20db
HORIZ GEOPHONES: NONE

Fig.7a. Seismic reflection records from the grid northwest corner of the Ross Ice Shelf and neighboring "ice plain". Records from stations E5, E6, and E7, all on the "ice plain"; note poor quality of the reflections.

(Robin and others 1970[a]) the radar-sounding records across it show the clutter echoes in the body of the ice stream and the marginal shear zones that are diagnostic of other ice streams (Rose 1979; Shabtaie and Bentley in press). Rose (1979) pointed out that the depth of burial of crevasses could give the date ice stream C was last active, but the resolution of his system allowed him only to say that that date was within the last 1000 years. A sounding program we carried out on the surface at UpC using a monopulse radar revealed numerous hyperbolae generated from the tops of crevasses 30-40 m below the surface. From measurements of snow accumulation rate and density, Shabtaie and Bentley (in press) estimate that the crevasses ceased to form about 250 years ago. As the crevasses might have been buried to some extent while ice stream C was still active, 250 years would then represent the upper limit to the time that has passed since ice stream C stagnated.

Since ice stream C is barely moving, its mass balance is strongly positive. Even if one assumes that the entire catchment area that once fed ice stream C is now diverted to ice streams B and D, the mass that falls on the surface of ice stream C itself is still eight times as large as the output flux (Shabtaie and Bentley in press). (If one includes

the nominal catchment area, the ratio is almost 40 : 1.) Clearly, then, ice stream C must be growing in elevation, and it seems likely that some time in the future it will become reactivated.

The map of surface elevations (Fig.5) (Shabtaie and others in press[b]) shows several minima in the surface of ice stream C. Flat spots were observed years ago by Robin and others (1970[b]), in an area that we now know to be on the sheet-flow ice grid southwest of ice stream C (Siple ice dome). They also noted that strong radar echoes were associated with these spots, and attributed both the flat surface and the strong echoes to "pseudo ice shelves" floating on trapped subglacial water without a hydrostatic connection to the sea. Detailed work by Shabtaie and others (in press[b]) has indicated that strongly reflecting areas are also found beneath ice stream C, and that in some places the hydraulic pressure gradient at the bed is negative, i.e such as to force water locally up-stream. Although a correlation between strong reflections and negative hydraulic pressure gradients has not been established, it is nevertheless interesting to speculate whether the Robin and Weertman (1973) instability mechanism, which depends upon reversals in the hydraulic pressure gradient, may become

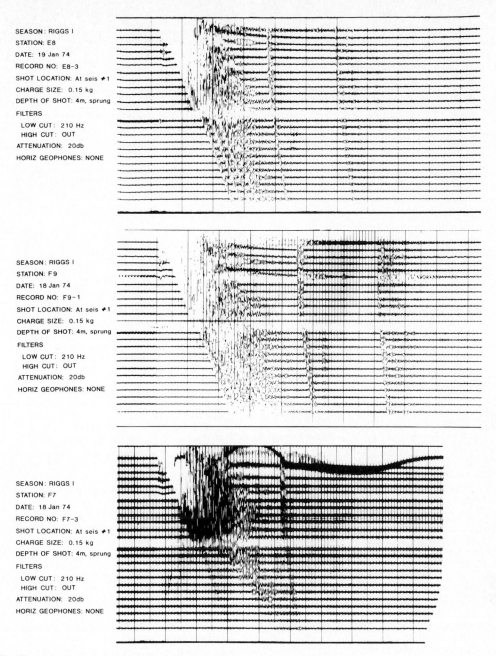

SEASON: RIGGS I
STATION: E8
DATE: 19 Jan 74
RECORD NO: E8-3
SHOT LOCATION: At seis ∓1
CHARGE SIZE: 0.15 kg
DEPTH OF SHOT: 4m, sprung
FILTERS
   LOW CUT: 210 Hz
   HIGH CUT: OUT
   ATTENUATION: 20db
HORIZ GEOPHONES: NONE

SEASON: RIGGS I
STATION: F9
DATE: 18 Jan 74
RECORD NO: F9-1
SHOT LOCATION: At seis ∓1
CHARGE SIZE: 0.15 kg
DEPTH OF SHOT: 4m, sprung
FILTERS
   LOW CUT: 210 Hz
   HIGH CUT: OUT
   ATTENUATION: 20db
HORIZ GEOPHONES: NONE

SEASON: RIGGS I
STATION: F7
DATE: 18 Jan 74
RECORD NO: F7-3
SHOT LOCATION: At seis ∓1
CHARGE SIZE: 0.15 kg
DEPTH OF SHOT: 4m, sprung
FILTERS
   LOW CUT: 210 Hz
   HIGH CUT: OUT
   ATTENUATION: 20db
HORIZ GEOPHONES: NONE

Fig.7b. Records from stations E8 and F9 on the floating ice shelf; note the excellent reflection quality. Also shown is a record from station F7, on the edge of the "ice plain".

operative beneath ice stream C and lead to its eventual re-activation, rather as suggested by Hughes (1975).

OTHER ICE STREAMS AND RIDGES

Ice stream A is different from ice streams B and C in its subglacial physiography (Shabtaie and others in press[b]). It lies against the wall of the Transantarctic Mountains on one side, and also shows a more pronounced subglacial margin on its other side than the other ice streams do. However, its surface looks very much like that of ice stream B, characterized by a ridge/trough structure parallel to ice movement with an amplitude of some 20 m (Fig.5). As on ice stream B, the surface slope diminishes downstream and out on to the "ice plain", but in contrast to ice stream B the slope continues to diminish right to the grounding line. There is nothing in the surface appearance of ice stream A that suggests that it is fundamentally different in dynamics from ice stream B.

Ridge AB is characterized by an irregular surface topography also (Shabtaie and others in press[b]). One might expect on a simple sheet-flow dome or ridge that the

surface contours would be smooth, indicating a simple pattern of outflow. Instead, the surface topography is characterized by a complex pattern of valleys and ridges (Fig.5), which suggests that simple equilibrium does not exist here. Ridge BC shows two domes (the larger around grid position 5.4°S, 4.6°W, and the smaller at grid 5.9°S, 2.7°W) separated by a transverse valley. Surface slopes are smooth downstream of the larger dome, which suggests to us an equilibrium form. Farther up-stream, however, the surface again is very irregular, perhaps reflecting a complex transient interaction with the upper parts of ice streams B and C. Siple ice dome, grid south of ice stream C, displays a regular, although asymmetric, dome-like shape.

ROSS ICE SHELF

Radar profiling has helped to delineate several features of the Ross Ice Shelf. Ice rises are of particular interest. Since an ice rise, by definition, is characterized by ice movement that is largely independent of the surrounding ice shelf and carries only the local accumulation in a largely radial outflow, longitudinal stresses are small and there are

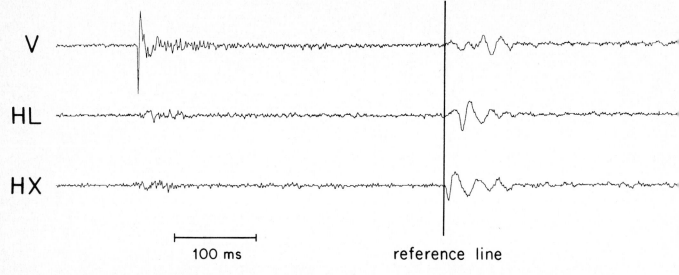

V

HL

HX

|← 100 ms →|

reference line

Fig.8. Passive seismogram of an event at the base of the ice stream near the recording array. The first trace (V) records vertical motion, the second (HL) horizontal motion along the axis of the ice stream, and the third (HX) horizontal motion normal to that axis. Note the close similarity between the wave forms on the HL and HV axes, and the 16 ms delay between the two traces.

no surface crevasses. Consequently, radar echoes from the surface of an ice rise exhibit little clutter; this characteristic can be used for mapping. On this basis Crary Ice Rise was found by Shabtaie and Bentley (in press) to extend about 70 km farther to the grid northwest than previously mapped, and to be flanked by two narrow, clutter-free strips with ill-defined boundaries on each side (Fig.3) that may also be ice rises.

Upstream from Crary Ice Rise there is another small area that shows a clutter-free zone on the radar record — "ice rise a" (Fig.3: grid position 5.3°S, 2.1°W). Although there are no boundary crevasses visible, either from the air or from the surface, the radar clutter is strong at the boundary, suggesting the presence of buried crevasses. Shabtaie and Bentley (in press) show "ice rise a" right on the suture zone that continues down-stream along the axis of ice stream B from the point of "ridge B1/B2" (Fig.3). Recent measurements by Bindschadler and others (in press) show that "ice rise a" is moving at the same speed as the surrounding ice. Following an idea of Whillans and others (in press), we suggest that "ice rise a" is a piece of undisturbed ice that broke off the point of "ridge B1/B2" and was carried on down the ice stream more or less un-deformed. It should be noted that "ice rise a" still lies up-stream of the grounding line, within the "ice plain". Because it is moving, "ice rise a" is not really an ice rise at all, nor is it an ice rumple, because the surrounding ice is grounded. We propose the term "glacial microplate" for such a feature, i.e. an "island" of undisturbed ice moving within an ice stream or "ice plain". It is possible that the clutter-free strips that flank Crary Ice Rise are "glacial microplates", and even that the up-stream end of Crary Ice Rise itself incorporates "glacial micro-terranes".

The strong boundary crevasse zones of the ice stream mark flow bands that can be followed far out on to the Ross Ice Shelf (Shabtaie and Bentley in press). Knowledge of the boundaries of the ice streams and their flow bands make it possible to calculate the ice discharge from each ice stream draining into the Ross Ice Shelf. Fluxes out of ice streams B and C have each been measured in two places. On ice stream B one line is at the head of the "ice plain" and the other near the grounding line. On the flow band from ice stream C, the up-stream gate is at the grounding line and the down-stream gate is about 200 km farther down-stream on the ice shelf. For neither of these pairs of gates is there a significant difference between influx and outflux (Shabtaie and Bentley, in press). Shabtaie and Bentley (in press) also compare the outflow from all the "Ross ice streams" to an estimate of input through snow accumulation on the whole interior of the West Antarctic

ice sheet up to the ice divide. They find for the entire system a net balance that is suggestively, but not conclusively (in terms of the errors in measurement), negative by about 10-20 km³ a⁻¹.

The ice flux into the Ross Ice Shelf from the West Antarctic inland ice can be compared with the output through a line about 100 km inland of the front of the ice shelf (Bentley 1985). If one assumes that the Ross Ice Shelf itself is in balance, that leads to an estimate for the average bottom melt rate of 0.11 ± 0.03 m a⁻¹ for the part of the ice shelf that lies inland of the output line. For the whole ice shelf, the estimated total volume loss by melting on the assumption of steady state is 60 ± 15 km³ a⁻¹. This is about half the estimate that one gets from the oceanographic measurements of Pillsbury and Jacobs (1985), which give 150 ± 75 km³ a⁻¹, after correction for the melt (30 km³ a⁻¹) in the high melt-rate region seaward of the output line (extrapolated from Giovinetto and Zumberge (1968)). In view of the quoted errors and the fact that the calculation by Giovinetto and Zumberge (1968) was based on only one line of measurements, these estimates of melt rate cannot be considered significantly different.

ACKNOWLEDGEMENTS

This work was supported by the National Science Foundation under grant DPP-8412404. This is contribution number 461 of the Geophysical and Polar Research Center, University of Wisconsin-Madison.

REFERENCES

Alley R B, Blankenship D D, Bentley C R, Rooney S T 1986 Deformation of till beneath Ice Stream B, West Antarctica. *Nature* 322 (6074): 57-59

Alley R B, Blankenship D D, Bentley C R, Rooney S T In press[a] Till beneath Ice Stream B. 3. Till deformation: evidence and implications. *Journal of Geophysical Research*

Alley R B, Blankenship D D, Rooney S T, Bentley C R In press[b] Till beneath Ice Stream B. 4. A coupled ice-till flow model. *Journal of Geophysical Research*

Bentley C R 1971 Seismic anisotropy in the West Antarctic ice sheet. *In* Crary A P (ed) *Antarctic snow and ice studies. II.* Washington, DC, American Geophysical Union: 131-177 (Antarctic Research Series 16)

Bentley C R 1972 Seismic-wave velocities in anisotropic ice: a comparison of measured and calculated values in and around the deep drill hole at Byrd Station, Antarctica. *Journal of Geophysical Research* 77(23): 4406-4420

Bentley C R 1985 Glaciological evidence: the Ross Sea sector. *In Glaciers, Ice Sheets, and Sea Level: Effect of a CO_2-induced Climatic Change. Report of a workshop held in Seattle, Washington, September 13–15, 1984.* Washington, DC, U.S. Department of Energy: 178-196

Bentley C R In press Antarctic ice streams: a review. *Journal of Geophysical Research*

Bindschadler R A, Stephenson S N, MacAyeal D R, Shabtaie S In press[a] Ice dynamics at the mouth of Ice Stream B, Antarctica. *Journal of Geophysical Research*

Bindschadler R A, MacAyeal D R, Stephenson S N In press[b] Ice stream-ice shelf interaction in West Antarctica. *In Proceedings of Utrecht Workshop on the Dynamics of the West Antarctic Ice Sheet.* Dordrecht, D. Reidel

Bindschadler R A, Stephenson S N, MacAyeal D R, Shabtaie S In press[c] Interaction between Ice Stream B and Ross Ice Shelf, Antarctica. *Antarctic Journal of the United States*

Blankenship D D Unpublished P-wave anisotropy in the high polar ice of East Antarctica. (Master's thesis, University of Wisconsin, Madison, 1982)

Blankenship D D, Bentley C R, Rooney S T, Alley R B 1986 Seismic measurements reveal a saturated porous layer beneath an active Antarctic ice stream. *Nature* 322 (6074):54-57

Blankenship D D, Anandakrishnan S, Kempf J L, Bentley C R 1987 Microearthquakes under and alongside ice stream B, Antarctica, detected by a new passive seismic array. *Annals of Glaciology* 9: 30–34

Blankenship D D, Bentley C R, Rooney S T, Alley R B In press Till beneath Ice Stream B. 1. Properties derived from seismic travel times. *Journal of Geophysical Research*

Giovinetto M B, Zumberge J H 1968 The ice regime of the eastern part of the Ross Ice Shelf drainage system. *International Association of Scientific Hydrology Publication 79 (General Assembly of Bern 1967 – Snow and Ice)*: 255-266

Gow A J 1970 Preliminary results of studies of ice cores from the 2164 m deep drill hole, Byrd Station, Antarctica. *International Association of Scientific Hydrology Publication 86 (ISAGE)*: 78-90

Gravenor C P, Meneley W A 1958 Glacial flutings in central and northern Alberta. *American Journal of Science* 256(10): 715-728

Hughes T 1975 The West Antarctic ice sheet: instability, disintegration, and initiation of ice ages. *Reviews of Geophysics and Space Physics* 13(4): 502-526

Kohnen H, Gow A J 1979 Ultrasonic velocity investigations of crystal anisotropy in deep ice cores from Antarctica. *Journal of Geophysical Research* 84(C8): 4865-4874

Pillsbury R D, Jacobs S S 1985 Preliminary observations from long-term current meter moorings near the Ross Ice Shelf, Antarctica. *In Jacobs S S (ed) Oceanology of the Antarctic continental shelf.* Washington, DC, American Geophysical Union: 87-107 (Antarctic Research Series 43)

Poulter T C 1950 *Geophysical studies in the Antarctic.* Stanford, CA, Stanford Research Institute

Robertson J D Unpublished Geophysical studies on the Ross Ice Shelf, Antarctica. (PhD thesis, University of Wisconsin, Madison, 1975)

Robertson J D, Bentley C R In press *Seismic studies on the grid – western Ross Ice Shelf.* Washington, DC, American Geophysical Union (Antarctic Research Series)

Robin G de Q, Weertman J 1973 Cyclic surging of glaciers. *Journal of Glaciology* 12(64): 3-18

Robin G de Q, Swithinbank C W M, Smith B M E 1970[a] Radio echo exploration of the Antarctic ice sheet. *International Association of Scientific Hydrology Publication 86 (ISAGE)*: 97-115

Robin G de Q, Evans S, Drewry D J, Harrison C H, Petrie D L 1970[b] Radio-echo sounding of the Antarctic ice sheet. *Antarctic Journal of the United States* 5(6): 229-232

Rooney S T, Blankenship D D, Alley R B, Bentley C R In press Till beneath Ice Stream B. 2. Structure and continuity. *Journal of Geophysical Research*

Rose K E 1979 Characteristics of ice flow in Marie Byrd Land, Antarctica. *Journal of Glaciology* 24(90): 63-75

Schultz D G, Powell L A, Bentley C R 1987 A digital radar system for echo studies on ice sheets. *Annals of Glaciology* 9: 206–210

Shabtaie S, Bentley C R 1986 Ice streams and grounding zones of West Antarctica and the Ross Ice Shelf. (Abstract.) *Annals of Glaciology* 8: 199-200

Shabtaie S, Bentley C R In press West Antarctic ice streams draining into the Ross Ice Shelf: configuration and mass balance. *Journal of Geophysical Research*

Shabtaie S, Bentley C R, Whillans I M, MacAyeal D R, Bindschadler R A In press[a] Partial collapse of the West Antarctic ice sheet: implications from mass balance studies of ice streams. *Journal of Geophysical Research*

Shabtaie S, Whillans I M, Bentley C R In press[b] Surface elevations on ice streams A, B, and C. *Journal of Geophysical Research*

Thomas R H, MacAyeal D R, Eiler D H, Gaylord D R 1984 Glaciological studies on the Ross Ice Shelf, Antarctica, 1973-1978. *In Bentley C R, Hayes D E (eds) The Ross Ice Shelf: glaciology and geophysics.* Washington, DC, American Geophysical Union: 21-53 (Antarctic Research Series 42)

Whillans I M, Bolzan J, Shabtaie S In press Velocity of Ice Stream B, Antarctica, and its mass balance. *Journal of Geophysical Research*

Annals of Glaciology 9   1987
© International Glaciological Society

MICROEARTHQUAKES UNDER AND ALONGSIDE ICE STREAM B, ANTARCTICA, DETECTED BY A NEW PASSIVE SEISMIC ARRAY

by

D.D. Blankenship, S. Anandakrishnan, J.L. Kempf and C.R. Bentley

(Geophysical and Polar Research Center, University of Wisconsin-Madison, Madison, WI 53706, U.S.A.)

ABSTRACT

A new seismographic array with a band width of 500 Hz per channel and a dynamic range of 96 dB was developed for detecting natural events on glaciers. It was first deployed on ice stream B during the 1985–86 austral summer. The network consists of nine solar-powered seismographs, each monitoring three components of ground motion. Each of the seismographs is connected by up to 4 km of fiber-optic cable to a central node where seismic events are both detected and recorded. During 85 h of passive seismic monitoring on ice stream B, 25 microearthquakes were observed. Sixteen of these events were associated with shallow crevassing, mostly near the margins, although not within the zones of extreme shearing that bound the ice streams. Nine microearthquakes were associated with low-angle thrusting near the base of the ice stream. The principal initial result of these passive seismic studies is the demonstration that virtually none of the energy dissipated beneath ice stream B takes place through brittle fracture near the base. Nevertheless, fracture associated with microearthquakes may play a significant role in sub-glacial erosion.

INTRODUCTION

The stability of the marine ice sheet of West Antarctica is one of the fundamental unsolved problems in glaciology. A first approach to understanding the dynamics of the West Antarctic ice sheet is to understand the dynamics of the so-called "Ross ice streams" (Bentley in press) that drain the majority of its ice (Fig.1). Recent advances in geophysical remote sensing (Bentley and others 1987) have assisted in comprehending the processes that control these ice streams as well as the nature of their interaction with the Ross Ice Shelf. To date, active seismology and radar have been the primary investigative tools; this paper discusses the first application of passive seismology to problems of ice-stream dynamics in West Antarctica.

The earlier use of passive seismology has been effective only to study the stress regime of Athabasca Glacier in North America (Neave and Savage 1970); no studies prior to ours have been undertaken in Antarctica. The work of Neave and Savage (1970) used a seismic array with a band width of several hundred Hz and a dynamic range of 35 dB; only the vertical component of ground motion was monitored. The microearthquakes observed in their investigation had a dominant period of 10 msec and were caused by crevassing within 60 m of the surface of the glacier.

Serious seismic studies of basal sliding have not been possible because seismic events associated with fracture or slip at the base of a glacier have not been identified conclusively. The seismic array deployed by Neave and Savage (1970) was theoretically large enough to locate microearthquake sources associated with basal sliding, but none was observed. It is possible that the area of rupture or slip associated with basal sliding of Athabasca Glacier was smaller than that associated with the crevassing near the

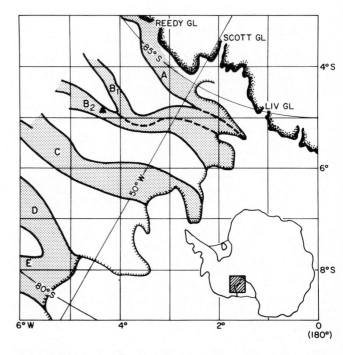

Fig.1. The Siple and Gould Coasts of West Antarctica. The "Ross ice streams" are lettered A, B_1, B_2, C, D, and E. The triangle denotes the location of "Upstream B" camp. Simplified from Shabtaie and Bentley (1986).

surface. Because the amplitude of the lower-frequency components of a seismic event is proportional to the size of the rupture area (Aki and Richards 1980, p.811) the limited dynamic range of the seismic array deployed by Neave and Savage might have made it difficult to detect seismic events associated with both near-surface crevassing and basal sliding. At the same time, because the upper-limit frequency of the seismic energy (i.e. the corner frequency) is inversely proportional to some power of the size of the rupture area (Aki and Richards 1980, p.821), most of the energy of any tremors caused by basal sliding might have occurred at frequencies above the upper limit of detectability for their seismic array.

To use passive seismology for studies of the internal stresses and basal sliding of the Ross ice streams, we have developed a seismic array that possesses much larger dynamic range (at least 96 dB) and band width (500 Hz) than previously used in glacier studies. In this paper the technical aspects of this newly developed passive seismic array are described and the results of its first deployment, on ice stream B, are reviewed.

DEVELOPMENT OF THE SEISMIC ARRAY

The increased dynamic range and band width required for the new passive seismic array were achieved by

incorporating recent advances in both analog-to-digital conversion and data-transmission technology. The array (Fig.2) comprises nine "remote" stations, at which three components of ground motion are digitized, connected by fiber-optic links to a "home" unit where the seismic information is reconstructed. The home unit is attached to a data logger (also developed at the Geophysical and Polar Research Center) where detection algorithms are applied to

REMOTE STATION

OPTICAL DATA LINK

HOME UNIT

Fig.2. Schematic representation of the three primary elements of the passive seismic array developed by the Geophysical and Polar Research Center. H_1 and H_x are the horizontal components of ground motion that are parallel and transverse to ice flow, respectively.

the digitized seismic signals and detected events are recorded.

Three geophones, one each for the vertical and two horizontal components of ground motion, are connected to each remote station. These geophones are characterized by a natural frequency of 28 Hz and a frequency response that is flat to at least 420 Hz. Low-noise pre-amplifiers (≈0.1 μV r.m.s.) are utilized at the input stages of the remote station. The pre-amplifiers are followed by amplifiers with a gain that can be varied between 36 to 72 dB in 12 dB steps. Four poles of a Butterworth low-pass filter with a corner frequency of 500 Hz are used to eliminate any spurious high-frequency noise. The amplified and filtered signal from each geophone is sampled at 8 kHz by a precision-integrated sample-and-hold chip. The sampled signal is digitized by a 16 bit analog-to-digital converter. As the sampling frequency is four octaves above the corner frequency of the filter, the signal is 96 dB down at the sampling frequency and aliasing is eliminated. The digitized sample (16 bits) plus a header of 8 bits containing gain information is multiplexed with the corresponding 24 bits from the other two channels in each remote station. The power for each remote station is derived from a 1 m² (10 W) solar panel; a 12 V lead-acid battery provides voltage regulation and back-up power.

Each optical link uses a 2 MHz clock and an optical transmitter with internal Manchester encoding to transmit data from the remote stations to the home unit. The bit stream produced by the remote stations modulates the light power of the transmitter. This modulated light is coupled to a fiber-optic cable with a step-index wave guide (100 μm core/140 μm cladding). The central wave guide in the cable is surrounded by a loose protective buffer, Kevlar braid runs the length of the cable for tensile strength, and the outer jacket is polyurethane. The optical attenuation in the wave guide is ≈4 dB/km and the loss in each coupling is about 2 dB, so for a 2 km optical link, there is a 12 dB

loss, and for a 4 km link, a 22 dB loss. An optical receiver with internal Manchester decoding receives the data at the home unit. The difference in light-power output by the transmitter and light-power sensitivity of the receiver (flux budget) is about 27 dB, i.e. a loss of up to 27 dB in the fiber-optic cable could be sustained without destroying the integrity of the link.

At the home unit, the optical receivers reproduce the bit stream as well as the clock from the remote stations. This bit stream is de-multiplexed into three channels and re-converted to analog form at 8 kHz by a 16 bit digital-to-analog converter. Four poles of a low-pass Butterworth filter with a corner frequency of 500 Hz are used for smoothing the analog wave form on each channel. A reproduction of the seismic information at each of the remote stations is then fed to a data logger, where the data are digitally monitored and events are detected and recorded.

DEPLOYMENT OF THE SEISMIC ARRAY

Commencing in 1983, the University of Wisconsin joined the Ohio State University, NASA, and the University of Chicago in an intensive study of the West Antarctic ice sheet near the Siple and Gould Coasts (Fig.1). As part of this cooperative investigation, the newly developed passive seismic array was deployed on ice stream B during the 1985–86 austral summer.

The array (Fig.3) was positioned near the center of the ice stream. The nine remote stations were distributed on a

UPSTREAM B PASSIVE SEISMIC ARRAY

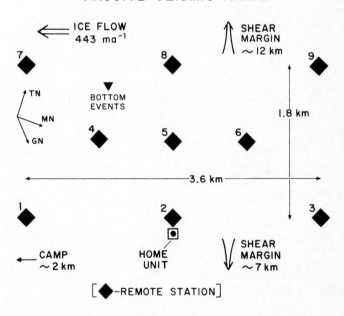

Fig.3. Schematic representation of the new passive seismic array as it was deployed on ice stream B. The numbered diamonds are the remote stations. TN, MN, and GN denote true north, magnetic north, and grid north, respectively.

rectangle measuring 1.8 km by 3.6 km, with the long dimension of the array parallel to the direction of ice movement. At each remote station, one horizontal geophone was placed parallel to the ice flow and the other transverse to the ice flow. The array was tested and calibrated by detonating 0.9 kg explosive charges at 30 surveyed locations within it.

The passive seismic array was maintained for 3 weeks during which 85 h of microearthquake monitoring were achieved (Fig.4). Because of the "cultural" noise emanating from the "Upstream B" camp, monitoring was generally possible only between midnight and 08.00 h (McMurdo local

MICROEARTHQUAKES AT UPSTREAM B

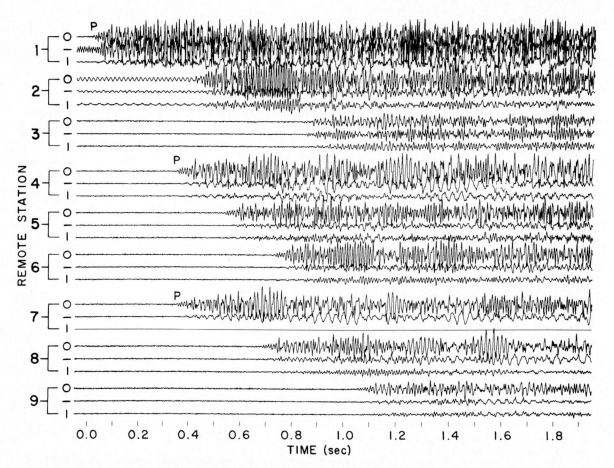

Fig.4. Calendar showing periods (stippled blocks) of seismic monitoring on ice stream B during 1985–86. Each individual segment is marked with the day/month and with tick marks at 00.00, 04.00, and 08.00 h McMurdo time. Seismic events are indicated by dots.

time). During the monitoring, the data logger continuously digitized each of the 27 output channels of the home unit; it also applied a threshold detector to one of these channels. A recording was triggered when the absolute value of the signal from the vertical channel of a pre-selected remote station exceeded a level set at about 0.5% of the total dynamic range of the array. Remote stations 2, 4, 6, and 8 were used at various times for event detection. When triggering occurred, digitized seismic signals from all remote stations for the 5 s interval starting 2 s before the triggering were recorded on a removable disk cartridge.

OBSERVATIONS OF MICROEARTHQUAKES

During the 85 h of seismic monitoring 25 micro-earthquakes were detected and recorded (Fig.4). Sixteen of these events were true "icequakes" associated with crevassing, presumably near the surface. Fig.5 depicts a typical icequake; the first arrivals are P-waves (compressional waves) that are characteristically emergent. The icequakes often occurred in groups of up to four events typically 1–5 min apart. Because of the frequent absence of clear S-wave (shear-wave) arrivals, precise locations of the sources for the icequakes (particularly their depths) will require careful numerical inversion of the P-wave travel times. Preliminary graphical analysis of the P-wave travel times for these icequakes shows that most of their sources were within 10 km of the seismic array and were associated with open crevasses near the shear margins of the ice stream that were observed on aerial photography by Vornberger and Whillans (1986). Surprisingly, none of the sources were located in the "chaotic zones" of intense shearing that bound the ice stream.

The most interesting phenomenon observed was a swarm of nine very small microearthquakes occurring just a few minutes apart on 22 December (Fig.4). The arrivals from one of these events are depicted in Fig.6; other microearthquakes in the swarm yielded very similar seismograms. All of the P-wave and many of the S-wave arrivals associated with these events are quite distinct.

Assuming the ice column is both vertically and laterally homogeneous, a simple iterative analysis of the P-wave and S-wave travel times places the foci of seven of these microearthquakes within several tens of meters of the base of the ice at the location indicated in Fig.3. The locations were calculated using a P-wave velocity in the ice of 3.83 km/s and an S-wave velocity of 1.94 km/s, both obtained by Blankenship and others (in press), as well as an ice thickness of 1080 m determined by Rooney and others (in press). The question of whether these foci are within the ice, sub-glacial till (reported by Blankenship and others

Fig.5.   Microearthquake associated with shallow crevassing near the shear margins of ice stream B. The vertical component of ground motion is denoted by O; − and | are horizontal components parallel and transverse to ice flow, respectively.   P-wave arrivals at the first of each group of three stations are indicated.

(1986; in press), and Alley and others (1986; in press)) or bedrock beneath the till, is important to the analysis of erosional mechanisms of ice stream B and will be addressed in future analyses.

Assuming these basal events arise from a double-couple source along some fault plane, the nodes and antinodes of the radiation pattern for the P-wave first motions (see Fig.6) can be used to infer the geometry of this fault plane. Aki and Richards (1980, p.82) show that a maximum compression and a maximum dilatation will occur at 45° angles to the fault-plane normal that has its origin at the source. These antinodes lie in a plane defined by the normal and slip vector; the intervening node lies along the fault-plane normal. The compressional antinode will lie on the side of the node that is in the direction of the slip of the near side of the fault; the dilatational antinode will lie on the opposite side. From Fig.6 we see that the P-wave first motion is up (representing compression) at stations 1 and 7 downstream of the bottom events (Fig.3), whereas at the other stations it is down (representing dilatation). The P-wave first motion at station 4 is quite small (the gain on the vertical channel at station 4 is higher than on the other vertical channels), implying that it lies close to a node in the P-wave radiation pattern. Because the source is known to lie nearly beneath station 4, we see from the description above that these observations are consistent with a fault plane that is nearly horizontal, and with a slip of the upper wall of the fault that is parallel to the direction of ice flow. The source location and fault-plane geometry for this event imply that it arises from low-angle thrust faulting near the base of the ice stream. Because we believe that this thrusting may be related to erosion at the base of the till as proposed by Rooney and others (in press), we are currently in the process of obtaining more precise fault-plane solutions for the nine-event swarm by formally inverting both the P-wave and S-wave radiation patterns.

Another interesting feature, best depicted on the seismogram for remote station 5 (Fig.6), is the possibility of

shear-wave splitting; the S-wave polarized transverse to ice flow travels approximately 2% faster than the S-wave polarized parallel to flow. One of the authors (DDB) has observed comparable anisotropy for S-waves recorded during seismic reflection experiments near "Upstream B" and a detailed analysis will be presented in a future paper.

A significant result of this first deployment of a passive seismic array on an active Antarctic ice stream is the paucity of microearthquakes whose sources are near the base of the ice. The seismic energy released by each of the nine bottom events can be calculated directly from the seismograms. Robin (1958) gives the energy per unit area (E_s) for a seismic pulse measured by a detector near the surface as

$$E_s = 1/8 \; \rho \; \lambda \; V^2$$

where ρ and λ are, respectively, the density of the medium and the seismic wavelength, and V is the maximum particle velocity. At the position of our detector ρ was $\approx 400 \, \text{kg/m}^3$ (Alley personal communication, 1986) and the P- and S-wave speeds were about 1000 and 600 m/s (Blankenship and others in press), respectively. For typical P- and S-waves arriving at stations 5, 6, and 8 (e.g. Fig.6), we obtained $V = 2 \times 10^{-7}$ m/s and $\lambda = 10$ m for the narrower initial P-wave pulse, $V = 1 \times 10^{-7}$ and $\lambda = 25$ m for the broader P-wave pulse, and $V = 1 \times 10^{-7}$ and $\lambda = 15$ for each S-wave component (i.e. one transverse and one parallel to flow). Thus for each microearthquake, the total E_s in the P- and S-waves is about $0.5 \times 10^{-10} \, \text{J/m}^2$. Since stations 5, 7, and 8 were each approximately 1 km from the epicenters (Fig.3), the total seismic energy for the nine-event swarm was $\approx 10^{-2}$ J. Using a typical seismic efficiency of 1% (Jaeger and Cook, 1970), the total energy released by these basal events seems to be about one Joule. However, because in the above calculation we have ignored both losses due to internal friction and likely asymmetry in the P- and S-wave radiation patterns, we believe that the total

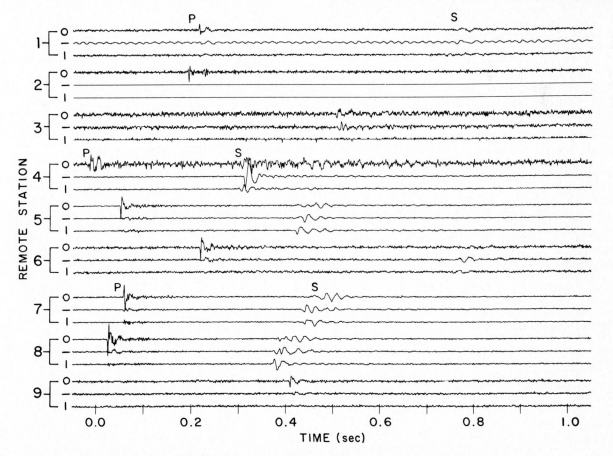

Fig.6. Microearthquake associated with low-angle thrusting near the base of ice stream B. The vertical component of ground motion is denoted by O; − and │ are the horizontal components parallel and transverse to ice flow, respectively. P-wave and S-wave arrivals at the first of each group of three stations are indicated. The gains of individual channels have been adjusted to increase the visibility of arrivals.

energy released by faulting at the base of the ice stream during our period of observation could have been as much as 10 Joules.

The energy released by basal faulting can be compared with the total energy dissipated by glacier sliding. The latter energy, E, is given by

$$E = u_s \tau_b TA$$

where u_s is the sliding velocity, τ_b is the basal shear stress, T is the elapsed time, and A is an appropriate basal area. If for A we take 3×10^6 m^2 (as an approximate area of basal coverage by the passive seismic array), $T = 0.01$ a (for the 85 hours of seismic monitoring), $u_s = 440$ m a^{-1} (Whillans and others in press) and $\tau_b = 2 \times 10^4$ Pa (from measurements of ice thickness and surface slope given by Shabtaie and Bentley (in press)), we obtain $E \simeq 10^{11}$ J. Thus the energy released by rupture at the base of ice stream B is at most the order of one part in 10^{10} of the energy dissipated in ice stream movement. Even though our short period of observation could have been unrepresentative, it seems clear that brittle fracture plays no direct role in the dynamics at the base of ice stream B. However, if the basal microearthquakes arise from fracture of the lithified sediments that have been shown to exist directly beneath the sub-glacial till (Rooney and others in press), they could be an important contributor to basal erosion.

CONCLUSIONS

A new passive seismic array that uses new technology to provide both a broad band width and a large dynamic range has been developed and deployed on an active Antarctic ice stream. Microearthquakes that occurred near the shear margins and close to the base of the ice stream were detected by this new array and digitally recorded. Although analysis of these events is still at a preliminary stage, a principal result of this deployment is that, at least for ice stream B, brittle fracture dissipates very little of the energy associated with basal sliding; however, it may play an important role in sub-glacial erosion.

ACKNOWLEDGEMENTS

We wish to thank L A Powell, B D Karsh, D R Novotny, and P Kinnerk for their engineering assistance, S T Rooney and K Kilalea for help with the field work, and A N Mares and J Gallagher for manuscript and illustration preparation. We are particularly indebted to B R Weertman, S T Rooney, and R B Alley for many helpful discussions. This work was supported by the US National Science Foundation under grant DPP-8412404. This is contribution No. 462 of the Geophysical and Polar Research Center, University of Wisconsin–Madison.

REFERENCES

Aki K, Richards P 1980 *Quantitative seismology*. San Francisco, W.H. Freeman

Alley R B, Blankenship D D, Bentley C R, Rooney S T 1986 Deformation of till beneath ice stream B, West Antarctica. *Nature* 322(6074): 57–59

Alley R B, Blankenship D D, Bentley C R, Rooney S T In press Till beneath ice stream B. 3. Till deformation: evidence and implications. *Journal of Geophysical Research*

Bentley C R In press Antarctic ice streams: a review. *Journal of Geophysical Research*

Bentley C R and 6 others 1987 Remote sensing of the Ross ice streams and adjacent Ross Ice Shelf, Antarctica. *Annals of Glaciology* 9: 20–29

Blankenship D D, Bentley C R, Rooney S T, Alley R B 1986 Seismic measurements reveal a saturated, porous layer beneath an active Antarctic ice stream. *Nature* 322(6074): 54–57

Blankenship D D, Bentley C R, Rooney S T, Alley R B In press Till beneath ice stream B. 1. Properties derived from seismic travel times. *Journal of Geophysical Research*

Jaeger J C, Cook N G W 1970 *Fundamentals of rock mechanics*. London, Chapman and Hall

Neave K G, Savage J C 1970 Icequakes on the Athabasca Glacier. *Journal of Geophysical Research* 75(8): 1351–1362

Robin G de Q 1958 Glaciology 3. Seismic shooting and related investigations. *Norwegian-British-Swedish Antarctic Expedition, 1949–52. Scientific Results* 5

Rooney S T, Blankenship D D, Alley R B, Bentley C R In press Till beneath ice stream B. 2. Structure and continuity. *Journal of Geophysical Research*

Shabtaie S, Bentley C R 1986 Ice streams and grounding zones of West Antarctica and the Ross Ice Shelf. (Abstract.) *Annals of Glaciology* 8: 199–200

Shabtaie S, Bentley C R In press West Antarctic ice streams draining into the Ross Ice Shelf: configuration and mass balance. *Journal of Geophysical Research*

Vornberger P L, Whillans I M 1986 Surface features of ice stream B, Marie Byrd Land, West Antarctica. *Annals of Glaciology* 8: 168–170

Whillans I M, Bolzan J, Shabtaie S In press Velocity of ice stream B, Antarctica, and its mass balance. *Journal of Geophysical Research*

Annals of Glaciology 9   1987
® International Glaciological Society

STAGNANT ICE AT THE BED OF WHITE GLACIER,
AXEL HEIBERG ISLAND, N.W.T., CANADA

by

Heinz Blatter

(Department of Geography, Swiss Federal Institute of Technology, Zürich, Switzerland)

ABSTRACT

A total of 400 soundings along 15 profiles were obtained on White Glacier, Axel Heiberg Island, N.W.T. with a monopulse radar equipment that was rebuilt according to a model of the US Geological Survey. The resulting data allowed maps to be compiled of the ice thickness for the glacier tongue. The radio echo-sounding data and englacial temperature measurements give some indication of the existence of stagnant ice in depressions of the glacier bed in the accumulation zone of White Glacier.

INTRODUCTION

Since 1959 glaciological and climatological data have been collected on the White Glacier, Axel Heiberg Island, N.W.T., Canada, by members of the McGill-Axel Heiberg Island Research Expedition of the late Professor F. Müller. In the first years, the main studies focused on mass balance (Müller 1963[a]), shallow depth ice temperature measurements (Müller 1963[b]) and the study of ice movement and fluctuations (Iken 1974). On 3 transverse profiles, ice thickness measurements with seismic (Redpath, 1965) and gravimetric (Becker 1963) soundings were made.

In the years 1974 to 1981, vertical profiles of the ice temperatures were measured at 32 sites, some reaching close to bedrock (Blatter 1985). A more detailed and accurate picture of the bedrock topography was gained by radio echo-sounding in May 1984.

White Glacier is a valley glacier situated at altitudes ranging from 80 to 1600 m a.s.l. with a correspondingly wide range of conditions. The mass balance, for example, varies from about −2.5 m water equivalent on the lowest part of the tongue area to about +0.4 m. w.e. in the higher parts. The mean annual temperature of the glacier is around 18°C; the annual precipitation is 170 mm, mainly falling during the spring and summer season. The melt season starts in June with strong radiation and usually ends in mid-August. However, the year-to-year variations may be rather high with the equilibrium line varying between 500 m and 1500 m a.s.l. and the mean altitude being at around 1100 m a.s.l. The glacier is cold except for a layer which is up to 40 m thick near the bedrock in the lowest tongue area (Blatter 1985). The surface ice velocities are up to 40 m a^{-1} in the central part of the glacier. During strong melt events or periods of rainfall, peak velocities may be up to 5 times faster than the mean annual value.

DEPTH SOUNDING

For the sounding campaign on the White Glacier, the US Geological Survey Monopulse Ice Radar (Watts and England 1976; Hodge unpublished; Watts and Wright 1981) was used. The equipment was partly built with the help of the Laboratory for Hydraulics, Hydrology and Glaciology at the Swiss Federal Institute of Technology, Zürich (Haeberli and others 1983).

The operating equipment consists of a transmitter and a receiver placed about 50 m apart on the glacier surface, both equipped with identical, resistively loaded, centre-fed dipole antennas. The antenna arms used were 20 m long, which gives the choice of 10 MHz radio frequency. The transmitter works with avalanching transistors, giving a pulse repetition rate of about 10 kHz (Hodge unpublished). The signal from the receiver antenna is fed directly into a battery-powered oscilloscope. The direct signal from the transmitter and the reflected signal from the glacier bottom are displayed on the screen, where the difference of the transmission time of the signals can be read to an accuracy of a 1/100 μs. The equipment allowed identification of reflected signals to a depth of 500 m. Soundings of glacier thicknesses of less than 50 m proved to be unsatisfactory. This may be caused by the length of the antenna which records a diffuse transmission time due to diffuse path lengths.

The sounding profiles were chosen transverse to the valley axis and the antennas were laid out orthogonal to the profile direction on the ice surface. The equipment, including theodolite and laser distomat for surveying the sounding locations, weighed less than 100 kg and could be handled and operated by three people.

BEDROCK TOPOGRAPHY

The evaluation of the sounding data was perfomed by direct geometrical means. It was assumed that the reflection points lie within the vertical plane of the surface profile. The possible locations for the reflection points, the ellipse with the focal points at the transmitter and at the receiver station could be closely approximated by circles for ice thicknesses of more than 100 m. The envelope of the circles of one profile is then assumed to give an upper limit to the elevation of the glacier bed, as is illustrated for example in Figs 3, 4, and 5.

The velocity of the radio waves in ice was chosen at 168 m/μs (Robin 1975, Glen and Paren 1975). The same value was used for firn and the refraction was neglected (Rasmussen 1986). Since firn layers are usually about 50 m thick, this error is negligible in the order of the reading accuracy. Finally, it is believed that the relative accuracy of the depth sounding is about 5 to 10% (Haeberli and Fisch 1984), owing to the uncertainty of wave velocity, transmission time reading, and the location of the reflection point at the bedrock.

The above method proved to work very well in the tongue area of the glacier where the bedrock forms a simple U-shaped valley. Usually, the radar soundings confirmed drilling depths within the limits of accuracy. On the other hand, the results obtained from earlier seismic soundings (Redpath 1965) and gravimetric soundings (Becker 1963) now show substantial differences (sometimes by as much as 50%) from the radar sounding thicknesses.

The data from 11 transverse sounding profiles have enabled compilation of a map of the glacier bed topography and thus for the glacier thickness in the tongue area (Fig.1), although smaller features (<50 m in the horizontal) of the bedrock cannot be spatially resolved.

The soundings performed in the acumulation zone, however, frequently produced multiple echoes. To explain

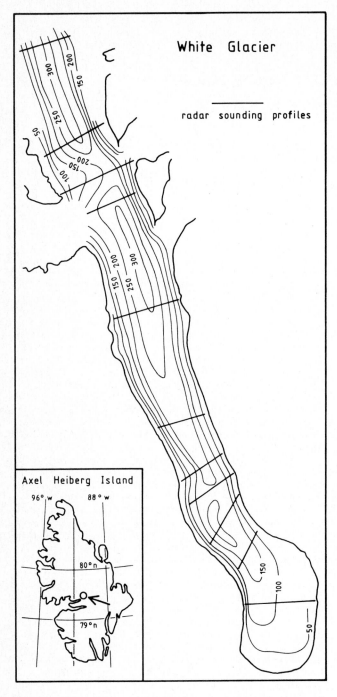

Fig.1. Map of the ice thickness in the tongue area of White Glacier with an inset map of Axel Heiberg Island.

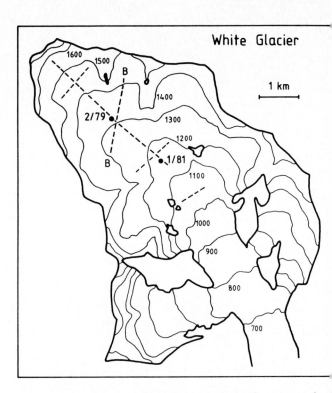

Fig.2. White Glacier accumulation area showing the surface topography, the location of drilling sites (labelled dots) and of the sounding profiles (dashed lines) (B----B indicates the transverse Beaver Profile).

those, additional information on the glacier and on the valley topography is needed.

A first indication for a rather irregular or uneven bedrock topography is given by the various rock ridges reaching into the glacier area that are partly visible from the glacier surface topography. Some transverse sounding profiles also reveal a relatively deep valley leading from the highest part down the centreline of the glacier accumulation basin. The locations of these profiles are given in Fig.2 and one example illustrating the transverse profile at the Beaver Profile shows the channel-like feature in the upper part of the glacier bed (Fig.3).

ENGLACIAL TEMPERATURE PROFILES

In the years from 1974 to 1981, a total of 32 holes was drilled on White Glacier (Blatter 1985). An open-system hot-water drill fed from a melt-water river or from water-filled plastic pools was used to melt the holes in the

glacier ice (Iken and others 1977). The drilling equipment could be handled and operated by three people. The deepest hole had a depth of 380 m and was drilled in about 6 hours. At the end of the drilling, a multi-core cable with a series of thermistors was inserted in the hole while it was still filled with water. The holes were usually frozen within 12 hours of the end of the drilling and the temperatures reached equilibrium after 3 weeks within the possible reading accuracy. This was confirmed for many data samples by repeated measurements during the adjustment period and even re-readings of the temperatures a year later. The readings were performed with a resistance bridge which proved to be quite reliable, since the voltage compensation showed little temperature dependence. The estimated accuracy of the temperature readings was $\pm 0.2°$ which is confirmed by the smoothness of measured profiles. This allows for an accuracy of better than ± 0.005 K m^{-1} for the temperature gradient if determined by 4 to 5 thermistors over a vertical distance of about 50 m thick.

In the U-shaped valley, the drilling depth usually lies within the 5 to 10% accuracy range of the radar sounding depth. Radio soundings also seem to underestimate the ice thickness, an observation which was also reported by Haeberli and Fisch (1984). However, at site 2/79 in the accumulation zone (Fig.2), the measured temperature profile was 350 m long and reached 25 m deeper than the deepest soundings and 100 m deeper than the most shallow sounding.

Several explanations of these differences are possible. First, a reflecting debris layer above the glacier bed can most likely be excluded. Some medial moraines in the tongue part of the glacier can be traced back to their origins. Otherwise, the glacier surface gives no indication of an extended debris content in the ice and the surrounding snow ridges give no hints of sources of debris in the highest parts of the glacier. The drilling was sometimes stopped before reaching the glacier bed, presumably because of meeting rocks within the ice, but only at sites where there was also debris on the surface. Secondly, a sharp turn in the direction of the drill hole could explain both the thickness difference and the sudden increase in the measured temperature gradients. Such abrupt changes occurred on rare occasions during drilling in Swiss Alpine

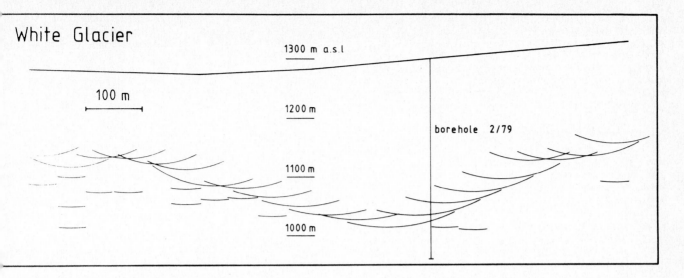

Fig.3. Radar sounding depth profile at the transverse Beaver Profile (B----B in Figure 2).

laciers (Röthlisberger, personal communication). To explain his change in the measured temperature gradient at site /79, the bend in the drill hole would have to be 60°, which is rather unlikely.

It also seems unlikely that the drilling did not reach he depth as indicated by the length of the drilling hose anging in the hole. It was usually possible to feel whether the drill was advancing or not. Since the drilling ose is rather stiff, it also seems unlikely that it became oiled in the lower part of the hole. There, the diameter of he hole is only about twice as large as that of the hose.

A third factor relates to the englacial temperature measurements. About 60 m above the bottom of the rilling, the temperature profile shows an abrupt bend Fig.4), giving a temperature gradient of 0.011±0.005 K $^{-1}$ below and 0.032±0.005 K m^{-1} above that level. The ower value seems to correspond to a geothermal heat flux f around 0.023 W m^{-2}, which is a very low value when ompared with reported values of 0.05 to 0.08 W m^{-2} for his region (Judge 1973). However, all vertical ice emperature profiles reaching close to bedrock in the ccumulation area show temperature gradients of the order f 0.02 to 0.032 K m^{-1}. This indicates a small local eothermal heat flux since these gradients also contain part

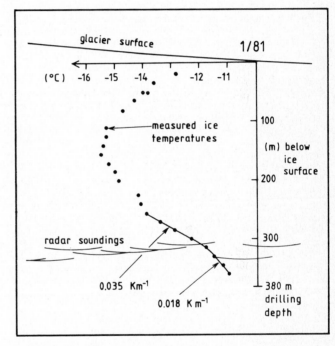

Fig.5. Comparison of the drilling depth and the radar sounding depths around the drilling site 1/81 with the measured temperature profile.

of the differential heat flux. The difference of 0.021± 0.007 K m^{-1} can be explained by the heat generated by the differential flow (W) at the level of the change which is given by:

$$W = 0.4(f\rho g \sin \alpha)^4 Ah^5 \qquad \text{where}$$

f = Nye shape factor
ρ = ice density
α = surface slope
A = flow law parameter
h = ice thickness

This equation is appropriate for a parallel-sided slab model with a flow law exponent of n=3. Although it contains the surface velocity from integration of the internal deformation over the whole thickness of the ice, it can be used for estimating the change of the temperature gradient within a relatively small layer of the bottom of the deforming part where most differential heat is released (Blatter 1985). Given the measuring accuracy of the temperatures, any gradient change can be located to a layer of 20-30 m thickness.

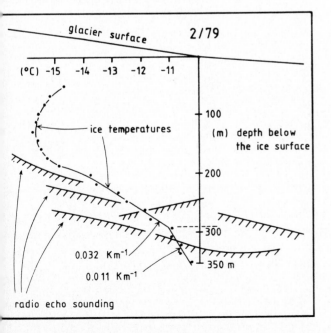

Fig.4. Comparison of the drilling depth and the radar sounding depths around the drilling site 2/79 with the measured temperature profile.

With the following values substituted into the equation: $f = 0.8$, $\rho = 900$ kg m^{-3}, $\sin \alpha = 0.08$, $A = 5.2 \cdot 10^{-25}$ s^{-1} Pa^{-3} (Paterson, 1981) and $h = 290$ m the heat generation $(W) = 0.0435$ W m^{-2}, it may thus be possible to explain the required increase of the temperature gradient of 0.021 K m^{-1}. Since most parameters in the equation have high powers (4 and 5) the numerical result should be considered an order of magnitude estimation of the heat generation.

The situation is similar at site 1/81 with a temperature gradient of 0.018±0.005 K m^{-1} below and 0.035±0.005 K m^{-1} above the bend in the measured temperature profile.

DISCUSSION AND CONCLUSIONS

Radio echo-sounding with the USGS Monopulse Ice Radar provided rather good results within the possible accuracy for the U-shaped glacier tongue bed. However, in the higher part of the glacier, in an area of a widened valley with several tributaries from the sides, the sounding consistently underestimated the ice thickness when compared with drilling depths by an amount to be explained by other factors.

The combined information of the soundings and the measured temperature profile at site 2/79 with a relatively sharp bend at about the depth given by the radar sounding indicates the possibility of the existence of stagnant ice about 60 m thick in a depression of the bed below site 2/79. This interpretation is rather speculative owing to the ambiguous sounding data and the accuracy of the temperature measurements of ± 0.2 K. However, the same features were observed further down the glacier at site 1/81, although not as clearly as at site 2/79, with a drilling depth of 380 m and a sounding depth of only 310 to 340 m (Fig.5).

ACKNOWLEDGEMENTS

All field activities on White Glacier were made possible thanks to the generous logistic support of the Polar Continental Shelf Project, Ottawa, Canada, and the financial support of the Swiss Federal Institute of Technology, Zürich, Switzerland, and McGill University, Montreal, Canada. Dr M Funk, Dr B Ott, and H J Frei did the laborious job of the sounding and K Schroff built the sounding equipment. Professor A Ohmura encouraged the project with stimulating discussions. Dr W Haeberli helped with the training on the sounding equipment and the evaluation method.

REFERENCES

Becker A 1963 Gravity investigations. (Interim Report.) *McGill University, Montreal. Axel Heiberg Island Research Reports. Preliminary Report 1961-1962*: 97-101

Blatter H 1985 On the thermal regime of Arctic valley glaciers; a study of White Glacier, Axel Heiberg Island, and the Laika Glacier, Coburg Island, Canadian Arctic Archipelago. *Zürcher Geographische Schriften* 22

Glen J W, Paren J G 1975 The electrical properties of snow and ice. *Journal of Glaciology* 15(73): 15-38

Haeberli W, Fisch W 1984 Electrical resistivity soundings of glacier beds: a test study on Grubengletscher, Wallis, Swiss Alps. *Journal of Glaciology* 30(106): 373-376

Haeberli W, Wächter H-P, Schmid W, Sidler C 1983 Erste Erfahrungen mit dem U.S.-Geological-Survey Monopuls-Radioecholot im Firn, Eis und Permafrost der Schweizer Alpen. *Zeitschrift für Gletscherkunde und Glazialgeologie* 19(1): 61-72

Hodge S M Unpublished USGS mono-pulse ice radar. Washington, DC, US Geological Survey (Internal Report)

Iken A 1974 Velocity fluctuations of an Arctic valley glacier; a study of the White Glacier, Axel Heiberg Island, Canadian Arctic Archipelago. *McGill University, Montreal. Axel Heiberg Island Research Reports. Glaciology* 5

Iken A, Röthlisberger H, Hutter K 1977 Deep drilling with a hot water jet. *Zeitschrift für Gletscherkunde und Glazialgeologie* 12(2), 1976: 143-156

Judge A 1973 Geothermal measurements in northern Canada. *In* Aitken J D, Glass D J (eds) *GAC-CSPG Proceedings of the Symposium on the Geology of the Canadian Arctic, Saskatoon, May 1973*: 301-311

Müller F 1963[a] Englacial temperature measurements. McGill University, Montreal. *Axel Heiberg Island Research Reports. Preliminary Report 1961-1962*: 81-89

Müller F 1963[b] Glacier mass budget and climate. *McGill University, Montreal. Axel Heiberg Island Research Reports. Preliminary Report 1961-1962*: 57-64

Paterson W S B 1981 *The physics of glaciers. Second edition*. Oxford etc, Pergamon Press

Rasmussen L A 1986 Refraction correction for radio echo sounding of ice overlain by firn. *Journal of Glaciology* 32(111): 192-194

Redpath B B 1965 Seismic investigations of glaciers on Axel Heiberg Island, Canadian Arctic Archipelago. *McGill University, Montreal. Axel Heiberg Island Research Reports. Geophysics* 1

Robin G de Q 1975 Velocity of radio waves in ice by means of a bore-hole interferometric technique. *Journal of Glaciology* 15(73): 151-159

Watts R D, England A W 1976 Radio-echo sounding of temperate glaciers: ice properties and sounder design criteria. *Journal of Glaciology* 17(75): 39-48

Watts R D, Wright D L 1981 Systems for measuring thickness of temperate and polar ice from the ground or from the air. *Journal of Glaciology* 27(97): 459-469

Annals of Glaciology 9 1987
© International Glaciological Society

NIMBUS-7 SMMR DERIVED GLOBAL SNOW COVER PARAMETERS

by

A.T.C. Chang, J.L. Foster and D.K. Hall

(Hydrological Sciences Branch, NASA/Goddard Space Flight Center, Greenbelt, MD 20771, U.S.A.)

ABSTRACT

Snow covers about 40 million km^2 of the land area of the Northern Hemisphere during the winter season. The accumulation and depletion of snow is dynamically coupled with global hydrological and climatological processes. Snow covered area and snow water equivalent are two essential measurements. Snow cover maps are produced routinely by the National Environmental Satellite Data and Information Service of the National Oceanic and Atmospheric Administration (NOAA/NESDIS) and by the US Air Force Global Weather Center (USAFGWC). The snow covered area reported by these two groups sometimes differs by several million km^2. Preliminary analysis is performed to evaluate the accuracy of these products.

Microwave radiation penetrating through clouds and snowpacks could provide depth and water equivalent information about snow fields. Based on theoretical calculations, snow covered area and snow water equivalent retrieval algorithms have been developed. Snow cover maps for the Northern Hemisphere have been derived from Nimbus-7 SMMR data for a period of six years (1978-1984). Inter-comparisons of SMMR, NOAA/NESDIS and USAFGWC snow maps have been conducted to evaluate and assess the accuracy of SMMR derived snow maps. The total snow covered area derived from SMMR is usually about 10% less than the other two products. This is because passive microwave sensors cannot detect shallow, dry snow which is less than 5 cm in depth. The major geographic regions in which the differences among these three products are the greatest are in central Asia and western China. Future study is required to determine the absolute accuracy of each product.

Preliminary snow water equivalent maps have also been produced. Comparisons are made between retrieved snow water equivalent over large area and available snow depth measurements. The results of the comparisons are good for uniform snow covered areas, such as the Canadian high plains and the Russian steppes. Heavily forested and mountainous areas tend to mask out the microwave snow signatures and thus comparisons with measured water equivalent are poorer in those areas.

INTRODUCTION

Remotely acquired microwave data in conjunction with essential ground observations will most likely lead to advanced extraction of snow properties beyond conventional techniques. Landsat visible and near-infrared data have recently become near operational for use in measurements of snow covered areas (Rango 1975, 1978). Operational NOAA satellites provide continuous global coverage with 4 km spatial resolution. Both Landsat and NOAA data acquisition are hampered by cloud cover, sometimes at critical times when a snowpack is ripe and ready to melt. Furthermore, information on water equivalent, free water content and other snowpack properties germane to accurate snow melt run-off prediction is not currently available using visible and near-infrared data because only surface and near surface snow contribute to the measured reflectances.

Microwave remote sensors which have the capability to penetrate the snowpack and respond to variations in snow properties, could provide information about snow depth and snow water equivalent (Rango and others 1979; Chang and others 1982). However, due to the coarse spatial resolution of the present microwave radiometers, combinations of vegetation, terrain and snow information within a pixel greatly complicate the retrieval algorithm development.

Algorithms need to be developed that are specific to physiographic areas like the Colorado River basin and the north slope of Alaska. These algorithms will take into account additional parameters related to microwave signatures. Until these algorithms are operational, the use of remotely collected microwave data for global quantitative snowpack analysis will not be operational due to the complexities involved in the data analysis.

MICROWAVE EMISSION FROM SNOW

Microwave emission from a layer of snow over ground consists of two parts: (1) emission by the snow volume and (2) emission by the underlying ground. Snow particles act as scattering centers for microwave radiation from a snowpack. The scattering effect which redistributes the upwelling radiation according to snow thickness and crystal size, provides the physical basis for microwave detection of snow. Mie scattering theory is used to account for the energy redistribution by snow crystals. Although the snow crystal usually is not spherical in shape, its ensemble scattering properties can be mimicked by spheres (Chang and others 1976). Theoretical computations indicate that scattering by individual snow crystals can be the dominant modification factor of upwelling 37 GHz (0.8 cm) radiation in the dry snow cases (Chang and others 1982). The effect of scattering is lessened by using the longer wavelengths. Fig.1 shows the calculated brightness temperatures versus snow water equivalent for SMMR frequencies. The effective microwave penetration depth into a dry snowpack, typically 10-100 times the wavelength, depends on the wavelength used and the characteristic crystal size of the snowpack. When the wavelength is much larger than the crystal size (> 5 cm), absorption will be the dominant effect. The brightness temperature will resemble the physical temperature of the snowpack. When the wavelength is comparable to the snow crystal size (< 1 cm), scattering becomes the dominant effect.

Nimbus-7 SMMR is a five-frequency, dual-polarized microwave radiometer which measures the upwelling microwave radiation at 6.6, 10.7, 18.0, 21.0, and 37.0 GHz (4.6, 2.8, 1.7, 1.4 and 0.8 cm) while scanning 25° to either side of the spacecraft (approximately 780 km swath width) with a constant incidence angle of approximately 50° with respect to the Earth's surface. The spatial resolution varies from 25 km for the 37 GHz (0.8 cm) to 150 km for the 6.6 GHz (4.6 cm). A detailed description of this instrument can be found in Gloersen and Barath (1977). The Nimbus-7 satellite was launched on October 24, 1978, into a sun-synchronous polar orbit with local noon/midnight equatorial crossing.

Using the multifrequency analysis approach, one may make inferences regarding not only the thickness of the snowpack, but the underlying soil (wet versus dry) condition. The shorter wavelengths, such as 0.8 cm (37 GHz), sense near surface (0-50 cm) temperature and emissivity, and

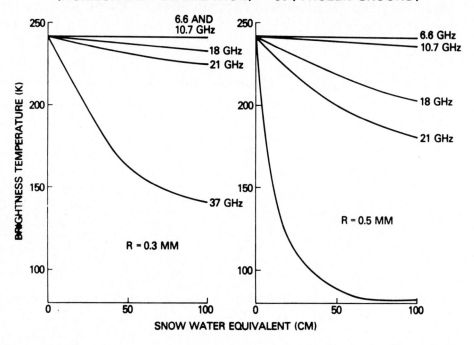

CALCULATED BRIGHTNESS TEMPERATURE AS A FUNCTION OF SNOW WATER EQUIVALENT
(HORIZONTAL POLARIZATION, $\theta = 50°$, FROZEN GROUND)

Fig.1. Calculated brightness temperatures versus snow water equivalent for five SMMR frequencies with horizontal polarization and 50° incidence angle.

surface roughness. At the intermediate wavelengths (1.4 and 1.7 cm), the radiance is less affected by the surface and more information is obtained on the internal characteristics of the snowpack. All of the above generalizations apply to snow conditions encountered by various satellite observations for different regions of the world.

The presence of liquid water content in a snowpack completely changes the observed microwave signatures (Chang and Shiue 1980; Mätzler and others 1980; Stiles and Ulaby 1980). A few per cent of liquid water in snow will cause a sharp increase in the brightness temperature (Chang and Gloersen 1975). This is because the emission of a small amount of liquid water within the snowpack alters the emitting radiance of dry snow.

The condition of the ground beneath the snow determines the intensity of the radiation emitted from below. Dry or frozen ground has a high emissivity (0.9 – 0.95) whereas unfrozen wet ground has a much lower emissivity (0.7). Knowledge of the condition of ground is important for the interpretation of observed brightness temperature of a shallow snowpack. The difference of vertical and horizontal polarized brightness temperature at 2.8 cm wavelength could reflect the frozen and thawed soil condition under a shallow snowpack.

SNOW-PARAMETER RETRIEVAL ALGORITHM

Kunzi and others (1982) reported an algorithm to retrieve snow cover parameters using the Nimbus-7 SMMR data. The brightness temperature gradients of 37 and 18 GHz were used to discriminate snow parameters. The snow water equivalent relationship was derived from the regressing brightness gradient and the measured snow water equivalent. Based on the SMMR data for the 1978-1979 winter season, encouraging results were obtained. Chang and others (1982) reported snow parameter retrieval results based on a theoretically derived algorithm and the results compared favorably with snow data taken from Eurasia and the Canadian high plains. These relationships will be used for the snow parameter retrieval algorithm used in this study. The snow depth–brightness temperature relationship for a uniform snow field can be expressed as follows (Chang 1986):

$$SD = 1.59 * (T_{18H} - T_{37H}) \text{ cm} \qquad (1)$$

where T_{18H} and T_{37H} are the brightness temperature at 18 and 37 GHz horizontal polarization. This equation was derived from linearly fitting the data shown in Fig.1 assuming a snow density of 0.3 Mg/m^3 and its application should be limited to snow depths less than 1 meter. Since the snow within a SMMR footprint (25 km × 25 km) can be quite variable, pixels with derived snow depth less than 2.5 cm will be assigned as no snow. The snow covered area is defined where derived snow depth is thicker than 2.5 cm.

INTERCOMPARISON OF SMMR-DERIVED SNOW MAPS AND OTHERS

Snow cover maps are routinely produced by the National Environmental Satellite Data and Information Service of the National Oceanic and Atmospheric Administration (NOAA/NESDIS) and by the U.S. Air Force Global Weather Center (USAFGWC). NESDIS monitors continental snow cover extent using NOAA satellites on a weekly basis (Dewey and Heim 1981). Skilled analysts hand prepare the snow cover chart based on imagery information. The Air Force charts rely on snow depth reported by weather stations. A complex computer program is then used to extrapolate the point measurements to area snow coverage. Kukla and Robinson (1981) reviewed the accuracy of these two products. Table I shows the variation and differences between them for the Northern Hemisphere snow covered area for the year of 1979. The snow covered area derived by NESDIS varied from 49.3 million km^2 in the 1978 winter season to 4.2 million km^2 in the summer of 1979. The differences between NESDIS and Air Force products are substantial in the fast-changing fall season. For example, in the 45th week of 1979, a difference of 12.5 million km^2 was observed between these two products. The major differences are in Europe and Asia (Fig.2). Persistent cloud cover probably was probably the reason for the differences.

A six-year snow cover/snow depth data set has been produced using Nimbus-7 SMMR data. It covers the period

TABLE I. NORTHERN HEMISPHERE SNOW-COVERED AREA COMPARISON FOR 1979 (UNITS: MILLIONS KM^2)

Week No.	NOAA	AF	NOAA–AF
1	48.4	47.3	1.1
5	49.3	47.6	1.7
10	45.3	42.4	2.9
14	38.0	34.9	3.1
18	28.7	20.5	8.2
23	17.3	7.9	9.4
27	7.1	2.2	4.9
31	5.8	4.2	1.6
36	4.2	5.0	−0.8
40	12.7	18.6	−5.9
45	24.1	36.6	−12.5
49	35.1	38.5	−3.4

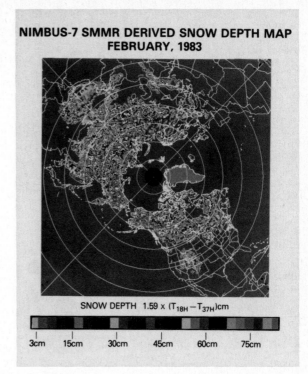

Fig.3. SMMR-derived snow map (February 1983). (Also reproduced in colour in appendix at end of volume.)

November 1978 to October 1984. Snow parameters were derived based on Equation (1) using calibrated SMMR brightness temperatures. Fig.3 shows a sample SMMR snow map (February 1983). In order to compare the different snow maps, both NESDIS and USAFGWC maps were projected onto a one degree by one degree SMMR map grid. Generally the major snowfield features matched well with those of NESDIS snow cover maps except in very shallow snow covered areas. Due to this limitation, total snow covered area derived from SMMR is usually about 10% less than the other two products. Fig.4 shows a typical

NESDIS AND USAFGWS SNOW MAP COMPARISON (1979 DAY 315/WEEK 45)

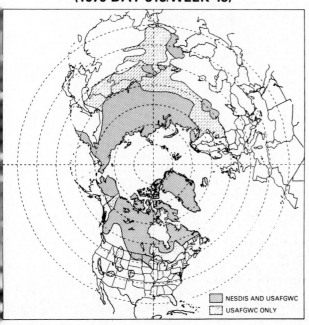

Fig.2. Differences in NESDIS and USAFGWC snow cover map (November 1979).

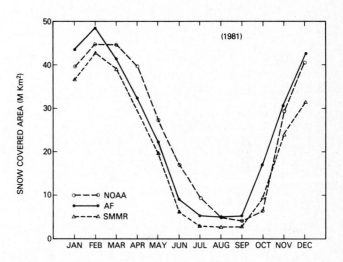

Fig.4. Comparisons of weekly snow covered area derived by NESDIS, USAFGWC and SMMR for 1981.

comparison of the Northern Hemisphere snow covered areas derived from these three methods for each month of 1981.

In order to make a preliminary evaluation of the accuracy of these products, the continental United States was chosen as the test area. The snow depths reported by weather stations were used as the reference point measurements. A subjective analysis similar to those used in the meteorological data analysis was performed to draw the snow covered area. Then the NESDIS and SMMR snow boundary lines were overlaid on the same chart. The snow cover maps for mid-January and mid-February of 1983 were chosen for comparison (Figs 5 and 6). In the western US, all three products gave very similar results for both January and February. Due to deeper snow conditions, mountainous terrain did not seem to affect the snow boundary determination. The January snow cover in the eastern US was very sparse and shallow. Only a small area in New England showed in the SMMR snow map. This is probably because the heavier vegetation cover and denser

41

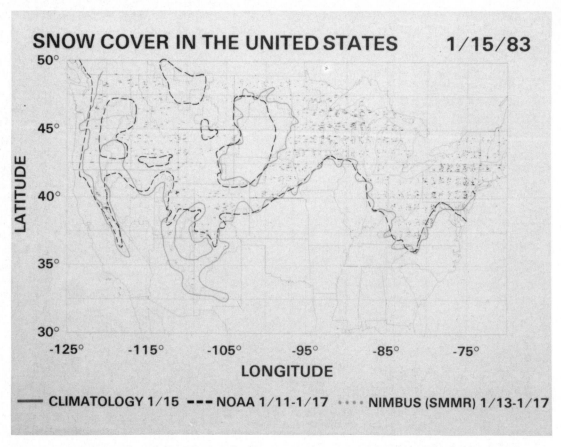

Fig.5. Comparisons of National Weather Stations reported snow depth with NESDIS and SMMR snow covered area, mid-January 1983.

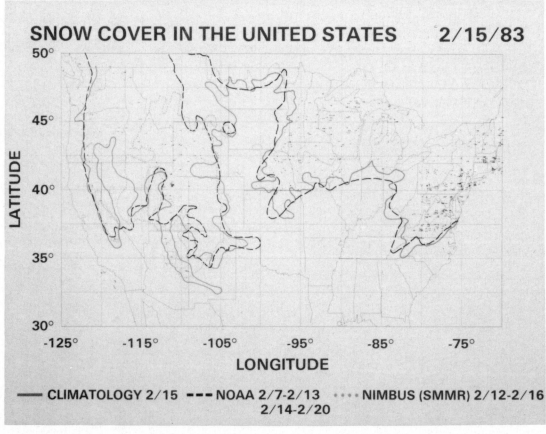

Fig.6. Comparisons of National Weather Stations snow depth with NESDIS and SMMR snow covered area, mid-February 1983.

rban development in the eastern US masked out the microwave signatures. The threshold for snow/no-snow iscrimination needs adjustment for this area. But in February the SMMR map matched better in the eastern US, ecause there was more snow on the ground. The outhernmost snow boundary is nearly always slightly ifferent in each map and the presence of shallow and atchy snow probably is the reason. No effort was made o compare the derived snow depth and those reported by veather stations.

CANADIAN HIGH PLAINS AND RUSSIAN STEPPES

Rango and others (1979) reported good correlations for rightness temperature and snow depth over the Canadian igh plains using Nimbus-5 and Nimbus-6 Electrically canned Microwave Radiometer (ESMR) data. Chang and thers (1982) reported similar results for the Russian steppes nd the Canadian high plains. The vegetation, topography, limate and latitude in these two areas are comparable. The egetation is predominantly a variety of grasses and the opography is generally flat. Both areas experience very cold vinters with snow covering the ground from December hrough March. Air temperatures for the study areas, before nd after satellite passes, were below $0°C$ with little chance f significant melting; dry snow conditions were assumed. Regression from both test sites is significant at the 0.005 evel with R^2 values of 0.75 for Russian and 0.81 for Canadian sites using only the 37 GHz vertically polarized rightness temperatures (Fig.7). Due to lack of snow water quivalent data for large area comparisons, snow depth nformation was utilized. A snow density of 0.3 Mg/m^3 was ssumed to relate the snow depth to snow water equivalent. The data displayed considerable scatter, which is probably lue to inhomogeneity within each footprint. The theo-etically calculated brightness curve for a mean snow grain radius of 0.35 mm fits well with the observed conditions. Snow depth and snow water equivalent for these areas could be inferred from microwave brightness temperatures measured by the SMMR spaceborne sensors.

COLORADO RIVER BASIN STUDY

In 1983 the snowfall for the Colorado River basin was well above average. Serious flooding in the western US was predicted for the spring melting season. A cooperative project between the US Geological Survey and the National Aeronautics and Space Administration was set up to explore the potential of using passive microwave data for determining snowpack properties in this mountainous region.

In order to analyze the relationship between microwave brightness temperature and snow parameters, a common geo-graphical grid was established to register all the available information, which includes elevation, slope, vegetation type, vegetation cover amount, snow course data, climatology data and SNOTEL data. The mesoscale footprint of the SMMR, an ellipsoid of approximately 40 km × 50 km for the 18 GHz (1.7 cm) radiometer, complicates the determination of snow water equivalent from microwave radiances in the mountainous regions. The R^2 between T_B and snow depth for a $1/4° \times 1/4°$ grid usually is only about 0.5 because of the difficulty in making the comparison between a point surface measurement and a mesoscale measurement from space. Analysis is now being concentrated on combining geographical information and microwave information to develop a better snow parameter retrieval algorithm for use in the mountainous regions.

ALASKA STUDY AREA

In northern Alaska the difference between the soil temperature and the air temperature can be as large as $40°C$. This steep temperature gradient in the snowpack promotes metamorphism and the formation of depth hoar crystals. The crystal size of depth hoar is usually larger than that of new snow crystals. Lower brightness temperatures were observed by the Nimbus-7 SMMR in this area, typically with a snowpack of 30–50 cm in depth, and can be explained by radiation interacting with a depth hoar layer (Hall and others 1986). In order to develop a retrieval algorithm for this area, information on the history of the thickness of the depth hoar layer and the characteristic grain size is needed. Based on this information and a snow energy and mass balance model, physical properties of the snowpack can be estimated.

DISCUSSION

Six year Nimbus-7 SMMR derived Northern Hemi-sphere snow parameter maps have been produced using a simple algorithm. The snow boundary lines derived by NESDIS, USAFGWC and SMMR were compared. Normally, the SMMR derived snow boundary maps showed less snow than those of the NESDIS and the USAFGWC snow maps. This is because microwaves can penetrate through a shallow snowpack without registering a noticeable scattering effect. However, when the microwave signatures indicate snow, the entire field of view is probably snow covered. Over the mountainous area, terrain effects tend to modify the microwave signatures. The sensitivity of microwave radiation

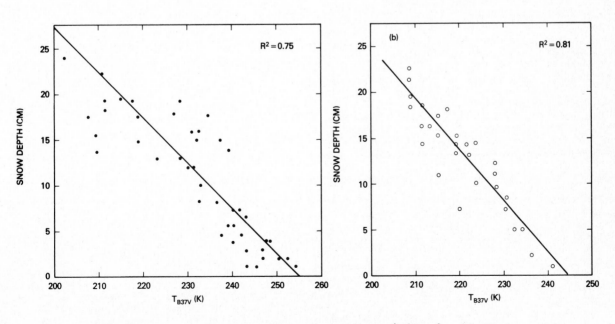

Fig.7. SMMR brightness temperature versus snow water equivalent for (a) Russian steppes, 15 February 1979 and (b) Canadian High Plains, 12 February 1983.

with respect to snow depth will be reduced, but the snow boundary can still be delineated. For heavily vegetated areas, the sensitivity of microwaves for detecting snow will also be reduced. The vegetation effect can be predetermined and classified by using visible and infrared imagery data. The depth hoar layer within a snowpack will greatly alter the microwave signatures. By taking into account all these factors, different algorithms will be developed to account for different snow conditions.

To evaluate the quality of retrieved snow water equivalent from passive microwave data, several areas were chosen for comparison. The best results were found in the Canadian high plains and the Russian steppe areas. The R^2 for these uniform and homogeneous areas are typically between 0.75 and 0.8. The results from mountainous areas, forested areas and snowpacks with depth hoar layers are not as good. The R^2 for these areas usually is only about 0.5 or less. Different algorithms are required to retrieve the snow parameters in these regions. Further tests are required before regional algorithms can be used in combination for global operational purposes.

REFERENCES

Chang A T C 1986 Nimbus-7 SMMR snow cover data. *Glaciological Data. Report* GD-18: 181–187

Chang A T C, Gloersen P 1975 Microwave emission from dry and wet snow. *In* Rango A (ed) *Operational applications of satellite snowcover observations.* Washington, DC, National Aeronautics and Space Administration: 399–407 (NASA Special Publication 391)

Chang A T C, Shiue J C 1980 A comparative study of microwave radiometer observations over snowfields with radiative transfer model calculations. *Remote Sensing of Environment* 10(3): 215–229

Chang A T C, Gloersen P, Schmugge T J, Wilheit T, Zwally H J 1976 Microwave emission from snow and glacier ice. *Journal of Glaciology* 16(74): 23–39

Chang A T C, Foster J L, Hall D K, Rango A, Hartline B K 1982 Snow water equivalent estimation by microwave radiometry. *Cold Regions Science and Technology* 5(3): 259–267

Dewey K F, Heim R Jr 1981 Satellite observations of variations in northern hemisphere seasonal snow cover. *NOAA Technical Report* NESS 87

Foster J L, Rango A, Hall D K, Chang A T C, Allison L J, Diesen B C III 1980 Snowpack monitoring in North America and Eurasia using passive microwave satellite data. *Remote Sensing of Environment* 10: 285–298

Gloersen P, Barath F T 1977 A scanning multichannel microwave radiometer for Nimbus-G and Seasat-A. *IEEE Journal of Oceanic Engineering* 2: 172–178

Hall D K, Chang A T C, Foster J L 1986 Detection of the depth-hoar layer in the snow-pack of the Arctic coastal plain of Alaska, U.S.A., using satellite data. *Journal of Glaciology* 32(110): 87–94

Kukla G, Robinson D 1981 Accuracy of operational snow and ice charts. *IEEE International Geoscience and Remote Sensing Symposium '81*: 974–987

Kunzi K F, Patil S, Rott H 1982 Snow-cover parameters retrieved from Nimbus-7 Scanning Multichannel Microwave Radiometer (SMMR) data. *IEEE Transactions on Geoscience and Remote Sensing* GE20: 452–467

Matson M, Ropelewski C F, Varnadore M S 1986 *An atlas of satellite-derived northern hemispheric snow cover frequency.* Washington, DC, US Department of Commerce

Mätzler C, Schanda E, Hofer R, Good W 1980 *Microwave signatures of the natural snow cover at Weissfluhjoch.* Washington, DC, National Aeronautics and Space Administration: 203–223 (NASA Conference Publication 2153)

Rango A (ed) 1975 *Operational applications of satellite snowcover observations.* Washington, DC, National Aeronautics and Space Administration (NASA Special Publication 391)

Rango A 1978 Pilot tests of satellite snowcover/runoff forecasting systems. *Proceedings of the Western Snow Conference*, 46th annual meeting: 7–14

Rango A, Chang A T C, Foster J L 1979 The utilization of spaceborne microwave radiometers for monitoring snowpack properties. *Nordic Hydrology* 10(1): 25–40

Stiles W H, Ulaby F T 1980 *Microwave remote sensing of snowpacks.* Washington, DC, National Aeronautics and Space Administration (NASA Contractor Report 3263)

nnals of Glaciology 9 1987
International Glaciological Society

NIMBUS-7 SMMR DERIVED GLOBAL SNOW COVER PARAMETERS

by

A.T.C. Chang, J.L. Foster and D.K. Hall

(Hydrological Sciences Branch, NASA/Goddard Space Flight Center, Greenbelt, MD 20771, U.S.A.)

(See pages 39–44)

Colour print of Fig.3, which appears as a black-and-white print on page 41

NIMBUS-7 SMMR DERIVED SNOW DEPTH MAP
FEBRUARY, 1983

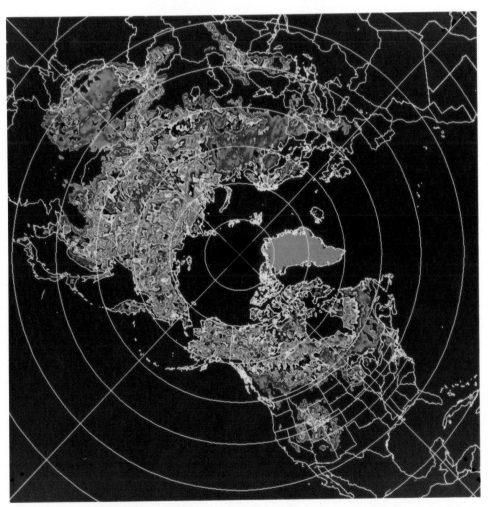

SNOW DEPTH $1.59 \times (T_{18H} - T_{37H})$ cm

3cm   15cm   30cm   45cm   60cm   75cm

Annals of Glaciology 9 1987
© International Glaciological Society

GROUND DATA INPUTS TO IMAGE PROCESSING FOR ESTIMATING TERRAIN CHARACTERISTICS FOR GLACIO-HYDROLOGICAL ANALYSIS

by

M.J. Clark, A.M. Gurnell and P.J. Hancock

(GeoData Unit, University of Southampton, Southampton SO9 5NH, U.K.)

ABSTRACT

Remote-sensing research in glacial and pro-glacial environments raises several methodological problems relating to the handling of ground and satellite radiometric data. An evaluation is undertaken of the use of ground radiometry to elucidate properties of relevant surface types in order to interpret satellite imagery. It identifies the influence that geometric correction and re-sampling have on the radiometric purity of the resulting data set. Methodological problems inherent in deriving catchment terrain characteristics are discussed with reference to currently glacierized and pro-glacial zones of south-western Switzerland.

INTRODUCTION

The methodological problems of ground and remotely sensed data handling are as important as the technical challenges of the field. Whilst most large-scale operational systems for monitoring snow, ice or water make sophisticated use of ground data and specify techniques through which they are integrated with remotely sensed data, the same is not always true of small-scale research applications. Simple procedures are required to incorporate ground data within the research design. For operational purposes, it is also necessary to specify precisely the processing of the remotely sensed data and its influence on the radiometric purity of the resulting data set.

Basic techniques in three interlinked contexts of relevance to glaciologists or alpine hydrologists are considered here. First, the use of ground radiometry in image interpretation and to provide a quantitative basis for classification. Secondly, the recognition of atmospheric and shadow effects. Thirdly, the comparison of geometric correction and re-sampling routines which are used to provide cartographic fit, but which may have an influence on the radiometric quality of the data. The techniques are evaluated using Landsat Thematic Mapper data for south-west Switzerland from a scene dated 7 July 1984, path 195/row 028. Ground radiometry was undertaken from 5 to 9 September 1985.

GROUND RADIOMETRIC DATA

Ground radiometry provides an insight into the radiometric response of specific surface types, and thus supports the interpretation of satellite data by providing a basis for selecting appropriate image indices (e.g. choice of bands, ratios, or other mathematical routines) and for guiding the quantitative calibration of supervised classifications. This assumes that the technique provides robust data, that the conditions of sensing are equivalent for ground and satellite data, that ground radiometric response is physically and environmentally interpretable, and that a meaningful relationship can be established between ground and satellite data.

Ground-data quality depends on the instrumentation and the radiometric stability of the terrain surfaces. The present study employed a field-portable Milton Multiband radiometer (Milton 1980). Reflected radiation was recorded in four bands, using a Kodak grey card reference to calculate an approximation to the bidirectional reflectance factor (BRF) (Robinson and Biehl 1979), and the conditions at times of survey were noted since the ground and satellite data were collected on different dates. The technique used a 2 m mast with a single sensor head which sequentially accessed the target (1 m diameter instantaneous field of view) and reference. The four bands represented Thematic Mapper bands 2–5.

Radiometry yields a coherent ground-data set but its relationship to satellite data requires discussion. Ideally, the two data sets should be acquired simultaneously. If this is not possible, the radiometric consequence of any differences must be appreciated. In the present investigation the distinction between the July satellite image and the September ground radiometry raises several concerns, even though weather conditions were similar on both dates. In the pro-glacial context, the difference introduces contrasts between river-flow and sediment transport conditions, and for vegetation in the growth and senescent phases. On the glaciers, major distinctions are the reduced snow cover and increased supraglacial water of the ground survey.

However, much can be gained from an analysis of the ground data (Table I). Pro-glacial vegetation represents significant catchment characteristics. Complex patterning of vegetation related to topoclimatic factors is superimposed on a regional altitudinally related zonation. In a similar field context, Frank and Isard (1986) found that a combination of remotely sensed data and ground-derived topoclimatic indices yields improved Alpine tundra vegetation discrimination. In the absence of such integrated data, the ground data suggest that a reasonable vegetation classification can be obtained from TM equivalent band 4/3 ratios.

The dry surfaces of roads, rock spreads, and continuous moraine cover on ice provide data that are radiometrically moderately stable and can to an extent be discriminated both by the pattern of the four-band reflectance values (which have similar form, but distinct absolute levels), and by their TM band 4/3 ratios, although there is partial overlap of the samples.

The water samples are clearly differentiated from other surfaces by high absorption in bands 4 and 5, but they display considerable variability associated with water shallowness, flow turbulence and suspended sediment concentration. Thus the radiometric performance of pro-glacial water is likely to be very sensitive to seasonal and diurnal flow and sediment variations, reinforcing the importance of synchronizing ground- and satellite-data collection.

Three sub-sets of glacier terrain were assessed: bare and generally dry ice, supraglacial water, and varying cover of rock debris on the ice. Ice yields a dual radiometric

TABLE I. GROUND RADIOMETRIC ATTRIBUTES OF GLACIAL AND PRO-GLACIAL TERRAIN IN SOUTH-WEST SWITZERLAND

Surface type	N		Equivalent TM band mean 2	3	4	BRF 5	TM 4/3 ratio	TM 5/2 ratio
Grass	12	m	9.6	8.0	35.4	32.2	4.63	3.38
		s	1.3	1.7	4.8	4.8	1.23	0.32
Shrubs	12	m	8.5	5.0	36.2	30.6	7.53	3.63
		s	1.5	1.1	4.8	4.8	1.45	0.38
Dense shrub	12	m	10.4	6.9	38.2	33.6	5.60	3.24
		s	1.8	1.6	5.8	5.7	0.49	0.27
Trees	12	m	10.0	7.9	41.9	35.7	6.05	3.91
		s	3.1	2.9	4.6	2.8	2.23	1.08
Roads	30	m	24.5	26.5	26.3	27.1	1.00	1.11
		s	1.8	2.3	1.7	2.3	0.03	0.08
Rocks	12	m	38.9	38.7	36.2	40.8	0.95	1.06
		s	7.4	7.9	5.9	6.2	0.14	0.14
Moraine	14	m	20.4	21.2	17.8	19.8	0.84	0.97
		s	2.8	3.1	2.9	3.3	0.03	0.07
Rivers A	24	m	32.6	25.9	6.1	3.8	0.26	0.14
		s	10.5	8.0	2.5	1.6	0.13	0.08
Rivers B	12	m	36.7	32.1	9.5	5.1	0.30	0.14
		s	2.4	2.3	2.2	1.6	0.06	0.04
Rivers C	12	m	28.6	24.0	7.1	5.3	0.30	0.20
		s	6.8	5.3	2.6	1.6	0.11	0.08
Pools	10	m	14.7	12.7	3.9	3.4	0.30	0.23
		s	3.0	2.9	2.1	1.1	0.12	0.05
Bare dry ice	12	m	27.5	28.4	17.6	10.4	0.62	0.38
		s	7.0	7.3	4.7	2.6	0.03	0.03
Water on ice	24	m	20.8	20.1	7.2	3.7	0.36	0.18
		s	4.5	4.4	2.0	1.0	0.08	0.04
Debris on ice	28	m	18.9	19.1	13.3	11.4	0.72	0.66
		s	5.7	5.6	2.6	2.2	0.09	0.24

m, sample mean; s, sample standard deviation.

population differentiated by presence or absence of water. The potential for quantitative radiometric evaluation of debris cover on ice also justifies comment. Aggregate values indicate less differentiation from bare ice than might have been expected, but this can be explained by the influence of varying percentage debris cover. Debris cover was assessed in 5% classes and compared with BRF. Band 2 provided the strongest (but non-linear) single-band relationship between reflectance and percentage debris cover, with increases of debris cover beyond about 30% inducing little further ground radiometric change. A stronger and linear relationship was found between the band 5/2 ratio and percentage debris cover. The ability of this ratio to discriminate water, ice, and debris-related terrain classes complements the band 4/3 ratio which is more useful in discriminating other terrain types (Table I).

This discussion has identified two areas of potential difference between ground and satellite data induced by the differences in the timing of data collection. First, some surfaces on the satellite image may not be represented in the ground data because of changes in their areal extent or radiometric characteristics (e.g. fresh and melting snow cover, lush meadow vegetation). Secondly, some surface types may have similar radiometric characteristics but their areal extent may have changed (e.g. water of given depth, turbulence, and suspended sediment concentration).

RELATING GROUND AND SATELLITE DATA ATMOSPHERIC AND SHADOW EFFECTS

Physically rigorous techniques for estimating atmospheric corrections to satellite data are notoriously complex but the use of band ratios in the present study meant that it was important to assess the variable atmospheric effects on different bands. A simple alternative to physically rigorous atmospheric correction uses base values from very strongly absorbing surfaces such as deep clear water bodies and areas of deep shadow, although it is stressed that this only tackles the additive component of the atmospheric influence, the path radiance (Kriegler and others 1969). Areas of deep shadow and a clear lake were employed to determine an atmospheric correction base value for each of the TM bands 2 to 5 and the following discussion of band ratios relates to data for which base value correction has been applied.

An additional problem in relating ground to satellite data arises from shadow effects, particularly in areas of strong relief. Algorithms for the removal of shadow effects are not yet well developed for TM data, but the use of band ratios can be quite effective. This discussion will concentrate upon their use when relating ground to satellite data, but it should be noted that, although much of the shadow effect was removed by this means, there were still

ear shadow influences remaining in the ratio data in areas
f very steep relief.

RELATING GROUND AND SATELLITE DATA: THE INFLUENCE OF GEOMETRIC CORRECTION AND RE-SAMPLING

The ground and satellite data were related in two 40 ×
0 pixel extracts centred, respectively, on the tongue of
Lower Arolla glacier and on the down-stream valley bottom.
These extracts incorporate the field areas in which the
ground data were collected, and thus the range of terrain
types to be studied. Plots of band 4 against band 3 digital
numbers for all the pixels in these two areas show the
marked contrasts between the predominantly ice- and
rock-covered terrain of the glacier extract and the
increasing vegetational influence in the extract of the
down-stream valley bottom (Fig.1).

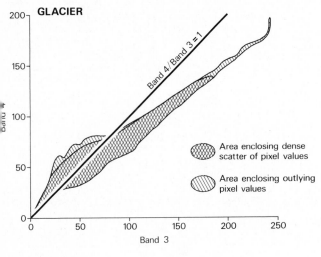

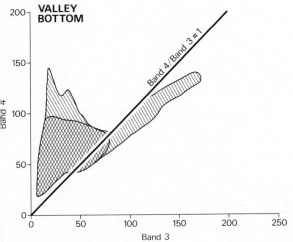

Fig.1. Digital numbers for bands 3 and 4 for pixels in
extracts covering the tongue of Lower Arolla glacier and
the valley bottom near Arolla village.

Before classification, satellite data are frequently geometrically corrected to a map base. The ground data were,
therefore, related to both uncorrected satellite data and
geometrically corrected and re-sampled data. An extract of
the TM scene, including the entire Arolla Valley, was
geometrically corrected by fitting both linear and quadratic
trend surfaces to ground-control points on the image and
their map coordinates on the 1 : 25 000 scale map. Geometrically corrected images with a 20 m pixel size were
created using nearest-neighbour and cubic-convolution re-
sampling algorithms. Identical 42 × 80 pixel extracts were
selected from each of the four re-sampled scenes to overlap
the glacier and valley-bottom extracts from the uncorrected
satellite data. Fig.2 shows the uncorrected 40 × 60 pixel

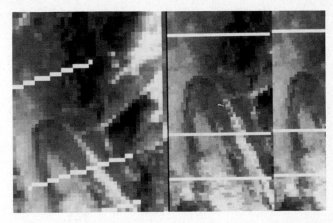

Fig.2. Extract of Lower Arolla glacier showing a composite
of bands 4, 3, and 2 for the raw data (left) and linear,
nearest-neighbour geometrical correction and re-sampling
(right). (Both photographs are presented at the same
ground scale and locate glacier profiles 1–3 from top to
bottom of the photographs.)

(pixel size 30 m) glacier extract and the geometrically
correct, quadratic, nearest-neighbour re-sampled 42 × 80
pixel (20 m pixel size) extract photographed at the same
scale.

Because of the different pixel size and orientation of
the uncorrected and corrected, re-sampled data, it is
difficult to draw quantitative comparisons. Nevertheless, the
influence of the order of trend surface and re-sampling
technique on the data was assessed by comparing pixel
values for six identical profiles (three across the glacier and
three across the valley bottom) on each of the four
geometrically corrected images. Table II expresses the
percentage deviation of the digital numbers for the same
pixels on the central profiles shown in Fig.2 for the linear,
cubic convolution; quadratic, nearest-neighbour; and
quadratic, cubic convolution re-sampled data in comparison
with the linear, nearest-neighbour re-sampled data. It was

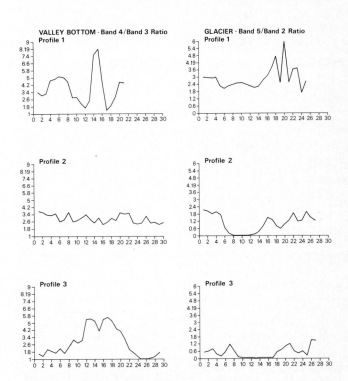

Fig.3. Raw-data pixel profiles of band ratios for extracts
across Lower Arolla glacier (band 5/2) and the valley
bottom near Arolla village (band 4/3). (The vertical axis
represents the band ratio and the horizontal axis represents
the pixel location on the profile.)

TABLE II. PERCENTAGE DEVIATION OF PIXEL VALUES FOR VARIOUS GEOMETRIC CORRECTION AND RE-SAMPLING PROCEDURES FOR THE SAME 42 PIXEL PROFILE ACROSS LOWER AROLLA GLACIER (all percentages are calculated as deviations from the pixel values for the linear, nearest-neighbour re-sampling)

Band	Trend surface	Re-sampling algorithm	Percentage Deviation			
			Mean	St. dev.	Range	
2	Linear	Cub. conv.	-1.4	8.7	14.8	-21.3
	Quadratic	Nrst. neigh.	0.2	12.5	62.3	-34.8
	Quadratic	Cub. conv.	-1.3	12.6	41.0	-38.3
3	Linear	Cub. conv.	-1.7	8.6	16.7	-22.2
	Quadratic	Nrst. neigh.	0.4	12.9	65.3	-35.5
	Quadratic	Cub. conv.	-1.7	12.6	43.1	-39.4
4	Linear	Cub. conv.	-1.9	8.2	17.0	-24.5
	Quadratic	Nrst. neigh.	0.8	14.0	73.6	-37.5
	Quadratic	Cub. conv.	-1.6	12.7	47.2	-38.2
5	Linear	Cub. conv.	-1.6	16.1	44.4	-39.4
	Quadratic	Nrst. neigh.	5.0	35.2	130.0	-54.8
	Quadratic	Cub. conv.	-0.9	23.9	77.8	-57.6

TABLE III. A COMPARISON OF THE CLASS BOUNDARIES FOR DIFFERENT GLACIAL TERRAIN TYPES BASED UPON GROUND DATA AND A DENSITY SLICE OF AN EXTRACT FROM A THEMATIC MAPPER SCENE

Terrain type	Band 4/3 ratio		Band 5/2 ratio	
	Ground data m+/-1s	Density slice class	Ground data m+/-1s	Density slice class
1. Bare ice, wet	0.28 − 0.44	0.00 − 0.72	0.14 − 0.22	0.00 − 0.20
2. Bare ice, dry	0.59 − 0.65		0.35 − 0.41	0.21 − 0.50
3. Debris on ice	0.63 − 0.81	0.73 − 0.75	0.42 − 0.90	0.51 − 0.80
4. Moraine	0.81 − 0.87	0.76 − 0.81	0.90 − 1.04	0.81 − 1.80
5. Weathered rock	0.81 − 1.09	0.81 − 1.49	0.92 − 1.20	1.81 − 3.20
6. Rock and sparse veg.	−	1.50 − 2.99	−	3.21 − 3.50
7. Thin grass and herbs	3.40 − 5.86 (grass)	2.99 − 3.52	3.02 − 3.70 (grass)	>3.50
8. Trees and dense shrubs	3.82 − 8.28 (trees)	3.53 − 7.00	2.83 − 4.99 (trees)	
9. Meadow	−	>7.00	−	

m, sample mean; s, sample standard deviation.

clear from these comparisons for all six profiles that substantial differences in digital number can arise for the same pixel according to the trend surface and re-sampling algorithm employed, and that these differences will affect band ratios with consequences for classification.

Tables III and IV indicate the implications of linking ground data to raw and re-sampled satellite data in terrain classification. Table III lists the class boundaries used to density-slice the band 4/3 and 5/2 ratios to classify terrain in the uncorrected valley bottom and glacier extracts. The correspondence between band ratios estimated from the ground-reflectance data and the satellite-radiance data for areas of known terrain type was found to be surprisingly good. Thus, the density-slice classes were estimated by starting with the ground data, inspecting the satellite data in the areas from which the ground data were derived, and creating class boundaries from these two sources to permit classification of all pixels in the extracts.

The greatest difficulties in establishing class boundaries were found in areas of meadow (for which there were no ground data) and in coarse rocky terrain with sparse vegetation (where the ground data related only to

TABLE IV. COMPARISON OF THE PERCENTAGE OF A PROFILE ACROSS LOWER AROLLA
GLACIER CLASSIFIED INTO DIFFERENT TERRAIN CLASSES ACCORDING TO THE
PROCESSING OF THE DATA

Terrain number (Table III)	Raw data	Linear nrst. ne.	Linear cub. con.	Quadratic nrst. ne	Quadratic cub. con.
Band 4/3					
1+2	30	29	33	29	29
3	15	14	7	7	14
4	22	36	38	40	26
5	33	21	21	24	31
Band 5/2					
1	26	19	19	17	17
2	4	12	14	12	10
3	7	5	2	7	10
4	41	38	33	38	31
5	22	26	30	26	31

components of the terrain). A small amount of vegetation across rocky terrain appears to have a large effect on the 4/3 and 5/2 ratios. Problems in classifying water, which had been anticipated from the ground data, arose more from the spatial scale than from the nature of the water surfaces. In the glacial extract, the high July suspended sediment concentration and small width (4–8 m) of the pro-glacial stream caused it to be absorbed into the moraine/bare-rock class of the surrounding flood plain. Fig.3 presents band-ratio values prior to geometric correction for the six profiles from the glacier and valley-bottom extracts. The ice- and water-covered surface of Lower Arolla glacier is clearly identifiable in the very low band 5/2 ratios, whereas the high band 4/3 ratios of profiles 1 and 3 of the valley bottom correspond to areas of forest and meadow.

The influence of data processing was assessed by considering the percentage of the six profiles on the valley bottom and glacier extracts that would be classified into different terrain types after different types of processing. Table IV shows the results of this analysis for profile 2 on the glacier extract. On ice and rock, the classification of the band 5/2 is slightly more stable than that of the 4/3 profile, but differences in both occur according to different processing techniques. On more vegetated terrain, comparison of the ratios is precluded by the poor definition of vegetation categories by the band 5/2 ratio.

GROUND DATA IN PERSPECTIVE
A methodology for terrain classification in Alpine areas has been described which begins with the collection of ground data and proceeds to use these data as the primary input to the interpretation and classification of satellite data. In spite of the differences in the timing of ground- and satellite-data collection, the ground data proved surprisingly useful. It appears that there is great potential for building up a data base of well-documented ground reflectance data on different dates and under different environmental conditions to aid in future image processing for the same field area. Ground data are both qualitatively and quantitatively useful. In the present study they not only provided a quantitative basis for classifying the satellite data but their analysis also indicated profitable approaches to the processing of the satellite data. The usefulness of a band 4/3 ratio in separating terrain categories might have been anticipated from the results of other studies, but the potential applicability of the band 5/2 ratio to terrain separation in areas of variable ice, rock and water cover was discovered from ground data analysis. In addition to stressing the great potential of ground data in increasing the rigour of glacial and pro-glacial remote sensing, the results of this study underline the need to adopt caution when interpreting re-sampled data.

ACKNOWLEDGEMENTS
The authors acknowledge the invaluable technical advice of Dr E J Milton, University of Southampton. Field radiometry was funded by the University of Southampton and assisted by undergraduates from the Department of Geography. Image data were provided by the Natural Environment Research Council.

REFERENCES
Frank T D, Isard S A 1986 Alpine vegetation classification using high resolution aerial imagery and topoclimatic indices. *Photogrammetric Engineering and Remote Sensing* 52(3): 381-388
Kriegler F J, Malila W A, Nalepka R F, Richardson W 1969 Preprocessing transformations and their effects on multispectral recognition. *In Proceedings of the 6th Michigan Symposium on Remote Sensing of the Environment. Ann Arbor, MI, University of Michigan:* 97-131
Milton E J 1980 A portable multiband radiometer for ground data collection in remote sensing. *International Journal of Remote Sensing* 1(2): 153-165
Robinson B F, Biehl L L 1979 Calibration procedures for measurement of reflectance factor in remote sensing field research. *Society of Photo-Optical Instrumentation Engineers* 196: 16-26

Annals of Glaciology 9   1987
© International Glaciological Society

INTERFACE TRACKING IN DIGITALLY RECORDED GLACIOLOGICAL DATA

by

A.P.R. Cooper

(Scott Polar Research Institute, University of Cambridge, Cambridge CB2 1ER, U.K.)

ABSTRACT

As more data are recorded in digital form the importance of automatically extracting parameters of glaciological significance increases. This paper addresses the problem, with particular reference to tracking bedrock or internal reflectors in digitally recorded Radio Echo Sounding (RES) data. It has been found that the simplest solution to this problem is a "supervised" system, where operator decisions may be added interactively, either on operator command or upon loss of track. Increasing internal decision making within the program may reduce the number of operator interventions required, but is unlikely to eliminate them completely.

Algorithms are presented and discussed for determining the position of the interface, for predicting the position of the interface in successive records, for determining the loss of track condition, and for re-acquiring track after loss of track.

INTRODUCTION

Tracking is a process whereby the position of an interface is determined by an automatic or semi-automatic method. It is applicable to many glaciological data, especially altimeter data, or Radio Echo Sounding (RES) data. The problem of tracking may be divided into several parts, as follows (Fig.1). First, an algorithm must be provided to enable the program to locate the interface in the digitized trace. Secondly, the program must be provided with a means of predicting where in the trace the interface is likely to be found on the following trace, in order to eliminate unnecessary searching, to eliminate glaciologically impossible points and, in the case of instruments with a narrow window (e.g. satellite altimeters), to position the window so that the surface lies within it. Thirdly, the program must be able to recognize when no point in the area of the trace defined above fulfils the criteria specified by the first algorithm. Finally, a means of re-acquiring track after its loss must be provided. In addition to these four processes that are specific to the tracking problem, it will almost certainly be necessary to pre-process the signal to reduce the noise level as far as possible, and to convert the data from instrumental units to geophysical units.

PRE-PROCESSING

Many techniques are available for reducing the noise level in a signal, and the final choice of method will be dependent on the characteristics of the data being processed. If the data contain isolated spikes, a spike-editing procedure such as the median filter (Sy 1985) should be used to remove them. Otherwise, a very useful technique is that of low-pass filtering (e.g. Albert 1986), provided it is known that there is a limiting frequency above which there is unlikely to be useful information in the signal, as is the case for pulsed radar systems. There are many other useful techniques such as deconvolution, Wiener filtering and maximum entropy deconvolution which are of use, especially when the signal being processed may be regarded as the convolution of a series of spikes with a wavelet, as in seismic processing.

It may also be useful at this point to convert the data from the instrument units to geophysical units. This is often of benefit if the relationship between the two is non-linear, as, for example, in the case where an amplifier with a logarithmic response is being used.

LOCATING THE TRACK POINT

Various algorithms are in use for specifying the point in the recorded trace that is to be regarded as representing the interface to be tracked. These depend on the characteristics of the surface being tracked, and also on the signal returned by the instrument in question. For example, the surface of the sea is tracked by satellite radar altimeters using a simple model of the expected radar return, which has the advantage of allowing various useful properties of the pulse to be extracted directly from the tracking algorithm. This approach is very successful for returns from ocean surfaces, which are statistically homogeneous and

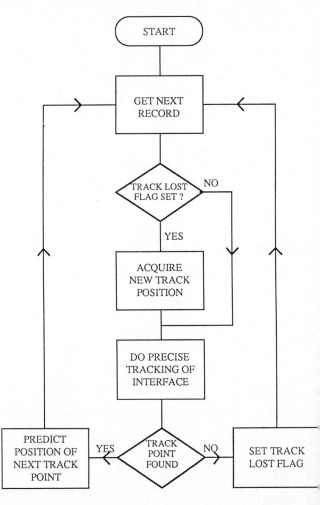

Fig.1. General flow chart of the tracking process

geometrically simple. Unfortunately, for ice surfaces the situation is not so clear cut, because the surface is geometrically complex and because the scattering mechanism from the ice surface is not well understood. This prevents the use of simple models of the return pulse shape, and other approaches have to be considered. The most useful point to track in the return pulse is the first arrival time, as this is the simplest point to interpret in terms of the overall shape of the surface being studied (Harrison 1970). This point can be detected either by setting a threshold level and testing for the first point in the pulse to exceed this level, or by differentiating the return pulse and testing for positive-going zero crossings, which represent minima in the returned signal. In either case, noise in the signal will produce spurious results, and secondary criteria such as testing the next maximum in the signal must be used to reject invalid points. The first method is suitable for signals where there is only one interface, or for signals where the interfaces are separated by intervals where the signal level returns to the noise level, and the second is more suitable where the returns from successive interfaces overlap, as is the case in RES when the bedrock return arrives during the wide-angle reflections from the surface. Fig.2 illustrates the difference between the two types of signal described above, showing possible alternate track points.

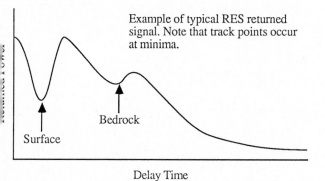

Example of typical RES returned signal. Note that track points occur at minima.

Bedrock

Surface

Delay Time

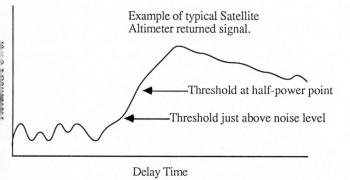

Example of typical Satellite Altimeter returned signal.

Threshold at half-power point

Threshold just above noise level

Delay Time

Fig.2. Examples of typical returned waveforms, illustrating the required tracking points.

PREDICTING THE NEXT TRACK POINT

Prediction of the position of the interface in the record succeeding the current one is the heart of the tracking problem. It is dependent on the geophysical characteristics of the interface being tracked, and also on instrument characteristics, especially the beamwidth in the case of the ranging instruments which are primarily being considered here. A further constraint on the type of algorithm to be used is the time available for the processing of the signal, which is clearly more limited in the case of a satellite altimeter which is tracking in real time than in the case of post-processing RES data.

At this point, it is necessary to discuss the exact meaning of the terms "broad" and "narrow" beamwidths. In this paper, the beamwidth of the instrument is considered to be broad if the nearest point of the surface lies within the antenna beam pattern, and narrow if it lies outside. If we consider a uniformly sloping surface, with undulations superimposed on it, then the beamwidth is broad if the height of the undulations above the sloping surface does not exceed

$$h_{max} = \frac{R.\mathrm{Sin}\left[\frac{\Theta}{2}\right].\mathrm{Sin}\left[\frac{\Theta}{2} - \phi\right]}{\mathrm{Cos}(\phi)} + \sqrt{R_e^2 + R^2.\mathrm{Cos}^2\left[\frac{\Theta}{2}\right]\mathrm{Sin}^2\left[\frac{\Theta}{2}\right]} - R_e \quad (1)$$

$$\text{for } 0 \geqslant \phi \geqslant \frac{\Theta}{2}$$

(This equation is derived in the appendix)

where h_{max} is the maximum height of an undulation above the sloping surface, R is the range of the instrument from the intersection of the slope with the nadir, ϕ is the angle of slope of the surface, Θ is the beamwidth of the instrument, and R_e is the radius from the centre of the Earth to the nadir point. Note that the last two terms of the equation correct for the curvature of the Earth, and are negligible for airborne instruments.

In the case of broad beam instruments, the maximum rate of change of range to the surface with distance along track is 1.0 (Harrison 1970). This means that it is only necessary to scan a portion of the digitized signal equivalent to ± the along track distance between successive returns from the instrument. In the case of airborne instruments where the signal is being digitized fairly frequently, this limits to manageable proportions the volume of data to be inspected for a particular interface. This constraint is not true for narrow beam instruments, but in these cases the digitized pulses returned by the instrument will be difficult to intepret, as the first arrivals will be coming from a region on the extremity of the antenna beam pattern.

For instruments which need a window to be positioned prior to the digitization of the returned signal, predictive tracking methods are commonly used, of which the most widely known is the alpha-beta tracker implemented in the Seasat altimeter (Rapley and others 1983). In this method, the range is predicted using the equations

$$H_{n+1} = H_n + \alpha.\Delta H_{n-1} + H_n'.\Delta t \quad (2)$$

$$H_n' = H_{n-1}' + \frac{\beta.\Delta H_{n-1}}{\Delta t} \quad (3)$$

where H_n is the range estimate at the nth, α is a constant, ΔH_n is the error in the range estimate at the nth point, H_n' is the estimate of the rate of change of range at the nth point, β is a constant and Δt is the time interval between pulses. The time taken by the tracker to respond to changes in the surface gradient, or to step changes in the surface elevation, may be adjusted by varying α and β. For Seasat, α was set to 0.25 and β was set to 0.015625, giving a response time of 0.28 s.

In these methods, assumptions are made concerning the form of the surface being tracked, and these assumptions are used to predict from past statistics the future position of the interface. Unfortunately, the major assumption of this method is that the differential of the surface observed varies smoothly, without discontinuities. Harrison (1971) showed that this is not a valid assumption for ice surfaces, or indeed for any undulating reflecting surface where the radius of curvature of the surface is less than the range from the sensor. Therefore such methods cannot be depended on over ice or other complex surfaces.

As the surface slopes of more than 90% of the Antarctic continent are less than the half beamwidth of altimeters like the Seasat altimeter (McIntyre and Drewry 1984), such altimeters are effectively operating in a broad beam mode and therefore returns may not change their range at a higher rate than 1.0. This gives rise to an

interesting possibility for tracking such surfaces without loss of lock. To outline one possible scheme, the position of the returned waveform within the window would be determined on a per pulse basis, and the position of the window would be changed so that the leading edge of the returned pulse is in the centre. Then, as long as the rate of advance per pulse of the satellite does not exceed the half width of the range window, the leading edge of the next pulse must appear within the window so placed, and the process may then be repeated. This method would require fast computing, as only about 1 ms would be available for the positioning algorithm. However, in the case of Seasat the rate of advance per pulse was ~7 m, and the range window was 30 m wide. Thus the range window was sufficiently wide to allow for more than twice the maximum possible shift in the position of the return, provided that the window was centred on the return from the previous pulse. The offset centre of gravity tracker developed by Wingham and others (1986) would be a very suitable algorithm for finding the position of the leading edge, as it is insensitive to noise and changes in the pulse shape, and operates linearly over a wide range of displacements from the centre of the instrument window. It should also be borne in mind that modern "off the shelf" microprocessors are capable of running at rates in excess of 1 million instructions per second, and that special purpose processors are much faster. In order to provide averaged pulses for the determination of significant wave height and back-scatter coefficient, it would still be necessary to maintain a value of the range rate, but this could be computed from a smoothed value of the corrections applied to the position of the range window.

DETERMINING LOSS OF LOCK

This can be a considerable problem for real-time algorithms, especially over complex surfaces such as ice sheets. Off-line, it is considerably simpler, as more time is available, and it is also possible to include human intervention in the loop. Indeed, for RES, human intervention is the only wholly reliable method of determining the loss-of-lock condition, as it is possible for tracking software to be misled by strong diffraction hyperbolæ, which are indistinguishable from a very steeply sloping interface and are often stronger reflections. In the case of satellite altimeters, criteria based on threshold values of the power difference between the early part of the return and the late part of the return are used. These can easily be invalidated by pulses of unusual shape, so that actual loss of lock occurs much earlier than it is detected by the instrument, as is seen in the case of Seasat data over Antarctica, or over sea ice (Rapley and others 1985).

ACQUISITION OF LOCK

The acquisition or re-acquisition of the track point after loss of lock is an essential part of any complete tracking system, and, in real-time systems, must be as fast as practicable, in order to minimize the volume of data lost during this phase. It must also be robust, in order to avoid spurious results. As with the loss-of-lock algorithm, the only method that is wholly reliable is human intervention, which can only be used in post-processing of data gathered in the field. In the case of satellite altimeters, it is usual to operate the altimeter in a completely different mode from the normal tracking mode (Rapley and others, 1983).

IMPLEMENTATION OF A RES TRACKING SCHEME

Following the acquisition of a large volume of digitally recorded RES data in 1983, with further data from 1985 and 1986 (Drewry and Liestøl 1985; Gorman and Cooper 1987), various tracking programs have been implemented by the author, both on a large mainframe computer and on Z80 based microcomputers. The mainframe has the advantage of processing the data much faster, so that relatively complex data processing can be carried out, but has the serious disadvantage of not allowing manual interaction with the data processing, so that direct human intervention in the case of loss of lock and re-acquisition of

lock cannot take place. In many ways, processing this data on a microcomputer has proved more successful, as an operator can intervene in the processing whenever appropriate, and hence simpler tracking algorithms can be used. The main disadvantage is the slow speed of the microcomputer, especially when floating-point arithmetic is to be carried out. Therefore little pre-processing has been carried out, as the rate of processing then drops to unacceptable levels. With more advanced microprocessors than a Z80, there is less of a problem, and experiments will be carried out in the near future using an Intel 80286 fitted with an 80287 maths co-processor.

The scheme followed on microprocessors has been that outlined in Fig.3. For each record, the data are first displayed. This has been one of the least satisfactory parts of the microcomputer implementation, as the resolution of

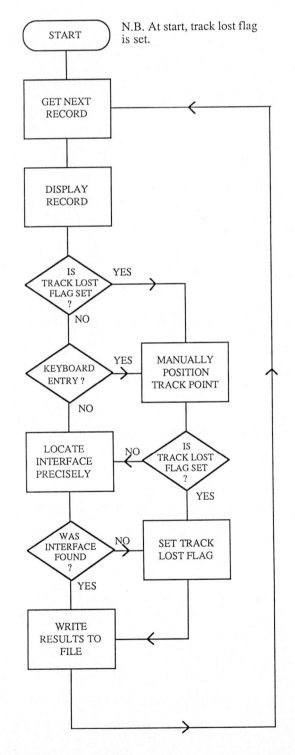

Fig.3. Flow chart of RES tracking program used for data gathered in Svalbard, 1985.

the graphical display has not been adequate to allow the data to be displayed as a profile along the track, with power modulating the intensity of the display as in a Z-scope display. The alternative adopted has been to display successive records as plots of delay time against power, as in an A-scope display. This allows the full resolution of the data to be displayed, and the location of the interfaces can be seen quite clearly. The disadvantage is that the continuity of interfaces cannot readily be visualized. Then the program checks whether an interface was found on the last record (the "track lost flag" referred to in Fig.3), and also whether there has been any keyboard input from the operator. In either case, the operator is given the opportunity to position the track point manually, or, alternatively, to set the "track lost flag" manually, thus preventing mis-tracking by the automatic routines. If the "track lost flag" is set, the results are written to an output file, with identifying data. Otherwise, the interface is located precisely by searching within a window, whose width is constrained as described above, for minima in the signal. In this particular implementation, an interface was defined as occurring when the returned power had risen by one twentieth of the difference in power between the minimum and the next maximum, in order to avoid difficulties with instrumental noise near the minimum. This approach was required because there was instrumental noise near the minimum, and could have been improved on if more pre-processing had been possible. The results were consistent, with a small systematic offset which was determined by inspection of the data. Various constraints were added to the simple test described here, such as a minimum difference in power between the minimum and the maximum, which improved the selectivity of the system. If no suitable interface was found, then the "track lost flag" was set. In either case, the result was written out to file.

ACKNOWLEDGEMENTS

Much of this study was funded by NERC grant GR3/4463 to D J Drewry. M R Gorman, J A Dowdeswell and W G Rees all contributed much helpful discussion during the development of tracking programs for RES.

REFERENCES

Albert D G 1986 FORTRAN subroutines for zero-phase digital frequency filters. *CRREL Special Report*: 86–4

Drewry D J, Liestøl O 1985 Glaciological investigations of surging ice caps in Nordaustlandet, Svalbard, 1983. *Polar Record* 22(139): 359–378

Gorman M R, Cooper A P R 1987 A digital radio echo-sounding and navigation recording system. *Annals of Glaciology* 9: 81–84

Harrison C H 1970 Reconstruction of subglacial relief from radio echo sounding records. *Geophysics* 35(6): 1099–1115

Harrison C H 1971 Radio echo sounding: focussing effects in wavy strata. *Geophysical Journal of the Royal Astronomical Society* 24: 383–400

McIntyre N F, Drewry D J 1984 Modelling ice-sheet surfaces for ERS-1's radar altimeter. *ESA Journal* 8(3): 261–274

Rapley C G and 22 others 1983 *A study of satellite radar altimeter operation over ice-covered surfaces*. Paris, European Space Agency (ESA Contract Report 5182/82/F/CG(SC))

Rapley C G and 16 others 1985 *Applications and scientific uses of ERS-1 radar altimeter data; final report*. Noordwijk, European Space Agency (ESA Contract Report 5684/83/NL/BI)

Sy A 1985 An alternative editing technique for oceanographic data. *Deep Sea Research* 32A(12): 1591–1599

Wingham D J, Rapley C G, Griffiths H 1986 New techniques in satellite altimeter tracking systems. *In IGARSS '86 Symposium, Zurich, September 1986.* Paris, European Space Agency: 1339–1344 (ESA Special Publication 254)

APPENDIX A

The equation describing the broad beam criterion is derived in this appendix.

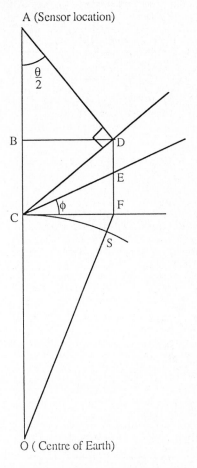

If A is the sensor location, C is the nadir point at the surface, and CE is the mean surface, with a slope of ϕ, then for a sensor beamwidth of θ, DE represents the maximum elevation of a point above the slope for the closest approach of the point to be within the beam-limited footprint of the sensor. We can calculate DE as follows:

$$CD = AC.Sin\left(\frac{\theta}{2}\right) \tag{A1}$$

$$DF = CD.Sin\left(\frac{\theta}{2}\right) \tag{A2}$$

therefore

$$DF = AC.Sin^2\left(\frac{\theta}{2}\right) \tag{A3}$$

$$BD = CD.Cos\left(\frac{\theta}{2}\right) \tag{A4}$$

therefore

$$BD = AC.Sin\left(\frac{\theta}{2}\right).Cos\left(\frac{\theta}{2}\right) \tag{A5}$$

$$CF = BD \tag{A6}$$

therefore

$$EF = AC.Sin\left(\frac{\theta}{2}\right).Cos\left(\frac{\theta}{2}\right).Tan(\phi) \tag{A7}$$

therefore

$$DE = AC.Sin\left(\frac{\theta}{2}\right).\left[Sin\left(\frac{\theta}{2}\right) - Cos\left(\frac{\theta}{2}\right).Tan(\phi)\right] \tag{A8}$$

therefore

$$DE = \frac{AC.Sin\left[\frac{\theta}{2}\right].Sin\left[\frac{\theta}{2} - \phi\right]}{Cos(\phi)} \qquad (A9)$$

In order to correct for the curvature of the Earth, we have to add the further correction SF:

$$OF^2 = OC^2 + CF^2 \text{ (by Pythagoras)} \qquad (A10)$$

but $$CF = AC.Cos\left[\frac{\theta}{2}\right].Sin\left[\frac{\theta}{2}\right] \qquad (A11)$$

therefore

$$OF^2 = OC^2 + AC^2.Sin^2\left[\frac{\theta}{2}\right].Cos^2\left[\frac{\theta}{2}\right] \qquad (A12)$$

and therefore

$$SF = \sqrt[2]{OC^2 + AC^2.Sin^2\left[\frac{\theta}{2}\right].Cos^2\left[\frac{\theta}{2}\right]} - OC \qquad (A13)$$

The total correction, DE + SF, is given by:

$$DF + SF = \frac{AC.Sin\left[\frac{\theta}{2}\right].Sin\left[\frac{\theta}{2} - \phi\right]}{Cos(\phi)} + \sqrt{OC^2 + AC^2.Cos^2\left[\frac{\theta}{2}\right].Sin^2\left[\frac{\theta}{2}\right]} - OC \qquad (A14)$$

This equation can now be converted from the geometric notation used here into the algebraic notation used in the main text (Equation 1), as follows:

$$h_{max} = \frac{R.Sin\left[\frac{\theta}{2}\right].Sin\left[\frac{\theta}{2} - \phi\right]}{Cos(\phi)} + \sqrt{R_e^2 + R^2.Cos^2\left[\frac{\theta}{2}\right].Sin^2\left[\frac{\theta}{2}\right]} - R_e$$

where h_{max} (= DE + SF) is the maximum height of an undulation above the sloping surface, R (= AC) is the range of the instrument from the intersection of the slope with the nadir, ϕ is the angle of slope of the mean surface, θ is the beamwidth of the instrument, and R_e (= OC) is the radius from the centre of the Earth to the nadir point.

Annals of Glaciology 9   1987
© International Glaciological Society

SEASAT ALTIMETER OBSERVATIONS OF AN ANTARCTIC "LAKE"

by

W. Cudlip and N.F. McIntyre

(Mullard Space Science Laboratory, University College London, Holmbury St. Mary,
Dorking, Surrey RH5 6NT, England)

ABSTRACT

The word "lake" in the context of Antarctica has been used to describe both surface features (10-12 km in size) and areas of liquid water on the bedrock beneath the ice sheet (typically a few kilometres in size and at a depth of over 3000 m). There has been however only limited evidence for a relationship between the two phenomena.

This paper reports an analysis of Seasat altimeter observations of an extremely flat area on the surface of the East Antarctic ice sheet, approximately 30 km in extent, centred at 68.6°S 136.0°E and close to the edge of the sub-glacial Astrolabe Basin. It has a regional slope of between zero and 0.01°, and non-random variations in height along track of about ± 1m on the scale of a few kilometres. The average radar backscatter coefficient is 5±2 dB in the region of the Astrolabe Basin, compared to a more usual value of about 10 dB for other areas of the ice sheet. A computer enhanced Landsat image of the region clearly shows the rougher steeper terrain to the North, with the surface in and around the flat area appearing totally smooth and featureless.

NSF/SPRI/TUD radio echo-sounding data from the region, although limited in extent, shows a relatively strong signal (indicative of ice at the pressure melting point) over a large region. The signal under the flat area, however, is particularly strong and smooth, confirming the association between the surface feature and a bedrock lake 3800 m below.

INTRODUCTION

Robinson (1964), senior navigator with the Soviet Antarctic Expedition of 1959, reported "........oval depressions with gentle 'shores' which are visible from an airplane over the plateau. The depth of the depressions usually does not exceed 20–30 m. These unusual depressions are sometimes called 'lakes' by pilots". At that time, however, there was no reason to associate these visual lakes with liquid water. Some years later, as the techniques of radio echo-sounding were developed, anomalously strong echoes at a depth of 4200 m near Sovetskaya (Robin and others 1970) were reported and a substantial area of water deep under the Antarctic ice was suggested as an explanation. Subsequent radio-echo observations by the NSF/SPRI/TUD radio echo-sounding programme have detected 57 examples of strong specular returns from smooth flat areas in the bedrock (Oswald and Robin 1973; Robin and others 1977; McIntyre unpublished). These authors conclude, in combination with other studies, that water pockets or lakes do exist under certain parts of the ice sheet. Most of the examples suggest lake size of the order of a few kilometres; however, there is evidence for a much larger lake near Vostok about 180 × 45 km in size (Robin and others 1977). A flight over this lake did in fact give visual confirmation of a correlation between surface and bedrock features, with the "shores" showing as areas of whiter snow which corresponded well in some cases with the edges of the radio echo lake. The visibility of the surface lake was attributed to the difference in appearance between a uniform flat surface (the lake) and a gently sloping one (the shore). This could be due either to changes in albedo, caused by differences in relative sun angle, or directly to changes in texture resulting from different snow accumulation. This is the only link so far that suggests that the earlier reported surface "lakes" are in fact related to sub-glacial lakes.

Satellite altimeters, because of their ability to provide data for accurate maps of surface elevation, can be used to help quantify the ice-sheet surface characteristics of bedrock lakes. Towards this end, this paper reports the detection, using Seasat altimeter data, of a particularly flat area in East Antarctica, about 30 km in extent, which can be linked with such a sub-glacial lake.

DATA AND ANALYSIS

Although designed for use over the open ocean, the altimeter on board the Seasat satellite (operational between 28 June and 10 October 1978) provided useful data over non-ocean surfaces. However, the analysis and interpretation of such data is usually more difficult due to the geometric and dielectric variability of the terrain and to the inability of the range·tracker to stay securely locked onto a return echo that can change rapidly in both shape and delay time. This latter problem can be overcome to some extent by recalculating the range estimate from the return echoes, a process called retracking (Martin and others 1983).

The ability to derive from Seasat Altimeter data detailed topographic maps over ice sheets has already been demonstrated (Brooks and Norcross 1983; Zwally and others 1983); over the smoother and flatter areas a potential precision of 0.25 m (for 10 Hz data) has been quoted, degrading to 1.6 m over rougher, sloping surfaces. (Zwally and others 1983). This precision of the Seasat elevation data can result in much detail remaining hidden in a contour map with, say, a 25 m contour interval.

Fig.1, taken from Brooks (1983), shows a 25 m contour map of a part of Wilkes Land in East Antarctica. Marked on this map are two sections of Seasat altimeter tracks A and B which intersect over a local ice divide separating ice flowing towards Frost and Mertz Glaciers. The retracked elevation profiles along these tracks, derived from Seasat Sensor Data Record (SDR) tapes of waveform data, are shown in Fig.2. The echoes received over this section of the ice sheet were sufficiently similar in shape to those received over the ocean to justify retracking the waveforms using a threshold of 50% of peak return power (Partington and others 1987, this conference). This was sufficient to provide measurements of the mean surface elevation to a precision of ±1 m within the pulse-limited altimeter footprint of about 1.75 km. Slight distortions of the waveforms observed in some locations did not warrant the use of a more sophisticated retracker.

Despite the frequent loss of lock of the altimeter, it can be seen from Fig.2, track A, that there is a marked transition from an upward slope to a very flat region near the point of intersection. Unfortunately, the beginning of the flat area is lost due to the loss of lock, and the altimeter again loses lock soon after leaving the flat area.

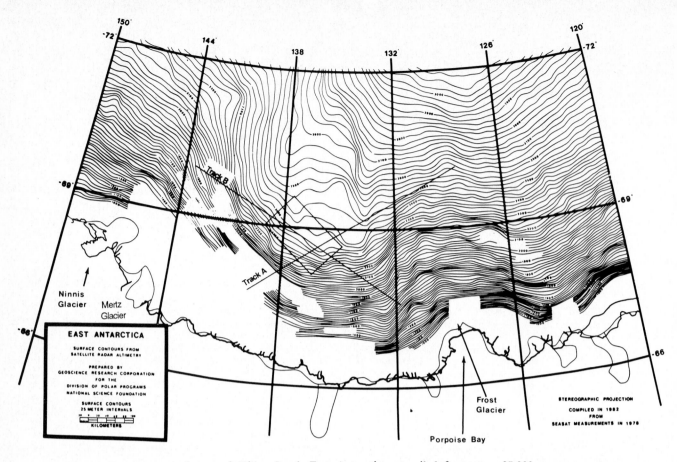

Fig.1. Contour map of part of Wilkes Land, East Antarctica compiled from over 27 000 measurements by the Seasat radar altimeter and contoured at a 25 m interval (Brooks 1983). Also shown are the Seasat ground tracks used in Fig.2 (A and B) and the area recontoured in detail in Fig.3.

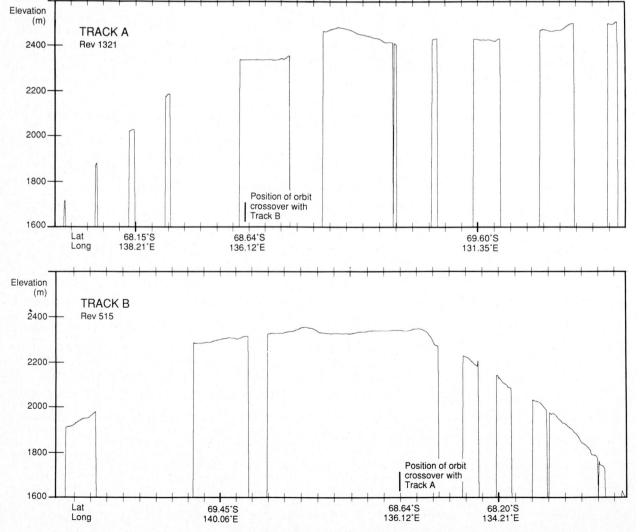

Fig.2. Two tracks of ice sheet surface elevations derived from Seasat altimetry data showing their intersection in the area of extremely level surface terrain. Gaps in the profiles are due to losses of lock by the altimeter's onboard tracking loop. Ground track locations are shown in Fig.1.

Track B also shows the surface to be flat where it intersects with track A, demonstrating that the flat area extends in two directions and is not merely the result of a track fortuitously aligned with a surface contour.

A linear least-squares fit to the elevation data over the flattest parts of tracks A and B gives the following gradients and root mean square deviations:

	length	gradient		rms deviation
Track A	36.4 km	<1:12 000	(<0.005°)	0.6 m
Track B	28.5 km	1:6000(±2000)	(0.01°)	0.6 m
Larsen Ice Shelf	30 km	<1:6000	(<0.01°)	1.5 m

For comparison, the table also shows values for a length of track from the central part of the Larsen Ice Shelf. The gradient and deviations are similar. The data over these two areas could be retracked with a more sophisticated retracker (e.g. a least-squares, five-parameter, ocean-like fit) (Martin and others 1983) in order to reduce random retracking errors, but this is unlikely to reduce the rms deviations given above, as close examination of the elevation profiles does show some non-random variation in height of ±1 m with a scale length of a few kilometres. Other comparable lengths of track can be found on the ice-sheet with similar gradients, but these have larger standard deviations (of ≈1.5 m or more) and appear to be chance alignments with contours.

Fig.3a shows a more detailed contour map of the region around the intersection of the tracks A and B. This has been compiled from 9 tracks of retracked Seasat altimetry data and contoured by hand to an interval of 5 metres. The surface-elevation values are with respect to the Seasat's orbital reference ellipsoid (a=6378.137 km, f=1/298.257) and use the orbital parameters from the SDR tapes.

As we are concerned only with relative changes in surface elevation, no atmospheric or other corrections have been applied to the data. Analysis of the 8 orbit crossovers in the area of Fig.3 revealed a mean difference of 0.6 m and a standard deviation of ±1.6 m. As this was of similar magnitude to the uncertainty in the individual elevation estimates, no correction for orbit errors was deemed necessary. Similarly, no correction for slope-induced error was carried out since for slopes of 0.1° this amounts to only 1 m; the flat area would thus be unaffected by this correction although other parts of the map would be subject to error. The portions of the altimeter tracks with valid data used in the construction of the contour map are shown in Fig.3b. Many more Seasat orbits crossed the area of Fig.3 but these were repeat or near repeat tracks. The density of tracks is mostly limited by the maximum repeat orbit of 17 days used by the Seasat satellite, together with the relatively short lifetime of the mission (3 months).

The flat area at the centre of Fig.3 is bounded to the south and west by higher ground, and by lower ground to the east and north. There does not appear to be a common surface feature with an elevation change of more than a metre which could be identified as a "lake shore" surrounding the flat area; however, the relatively rapid change in gradient around the flat area might make it distinguishable to an airborne observer. Similarly, although changes of a factor 2 in return power in this region of the Astrolabe Basin can be seen in the return waveforms, there is no marked difference between the flat section and the surrounding area. The normal-incidence backscatter coefficient (Moore and Williams 1957) over the flat area is ±2dB, which is relatively low when compared with the more usual value of 10 or 11dB seen over other areas of the ice sheet. An analysis of the backscatter coefficient over a much larger area of the ice sheet will have to be carried out before this lower value can be explained.

COMPARISON WITH LANDSAT

One of the few Landsat images of the area (path 90, row 109, taken on 29 January 1973) was purchased in order

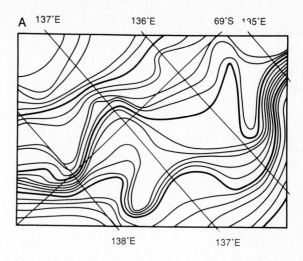

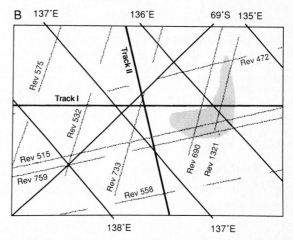

Fig.3. A: Contour map of ice sheet surface elevations in the boxed area of Fig.1. Contour interval is 5 m with every sixth contour drawn in bold. From the bottom, the bold contours are 2312, 2342 and 2372 m above Seasat's orbital reference ellipsoid. The estimated relative error on these contours is ±2 m.

B: Location of the 9 tracks of Seasat altimetry data (dotted lines) used to compile the contour map in A. Two NSF/SPRI/TUD radio echo flight lines (I and II) are shown as bold continuous lines. The very flat terrain identified from the altimetry is shown as a stippled zone.

to compare the visible and radar observations. Fortunately, although half the area of the image was obscured by cloud, the lower half of the area of Fig.3 was cloud-free. Bands 4 and 5 were saturated, as expected for many Landsat images over the ice sheet (Dowdeswell and McIntyre 1986). In bands 6 and 7, even after digitally processing the image with an extreme contrast stretch, no surface features were discernible on the flatter central area of Fig.3. However, the steeper area to the lower end of Fig.3a was clearly discernible as an apparently much rougher, pock-marked surface. There were no discernible surface features which could be attributed to the presence of a surface "lake" or "lake shore".

COMPARISON WITH RADIO ECHO-SOUNDING OBSERVATIONS

Fig.4 shows the ice-surface and bedrock profiles taken from two radio echo-sounding flights obtained during the NSF/SPRI/TUD campaign of 1974-75. The data are plotted for the lengths of the flight lines which intersect the area of Fig.3a, and the locations of the flight lines are shown in Fig.3b. These two flights provide the only radio echo-sounding data directly over or close to the area of the flat surface. As can be seen in Fig.3b, neither of the lines are exactly coincident with Seasat ground tracks and so care

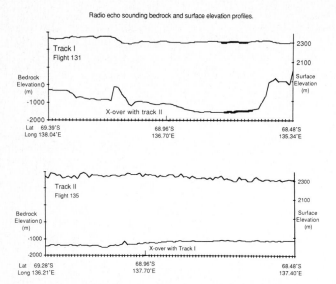

Radio echo sounding bedrock and surface elevation profiles.

Fig.4. Surface and bedrock profiles from two NSF/SPRI/TUD radio echo-sounding flights (I and II) with ground tracks shown in Fig.3b. Note the different vertical scales for bedrock and surface profiles. The point of intersection of each is marked. The section of exceptionally bright and uniform return in track I, identified as a sub-glacial lake, is shown in bold on the bedrock profile, and the intersection of track I with the stippled lake area of Fig.3b is shown in bold on the surface profile.

must be taken in any comparison between these and Seasat data.

Track I is from a flight line along the length of the Astrolabe Basin, which has been described as a valley similar to the Great Rift Valley in East Africa. The surface and bedrock profiles along the whole flight line can be found in Steed and Drewry (1982). Almost the whole of Fig.3 appears to lie over a relatively flat bedrock shelf about 70 × 70 km at the northern end of the basin. The ice thickness in this area is between 3800 and 4000 m, while over the deepest part of the basin it increases to 4670 m.

A study of the original photographic record of these flights shows the echo returns to be relatively strong over the whole area but with a short section of track II being particularly strong and smooth. This short section, 16.5 km in length (shown highlighted in Fig.4b), shows good agreement with the flat surface area mapped with the Seasat altimeter (Fig.3). The strong return over the whole area suggests the presence of ice at the pressure melting point (Oswald 1975) over most of the shelf area, and the particularly strong section suggests a body of water a metre or more in depth i.e. a subglacial lake directly beneath the area of particular surface flatness. The presence of a lake at this location was suggested by McIntyre (unpublished) on the basis of the radio echo-sounding data alone.

A linear least-squares fit to the bedrock elevation over the 16.5 km of strongest return in track II gives a gradient of 1:160±20, which compares with values of 1:70±10 and 1:50±10 determined for some other sub-glacial lakes (Oswald 1975). The mean surface elevation derived from radio echo-sounding data over the same length of track is 2310 m with a standard deviation of ±4 m. This compares with a geoid-corrected value derived from the Seasat data of 2295±2 m. (The geoid correction using the GEM 10b model is 45.1 m at this location, and the local Geoid gradient is 0.001°). This residual difference of about 15 m is well within the absolute calibration error of the radio echo-sounding data which has been estimated to be ±50 m (Steed and Drewry 1982).

DISCUSSION AND SUMMARY

The removal of basal friction over an area of tens of kilometres must be expected to produce some effect on ice flow and hence the upper surface of an ice sheet (Whillans

and Johnsen 1983), even though local ice thickness may be 3000 m or more. This effect was predicted by Robin and others (1977) for the Vostok Lake, but sufficiently accurate data were not available to provide more than visual confirmation. With Seasat altimetry data we have shown these predictions to be correct, even though applying to a much smaller area of basal water.

Most of the sub-glacial lakes found to date are in areas of low surface slopes and low ice velocity; there are high concentrations near Dome Circe and Ridge B. The location of the lake observed here is, however, unusual in that it is much closer to the Wilkes Land coast (250-300 km) than most previous observations of pressure melting or sub-glacial lakes (about 700 km at Dome Circe). The surface conditions for its occurrence are provided by the local ice divide shown in Fig.1, the location of which was not clearly demonstrated until production of the surface-elevation map with Seasat data. This would suggest first the importance of generating accurate topographic maps using satellite altimetry data as an aid to studies of ice-sheet dynamics; and, secondly, that prerequisites for the formation of lakes (such as high geothermal heat flux (Oswald 1975) may exist over wider areas than previously thought.

The discovery of a large flat area on the ice-sheet which is even flatter than the major portion of the central ice sheet is also important from the point of view of future satellite missions. It has been suggested that the level topography of the central areas of ice sheets be used as reference surfaces for the calibration of the satellite orbit (Gorman and Drewry 1984), since they have advantage over tropical sites in terms of atmospheric corrections, smooth terrain and tidal variations. It is possible that the surface of ice sheets in the region of subglacial lakes will offer exceptionally smooth sites for altimeter calibration. The large subglacial lake near Vostok and other smaller ones near Dome Circe and Ridge B will be within the orbital limit of the European Space Agency's ERS-1 satellite (82°N and S) and could be considered for this purpose.

ACKNOWLEDGEMENTS

We are most grateful to Dr D J Drewry of the Scott Polar Research Institute for providing access to the radio echo-sounding data. WC and NFM acknowledge the receipt of NERC and SERC PDRAs respectively.

REFERENCES

Brooks R L 1983 Scientists use satellite data to map ice-sheet contours. *Antarctic Journal of the United States* 18(2): 17-18

Brooks R L, Norcross A N 1983 *East Antarctic ice sheet surface contours from satellite radar altimetry — a demonstration.* Salisbury, MD, Geoscience Research Corporation

Dowdeswell J A, McIntyre N F 1986 The saturation of LANDSAT MSS detectors over large ice masses. *International Journal of Remote Sensing* 7(1): 151-164

Gorman M R, Drewry D J 1984 Ice sheets as invariant surfaces for radar altimeter calibration and orbit determination. *In* Guyenne T D, Hunt J J (eds) *ERS-1 Radar Altimeter Data Products. Proceedings of an ESA workshop held at Frascati, Italy on 8-11 May, 1984.* Paris, European Space Agency: 173-176 (SP221)

McIntyre N F Unpublished *The topography and flow of the Antarctic ice sheet.* (PhD thesis, University of Cambridge, 1983)

Martin T V, Zwally H J, Brenner A C, Bindschadler R A 1983 Analysis and retracking of continental ice sheet radar altimeter waveforms. *Journal of Geophysical Research* 88(C3): 1608-1616

Moore R K, Williams C S 1957 Radar terrain return at near vertical incidence. *Proceedings of the Institute of Radio Engineers* 45(2): 228-238

Oswald G K A 1975 Investigation of sub-ice bedrock characteristics by radio-echo sounding. *Journal of Glaciology* 15(73): 75-87

Oswald G K A, Robin G de Q 1973 Lakes beneath the Antarctic ice sheet. *Nature* 245(5423): 251-254

Partington K C, Cudlip W, McIntyre N F, King-Hele S 1987 Mapping of Amery Ice Shelf, Antarctica, surface features by satellite altimetry. *Annals of Glaciology* 9: 183–188

Robin G de Q, Swithinbank C W M, Smith B M E 1970 Radio echo exploration of the Antarctic ice sheet. *International Association of Scientific Hydrology Publication* 86 (*ISAGE*): 97-115

Robin G de Q, Drewry D J, Meldrum D T 1977 International studies of ice sheet and bedrock. *Philosophical Transactions of the Royal Society of London* Ser B 279(963): 185-196

Robinson R V 1964 Experiment in visual orientation during flights in the Antarctic. *Soviet Antarctic Expedition. Information Bulletin* 2: 233-234

Steed R H N, Drewry D J 1982 Radio-echo sounding investigations of Wilkes Land, Antarctica. *In* Craddock C (ed) *Antarctic geoscience. Symposium on Antarctic Geology and Geophysics, Madison, Wisconsin, U.S.A., August 22-27, 1977.* Madison, WI, University of Wisconsin Press: 969-975

Whillans I M, Johnsen S J 1983 Longitudinal variations in glacial flow: theory and test using data from the Byrd Station strain network, Antarctica. *Journal of Glaciology* 29 (101): 78-97

Zwally H J, Bindschadler R A, Brenner A C, Martin T V, Thomas R H 1983 Surface elevation contours of Greenland and Antarctic ice sheets. *Journal of Geophysical Research* 88(C3): 1589-1596

Annals of Glaciology 9   1987
© International Glaciological Society

A MULTI-SENSOR APPROACH TO THE INTERPRETATION OF
RADAR ALTIMETER WAVE FORMS FROM TWO ARCTIC ICE CAPS

by

Mark R. Drinkwater and Julian A. Dowdeswell*

(Scott Polar Research Institute, University of Cambridge, Cambridge CB2 1ER, U.K.)

ABSTRACT

Data collected over Svalbard on 28 June 1984 by a 13.81 GHz airborne radar altimeter enabled analysis of signals returned from two relatively large ice masses. Wave forms received over the ice caps of Austfonna and Vestfonna are analysed with the aid of existing aerial photography, radio echo-sounding data, and Landsat MSS images acquired close to the date of the altimeter flight. Results indicate that altimeter wave forms are controlled mainly by surface roughness and scattering characteristics. Wet snow surfaces have narrow 3 dB back-scatter half-angles and cause high-amplitude signals, in contrast to relatively dry snow surfaces with lower-amplitude diffuse signals. Metre-scale surface roughness primarily affects wave-form amplitude and leading-edge slope, this becoming apparent over ice streams on Vestfonna.

INTRODUCTION

The European Space Agency's ERS-1 satellite will carry a suite of Earth remote-sensing instruments on a dedicated polar and oceanographic mission. For studies of large ice masses, an important component of its payload will be a Ku-band radar altimeter with the capability of providing precise surface elevations of Arctic and Antarctic ice masses (cf. Zwally and others 1983). It will yield synoptic information on the topography of major ice-sheet and ice-cap surfaces, enabling morphological and mass-balance studies to be undertaken. In addition to the radar altimeter's capability of returning accurate range data, however, it also returns information in the record of energy back-scattered from the surface. Few investigations have been made of the resulting wave forms, and their relationship to ice-mass surface character.

The aims of this paper are, first to analyse the nature of radar altimeter wave forms received from two Arctic ice caps and, secondly, to interpret these data with the aid of other information on ice-mass elevations and characteristics of the ice surface. Data were collected during the 1984 Marginal Ice Zone Experiment on 28 June 1984 using an aircraft-mounted 13.81 GHz radar altimeter, designed and constructed by the Rutherford Appleton Laboratory (RAL) (McIntyre and others 1986). This provided an opportunity to examine wave forms obtained over relatively large terrestrial ice masses.

Some 400 track kilometres of radar-altimetric data were acquired over the ice caps of Austfonna (8105 km^2) and Vestfonna (2510 km^2) in Nordaustlandet, Svalbard (Fig.1). Three sections of the altimetric data are analysed in detail to illustrate a number of points concerning ice-surface characteristics and topography. Information from other sensors is also used to aid interpretation of the altimeter wave forms. Ice-surface topographic data from airborne radio echo-sounding (RES) of the ice caps are available (Dowdeswell and others 1986), and the character of the ice surface has also been investigated using a combination of

aerial photographs and Landsat images acquired close to the date of the flight.

BACKGROUND: SCATTERING FROM SNOW AND ICE AT NEAR-NORMAL INCIDENCE

The main factors influencing scattering of radar pulses from snow and ice surfaces are outlined prior to analysis and interpretation of altimetric data from the Nordaustlandet ice caps. Moore and Williams (1957) suggested that the mean pulse returns received by an altimeter correspond mainly to incoherent back-scattering from a rough surface. If volume scattering is considered negligible, a model mean pulse return may be expressed as a convolution of two parts: the transmitted pulse shape and a function including the antenna-beam pattern, ground properties, and range to target. Brown (1977) incorporated and refined these ideas in a model of rough-surface scattering by including an explicit expression for the antenna pattern and variation of back-scatter with the angle.

Passive microwave radiometric data were obtained during the same flight by imaging and non-imaging instruments operating at wavelengths from 0.3 to 1.7 cm. Gloersen and others (1985) concluded from brightness temperatures recorded that the surface layer was close to melting point throughout most of Nordaustlandet. Many areas will therefore have had a relatively high volumetric free-water content in the surface layers. The resulting dielectric discontinuity between the air and snow is large, preventing significant penetration of pulses. From this, we may deduce that scattering of the transmitted altimeter pulses occurs almost exclusively at the surface over Nordaustlandet. Even when snow is dry, back-scatter at nadir will be dominated by surface scattering, using typical values for snow parameters taken from Ulaby and others (1982). This enables the use of a model incorporating rough-surface theory to understand variations in wave forms obtained in Nordaustlandet.

Ulander (unpublished) discussed the use of Brown's model in understanding mean pulse returns over planar ice sheets, with the proviso that its assumptions are only interpreted as first-order approximations of surface character. The main drawback of the model is that ice masses are variable across the altimeter footprint. A series of predicted mean RAL altimeter pulse shapes, which Ulander generated using Brown's model, is shown in Fig.2. This illustrates the effects of different values for RMS surface roughness (Fig.2a) and for the "surface polar-scattering diagram" or angular dependence of scattering (Fig.2b), which is quantified in terms of the 3 dB back-scatter half-angle (i.e. the angle from normal incidence to the point at which the back-scatter coefficient is reduced to half its maximum value, assuming a Gaussian back-scatter function and which is here abbreviated to TBH). Increasing RMS roughness at the metre-scale reduces peak power and increases rise time, and increasing back-scatter half-angle causes more diffuse returned signals with progressively longer durations. These idealized wave forms are presented for comparison with the data obtained from Nordaustlandet shown in the subsequent analysis.

---

*Present address: Department of Geography, University College of Wales, Aberystwyth, Penglais, Aberystwyth, Dyfed SY23 3DB, UK.

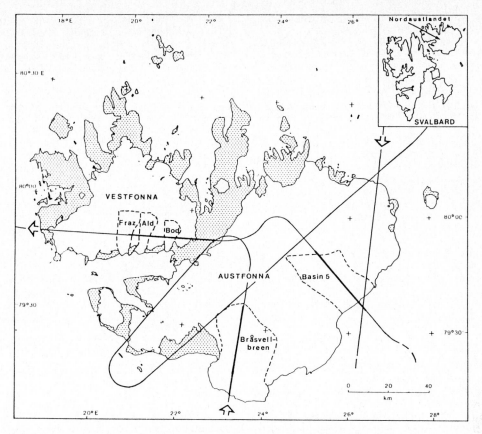

Fig.1. Map of Nordaustlandet ice caps illustrating the full radar-altimeter flight track and sections analysed in detail (heavy). The location of Nordaustlandet within Svalbard is inset.

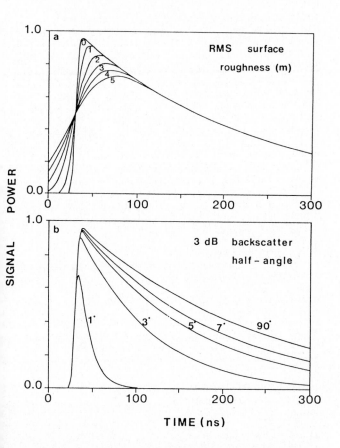

Fig.2. Predicted pulse shapes for variations in (a) surface roughness and (b) TBH, using a 7.3 ns pulse duration. All have been normalized assuming constant back-scatter coefficient at normal incidence and antenna gain, and Gaussian variation of back-scatter coefficient with angle (adapted from Ulander unpublished).

Various approaches to modelling have demonstrated that for comparatively rough surfaces, such as glacier ice, the coefficient of back-scatter becomes independent of radar frequency and depends on surface RMS slope, reflectivity, and incidence angle (Ulaby and others 1982). For locally smooth surfaces with only gentle undulations, such as snow surfaces, surface scattering may be separated into a coherent (specular) component, and a non-coherent (diffuse) component (Fung 1981).

Properties affecting scattering include:

i. Water content;
ii. Surface roughness;
iii. Reflectivity;
iv. Crystal structure;
v. Density of snow-pack.

The presence of small amounts of free water in snow is known to affect the loss tangent of the snow significantly. For example, a free-water content of 1% by weight will increase the loss tangent from 0.001 to 0.01 in dry snow (Cumming 1952), thereby substantially reducing signal penetration. The penetration depth through snow is reduced to about one wavelength for free-water contents of only 3–4% by volume. Temperature variations also influence snow-crystal metamorphism and alter the texture of the surface, thus affecting scattering. Reflectivity is determined by the permittivity of the medium and controls reflection from and transmission across the air/medium interface. Finally, the crystal structure and density of the medium are most likely to influence the directionality and strength of scattered energy. Intuitively, because of differences in material properties between wet snow, dry snow, and bare ice, we may expect each to have different scattering signatures.

METHODS
Altimetric data collection and navigation

The RAL altimeter was mounted in a NASA CV-990 aircraft. Measurements were made with a nadir-pointing

horn antenna between 07.34 and 08.30 h on 28 June 1984 from an altitude of 10 km. A full 3 dB beam width of 10° determined a pulse-limited footprint diameter of approximately 1.75 km. Flight lines were planned to follow tracks previously surveyed during an airborne RES campaign over Nordaustlandet (Dowdeswell and others 1986). Additional instruments aboard the CV-990 included a metric camera, providing overlapping vertical photographic coverage. However, cloud cover during the flight precluded its use.

Signals were recorded using a Biomation wave-form digitizer, in which one biomation amplitude unit of power (bau) equals 15.6 mV. Wave forms were digitized at delay intervals in range bins (or range gates), where one bin equals 3.33 ns. Subsequent use of the "bau" and "bin" units in the text considerably simplifies the discussion of data. A background d.c. voltage from the detector introduces an offset in mean wave forms, equal to 4 bau. Before data were interpreted this background level was subtracted. Over the ice sheet, signal amplitude varied considerably but the instrument had a fixed receiver gain. This had important implications, since at one extreme very strong returns saturated the receiver while, at the other, returns were not separable from background noise. Instrument calibration was also necessary for the calculation of back-scatter coefficients (Ulander unpublished). External calibrations to determine the instrument parameters were undertaken over retro-reflectors located on the Norwegian island of Andøya, and internal calibration was possible during the flight using the instrument's in-built "calibration mode".

The aircraft's flight path was recorded on magnetic tape by the Airborne Digital Data Acquisition System and was derived from the on-board Inertial Navigation System (INS). First, corrections were applied to original records of latitude and longitude to eradicate biases caused by drift of the gyros in the INS mechanism, and the cumulative error was linearly redistributed throughout the flight. Secondly, the flight line was re-positioned by fixing its output to the known locations of timed crossings on to the ice caps. Navigational data recorded throughout the flight were used to give an accurate flight line, which was superimposed on satellite images (Fig.3).

Supporting data sets

We selected Landsat 5 Multispectral Scanner (MSS) imagery from 24 June, and 12 and 17 July 1984, since these dates bracketed the altimetric data set and were the nearest days on which partially cloud-free digital scenes were available. Digitally enhanced Landsat sub-scenes corresponding to each flight section are shown in Fig.3. These images provide information on the surface character of the ice caps. Even where slopes are less than 1° or 2°, relatively small changes in ice-surface topography are indicated by differences in radiance (Dowdeswell and McIntyre 1986). Changing ice-surface characteristics, for example snow density, crystal size, and water content, will also affect radiance recorded by the MSS. Digital enhancement of these MSS data recorded on Computer Compatible Tapes (CCTs) therefore reveals considerable detail of the ice-cap surface (Fig.3). Aerial photographs taken of the ice caps in other years, and field seasons on the ice caps in

May 1983 and May 1986, provide further background to the interpretation of altimeter wave forms. Airborne RES data, including both surface and bedrock elevations (Dowdeswell and others 1986), were also important in planning the CV-990 flight lines.

Wave-form tracking and averaging

Algorithms have been designed by Ulander (unpublished) which derive mean statistics from RAL altimeter wave forms. Such methods are akin to tracking and averaging routines used to sum and average raw individual pulses on previous altimetric missions (MacArthur 1978). They also act as a data quality check, and include sorting routines which guard against spurious data resulting from instrument malfunctions, and data which are invalid geophysically, becoming incorporated in the analysis. For an altimeter flying over a variable surface, such as a large ice mass, rapid changes in the form of scattered radar energy may be encountered due to variations in surface geometry and material properties. There is a temptation, therefore, to decrease the number of wave forms averaged in order to minimize the surface area sampled. However, confidence in the mean statistical values is determined by sample size. The control which this has on confidence limits, and the corresponding integration times and along-track integration distances, are presented in Table I.

The statistics in Table I are calculated assuming a constant aircraft velocity of 200 m s^{-1} and a known pulse-repetition frequency of 100 Hz. However, with an unstable moving platform such as an aircraft, these integration distances may only be used as an approximate guide, since the effects of pitch and variable velocity either compress or extend along-track integration distance. Rejecting pulse wave forms during the sorting procedure also extends the sampling interval. For a sample size of 50, the integration distance has been observed to expand, at worst, by 100 m. Normally, the use of 100 pulses is statistically adequate to construct a mean wave form. However, for surfaces of high spatial variability, we may have to accept increased variance, due to inadequate averaging out of "fading" effects (Ulaby and others 1982), in order that the terrain characteristics do not change significantly during along-track averaging. Thus, for a mean wave form composed of typically 50 individual pulse wave forms, the 95% confidence limits on the mean power $\overline{(P)}$ lie between $0.77\overline{P}$ and $1.35\overline{P}$. This corresponds to a range of approximately 2.5 dB. Such uncertainty in a single estimate of mean power, for example, must be taken into account when contiguous mean pulse wave forms are compared or geophysical parameters are extracted. The 2.5 dB range, although large, may still enable us to distinguish between: (a) bare ice and dry snow, which have a possible relative difference in Fresnel reflection coefficient of up to 7 dB, and (b) wet snow and dry snow, which have a possible relative difference of up to 11 dB in reflection coefficient (Evans 1965, Smith and Evans 1972).

Wave-form descriptors

Several parameters are extracted to describe mean pulse wave forms, and have been adapted from Ulander (unpublished):

TABLE I. CONFIDENCE LIMITS FOR POWER VALUES (P) OF MEAN WAVE FORMS CONSTRUCTED USING VARYING SAMPLE SIZES. CORRESPONDING INTEGRATION DISTANCES AND TIMES ARE INCLUDED

N	Integration distance m	Integration time s	95% Confidence limits for 2N degrees of freedom
10	40	0.2	$0.59\overline{P} <$ mean $< 2.10\overline{P}$
20	80	0.4	$0.67\overline{P} <$ mean $< 1.64\overline{P}$
50	200	1.0	$0.77\overline{P} <$ mean $< 1.35\overline{P}$
100	400	2.0	$0.81\overline{P} <$ mean $< 1.20\overline{P}$

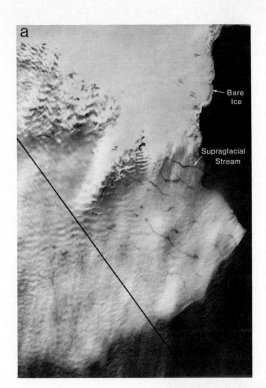

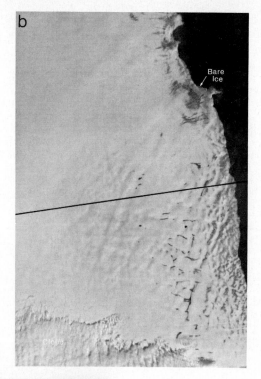

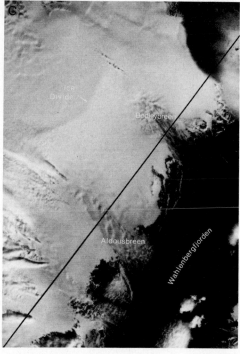

Fig.3. Digitally enhanced Landsat 5 images with the three flight tracks marked. Each image measures 28 km by 41 km.

(a) Bråsvellbreen on 12 July 1984 (Landsat path/row 214/03);

(b) Basin 5, Austfonna, on 17 July 1984 (217/02);

(c) Vestfonna on 24 June 1984 (216/02).

(1) <u>Leading-edge slope</u> (L_e). This is calculated by fitting a function to five points around the half-peak power point in a least-squares manner. It is expressed in dB bin^{-1} (where one range bin equals 3.33 ns).

(2) <u>Calibrated coefficients of back-scatter</u> ($\sigma°$). Each of the two coefficients expresses the gain (in dB) of an average scatterer on the ice-mass surface, assuming that the system point-target response is Gaussian with time.

(a) In the first algorithm the back-scatter coefficient ($\sigma°_p$) is calculated using the peak power of the received signal. This gives a measure of back-scatter at normal incidence (i.e. from within the first few range bins of the footprint). Since peak value is, in most cases, attained within the first few bins of the wave form, it is independent of the polar-scattering diagram of the surface. On the other hand, this value is sensitive to saturation effects in high-amplitude returns.

(b) An alternative algorithm calculates a weighted value of the coefficient of back-scatter (σ_i°). This method assumes a "diffuse" return and so utilizes the sum or integral of power in the range bins occupied by the returned wave form. It is weighted to account for the antenna-beam attenuation pattern.

Accurate calculation of the coefficient of back-scatter necessitates that the aircraft tilt relative to the surface is small. A pitch, for example, of 1.5° would reduce back-scatter by about 0.5 dB, and so only flight sections where pitch and roll and regional slope are small ($<1.5^{\circ}$) at all times were analysed.

(3) <u>Trailing-edge attenuation coefficients</u> (X). Since both coefficients of back-scatter are sensitive to pointing errors and surface-slope variations, and because they do not take into account variation in back-scatter with incidence angle, Ulander (unpublished) also used two trailing-edge attenuation coefficients to measure rates of decay in mean wave forms. They are applicable to wave forms with longer delays where power values are limited by antenna-beam attenuation.

(a) The early trailing-edge attenuation coefficient (X_e) is used to measure attenuation of power in the interval between 6 and 25 range bins after the wave-form peak. If we assume that signal power falls exponentially, then the following equation is used to calculate the coefficient of exponential decay:

$$y = ae^{-x} \qquad (1)$$

where y is power in units of bau, x is the coefficient of attenuation X_e, and a is a constant (300), scaling X_e to units of μs^{-1}.

(b) A late trailing-edge coefficient (X_1) is calculated in a similar manner for the interval 50 to 89 range bins after the wave-form peak.

The early trailing-edge attenuation coefficient gives good differentiation between mean returns of short duration, while the late attenuation coefficient gives better separation of longer-duration pulse returns. They are chosen to distinguish between surfaces having small 3 dB half-angles (i.e. $1-3^{\circ}$), and surfaces having large 3 dB half-angles (i.e. $>3^{\circ}$). The latter coefficient is also independent of surface-height distribution effects on the received signals.

The above parameters are calculated for three sections of the altimeter flight over Austfonna and Vestfonna.

VARIATIONS IN WAVE-FORM PARAMETERS: BRÅSVELLBREEN, AUSTFONNA
Results

Mean wave forms from a long profile of Bråsvellbreen (Fig.1) are illustrated in Fig.4, with corresponding wave-form parameters in Fig.5. RMS surface roughness is also calculated from the standard deviation of pulse delays (Fig.5f). Fig.5a–g reveals three zones, after initial wave-form disruption caused by the inbound coastal crossing. Note that the boundaries between ice-surface zones discussed here are in reality transitional.

The first identifiable region lasts from approximately 3 to 17 km, and is characterized by the mean returns shown in Fig.4a and b. It is a period of relative stability and σ_p° and σ_i° show only slight steady increases, from 12 to 14 dB and 6 to 8 dB, respectively. X_e stays fairly constant at around 25 μs^{-1}, while an increase of X_1 from 2 to 10 μs^{-1} picks out the extended tails of wave-form trailing edges. Leading-edge gradient L_e (Fig.5c) fluctuates between 1.5 and 2.0 dB bin^{-1}, after having been stretched to 0.5 dB bin^{-1} at the 3 km stage. RMS surface roughness varies between 0.5 and 2.0 m (Fig.5f).

The second region lasts from 17 to 25 km, and is characterized by the mean wave forms in Fig.4c. Peak signal amplitudes reach their maximum, with σ_p° attaining a plateau at around 15 dB. Wave-form duration increases to 300 ns, resulting in a fall of X_e from 30 to 20 μs^{-1}.

The final region continues from 25 km until the end of the profile, and mean signals received at 31 km inland are displayed in Fig.4d. A marked reduction in signal amplitude occurs over the space of 5 km, and the coefficients of back-scatter σ_p° and σ_i° fall by 5 dB to 9 dB, and by 6 dB to 4 dB, respectively. Additionally, L_e (Fig.5c) falls to minima of below 1.0 dB bin^{-1}. More important, however, is the change occurring at around 29 km inland, after which both trailing-edge attenuation coefficients become markedly more variable. Beyond 35 km inland, signal amplitudes fall to their lowest values, and mean returns observed at 38 km

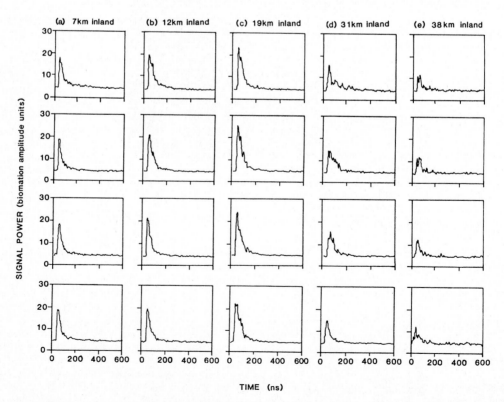

Fig.4. Four contiguous mean wave forms taken at five intervals along the Bråsvellbreen flight section (Fig.1) illustrating varying wave-form characteristics.

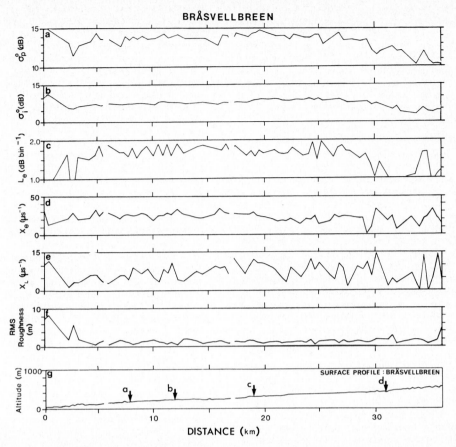

Fig.5 Traces of varying wave-form parameters and ice-surface profile along Bråsvellbreen flight section. The arrows labelled a to d in part g of the figure represent the locations of the sets of wave forms in Fig.4.

inland are displayed in Fig.4e. At this point the proportion of spurious delays increases, as many signals are too weak to be detected by the receiver. This situation is characteristic of the central parts of the ice cap.

Interpretation

Bråsvellbreen surged between 1936 and 1938, and is now in the quiescent period between surge activity (Schytt 1969). It has a regional gradient of less than $1°$, and this minimizes inaccuracies in back-scatter calculations.

Comparison of the mean returns in Fig.4 with the model wave forms in Fig.2a and b indicates that over the sections identified progressive changes in wave-form geometry occur. Wave forms in Fig.4a, b, and c illustrate the increasing mean signal amplitudes and durations between 7 and 19 km. Throughout, signals have a specular component which dominates over a diffuse scattered component. As the bulk of each wave form is composed of reflected energy, they are termed "quasi-specular". Since signals do not principally result from diffuse scattering mechanisms, the two values of $\sigma°$ (Fig.5a and b) show a consistent discrepancy, with $\sigma_p°$ continually greater by 5 dB. Only in places where diffuse scattering becomes dominant will the two values meet and follow similar trends. Otherwise, although incorrect as an absolute value of $\sigma°$, the changes in $\sigma_i°$ indicate variations in scatter contributions from angles off-nadir (Fig.5b). At this point, the surface has a narrow TBH, attaining a value of $3°$ at 19 km inland. Tonal variations on Landsat imagery indicate that most of the ice surface is snow-covered, apart from small coastal areas where bare ice is visible (Fig.3a). At this time of year the snow depth is likely to be significantly less than 1 m in depth within 25 km of the coast, and surface melting results in high free-water content in the upper layers. Snow depths are based on field surveys on Austfonna by Scott Polar Research Institute (SPRI) parties in spring 1983 and 1986. We suggest that in relatively wet snow areas the main mechanism causing near-specular signals is probably that of

reflection from flat surfaces orientated perpendicular to incident pulses. Additional important effects are thought to be caused by reflections from supraglacial streams which cross the lower parts of the wet-snow zone. These streams are visible beneath thin cloud on the Landsat scene from 12 July (Fig.3a).

Maximum observed signal amplitudes occur between 17 and 25 km, and large TBH (over $3°$ in places) causes longer-duration wave forms (Fig.4c). More dense refrozen or glazed surface crusts, or shallow ice lamellae or lenses in this region (Schytt 1964), would explain the continuing dominance of the specular component of wave forms over the diffuse component.

From 25 km onwards signal amplitude falls considerably. This may be a response to moving to progressively higher elevations, where the effects of melting are reduced (Fig.4c). Surface melting and the formation of ice lenses has been observed even on the high crest of Austfonna (e.g. Schytt 1964, and SPRI party observations). However, whether or not surface melting had taken place high on Austfonna by late June 1984 is not certain. Trailing-edge attenuation remains fairly stable until 30 km, whereupon X_e and X_1 begin to fluctuate, in response to TBH varying between $3°$ and $7°$. Variability in polar-scattering characteristics of dry snow surfaces has been observed by Ulander (unpublished). He derived a similar range of values from altimetry over central Greenland on 1 July. The cause of this variability is not understood, but it is most likely to be a combination of small-scale roughness and the reduced effects of snow-grain metamorphism.

In relatively dry snow some penetration may occur, and volume scattering is possible. However, at near-normal incidence surface scattering continues to dominate. If we assume a Gaussian height distribution, then the value of 1.5 dB bin^{-1} for L_e (Fig.5c) at 33 km corresponds with a value of 1 m surface roughness predicted by Brown's model for a similar leading-edge gradient. This matches our calculation of RMS surface roughness at 33 km (Fig.5f), and

is similar to the height of wind-generated snow features such as sastrugi, dunes and ridges. Snow-feature dimensions observed by Ivanov (1968) range from a few centimetres to 2 m amplitude, with long axes from metres to several tens of metres. Beyond 30 km, peaks in RMS roughness are observed (Fig.5f), indicating larger-scale roughness than aeolian features (2 and 3 m). These account for some reduction in leading-edge gradient. Signals in Fig.4e have more rounded peaks and considerable variability in late bins.

VARIATIONS IN WAVE-FORM PARAMETERS: BASIN 5, AUSTFONNA
Results
Three zones are identifiable in this section of data; 0–7, 7–25, and 25–33 km (Fig.6). Between 0 and 7 km inland σ_p° is consistently the highest value of back-scatter by 10 dB (Fig.6a and b). L_e remains steady at 1.3 dB bin^{-1} (Fig.6c), but the attenuation coefficients fluctuate: X_e between 0 and 35 μs^{-1}, and X_1 between 0 and 15 μs^{-1}. The surface profile in Fig.6f has discontinuities indicating periods when the altimeter records spurious delays. These result from instrument malfunction or received signals being too weak to cross the triggering power threshold.

From 7 to 25 km, L_e becomes greater, attaining values comparable to those observed over Bråsvellbreen. There are increases in signal amplitude indicated by peaks in σ_p°, but σ_i° rises dramatically by over 10 dB in places following similar trends to 0_i°. In contrast, X_e falls to low values between 15 and 20 km, periodically reaching zero. X_1 attains values similar to those given over the section of several kilometres before the inbound coastal crossing indicated on the surface profile (Fig.6f).

In the final section (25 km onwards), parameters return to similar levels observed for the first 7 km over land. Once again, breaks in the profile indicate a large proportion of spurious altimeter delays.

Interpretation
Signals from 0 to 7 km are quasi-specular and comparisons made with Fig.2b yield a mean TBH of less than 3°. Despite the tall narrow peaks, however, some mean wave forms have shallower leading edges, resulting from

large-scale surface roughness. The value of L_e of 1.2 dB bin^{-1} at 4 km inland corresponds with a surface-roughness prediction using Brown's model of the order of 3 m amplitude. Aerial photographs from previous field campaigns reveal that there is a high degree of crevassing and undulating serac ice in this coastal region and this would explain the loss or severe attenuation of many signals, and the resulting large number of spurious delays. However, the specular component of signals is likely to occur by reflections from wet snow and/or large planar ice surfaces between crevasses.

Aerial photography indicates that the surface is markedly less broken between 7 and 25 km inland despite large-scale surface perturbations of over 10 m height and several kilometres in wavelength. The profile in Fig.6f indicates that signals are strong in this part of the basin. The rise in σ_i° and similarity of the two back-scatter coefficients confirms that returns are diffuse, wave forms periodically having durations of over 300 ns and 3 dB half-angles between 5° and 7°.

It is useful at this point to make comparisons between diffuse land-ice signals and ocean returns, and the corresponding responses of the calculated wave-form parameters. Ocean returns (diffuse) are a useful yardstick since they result from Lambertian or isotropic surface scatter, tending to a 10° situation in Fig.2b (Gatley and Peckham 1983). Diffuse ocean returns characteristically exhibit high values of σ_p°, σ_i°, and X_1, with X_e returning zeros or very low values. By contrast, quasi-specular returns show high values for σ_p° and X_e with low σ_i° and X_1. Surface returns at 14 and 16 km show similar characteristics to ocean signals observed between −5 and 0 km, and L_e and X_1 return similar values, despite reduced back-scatter values. High-density snow with a refrozen crust or high water content is a known cause of high reflectances, while small-scale surface roughness is the suggested cause of wider polar scattering.

Between 20 and 25 km, mean signals return to values similar to those recorded in the initial 7 km over the ice cap, the TBH falling to around 3°. Scattering contributions are limited to highly reflective surfaces within the pulse-limited footprint, antenna-beam attenuation effectively suppressing weak returns from areas on the fringe of the

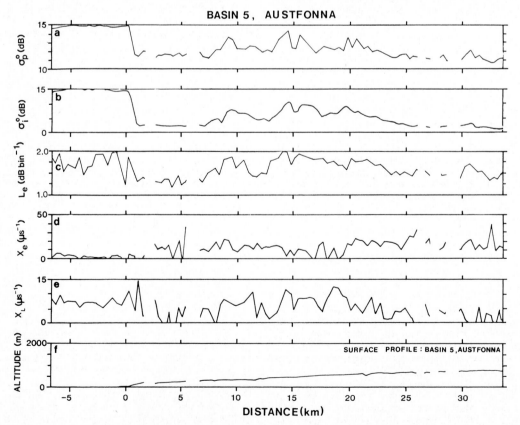

Fig.6. Traces of varying wave-form parameters and ice-surface profile along Basin 5 flight section, Austfonna.

eam-limited footprint. These signals are probably from eas where the snow surface is still relatively dry, and no revasses have been observed in the upper part of Basin 5. igh back-scatter values for drier snow have previously een explained by the effects of a complex snow ratigraphy. Snow layering, ice lamellae, and ice glands or nses produce a component of specular reflection with ypical reflectivity of 0.1, this explaining strong back-scatter mited to near-normal incidence. Snow-pit studies and hallow drilling in Basin 5 during spring 1986 show that uch density variations are a common feature of the strati- raphy in this area.

ARIATIONS IN WAVE-FORM PARAMETERS: ESTFONNA

Results

Wave-form parameters from a track across Vestfonna nd the central, ice-free valley in Nordaustlandet (Fig.1) ndicate three main zones (Fig.7). The first is for the initial 2 km, and is from non-glacierized land. The two values of σ° show several distinct peaks at irregular periods, which re accompanied by rises in L_e of up to 2.0 dB bin^{-1} from level of around 13 dB bin^{-1}. Both attenuation coefficients re extremely variable, X_e registering a maximum of over 0 μs^{-1} at 6.5 km, but when no signal is evident in late-gate ins X_l returns zeros. Throughout this period spurious elays are recorded, breaks in traces a to f in Fig.7 ndicating these occurrences.

The second zone, comprising three short sections, is etween approximately 15 and 20 km, 24 and 33 km, and 39 nd 42 km. All have the same characteristics with relatively teady values of X_e, at around 20 μs^{-1}. Although back- catter at nadir (σ_p°) increases by only 1 or 2 dB, σ_i° rises nstead by several dB, to plateaus of between 5 and 8 dB. $_e$ behaves similarly (as at 16 km) by rising by 0.6 dB bin^{-1} o 1.9 dB bin^{-1}.

The final set of short sections last from approximately 20 to 24 km, 33 to 39 km, and from 42 km onwards. σ_p° alls by 1 or 2 dB in places to 12 dB, while σ_i° shows a orresponding decrease to a level of 3 dB. Wave-form eading edges have markedly reduced gradients, with minima t 23 and 35 km of 12 and 11 dB bin^{-1}, respectively.

During these periods both trailing-edge coefficients record a large proportion of zeros (Fig.7d and e).

Interpretation

Before crossing the ice streams Bodleybreen, Aldousbreen, and Frazerbreen on Vestfonna, the altimeter traverses Helvetesflya, an area of non-glacierized terrain with several melt lakes. At 1 km, over Flysjøen (the largest lake) signal amplitudes are the highest observed. Although σ_p° only picks out two definite peaks over lakes (at 1 km and 11 km), σ_i° has additional increases at 4 and 5 km. The diffuse nature of signals from Flysjøen, and the TBH of 5°, indicate that the lake surface does not act as a plane reflector, despite minimal metre-scale surface roughness indicated by values for L_e of almost 2.0 dB bin^{-1}. Analysis of Landsat imagery showed that Flysjøen was still ice-covered on 12 July. It is therefore inferred that lake ice, accompanied by surface snow dunes, is a likely cause of the relatively wide polar-scatter diagram and longer-duration returns. Similar wave-form characteristics are observed at 4, 6.5, and 11 km, but at 6.5 km X_e exceeds 50 μs^{-1}. This occurs because mean signals have markedly shorter duration, due to reflection from a more specular surface such as a stream. Otherwise, between the lakes there is a high pro-portion of spurious delays and rejected wave forms. This, in combination with stretched leading edges, is a response to partially snow-covered terrain having metre-scale roughness of the order of 3 m. Such intermittently snow-covered land was observed on the Landsat scene for 12 July.

The second distinguishable set of areas is beyond about 12 km, where the altimeter begins to traverse Vestfonna. This is the first of several snow-covered and relatively flat ridges with minimal gradient and metre-scale surface roughness (less than 1 m). They return high-amplitude signals. Corresponding 3 dB back-scatter half-angles are approximately 3°.

The last group of areas is associated with ice streams flowing orthogonal to the altimeter flight track and between the series of ridges mentioned above (Fig.7f). They are identifiable both by their distinct surface character and wave-form response. Large-amplitude (15–25 m) kilometre-scale surface roughness has been identified in previous work by Dowdeswell (unpublished), but the L_e minimum of 12 dB

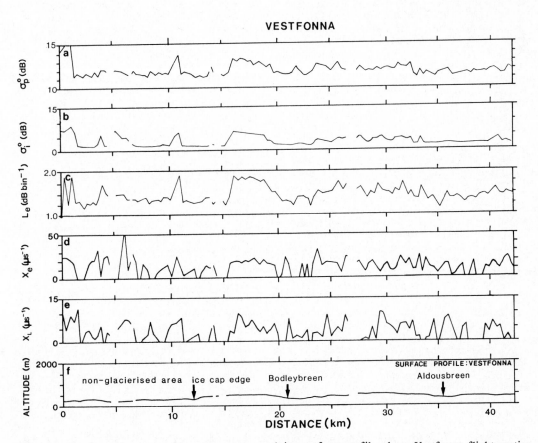

Fig.7. Traces of varying wave-form parameters and ice-surface profile along Vestfonna flight section.

bin^{-1} at 23 km (Fig.7c), for instance, also suggests metre-scale roughness of 3 m or more when compared with predictions for leading-edge gradient made by Brown's model. Bodleybreen (Fig.7f) has surged since 1976 (Dowdeswell 1986). Its boundaries are well defined on Landsat and aerial photographic images of the area by heavy crevassing truncated by shear zones (Fig.3c). The combined roughness and steep lateral margins of Bodleybreen cause severe attenuation of pulses, and loss of the weakest returns. In regions of broken ice the slope distribution is likely to be markedly different than elsewhere and results in TBH of around 5° over parts of Bodleybreen. The other areas of similar wave-form characteristics occur from approximately 33 to 39 km over Aldousbreen, and over the beginning of Frazerbreen from 42 km onwards. Neither has been observed to surge but both are highly crevassed and are faster flowing than the surrounding ridges (Dowdeswell 1986). This surface roughness has a significant effect upon wave forms, in contrast with the signals resulting from surrounding ridges. Small-scale surface roughness is again liable to cause wider scattering and thus the observed longer-duration wave forms.

CONCLUSIONS

Our results indicate that significant variations in returned altimeter signals occur through changes in the terrain type and surface character of Nordaustlandet ice caps. A number of conclusions may be drawn from this work:

1. When wet snow or ice surfaces are encountered Fresnel reflection coefficients are large and a specular component is observed to dominate wave forms.
2. Brown's rough-surface scattering model appears valuable as an indicator of the effects of varying surface character on wave forms.
3. Variations in observed polar scattering suggest that the scattering properties of snow and ice vary across the ice caps. Mean 3 dB back-scatter half-angles of 3° over flat, and relatively less wet snow surfaces correspond with experimental results of Fung and others (1980).
4. Significant changes in surface reflection occur in the early stages of snow surface melt, when free water produced by melting is mainly suspended by necks between snow grains. Previous authors such as Suzuki and others (1983) have noted its importance in influencing the scattering characteristics of the snow surface.
5. Where large-scale surface roughness and crevassing is minimal, as on Bråsvellbreen, differences between the polar-scattering properties of wet and relatively dry snow lead to a significant change in the type of altimetric returns.

Ambiguities remain in the understanding of mechanisms influencing scattering from ice-mass surfaces, and how these in turn affect the general characteristics of mean altimeter wave forms. Further work is needed at 13.6 GHz and at near-nadir into the effects of particular surface properties on radar-altimeter wave forms. This should enable development of algorithms to extract data on key glaciological parameters from the forthcoming generation of satellite radar altimeters (e.g. ERS-1). The delineation of dry and wet snow, and bare-ice zones on large ice masses from altimeter wave-form characteristics should in the future provide valuable information for mass-balance studies.

ACKNOWLEDGEMENTS

We thank Andrew Birks of the Rutherford Appleton Laboratory for providing the Svalbard radar-altimeter data set, and for his assistance with tape handling and back-scatter calibrations. Dr C Rapley and the Algorithm Development Facility, Royal Aircraft Establishment, Farnborough, kindly funded digital image acquisition and processing. Sue Pilkington assisted by drafting the figures. M.R.D. acknowledges support from a UK Natural Environment Research Council studentship. L M Ulander, and Drs N F McIntyre, V A Squire, and G Rees kindly commented on drafts of this paper.

REFERENCES

Brown G S 1977 The average impulse response of a rough surface and its applications. *IEEE Transactions on Antennas and Propagation* AP-25 (1): 67–74
Cumming W A 1952 The dielectric properties of ice and snow at 3.2 centimeters. *Journal of Applied Physics* 23(7): 768–773
Dowdeswell J A 1986 Drainage-basin characteristics of Nordaustlandet ice caps, Svalbard. *Journal of Glaciology* 32(110): 31–38
Dowdeswell J A Unpublished Remote sensing studies of Svalbard glaciers. (PhD thesis, University of Cambridge 1984)
Dowdeswell J A, McIntyre N F 1986 The saturation of LANDSAT MSS detectors over large ice masses. *International Journal of Remote Sensing* 7(1): 151–164
Dowdeswell J A, Drewry D J, Cooper A P R, Gorman M R, Liestøl O, Orheim O 1986 Digital mapping of the Nordaustlandet ice caps from airborne geophysical investigations. *Annals of Glaciology* 8: 51–58
Evans S 1965 Dielectric properties of ice and snow – a review. *Journal of Glaciology* 5(42): 773–792
Fung A K 1981 A review of surface scatter theories for modeling applications. In *Coherent and incoherent radar scattering from rough surfaces and vegetated areas. Proceedings of an ESA EARSeL Workshop, held at Alpbach, Austria, 16-20 March 1981.* Paris, European Space Agency: 71–82 (ESA SP-166)
Fung A K, Stiles W H, Ulaby F T 1980 Surface effects on the microwave backscatter and emission of snow. In *Institute of Electrical and Electronic Engineers International Conference on Communications '80 Conference Record* 3: 49.6.1–49.6.7
Gatley C, Peckham G E 1983 *Radar altimetry and scatterometry from aircraft over the oceans.* Edinburgh, Heriot-Watt University. Department of Physics
Gloersen P, Mollo-Christensen E, Wilheit T, Dod T, Kutz R, Campbell W J 1985 *MIZEX '84 NASA CV-990 flight report.* Greenbelt, MD, Goddard Space Flight Center (NASA Technical Memorandum 86216)
Ivanov V B 1968 Eolovyye formy mikrorel'yefa snezhnoy poverkhnosti na lednikakh Antarktidy i Arktiki [Aeolian forms of microrelief of the snow surface of Antarctic and Arctic glaciers]. *Trudy Sovetskoy Antarkticheskoy Ekspeditsii* 38: 22–38
MacArthur J L 1978 *Seasat-A radar altimeter design description.* Baltimore, MD, Johns Hopkins University (Report SDO-5232)
McIntyre N F and 9 others 1986 *Analysis of altimetry data from the Marginal Ice Zone Experiment. Final report.* Paris, European Space Agency (Report 5948/84/NL/BI)
Moore R K, Williams C S 1957 Radar terrain return at near-vertical incidence. *Proceedings of the Institute of Radar Engineers* 45: 228–238
Schytt V 1964 Scientific results of the Swedish glaciological expedition to Nordaustlandet, Spitsbergen, 1957 and 1958. *Geografiska Annaler* 46(3): 243–281
Schytt V 1969 Some comments on glacier surges in eastern Svalbard. *Canadian Journal of Earth Sciences* 6(4, Pt 2): 867–873
Smith B M E, Evans S 1972 Radio echo sounding absorption and scattering by water inclusion and ice lenses. *Journal of Glaciology* 11(61): 133–146
Suzuki M, Matsumoto T, Kuroiwa D, Fujino K, Wakahama G 1983 Research on the interaction of microwaves with snow and ice. Part 1. A study on the microwave backscattering from melting snowpack. *Memoirs of National Institute of Polar Research.* Special Issue 29: 166–175
Ulaby F T, Moore R K, Fung A K 1982 *Microwave remote sensing active and passive. Vol 2. Radar remote sensing and surface scattering and emission theory.* Reading, MA, Addison Wesley
Ulander L Unpublished Airborne radar altimetry over the Greenland ice sheet. (MSc thesis, University of London 1985)
Zwally H J, Bindschadler R A, Brenner A C, Martin T V, Thomas R H 1983 Surface elevation contours of Greenland and Antarctic ice sheets. *Journal of Geophysical Research* 88(C3): 1589–1596

Annals of Glaciology 9 1987
© International Glaciological Society

A METHOD TO ESTIMATE OPEN PACK-ICE THICKNESS
FROM TWO-DAY SEQUENCES OF SIDE-LAPPING SATELLITE IMAGES

by

Uri Feldman

(Department of Geography, Bar-Ilan University, Ramat Gan 52100, Israel)

ABSTRACT

A method to estimate open pack ice thickness drifting in a marginal ice zone (MIZ) is presented. The estimates are obtained from two-day sequences of sidelapping Landsat-1 MSS images and two-day sequences of wind field data by four steps: estimating the surface wind speed, estimating the angle of sea ice deflection, estimating three ratios between ice parameters and estimating the lower and upper limits of pack ice thickness. The method has been applied to six groups of open pack ice floes drifting in the MIZ of the Beaufort Sea during 1973-1975. In the absence of simultaneous in-situ observation, the results have not been tested. The method presented may be applied to any MIZ. Rather than using Landsat-1 MSS images, data from a high resolution active microwave remote sensing system should be employed in the future as its data will be independent of sun illumination and cloud cover.

INTRODUCTION

A knowledge of pack ice thickness is vital for studying pack ice dynamics (Hibler 1979) and its impact on the environment (Weller and others 1983). Since in situ or remotely sensed measurements of pack ice thickness are not routinely available for the polar oceans, a method to estimate thickness of open pack ice floes, drifting in a marginal ice zone (MIZ), is suggested.   The estimates are obtained by solving a reduced form of the general equation of motion for drifting pack ice, using data from four readily available sources. The estimates are obtained by four steps:

1. Estimating the surface wind speed.
2. Estimating the angle of sea ice deflection.
3. Estimating three ratios between ice parameters.
4. Estimating the lower and upper limits of pack ice thickness.

The method has been applied to six groups of open pack ice floes drifting in the MIZ of the Beaufort Sea during 1973-75.

ESTIMATING OPEN PACK ICE THICKNESS

The general equation of motion for drifting pack ice (Campbell 1965) may be reduced to a simple steady state equation of motion if applied to weekly drifts of open pack ice, since impact of ocean currents on pack ice drifting in the Beaufort Sea is negligible for weekly drifts (Feldman and others 1981). In this equation, wind stress, water drag and the horizontal Coriolis force per unit area of ice are assumed to be at equilibrium. If a Cartesian coordinate system, in which +x is chosen to be the direction of ice motion, is employed then wind stress is always positive, −x is direction of water drag and −y is direction of the Coriolis stress.   The x and y components of this equation are written as:

$$\rho_a \ C_d^a \ U^2 \cos \Delta_\gamma = \rho_w C_d^w \ V^2 \qquad (1)$$

and

$$\rho_a \ C_d^a \ U^2 \sin \Delta_\gamma = \rho_{ice} \ 2\omega \sin \phi \ h \ V \qquad (2)$$

where ρ_a is air density, C^a is drag coefficient at the air/ice interface, U is horizontal wind speed at 10 m above the ice surface, Δ_γ is the angle of sea ice deflection, ρ_w (= 1.03 10^3 kg m^{-3}) is ocean water density, C_d^w is drag coefficient at the water/ice interface, V is speed of center of gravity of a group of open pack ice floes, ρ_{ice} (= 0.91 10^3 kg m^{-3}) is ice density, ω (=7.292 10^{-5} s^{-1}) is Earth's angular speed of motion, ϕ is latitude and h is mean thickness of a group of open pack ice floes.

Position of ice floes has been measured from 1 : 1 000 Landsat-1 MSS images, by a Ruscom digitizer, using control points located on land. Assuming linear motion for the floes, during a two-day sequence, accuracy of positional measurements is ±100 m. Hence, errors in position may vary between ±0.25 and 5% relatively to the motion of the center of gravity of a group of ice floes. Location of ice floes drifting away from land may be determined by means of space triangulation. Errors of ice velocity are identical to errors in position, since temporal data are highly accurate.

Means of ϕ in Equation (2) have been determined for groups of drifting open pack ice from Landsat-1 MSS images. V in equations (1) and (2) has been derived for the intermediate scanning time $t_{1 1 i+1}$ (=t_{12}, t_{23}, t_{34}) from two-day sequences of sidelapping Landsat-1 MSS images, scanned at t_i and t_{i+1} respectively. ρ_a in equations (1) and (2) has been derived and time adjusted to $t_{1 1 i+1}$ by means of linear interpolation from two-day sequences of six-hour surface weather charts. Estimates of U in Equations (1) and (2) have been obtained for $t_{1 1 i+1}$ from the geostrophic wind speed G by the formula of Feldman and others (1979). Values of G have been derived and time adjusted to $t_{1 1 i+1}$ by means of linear interpolation from two-day sequences of six-hour surface weather charts. Estimates of Δ_γ in equations (1) and (2) have been obtained for $t_{1 1 i+1}$ following the work of Feldman and others (1981). The variables h, C_d^a and C_d^w cannot be derived from the available data. Hence, they must be estimated.

To determine the ratios M, N and B, Equation (2) has been rewritten as

$$M = h/C_d^a = (\rho_a \ U^2 \sin \Delta_\gamma)/(\rho_{ice} \ 2\omega \sin \phi \ V) \qquad (3)$$

Equation (1) has been written as

$$N = C_d^w/C_d^a = (\rho_a \ U^2 \cos \Delta_\gamma)/(\rho_w \ V^2) \qquad (4)$$

TABLE I. EXTREME VALUES OF h, C_d^a AND C_d^w

Variable	Min.	Max.
h	0.00	3.00
$10^3\ C_d^a$	0.95	4.00
$10^3\ C_d^w$	3.32	57.17

TABLE II. ICE THICKNESS ESTIMATED FROM TWO-DAY SEQUENCES OF LANDSAT-1 IMAGES

Data Set	LS-1 Cycle	UT	Date	h_{LL} m	h_{UL} m	$\bar{h}$ m
1.1	21	0810:25	July 25, 1973	1.81	3.00	2.43
1.2	21	0816:10	July 26, 1973	2.64	2.96	2.80
2.1	26a	0814.50	Oct. 24, 1973	3.09	2.36	2.73
2.2	26a	0820:20	Oct. 25, 1973	2.93	2.83	2.88
2.3	26a	0826:05	Oct. 26, 1973	2.22	3.00	2.61
3.1	26b	0827:30	Oct. 26, 1973	3.62	2.07	2.84
3.2	26b	0832:15	Oct. 27, 1973	2.14	3.00	2.59
3.3	26b	0837:50	Oct. 28, 1973	1.81	3.00	2.40
4.1	41	0831:20	July 25, 1974	4.08	2.05	3.06
4.2	41	0837:00	July 26, 1974	1.80	3.00	2.40
4.3	41	0842:40	July 27, 1974	2.07	3.00	2.53
5.1	43	0801:25	Aug. 25, 1974	5.00	3.00	4.00
5.2	43	0807:10	Aug. 26, 1974	7.14	3.00	5.07
6.1	60	0737:00	June 25, 1975	1.81	3.00	2.41
6.2.	60	0742:35	June 26, 1975	2.31	3.00	2.66

and B, defined from (3) and (4), has been written as

$$B = h/C_d^w = M/N \qquad (5)$$

The observed minimum and maximum of h, C_d^a and C_d^w, (Table I), summarized by Feldman and others (1981) from previous observations, have then been employed to rewrite Equation (5) as

$$h = C_d^w\ B \geqslant 3.32\ 10^{-3}\ B\ m \qquad (6)$$

Equation (3) as

$$h = C_d^a\ M \geqslant 0.95\ 10^{-3}\ M\ m \qquad (7)$$

and h as

$$h = h \geqslant 0.00\ m \qquad (8)$$

This allows Equations (6), (7) and (8) to be rewritten as

$$h \geqslant \max \{3.32\ 10^{-3}\ B,\ 0.95\ 10^{-3}\ M,\ 0.00\} = h_{LL}\ m \qquad (9)$$

where h_{LL}, the lower limit of h, is the maximum value among the three values in Equation (9). In the same way h_{UL}, the upper limit of h, is the minimum value among the three values in Equation (10)

$$h \leqslant \min \{57.17\ 10^{-3}\ B,\ 4.00\ 10^{-3}\ M,\ 3.00\} = h_{UL}\ m \qquad (10)$$

B and M, which are employed to estimate h in Equations (9) and (10), are linearly related to V in Equations (3) and (5). Hence, errors in V will linearly affect B and M.

The method has been applied to six groups of open pack ice floes, drifting in the MIZ of the Beaufort Sea. The lower limit, upper limit and mean thickness estimated by the method for the six groups are presented in Table II.

CONCLUSIONS

An estimate of open pack ice thickness is considered unacceptable if its lower limit is larger than its upper limit ($h_{LL} > h_{UL}$).

In the absence of simultaneous in situ observations of pack ice thickness, accuracy of the results cannot be tested.

The method may be applied to estimate thickness of open pack ice drifting in any MIZ, by using two-day sequences of sidelapping images.

The required wind field data should be obtained directly from weather data banks rather than from surface weather charts.

Rather than using Landsat-1 MSS images, data from a high resolution active microwave remote sensing system should be employed in the future, as its data will be independent of sun illumination and cloud cover.

REFERENCES

Campbell W J 1965 The wind-driven circulation of ice and water in a polar ocean. *Journal of Geophysical Research* 70(14): 3279-3301

Feldman U, Howarth P J, Davies J A 1979 Estimating the surface wind speed over drifting pack ice from surface weather charts. *Boundary Layer Meteorology* 16(4): 421-429

Feldman U, Howarth P J, Davies J A 1981 Estimating surface wind direction over drifting open pack ice. *Journal of Geophysical Research* 86(C9): 8117-8120

Hibler W D III 1979 A dynamic thermodynamic sea ice model. *Journal of Physical Oceanography* 9(4): 815-846

Weller G, Carsey F, Holt B, Rothrock D A, Weeks W F 1983 *Science program for an imaging radar receiving station in Alaska. Report of the Science Working Group.* Pasadena, CA, NASA Jet Propulsion Laboratory

Annals of Glaciology 9   1987
© International Glaciological Society

COMPARISON OF THE SURFACE CONDITIONS OF THE INLAND ICE SHEET, DRONNING MAUD LAND, ANTARCTICA, DERIVED FROM NOAA AVHRR DATA WITH GROUND OBSERVATION

by

Yoshiyuki Fujii, Takashi Yamanouchi

(National Institute of Polar Research, Kaga 1-chome, Itabashi-ku, Tokyo 173, Japan)

Kazuya Suzuki

(University of Electro-Communications, Chofugaoka 1-chome, Tokyo 182, Japan)

and

Shinya Tanaka

(Space Systems Department, Fujitsu Ltd, Kamata 1-chome, Ohta-ku, Tokyo 144, Japan)

ABSTRACT

The surface conditions of the inland ice sheet in east Dronning Maud Land, Antarctica, are derived from the NOAA-7 AVHRR data received at Syowa Station and then compared with the ground observations which were collected in November 1984 along a 243 km long traverse route at altitudes ranging from 2700 to 3400 m a.s.l. The variations in the AVHRR data are well related to the distribution of glazed surfaces. The areas with lower albedo, higher surface temperature, lower ratio of channel 2/channel 1, and lower T4–T5 coincide with the areas where a glazed surface has developed. This result is attributed to the fact that the glazed surface is composed of a multi-layered ice crust and that its radiative and thermal properties are closer to ice than to snow. The present study shows that the NOAA AVHRR data are useful for distinguishing bare ice, glazed surfaces, and snow surfaces of the Antarctic ice sheet.

INTRODUCTION

Surface conditions of the Antarctic ice sheet show regional characteristics from the coast towards the interior. The area between 1800 and 3200 m a.s.l., where precipitation is low and the katabatic wind is strong, is characterized by the extensive distribution of a flat, glazed surface with polygonal cracks formed by thermal expansion and contraction of the hard surface (Watanabe 1978). The flat glazed surface is a long-term (more than a few years) accumulation-free surface as identified by Watanabe (1978), and mass loss exceeding 5 g cm^{-2} per year occurs at the surface by sublimation (Fujii and Kusunoki 1982). The determination of the distribution of glazed surfaces is of great importance for the study of the mass and heat balance of the Antarctic ice sheet.

In this paper, the distribution of glazed surfaces observed during an oversnow traverse in November 1984 in the inland plateau of east Dronning Maud Land is compared with the NOAA-7 AVHRR visible and infra-red images. The satellite images used in this paper were received at Syowa Station (lat. 69°00'S., long. 39°55'E.) during the period of the oversnow traverse.

GROUND TRUTH

Surface conditions of the inland ice sheet of east Dronning Maud Land were observed visually from an

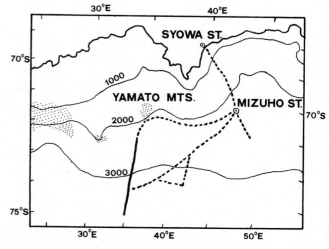

Fig.1. Glaciological traverse route (dashed line) for October to December 1984. The solid line indicates the present study line where surface conditions were derived from both ground and satellite observations.

oversnow vehicle along the traverse route as shown in Fig.1. The oversnow traverse was conducted from October to December 1984 as a part of the 5-year glaciological research programme called East Queen Maud Land Glaciological Program (EQGP) 1982–86.

Surface features are morphologically classified into three categories for the *in-situ* observations as described below.

Rough surface: composed of sastrugi greater than 30 cm in height.

Smooth surface: composed of a smooth surface, dune, erosional pit, and/or very small-scale sastrugi (less than 30 cm in height).

Glazed surface: composed of a multi-layered ice crust as shown in Fig.2.

Each surface feature changes successively; for example, sastrugi, dune, smooth surface, and glazed surface in accordance with the active, inactive, and equilibrium stages in the deposition–erosion process, respectively (Watanabe

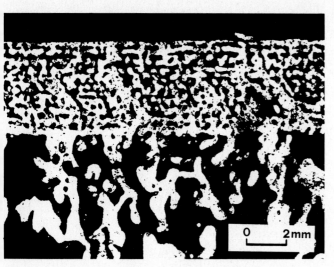

Fig.2. A thin section of a glazed surface with a multi-layered ice crust photographed in crossed polarized light.

1978). However, these stages in the depositional-erosion process may be affected by a large-scale step such as an undulation of the ice-sheet surface, i.e. a glazed surface develops at the relatively steep slope where the katabatic wind accelerates and snow deposition does not occur readily.

As these surfaces often co-exist, the areal ratio of each surface was described. The distributional boundaries were recorded with a resolution of 10 m using the distance meter of an oversnow vehicle, and their locations were interpolated from the positions of the nearest two overnight spots where exact satellite positioning was observed with an accuracy of ±30 m using a JMR-4A system.

The length of the present study line is 243 km along long. 35°E. from the southernmost route station (lat. 75°00'S., long. 35°01'E., 3396 m a.s.l.) to the station located at lat. 72°50'S., long. 35°09'E., and 2692 m a.s.l. south of the Yamato Mountains (Fig.1). In the illustrations for this paper, the southernmost station is located at km 0.

Fig.3 shows the observational results on the surface features along the study line, i.e. the glazed surface develops widely with a zonal width of 1 km to more than 10 km, and is the most predominant surface feature along the study line. The area where the glazed surface predominates amounts to about 45% of the area along the study line.

NOAA AVHRR DATA

For the present study, we used NOAA-7 AVHRR data received on 20 and 30 November 1984 when there were no clouds over the study line and the ground observation was carried out.

The AVHRR sensor of NOAA-7 provides imagery in five different spectral bands with a ground resolution of about 1.1 km at nadir. This ground resolution is comparable to the minimum width of the zonal distribution of the glazed surface. Channel 1 (0.58-0.68 μm) and channel 2 (0.725-1.10 μm) measure radiances of reflected solar

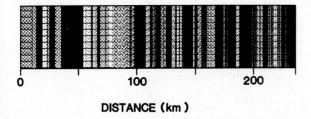

Fig.3. Distribution of surface features along the study line. Rough and smooth surfaces are illustrated as a wave-like dotted line and lateral stripes. Glazed surfaces are shown as black and darker parts for the features with an areal ratio of more than 70 and 50%, respectively.

radiation, and channel 3 (3.55-3.93 μm), channel 4 (10.3-11.3 μm), and channel 5 (11.5-12.5 μm) measure radiances of emitted terrestrial radiation and some minor parts of the reflected solar radiation.

Calibrations and corrections of the map projection were made by affine transformation using the ground-control positions of distinctive features on the bare-ice field near the Sør-Rondane and the Yamato Mountains obtained on 30 November, as shown in Fig.4. The ground-control points

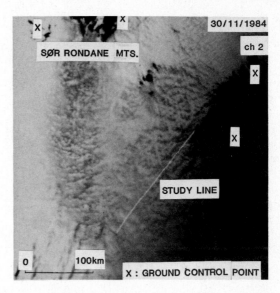

Fig.4. A NOAA-7 AVHRR image of channel 2 along the present study line, 243 km in length south of the Yamato Mountains.

were identified on the 1 : 250 000 scale Landsat-image maps issued by the Geographical Survey Institute of Japan in 1984 and 1985.

Fig.5 shows the data from channels 1, 2, 4, and 5 along the study line obtained on both 20 and 30 November. Channel 3 was not used because of noise in the data. Though the absolute values from each channel on both days differ from one another, the variation patterns show a good correlation. This may indicate that the calibrations and corrections of the map projection of the images for both days were made with almost the same accuracy, and these patterns from AVHRR data do not reflect temporal phenomena, such as clouds and drifting snow, but relatively stable surface conditions.

Reflected radiances from channels 1 and 2 decrease with distance because of the decrease in solar altitude with the distance. The spectral albedo depends on the surface conditions and the azimuth angle of the Sun as well as the relative angle between the Sun and the satellite. The albedo of bare ice is lower than that of dry and clean snow in the short-wave bands and is particularly low in the near infra-red bands (channel 2) (Warren 1982). Furthermore, the albedo depends on snow grain-size, i.e. it decreases with increasing grain-size of the snow. The dependence of the reflectivity on grain-size is more sensitive in the near infra-red region (channel 2) than in the visible region (channel 1) (Dozier and others 1981). Since the multi-year ice crust of a glazed surface has a higher density (0.69 Mg m^{-3}) than a snow surface (Fujii and Kusunoki 1982), the grain-size of a glazed surface can be regarded as large compared with that of a snow surface.

Brightness temperatures for channels 4 and 5 along the study line increase with distance, as shown in Fig.5, because of the decrease in altitude of the ice-sheet surface with distance.

DISCUSSION

The NOAA-7 AVHRR data from the different spectral bands are compared with each other in order to derive the surface conditions, which can then be compared with the

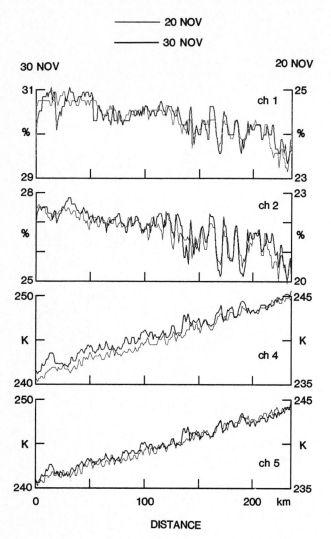

Fig.5. NOAA-7 AVHRR data from channels 1, 2, 4, and 5 along the study line on 20 and 30 November 1984.

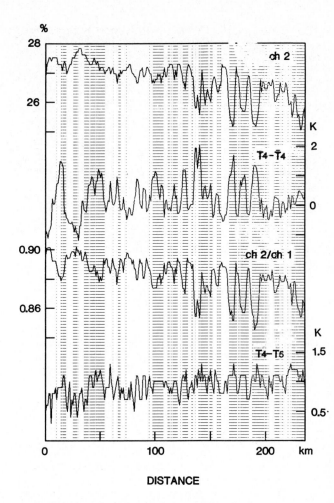

Fig.6. Comparison of NOAA-7 AVHRR data along the study line on 30 November, with the distribution of the glazed surface zone shown as the hatched areas.

surface conditions based on ground observation. Fig.6 shows the AVHRR data and distribution of the area where a glazed surface exceeds 70%. The glazed surface zone is hatched in Fig.6.

Since the satellite data for both 20 and 30 November have a similar variation along the study line but a larger amplitude in the variation of each channel on 30 November than on 20 November, the data for 30 November are chosen for discussion. Reflected radiance from channel 2 is given at the top of the figure because the variation is larger than that for channel 1. For brightness temperatures, channel 4 data are used because of the lower effect of atmospheric vapour content in comparison with channel 5.

Comparing the reflected radiance from channel 2 with the relative surface temperature, given as the deviation from the linear relation between surface temperatures and altitudes to eliminate the effect of the surface profile on the surface temperature, they have a high negative correlation. As the lower albedo zone coincides with the higher surface-temperature zone, the lower albedo is not due to the shadow of the surface relief but due to surface conditions. Fig.7 indicates the relationship between altitude and surface temperature (T4) for glazed and snow surfaces where each surface has an area greater than 90% along the study line. The surface temperature is about $0.5°K$ higher for a glazed surface composed of a multi-layered ice crust than for a snow surface, probably because of the difference in the thermal (or radiative) property of these surfaces, as was suggested by Mae and others (1981).

Since the albedo of bare ice is lower than that of snow, and particularly low in the near infra-red region (channel 2), the ratio of channel 2/channel 1 should be lower at the surface with properties more similar to those

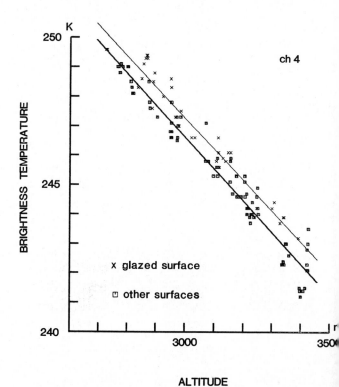

Fig.7. Relationship between altitudes and surface temperatures, T4, for glazed and snow surfaces along the study line.

of ice. As is shown in Fig.6, both channel 2/channel 1 and the albedo of channel 2 show a high positive correlation. This suggests that variations in the ratio of channel 2/channel 1 and the albedo of channel 2 depend on surface conditions such as snow or ice crust (properties more similar to those of ice).

Furthermore, as the difference in emissivity between channels 4 and 5 increases with increasing grain-size (Dozier and Warren 1982), the temperature difference between channels 4 and 5 given in the bottom of the figure should be larger for the ice crust of a glazed surface than on snow. Fig.6 shows a good negative correlation between T4–T5 and the channel 2 albedo. This may suggest that the temperature difference (T4–T5) along the study line depends on surface conditions such as snow or ice crust.

The NOAA AVHRR data are compared with the distribution of the glazed surface zone shown as hatched in Fig.6. Having taken the accuracy of the map projection of AVHRR imagery into account, it can be said that the variations of the AVHRR data are well related to the distribution of glazed surfaces. That is, the part with a lower albedo, higher surface temperature, lower ratio of channel 2/channel 1, and larger T4–T5 coincides with the area where a glazed surface develops. Therefore, the imagery of channel 2 of AVHRR shown in Fig.4 can be interpreted as follows. The relatively dark mottled parts indicate a glazed surface and the black-and-white parts (except the edges) indicate a bare ice (or rock) and snow surface.

The good correlation between NOAA AVHRR data and the distribution of glazed surfaces may be explained by the fact that a glazed surface is composed of a multi-layered ice crust, as shown in Fig.2, and the radiative and thermal properties of a glazed surface are closer to those of ice than snow.

CONCLUSIONS

The present study shows that with NOAA AVHRR data it is possible to distinguish bare ice, glazed surfaces, and snow on the Antarctic ice-sheet surface. This approach can therefore be applied to the surface mass and heat balance of the Antarctic ice sheet where they depend solely on surface conditions.

ACKNOWLEDGEMENTS

The authors wish to express their thanks to Professor Takeo Hirasawa of the National Institute of Polar Research, leader of the 25th Japanese Antarctic Research Expedition, and its members for receiving satellite data and supporting the field observations. Professor Atsumu Ohmura and Dr Konrad Steffen of the Department of Geography, Swiss Federal Institute of Technology, made valuable comments on the original manuscript. All computations were carried out at the Information Processing Center, National Institute of Polar Research.

REFERENCES

Dozier J, Warren S G 1982 Effect of viewing angle on the infrared brightness temperature of snow. *Water Resources Research* 18(5): 1424–1434

Dozier J, Schneider S R, McGinnis D F Jr 1981 Effect of grain size and snowpack water equivalence on visible and near-infrared satellite observations of snow. *Water Resources Research* 17(4): 1213–1221

Fujii Y, Kusunoki K 1982 The role of sublimation and condensation in the formation of ice sheet surface at Mizuho Station, Antarctica. *Journal of Geophysical Research* 87(C6): 4293–4300

Mae S, Yamanouchi T, Wada M 1981 The measurement of the surface temperature at Mizuho Station, East Antarctica. *Memoirs of National Institute of Polar Research. Special Issue* 19: 40–48

Warren S G 1982 Optical properties of snow. *Reviews of Geophysics and Space Physics* 20(1): 67–89

Watanabe O 1978 Distribution of surface features of snow cover in Mizuho Plateau. *Memoirs of National Institute of Polar Research. Special Issue* 7: 44–62

Annals of Glaciology 9 1987
© International Glaciological Society

PATTERN RECOGNITION OF AIR PHOTOGRAPHS FOR ESTIMATION OF SNOW RESERVES

by

W. Good and J. Martinec

(Swiss Federal Institute for Snow and Avalanche Research, 7260 Weissfluhjoch/Davos, Switzerland)

ABSTRACT

A certain snow-covered area determined from air photographs in different years can contain different volumes of snow. The pattern recognition of snow fields reveals whether the snow reserves are relatively high or low for the given snow coverage. It also signals a short term increase of the snow-covered area by summer snowfall.

INTRODUCTION

A typical feature of the seasonal snow cover in the mountains is its diminishing areal extent during the melt season. This phenomenon greatly reduces the risk of floods from snow melt in mountainous basins. Since the advent of remote sensing, the snow cover can be efficiently monitored. Efforts are in progress to improve the methods of snow-cover mapping with the use of orthophotographs from aircraft and of satellite data. This paper deals with estimating the snow reserves in terms of the water equivalent from this two-dimensional information by the pattern recognition of structured snowfields.

SNOW-COVERED AREA WITH REGARD TO SNOW ACCUMULATION

The snow-covered area had already been used as an index of the snow accumulation in terms of water equivalent 50 years ago (Potts 1937). The subsequent run-off volumes in various years deviated sometimes from forecasts based on such relations. These discrepancies were at first attributed to the varying summer precipitation in the respective years. Another source of error was the crude method of estimating the snow–covered area just from panoramic terrestrial photographs of a mountain range.

In the following decades, the accuracy and efficiency of snow – cover mapping were improved by terrestrial observations (Garstka and others 1958), air photographs and in particular by remote sensing from satellites. However, the old assumption linking the snow-covered area with the water volume stored in the snow cover remained (Ødegaard and others 1980; Shafer and Leaf 1980).

At the same time, detailed measurements in well-equipped representative basins (Martinec 1980) have shown that there is no unequivocal relation between the areal extent of the seasonal snow cover in a mountain basin and the snow accumulation in terms of water equivalent. As is explained elsewhere (Hall and Martinec 1985) the gradually decreasing area of the snow cover reflects not only the initial snow accumulation, but also climatic conditions during the snow-melt season of the given year. In order to overcome this problem, a method of evaluating the snow accumulation from a sequence of satellite or aircraft images of the snow cover has been developed (Martinec and Rango 1987). In this paper, the pattern of snow-cover boundaries is used to estimate whether a certain areal extent of the seasonal snow cover has a high or low snow–water equivalent.

IMAGE ANALYSIS AND PATTERN RECOGNITION
Working hypothesis

Ideally, a horizontal, two-dimensional surface is either covered with snow or it is void. In reality the transition time between these two states during ablation is short and becomes longer as the roughness of the surface increases. In mountainous areas where the topography is very complicated, the delay is even more accentuated because of the elevation range and temperature lapse rate.

During the process of disintegration, a contiguous snow cover successfully reveals details from the subjacent terrain. A given area of snow represents smaller amounts of water if more topographic details are visible (Good 1983). These signatures seem to be typical and reproducible for given areas and describe not only a state of the snow cover during melting, but they also represent the quantity of snow. In the present paper we try to quantify this hypothesis.

Methods used

It soon became evident that the resolution of the images used was one of the most critical conditions in defining the relevant structures. The resolution that is necessary to reveal the required topographic details, is of the order of a few metres.

Resolution, however, is not of itself sufficient; it has to be considered in relation to the areal extent of the test site under investigation (e.g. Dischma Valley, Swiss Alps, European Alps), and to the point resolution of the image analysis system. For the Dischma Valley, air photographs appear to be adequate. We work with a HAMAMATSU slow-scan video camera with a typical point resolution of 512 × 512 pixels (maximum 1024 × 1024 pixels) corresponding to a pixel size of 11.2 m. If the relevant topographic structures are finer or the resolution due to the pixel size is too coarse, no significant analysis is possible. Below this limit the patches without snow have only the effect of reducing the albedo. The snow appears to be slightly darker which influences the selection of the discriminant level. This does not, however, reveal the required signatures. These limitations are found in photographs which represent typical new snow situations such as that on 29 August 1985, when about 85% of the test area was covered with new snow. In spite of its large extent, this short-lived snow cover contained on the average only several centimetres of water equivalent. We tried various methods to quantify the signatures (structures) and to correlate the resulting parameters with the quantity of water stored in the corresponding snow cover.

Data set

Orthophotographs (1 : 50 000) from the Dischma Valley (flights between 1972 and 1981) were separated into an upper and a lower part in order to improve the point resolution and also to increase the number of samples (22). The upper part with an area of 21.5 km^2, ranging from 1900 to 3150 m a.s.l., is examined in this study. Care

TABLE I. POINT COUNT PARAMETERS

SYMBOL	NAME	DEFINITION
RHO	point density (mean)	$RHO = P_s / (P_s + P_v)$
P_s	fraction of snow(s)-pixels	
P_v	fraction of void(v)-pixels	
INS	intersections (mean)	transitions s/v and v/s
DMI	snow intercept length (mean)	distance between v/s and s/v
FRE	void intercept length (mean)	distance between s/v and v/s
F_1	form factor 1 (mean)	$F_1 = INS/2 \times FRE$
F_2	form factor 2 (mean)	$F_2 = 1 - DMI/FRE$

Histograms, mean (X..), variance (XSD), minimum (XMI) and maximum (XMA) are available per parameter [X is for R, I, D, F]

must be taken to avoid those samples where new snow would cause a bias in the evaluation.

A/D conversion

The 22 samples were digitized with the slow-scan video camera linked to a PDP 11/44. The grey level of each pixel is evaluated in an 8 bit scale. The 262144 intensities of each picture are stored on magnetic tapes. The conversion of one frame (512 × 512 pixel) takes roughly 90 sec.

Preprocessing

The only subjective step, and perhaps the most important one in the processing is the selection of the adequate limit where the intensity of a pixel signal is either a snow-covered or a snow-free area. In an ideal picture a single discriminant level may exist, but even here it has to be selected so that the original and the binary picture are as similar as possible. In real pictures, a dramatic amount of information is lost if one discriminant level only is used to split the pixels into a snow/no-snow binary picture. The reason is the slow and continuous variation of intensities over the whole picture and the local differences due to changing lighting conditions, snow quality, vegetation etc.

Local image operations of the edge-detection type were applied. The enhancement algorithm checks every pixel against its four nearest neighbours. A transition white-black is initiated if the pixel in the centre is slightly darker, within a discrimination band, than the surrounding ones, or if, in an absolute sense, it is darker than the lower limit of the discrimination band. In the enhanced pictures many of the required details reappear (Fig.1). The reproducibility of the areal extent of the snow cover could be kept within ±6%.

Parametrization

A set of pattern recognition parameters and a set of simple point count parameters were computed after image enhancement. The first group normally represents the geometry of the structures in a more comprehensive way. Because of the small extent of many structural details, the pattern recognition tended to reflect the coordinate frame rather than the structures. Therefore we worked with the less critical set of point count parameters.

NUMERICAL ANALYSIS

The extent of the snow cover in the analysed situations represented 57% and 93% respectively of the total area. We expect certain parameters describing the snow-free patches (topography) to be the important ones. A factorial analysis (Cooley and Lohnes 1973; SSP, ISPAHAN) yields an orthogonal coordinate frame where the axes (principal components) are linear combinations of the initial parameters. Axis number one, belonging to the first (largest) eigenvalue, is selected in such a way that the total

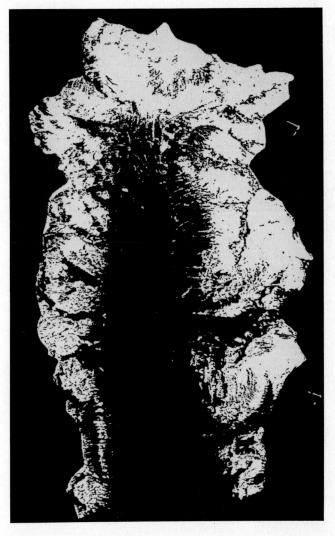

Fig.1.   Digitally enhanced orthophotograph of Dischma Valley. Pixel size is 11.2 m (situation from 5 July 1974).

dispersion of the points is maximized. The second one, associated with the second largest eigenvalue and all the following ones are mutually orthogonal.

Comparing the measured quantities of water from the respective snow covers with the linear combinations of parameters, we found reasonable descriptions of the two situations.

Models

The first principal component contains the information about the areal extent of the snow cover (RHO, RSD, RMA). (Table I). The following combination of number of intersections and distance between snow patches relates to some extent to the run-off volumes of water (ranging from 4 to 16×10^6 m^3) from the snow cover (Fig.2).

$$PRCO2 = 0.4 \times INS + 0.4 \times INS/FRE - 0.38 \times FRE + ...$$

The respective water volumes stored in snow are represented by the measured discharge volumes from the test area. The snow reserves are slightly higher in view of the run-off losses and problems in separating the snow melt and rain component, as explained in the next section. The linear relationship between the water volume and PRCO2 plotted in Fig.2 has a coefficient of determination of $R^2 = 0.89$. The differences in discharge volumes corresponding to approximately equal snow−covered areas, however, are significant. In the group of 56.63% ± 3.63% snow coverage, a discharge of 3.9×10^6m^3 corresponds to the individual with 58.2% of snow, whereas 6.8×10^6m^3 corresponds to the individual with 51.9% of snow. This contradiction to the areal extent can only be resolved with another combination of parameters describing the variability of geometric features of the snow cover: minimum of snow patch diameters and of covered area and difference of standard deviations of diameters of patches of snow and void. This finer discrimination is accomplished by plotting Q versus PRCO5 (see Fig.3).

$$PRCO5 = 0.58 \times DMI + 0.47 \times RMI + 0.31 \times FSD - 0.32 \times DSD + ...$$

The regression coefficient is 0.987. The same model is not directly applicable to the 93% group; apparently the 7% of the area that is snow−free has not enough structural details.

COMPARISON OF ESTIMATED SNOW RESERVES WITH MEASUREMENTS

The years 1972, 1974, 1976, and 1978 have been selected to show that an approximately equal snow-covered area in the alpine basin Dischma can contain either a high or a low snow volume. The years 1972 and 1976 are evaluated in the previous section as "low", and the years 1974 and 1978 as "high". This is in agreement with the direct point measurements at testsite, 2540 m a.s.l., which can be related to the examined test section of the Dischma basin.

The estimated snow reserves can also be compared with the resulting run-off volumes. To this effect, the run-off from the test area of 21.5 km^2 was derived from a recording water gauging station situated up-stream (catchment area 12.5 km^2) and from another recording water gauging station at the outlet of the Dischma basin (representing 43.3 km^2). Data are summarized in Table II, where S is the snow−covered area evaluated from an aircraft orthophotograph on the given date (100 × RHO), H_w is the point water equivalent of the snow cover measured at F Weissfluhjoch. R is the run-off depth from the test area of 21.5 km^2 and P×0.7 is precipitation during run-off period multiplied by a run-off coefficient.

Since there is no clear-cut time limit of snow melt run-off, the daily run-off volumes were added in each year until a total of 150 degree-days was reached from temperature measurements at Weissfluhjoch, 2693 m a.s.l. By subtracting the run-off from precipitation from the total run-off, run-off depths related to the initial snow reserves are obtained.

The years 1974, 1975, 1979, and 1980 show again a similar snow coverage at certain dates, considerably higher than in the previous example. In this group, however, it was not possible to discriminate between the respective years by the axis PRCO5. As shown in Table III, the snow reserves are very similar in 1974, 1975, 1980 and distinctly lower in 1979. Also, as already mentioned, the snow-free

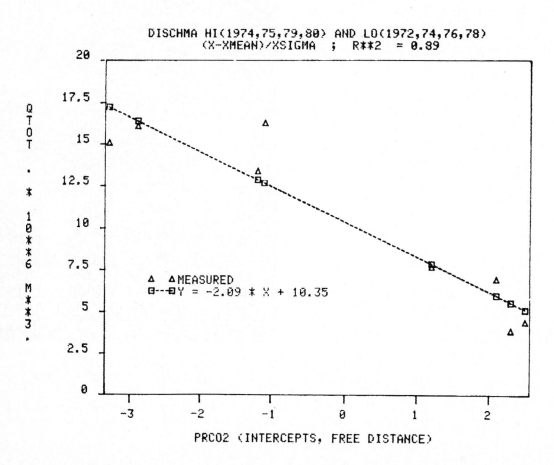

Fig.2. Run-off Q as a function of the number of snow/void and void/snow transitions and of the free distance (PRCO2).

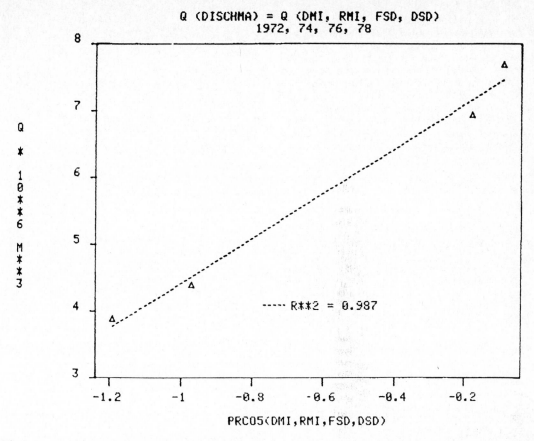

Fig.3. Run-off Q as a function of PRCO5. For a similar extent of the snow cover, two typical situations ("high", "low") can be resolved.

TABLE II. MEASURED WATER EQUIVALENTS OF SNOW AND RUN-OFF VOLUMES FOR COMPARISON WITH THE ESTIMATED SNOW RESERVES

DATE	S %	H_w F cm	R cm	P × 0.7 cm	R − P × 0.7 cm	Snow reserves
21 June 1972	56	33.7	30.7	10.3	20.4	low
8 June 1976	58	34.8	20.5	2.4	18.1	low
5 July 1974	60	70.4	45.5	9.7	35.8	high
17 July 1978	52	62.4	41.0	8.7	32.3	high

TABLE III. MEASURED WATER EQUIVALENTS AND RUN-OFF VOLUMES FOR COMPARISON WITH THE ESTIMATED SNOW RESERVES

DATE	S %	H_w F cm	R cm	P × 0.75 cm	R − P × 0.75 cm
16 May 1974	96	101.6	100.4	30.2	70.2
20 May 1975	94	117.1	102.1	26.4	75.7
5 June 1980	92	111.7	97.1	22.3	74.8
22 May 1979	93	95.7	79.6	17.2	62.4

area is too small to provide enough details for the analysis. The symbols are the same as in Table II. The run-off coefficient for precipitation was increased slightly to 0.75 because the snow-covered area is larger than in the previous case so that smaller losses are indicated. Since the run-off period starts at an earlier stage of the snow melt season than in the previous comparison, the number of degree-days measured at Weissfluhjoch was increased to 225 to obtain the respective final dates in each year. The run-off depths are again computed for the periods thus defined and run-off related to precipitation is subtracted in order to obtain values referring to the estimated snow reserves.

CONCLUSION

By the above method it is possible to recognize whether a certain areal extent (40–70%) of the seasonal snow cover contains a high or a low volume of snow. In this study the required spatial resolution of snow cover images was of the order of 10 m. If new snow is superimposed on the seasonal snow cover during the snow melt season, the fine structure of snow fields and the complex boundaries prevent the snow-covered and snow-free patches from being recognized. The identification of such scenes helps to avoid errors in periodical snow-cover mapping.

REFERENCES

Cooley W W, Lohnes R R 1973 *Multivariate data analysis.* New York, John Wiley and Sons

Garstka W U, Love L D, Goodell B C, Bertle F A 1958 *Factors affecting snowmelt and streamflow.* Washington, DC, Government Printing Office

Gelsema E S (ed) 1981 *ISPAHAN users' manual.* Amsterdam, Free University. Department of Medical Informatics

Good W 1983 *Estimation par des méthodes de traitement d'images de la quantité d'eau stockée dans un bassin versant.* Weissfluhjoch/Davos, Institut Féderal pour l'Etude de la Neige et des Avalanches (Rapport Interne 609)

Hall D K, Martinec J 1985 *Remote sensing of ice and snow.* London and New York, Chapman and Hall

IBM Corporation 1968 *System /360 Scientific Subroutine Package, version III. Programmers' manual.* New York, IBM Corporation

Martinec J 1980 Limitations in hydrological interpretation of the snow coverage. *Nordic Hydrology* 11(5): 209-220

Martinec J, Rango A 1987 Interpretation and utilization of areal snow-cover data from satellites. *Annals of Glaciology* 9: 166–169

Ødegaard H A, Andersen T, Østrem G 1980 Application of satellite data for snow mapping in Norway. *In* Rango A, Peterson R (eds) *Operational Applications of Satellite Snowcover Observations. Proceedings ... Nevada ... 1979.* Greenbelt, MD, NASA: 93-106 (NASA Conference Publication 2116)

Shafer B A, Leaf C F 1980 Landsat derived snowcover as an input variable for snowmelt runoff forecasting in south central Colorado. *In* Rango A, Peterson R (eds) *Operational Applications of Satellite Snowcover Observations. Proceedings of a final workshop ... held at Sparks, Nevada April 16–17, 1979.* Greenbelt, MD, NASA: 151-169 (NASA Conference Publication 2116)

Annals of Glaciology 9 1987
© International Glaciological Society

A DIGITAL RADIO ECHO-SOUNDING AND NAVIGATION
RECORDING SYSTEM

by

M.R. Gorman and A.P.R. Cooper

(Scott Polar Research Institute, University of Cambridge, Cambridge CB2 1ER, U.K.)

ABSTRACT

The Scott Polar Research Institute (SPRI) Mk IV 60 MHz radio echo-sounding (RES) system has proven itself to be a most effective and versatile tool in glaciology. During the last 15 years, it has been used from a variety of platforms, both surface and airborne, and over a range of ice thicknesses from 4000 m to 100 m. However, the photographic recording methods used during this period were felt to be increasingly outdated in the context of modern data handling procedures. Accordingly, in late 1982 the Mk IV system was modified to incorporate fast digitizing of the RES receiver output, with microcomputer-controlled magnetic-tape recording of both the radar data and navigational inputs (Drewry and Liestøl 1985). The new system will be described, along with the improvements in data processing which have resulted from its use.

INTRODUCTION

The fundamental principles of radio echo-sounding are well established. They rely on the fact that polar ice masses are relatively transparent to electromagnetic radiation in the Very High Frequency (VHF) range, *c.* 30–300 MHz. A simple low-frequency radar is flown over the surface, and short pulses are emitted directly downward. These propagate through the air, are partially reflected from the surface of the ice, continue to propagate (with some loss of energy) through the ice mass, and are finally reflected from the ice–bedrock interface. These reflections (together with faint echoes from any internal layering within the ice) return to the radar, where timing of the reflections enables the location of the different interfaces to be determined. By this means, a cross-section through the ice may be built up along any desired track.

Fig.1 is a block diagram of the elements of the complete RES system. The system includes the Mk IV radar, navigation, and recording systems, other avionics, and software for a digital RES system; each will be described in the following sections.

RADAR

The radar used is the well-proven SPRI Mk IV 60 MHz system, which is a development of the system described by Evans and Smith (1969). The Mk IV system specifications are:

Carrier frequency	60 MHz
Pulse length	300 ns
Receiver bandwidth	15 MHz
Transmitter power	300 W peak
System performance	160 dB (not including antenna gain)
Antenna type	2 half-wave dipole with reflector
Half-power beamwidth	90°
Forward gain	8 dB

A logarithmic receiver provides about 60 dB of dynamic range. The receiver output is digitized by a fast sample-and-hold circuit, using the sampling scheme discussed below.

The antenna system comprises two half-wave dipoles, one mounted under each wing of the DeHavilland Twin Otter aircraft. In this configuration, the wing acts as a reflector, giving a forward gain of about 8 dB. One antenna is used for transmitting, the other for receiving. Isolation between the two is found to be sufficient (>40 dB) to make a transmit–receive switch unnecessary.

The radar and recording systems are shock-mounted in a purpose-built aluminium-alloy rack secured in the main cargo compartment of the aircraft.

NAVIGATION SYSTEMS

The general approach to navigation systems which has been adopted might be termed *post hoc*. That is, great emphasis is placed on recording sufficient data to accurately reconstruct the precise track of the aircraft after the survey has been completed, rather than attempting to navigate very accurately during the flight. On occasion, however, the facilities provided by modern navigation systems have been used to fly, for example, a pre-determined grid over featureless terrain.

Primary navigation systems of two types have been used with the new system. In 1983, a Doppler navigator, linked through a Tactical Air Navigation System (TANS) navigation computer, provided a continuous read-out of latitude and longitude. Data from the TANS is provided in a serial stream, which is recorded directly on to tape, as described below.

During the 1986 season, a Litton 3000 Omega navigation system was used, which provided a similar data stream for recording. Omega uses low-frequency transmissions from a number of world-wide beacons to obtain a geographic position. However, the accuracy of this system is inferior to the Doppler–TANS combination. The normal accuracy of the Omega system is 2–3 km. This can be significantly improved (by up to an order of magnitude) by the use of along-track position fixes. During the radio echo-sounding operations in mountainous terrain there were frequent opportunities to acquire ground control when crossing nunataks, rock ridges, etc. Vertical aerial photographs were taken to precisely locate the aircraft over such features.

The serial data stream from the primary navigation system (doppler or Omega) includes the following data:

1. Present position (latitude and longitude),
2. Time (GMT),
3. Date (day, month, year),
4. Raw north wind and raw east wind,
5. X-track error,
6. Track angle,
7. True heading,
8. Drift angle,
9. Ground speed, and
10. True air speed.

These data, plus ranges from the Motorola system, compass and air-data outputs, as well as other sensors, such

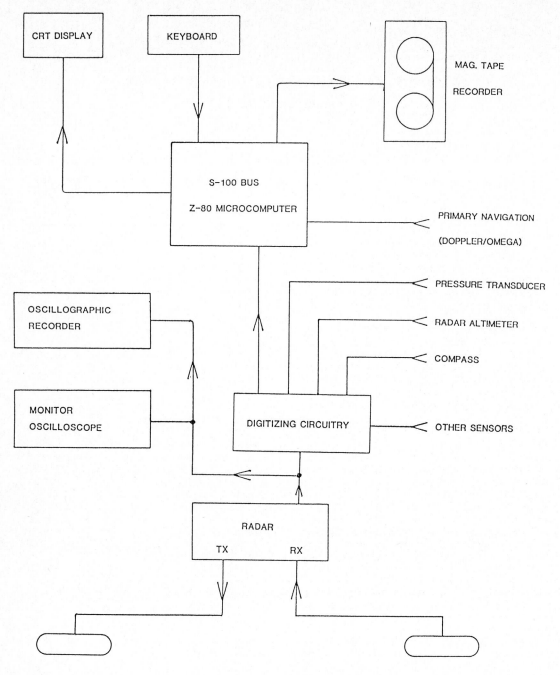

Fig.1. Block diagram of equipment.

as a precise pressure altimeter, are recorded on magnetic tape contiguously with the radar waveform to which they relate. This direct association of radar and navigational data represents a considerable advance over previous systems in which separate records have to be correlated.

Additionally, in 1983, a Motorola Mini-Ranger distance-measuring system was installed in the aircraft, operating in conjunction with ground-based transponders. These were located by satellite geoceiver. This provided much-improved precision over the centre of the Nordaustlandet ice cap, where errors of only ±30 m are estimated.

OTHER AVIONICS

Other avionics which are of importance to the RES system include the aircraft radar altimeter and gyro-stabilized compass. The radar altimeter is an S-band radar instrument which indicates the terrain clearance, i.e. the vertical clearance between the aircraft and the terrain below. The output of this unit is recorded, and may prove helpful in subsequent data analysis. The same information is usually available from the radio echo-sounder output itself,

and there is some indication that the RES terrain clearance figure exhibits lower variance than the aircraft's radar-altimeter output. Nevertheless, the radar-altimeter data provides an important check, especially when flying with small terrain clearances, when the surface return may tend to merge with the tail of the transmit pulse. It is important that an accurate figure for terrain clearance be obtained along the survey flight line, to enable the absolute elevation of the surface to be determined by comparison with the precise pressure-altimeter output.

A five-wire synchro output is available from the aircraft's gyro-stablized compass. Reconstructing the aircraft's heading may be quite important, and magnetic compasses are of little value over much of the polar regions. In practice, the same information is available in the data stream from the primary navigation system (Doppler–TANS or Omega), and this source has proved to be reliable.

RECORDING

Recording of both radar and navigational data is controlled by an onboard S-100 bus Z80 based micro-computer, which has proved to be amply powerful. The

recording media chosen initially were data cartridges of type DC 300XL. These cartridges hold 3.2 Mbytes of data, which represent about 4 h of airborne recording. For the 1986 season, the recording medium was changed to IBM-compatible magnetic tape of 600 foot (183 m) reel size. In each case, commercial recorders of the appropriate type were used, communicating with the microcomputer over an RS-232 link in the first instance, and a fast parallel link in the case of the IBM tape. The IBM tape option has greatly eased post-survey data reduction work in the laboratory.

The basic record cycle during airborne work is 2 s. During the first second of each cycle, radar data is gathered, Motorola ranges are acquired, other navigational data read, etc. During the last second, the data are written to tape. Sufficient memory is available to buffer all data with no loss during the required rewind time.

For digitizing the radar output, a sampling scheme was chosen. Using this scheme, the entire received waveform of 256 points is built up during the course of 256 transmit–receive cycles. Fast analogue-to-digital converters are available which could provide all 256 points in the time of one cycle (60 μs), but calculations suggest that the advantages to be gained are not worth the added circuit complexity. The aircraft moves only about 0.95 m during the 256 cycles required for a complete waveform.

The scheme which was chosen samples at 100 ns intervals in range, to build up a plot of log power against range over 25.6 μs. This total range has proved to be more than adequate for work in Svalbard. Each sample consists of an 8 bit value, which implies a resolution of about ±0.5 dB in received power. This is a considerable improvement on the previous photographic recording systems.

In the first version of this equipment, a single average of 8 waveforms, which were uniformly spaced over the 1 s radar interval, was recorded. Although effective, and with a useful increase of signal-to-noise ratio (c. 9 dB), it was realized that information about the variance of the signal was lost by this process. The latest version allows a choice of 2, 4, or 8 waveforms (unaveraged) or the single averaged waveform to be recorded. As the radar waveform comprises about half of each record, there is, of course, a corresponding increase in the volume of data; this will result in a rather short (c. 1 h) recording time per tape for the largest number.

The digitization interval of 100 ns, compared to the frequency content of the data, implies that valid interpolations should be possible between the data points. This was found to be the case, and the leading edge of the strong ice-surface return has been found to be interpolable to an accuracy approaching 2 m. The weaker bottom echoes can also be interpolated, although not to the same accuracy.

The digitized radar data are displayed in the aircraft on a small CRT display, which can be operated in the A-scope mode, or, more usually, in an intensity-modulated mode (Z-scope). This latter mode provided a very useful display, which appears visually as a cross-section through the ice during the last 8 min of flying time.

To provide a back-up in case of failure of the digital system, an oscillographic recorder is flown. This provides a Z-scope type of record on heat-developed paper. The dynamic range of this analogue record is rather limited, but it does provide a handy "quick-look" facility, as well as fulfilling a back-up role.

In summary, the new SPRI digital RES represents a considerable advance over previous systems. The direct association of radar and navigational data on the same record, and the directly computer-compatible form of the radar, have greatly eased and speeded data reduction. At the same time, the high density and wide dynamic range of the radar record have made possible investigations (e.g. power reflection coefficient of the bed) which have not previously been attempted on a large scale. An example of a processed digital record is shown in Fig.2.

SOFTWARE FOR DIGITAL RES SYSTEM

The software installed in the SPRI digital logging systems has been under continuous development since 1983 and has markedly increased in complexity and sophistication

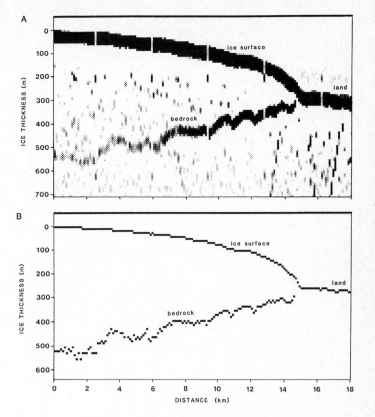

Fig.2. Differentiated digital data.

during this time. It is written in Z80 assembly language, both to ensure that the program executes as rapidly as possible, and also to overcome the constraints imposed by the limited space available in the Erasable Programmable Read-Only Memory (EPROM) for the program. All versions have had a similar basic design philosophy, in that all consist of a simple loop which continuously checks if data are available in the output buffer and, if they are, dumps them to an output device. The radar digitizing system, and in later versions other devices as well, cause this basic loop to be interrupted. The routine(s) servicing these interrupts read data from the input devices and place them in the output buffer. This basic interrupt-driven scheme allows data to be gathered as fast as they are available, without time being lost by the software in polling devices to see if data are available.

The keyboard interrupt collects characters from the keyboard into a buffer and, when a carriage return is pressed, passes the contents of this buffer to a simple command interpreter. The interpreter is written so as to allow new commands to be included in a straightforward way, giving greater flexibility of use. The current command set includes commands to alter the monitor display, to change the number of A-frames recorded at each radar interrupt, and to fire an external device either on command or at pre-set intervals.

The radar interrupt is the central part of the program, because it causes data to be assembled into a record in the output buffer, as well as reading data from the radar digitizing system, and other inputs such as a pressure transducer. If requested, the software will average the 8 A-frames digitized during the 1 s of the digitizer's activity. Otherwise, the number of A-frames that have been requested are placed in the output record.

The timer interrupt occurs every 10 ms, and is used to time the firing of an external device which during the current field season was a vertical camera. The camera was used for navigation and to acquire stereo pairs of photographs for studies of surface roughness.

The final interrupt is generated by the aircraft's Omega navigation system each time a byte of data is available. These data are screened to remove that part of the data stream which is not required. The remainder is assembled into a buffer ready for inclusion in the output record.

Because of the highly modular nature of the software, it should prove to be a straightforward task to modify it for future field seasons, when different data inputs are available.

REFERENCES

Drewry D J, Liestøl O 1985 Glaciological investigations of surging ice caps in Nordaustlandet, Svalbard, 1983. *Polar Record* 22(139): 359–378

Evans S, Smith B M E 1969 A radio echo equipment for depth sounding in polar ice sheets. *Journal of Scientific Instruments (Journal of Physics* E*)* Ser 2, 2: 131–136

Annals of Glaciology 9 1987
© International Glaciological Society

CHARACTERISTICS OF THE SEASONAL SEA ICE OF EAST ANTARCTICA AND COMPARISONS WITH SATELLITE OBSERVATIONS

by

T.H. Jacka, I. Allison, R. Thwaites

(Antarctic Division, Department of Science, Kingston, 7150 Australia)

and

J.C. Wilson

(Bureau of Meteorology, Department of Science, Melbourne, 3000 Australia)

ABSTRACT

A cruise to Antarctic waters from late October to mid December 1985 provided the opportunity to study characteristics of the seasonal sea ice from a time close to that of maximum extent through early spring decay. The area covered by the observations extends from the northern ice limit to the Antarctic coast between long. $50°$E and $80°$E. Shipboard observations included ice extent, type and thickness, and snow depth. Ice cores were drilled at several sites, providing data on salinity and structure.

The observations verify the highly dynamic and divergent nature of the Antarctic seasonal sea-ice zone. Floe size and thickness varied greatly at all locations, although generally increasing from north to south. A high percentage of the total ice mass exhibited a frazil crystal structure, indicative of the existence of open water in the vicinity.

The ground based observations are compared with observations from satellite sensors. The remote sensing data include the visual channel imagery from NOAA 6, NOAA 9, and Meteor 11. Comparisons are made with the operational ice charts produced (mainly from satellite data) by the Joint Ice Center, and with the analyses available by facsimile from Molodezhnaya.

INTRODUCTION

Because Antarctica has no permanent population and is distant from international travel routes and areas of major strategic importance, there are few observations of sea-ice characteristics from within the seasonal sea-ice zone. Our knowledge of winter ice conditions comes almost solely from relatively coarse resolution passive microwave and infrared satellite imagery.

Most sea ice in the Antarctic is believed to be of one type (first-year ice) and hence it has been assumed that the ice concentration can be obtained from passive microwave data, analysed using simple algorithms based on the difference between the microwave emissivity (at 19GHz) of water and a unique value for ice (e.g. Comiso and Zwally 1982; Zwally and others 1983). There are, however, few ground-based data to verify these algorithms. Comiso (1983) suggested that using multi-spectral cluster analysis techniques, the satellite data themselves can be used to identify the microwave characteristics of different ice types (including new ice), but there still remains a need for *in-situ* validations.

Ackley and others (1980) made observations of ice characteristics in the Weddell Sea from USCGC *Polar Sea* in February/March 1980. Comprehensive observations in the same area were made in late spring (October/November, 1981) from NES *Mikhail Somov* during the Weddell Polynya Experiment. The latter study produced detailed data on sea-ice conditions (Ackley and Smith 1983) and on the physical, chemical and biological properties of sampled ice

floes (Ackley and others, 1983). Comiso and others (1984) compare the *in-situ* data from *Mikhail Somov* with Nimbus-7 SMMR data and conclude that the 18 GHz channel gives good discrimination of the ice edge but that in late spring and summer, surface wetness leads to differences identifiable by multi-spectral cluster analysis, particularly in the 37 GHz emissivity. These differences would cause errors in the ice concentrations if a unique ice emissivity was assumed. Comiso and others (1984) stress the importance of surface observations to aid interpretation of satellite data but, while there is an ongoing winter sea-ice program in the Weddell Sea, no previous winter data have been reported from other areas of Antarctica.

The data presented in this paper were collected on board MV *Nella Dan* between early October and mid December 1985, in the Indian Ocean sector of the seasonal sea-ice zone. Climatic ice data (e.g. Zwally and others 1983; Jacka 1983) show that the maximum extent of the ice in this sector occurs in October. In contrast, the maximum extent in the Weddell sector occurs from mid September to mid October and the data reported to date from the Weddell Sea have been from the period of decay of the annual ice cover.

The cruise of *Nella Dan* was primarily for biological studies (Antarctic Division Biomass Experiment III; ADBEX III) but included a sea-ice subprogram. Observations of sea-ice conditions were made about every 5 km. A number of vertical photography runs over the pack ice were made from a helicopter, and cores were collected for study of the physical and structural properties of the ice. Automatic Picture Transmission (APT) data were received on board from NOAA 6, NOAA 9 and Meteor 11 satellites.

The cruise track of *Nella Dan* is shown in Fig.1. The ship entered the pack on 8 October at long. $60°$E and worked southwards until 17 October but was unable to penetrate beyond about $64°$S. A region of lower concentration ice to the north of the Enderby Land peninsula was exploited to then work south towards the coast. *Nella Dan*'s track south was approximately along $55°$E until poor weather, strong winds and south-west running leads eventually led to the ship approaching the Antarctic coast near $50°$E. In this area, *Nella Dan* became beset and remained so, in a large consolidated floe, drifting slowly west for 7 weeks. MV *Icebird* was unable to approach closer than 18 km to *Nella Dan* in early December because of severe ice conditions in a shear zone. *Nella Dan* was eventually freed on 15 December by the Japanese ice breaker *Shirase*.

ICE CONDITIONS IN THE MARGINAL ICE ZONE

The marginal ice zone (MIZ) is the outer region of a sea-ice field, where oceanic and atmospheric interactions with pack ice are at their most dynamic. In the area close

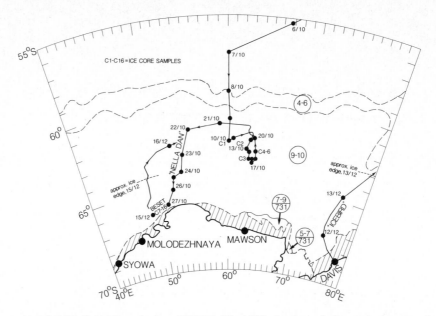

Fig.1. Cruise track of MV *Nella Dan* during ADBEX-III, 1985. Ice boundaries and concentrations (shown in the standard "egg" code) are from the US Navy—NOAA Joint Ice Center ice chart for 10 October 1985

to the open ocean the ice has properties quite different from those of ice in the interior pack (typically more than 150 km from the ice edge) and special techniques may be required to obtain estimates of ice characteristics from remote-sensing data (NASA 1984).

The change in ice conditions across the late winter Antarctic MIZ along 60°E longitude is summarized in Fig.2 from observations made from *Nella Dan*. In the first 100 to 150 km from the ice edge (at 59°32'S), the floe size and ice thickness generally increase and the ice type becomes increasingly more mature. The floe size affects such large-scale properties as the mechanical strength of the ice as a continuum and summer melt from the sides of the floes. Rothrock and Thorndike (1984) discuss the detailed measurements required to determine the floe size distribution. The data of Fig.2 give the typical range of floe sizes

as estimated from the ship. Similarly, the ice thicknesses given are estimates only and, where the pack consists of a mixture of new and more mature ice forms, the range of thicknesses given is typical of that of the older floes. These shipboard observations were made by three observers. To ensure consistency of their estimates of thickness, floe size and concentration, the initial observations during the cruise were made simultaneously but independently by all observers and single observer estimates were used only after the observers had achieved a consistent standard. Throughout the MIZ, newly formed ice (nilas, pancakes, young ice, etc.) constitutes a significant percentage of the total ice cover. The structure of the pack ice in the late winter MIZ contrasts with that seen during the summer. Figs 3 and 4 show north-south transects from the interior ice near the coast to the ice edge, observed from MV *Icebird* and MV

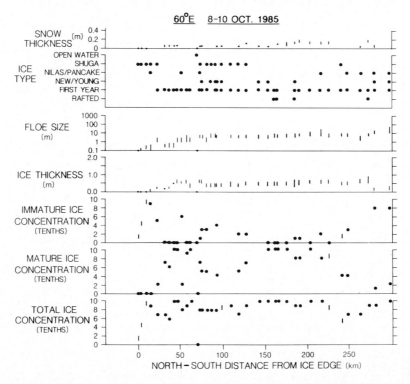

Fig.2. Sea-ice conditions observed on board *Nella Dan* along a north-south transect through the MIZ at 60°E, October 1985. Ice concentration is shown for mature and immature ice forms as well as a total. Immature ice includes nilas, pancakes and new ice and is typically less than 0.4 m thick.

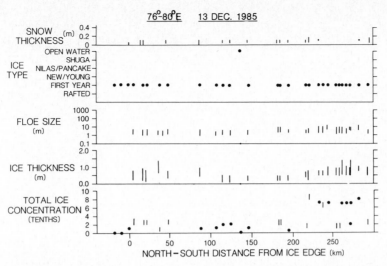

Fig.3. Sea-ice conditions observed on board *Icebird* along a north–south transect through the MIZ at 76–80°E, December 1985.

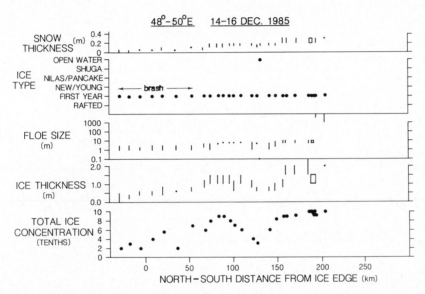

Fig.4. Sea-ice conditions observed on board *Nella Dan* along a north–south transect through the MIZ at 48–50°E, December 1985.

Nella Dan respectively in December 1985. At this time the pack is much more uniform in ice type and floe size since all the thin ice forms have melted, and the low ice concentrations in the outer 200 km allow waves to break up the remaining first-year floes.

ICE CONDITIONS IN THE INTERIOR ICE ZONE

The ice conditions deeper within the seasonal ice zone are summarized in part of Fig.2 (beyond 150 km) and in Fig.5, which shows a transect in mid October along approximately 55°E. The floe sizes in this region are determined more by constructive processes (freezing together of smaller floes and brash ice) than by the destructive process of fracture. *Nella Dan* eventually became entrapped in a huge consolidated floe of approximately 20 km diameter. Even within the interior pack, 10-20% of the ice typically consisted of thin new ice types but this new ice was concentrated in wide leads, rather than being spread fairly uniformly as in the MIZ. *Nella Dan* was, however, following a route through systems of leads and polynyas. Extensive rafting of new and young ice up to 0.5 m thick occurs throughout the interior ice zone, and this contributes to the thickening of the ice cover. Only in the strong shear zones nearer to the Antarctic coast, do pressure ridges of up to 3 m in height develop due to rafting of thicker ice.

A 7 km wide band of landfast ice west of 51°50'E

longitude was separated from the pack by shore polynyas several kilometres wide, and a large polynya of tens of kilometres diameter lay to the north of Enderby Land, around 65°40'S, 52°20'E.

COMPARISON WITH SATELLITE-DERIVED ICE CONDITIONS

The value of remote sensing, particularly with passive microwave and active Synthetic Aperture Radar systems, to the delineation of ice extent and to the study of sea–ice processes has been well demonstrated. However, most researchers investigating the climatic role of sea ice, or studying ice–weather relationships, rely not on raw data from these systems but on derived products (such as the ice charts produced by the U.S. Navy/NOAA Joint Ice Center), on time-averaged compilations of satellite passive microwave data (e.g. Zwally and others 1983) or on the more easily obtainable visible and infrared imagery from polar orbiting weather satellites. Ice conditions from these sources are compared with the surface observations from *Nella Dan*.

APT data (in low resolution format) from the AVHRR on NOAA 6 and NOAA 9 and from Meteor 11 were routinely received on board *Nella Dan*. These data were limited by cloud which obscured surface conditions and by the extremely oblique viewing angles of some images. For good orbit geometry and largely cloudless conditions, the

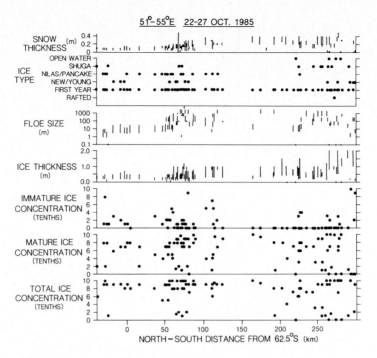

FIGURE CAPTIONS

Fig.5.   Sea-ice conditions observed on board *Nella Dan* along a north–south transect through the interior pack at 51-55°E, October 1985. Ice concentration is shown for mature and immature ice forms as well as a total.

Meteor pictures were superior for ice interpretation to those from NOAA. Fig.6 shows a Meteor 11 (visual band) picture for 10 October 1985 and an ice chart interpreted from this image. Also shown is a map drawn from the ice chart for the period 11-20 October 1985 which was transmitted by facsimile from the Soviet Antarctic Station, Molodezhnaya. These operational charts are prepared on the 1st, 11th, and 21st of each month, and are largely based on APT data. The ice concentration contours from the JIC chart of 10 October 1985 are plotted on Fig.1 along with the track of *Nella Dan*.

Although Fig.6b and 6c show a very similar fast ice extent near the coast, the operational chart (Fig.6c) shows an average ice concentration in the interior pack higher than that in Fig.6b. Fig.6c represents the mean ice conditions for the ten-day period, and hence does not show fine detail, and some of the difference may be due to changing ice conditions during the analysis period (e.g. the ice concentration around 50°E increased considerably in the latter part of the period). The interpretation of ice concentration from the satellite imagery may, however, also differ between charts.

The system of leads and polynyas extending south towards the coast can be seen in the Meteor image. Strong west to south–west winds in the rear of a series of deep low pressure systems helped produce these open conditions. It would appear from satellite imagery and aerial observations over a number of years that this area of lower concentration is a semi-permanent feature (Knapp 1966).

Thorough analysis of the Meteor 11 picture can provide a detailed description of ice concentrations, and three categories are shown in Fig.6b. The general pattern of this detailed analysis agrees with the variable nature of the ice recorded on board *Nella Dan*. This detail is not displayed by either of the operational charts, but the more general concentration of 8-10/10 throughout the interior ice zone is not incompatible with the ground–based observations. The sea-ice data derived from ESMR data for 1973-1976 (Zwally and others 1983) show typical October ice concentrations in this region of 8-9/10. Both the Molodezhnaya and JIC charts show an outer ice band, of about 100 km width, of 4-6/10 concentration. The actual concentration in this zone was greater than 7/10, although

much of this cover consisted of ice less than 5 cm thick, and thus was not detectable by the satellite sensors.

The winter ice charts are based exclusively on satellite data (IR and microwave) but unfortunately only an estimated ice edge is shown on the JIC chart for 10 October. The extent of fast ice from the analysis of the Meteor 11 picture, the Molodezhnaya chart and the JIC chart show good agreement. The northern ice limit interpreted from the Meteor 11 picture is further south than shown on both the operational charts, which, however, agree to better than the resolution of the microwave sensor (~30 km) with the ship-based observations.

The Meteor 11 picture was scaled by using known latitudes and longitudes of identifiable locations on the picture. Small errors in scaling are possible due to the projection of the picture. In addition, the picture is an instantaneous record of sea–ice conditions while the operational charts are based on data collected over a period of about a week. These may account for the above discrepancy.

The ice concentrations observed in the marginal ice zone in December along 80°and 50°E longitude (Figs 3 and 4) were, in both cases, lower than the 4-6/10 concentrations shown in the JIC chart for 12 December. Zwally and others (1983) also show monthly mean concentrations derived from the ESMR data that are somewhat higher at this time of year than reported here. It is difficult to resolve this apparent anomaly. Changes in the ice emissivity due to melt processes can result in errors in ice concentrations (Comiso and others 1984) of 3 to 4% at 19 GHz. Sea–ice concentrations derived from ESMR data, however, are only made to the nearest 10%. In addition, no large-scale surface melt was observed, and thus it seems unlikely that melt processes could account for the total anomaly. Streten and Pike (1985) noted from shipboard observations in the same region and during the same season that satellite visual imagery frequently did not resolve low concentrations in the range 0-2/10.

The ice edge shown on the JIC ice chart for 12 December is in general agreement with that observed from *Nella Dan* and from *Icebird* (Figs 3 and 4), although at this time of year direct comparisons should not be made. The JIC reports the mean ice conditions determined over a one-

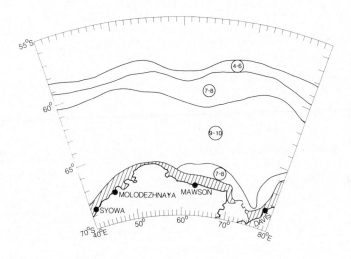

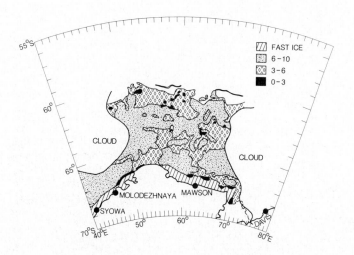

Fig.6. (a) Meteor 11 visual image of the observational area on 10 October 1985; (b) a detailed analysis of the image showing three levels of concentration; and (c) the analysis available by facsimile from Molodezhnaya for the period 11–20 October 1985.

week period, and in mid December the ice edge retreats at up to 200 km per week. Bands of small floes and brash ice, from 100 to 1000 m wide, were observed at distances up to 100 km north of the ice edge. Martin and others (1983) have studied the movement and decay of similar bands in the Bering Sea and consider them to play a major role in controlling the ice-edge position.

PHYSICAL AND STRUCTURAL PROPERTIES OF THE PACK ICE

Ice cores were taken from floes at sites shown in Fig.1, and these are described in Table I. The cores were examined structurally and sampled every 0.1 m for salinity measurements. Core samples were melted shortly after collection, and salinities were determined from measurements of the specific gravity at constant temperature. The resulting profiles are shown in Fig.7. Few of the cores exhibited the characteristic C-shaped salinity profile typical of first-year fast ice, but cores 10 and 12 in particular show "stacked C" profiles indicative of multiple rafting of floes, each typically 0.6–0.8 m thick.

The ice structure was determined from examination of thick longitudinal sections between crossed polarizing filters. Three major ice types were identified from the examination: ice with long columnar crystals originating from congelation growth; fine-grained frazil crystals; and infiltration ice formed from freezing of water-saturated snow. Almost all the examined cores show a very high fraction of frazil ice, which is similar to what was found in the Weddell Sea by Gow and others (1982), where many floes contained more than 50% frazil ice. They suggested that much of this frazil originates as a result of brine convection induced by rapid freezing of sea water in leads and polynyas, and that the widespread distribution of frazil in the Weddell Sea indicates the importance of ice deformation.

The area of ice sampled in cores 1 to 6 was quite different from that sampled in cores 7 to 16. Cores 1 to 6 were obtained from young ice floes less than 1 m thick and within 300 km of the ice edge. Cores 7 to 16 were obtained when the ship was beset near the coast in ice more than 2 m thick. Notwithstanding these differences, the average fraction composition of the ice at both sites (Table I) was remarkably similar, with more than 50% of the total column being composed of frazil ice.

Studies of sea-ice structure at Casey in 1983 (Allison and Qian 1985) showed that frazil ice only contributed significantly to the total ice fraction when there was open water within 100 m or so of the fast ice. The large frazil fractions found in all cores hence suggest that divergent motion and lead formation is common in this region of the Antarctic seasonal sea-ice zone.

TABLE I. AVERAGE COMPOSITION OF ICE CORES

Core number	Typical depth (m)	Weighted average composition (%)		
		Infiltration ice	Congelation ice	Frazil ice
1-6	~ 0.8	7	37	56
7-16	> 2.5	2	40	58
TOTAL		3	40	57

SUMMARY AND CONCLUSION

The observations from *Nella Dan* show that there is a wide mixture of ice type, age and thickness throughout the late winter Antarctic seasonal sea-ice zone in the region studied. The total ice concentrations are generally 9–10/10, but a high fraction of this ice cover is very young and often less than 10 cm thick. This has important implications for the energy exchange between the ocean and atmosphere. For example, in the central Arctic the total heat input to the atmosphere from regions of young ice may be equal to or greater than that from regions of open water or thick ice (Maykut 1978).

More than half the total ice column in the cores examined was composed of frazil ice, a similar fraction to that found in the Weddell Sea by Gow and others (1982). If frazil formation is widespread in the Antarctic seasonal sea-ice zone, there are important implications both for dynamic sea ice models and for the energy exchange. The dynamic models require a theoretical ice thickness distribution which presently does not allow for ice growth by frazil accretion. Similarly, the shift of frazil formed in leads and polynyas to other parts of the pack keeps the leads open as areas of very high ice formation rates and as areas of high heat loss. Current energy calculations do not take account of persistent leads.

Rafting of ice less than about 0.5 cm thick also contributes to the ice thickness development. Extensive rafting was observed in the interior ice zone. Large ridge building only occurred in a few areas near the coast.

The observed nature of the pack, and the ice core structure found, result from the relatively rapid and generally divergent motion of the Antarctic pack ice. This stresses the importance of ice dynamics as well as thermodynamics to large-scale sea-ice interaction in the Antarctic (Hibler and Ackley 1982).

The JIC and Molodezhnaya ice charts, derived from satellite imagery, give a reasonable representation of the Antarctic ice extent and broadscale concentration within the interior pack. In the marginal ice zone, however, thin new ice is not detected and the charts underestimate total ice concentration. Neither do they indicate the considerable percentage of young ice in the interior pack. These ice charts are widely used by many researchers as a primary sea-ice data source for ice and climate interaction studies.

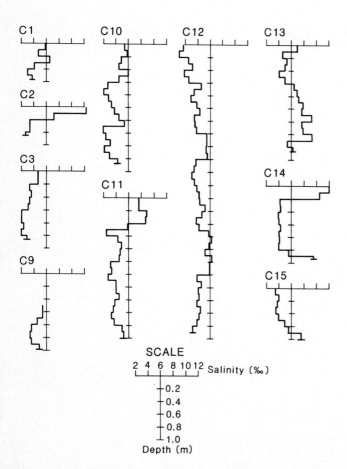

SCALE

2 4 6 8 10 12 Salinity (‰)

0.2
0.4
0.6
0.8
1.0

Depth (m)

Fig.7. Salinity profiles from cores drilled at locations shown in Fig.1.

Because of the importance of thin ice to the surface energy balance, their use for this purpose should be treated with caution.

REFERENCES

Ackley S F, Smith S J 1983 Weddell Polynya Expedition. Sea ice observations. *Reports of the U.S.-U.S.S.R. Weddell Polynya Expedition, October-November, 1981. Vol 5.* Hanover, NH, US Army Cold Regions Research and Engineering Laboratory

Ackley S F, Gow A J, Buck K R, Golden K M 1980 Sea ice studies in the Weddell Sea aboard USCGC *Polar Sea. Antarctic Journal of the United States* 15(5): 84-86

Ackley S F, Clarke D B, Smith S J 1983 Weddell Polynya Expedition. Physical, chemical and biological properties of ice cores. *Reports of the U.S.-U.S.S.R. Weddell Polynya Expedition, October-November, 1981. Vol 4.* Hanover, NH, US Army Cold Regions Research and Engineering Laboratory

Allison I, Qian Songlin 1985 Characteristics of sea ice in the Casey region. *ANARE Research Notes* 28: 47-56

Comiso J C 1983 Sea ice effective microwave emissivities from satellite passive microwave and infrared observations. *Journal of Geophysical Research* 88(C12): 7686-7704

Comiso J C, Zwally H J 1982 Antarctic sea ice concentrations inferred from Nimbus 5 ESMR and Landsat imagery. *Journal of Geophysical Research* 87(C8): 5836-5844

Comiso J C, Ackley S F, Gordon A L 1984 Antarctic sea ice microwave signatures and their correlation with in situ ice observations. *Journal of Geophysical Research* 89(C1): 662-672

Gow A J, Ackley S F, Weeks W F, Govoni J W 1982 Physical and structural characteristics of Antarctic sea ice. *Annals of Glaciology* 3: 113-117

Hibler W D III, Ackley S F 1982 On modelling the Weddell Sea pack ice. *Annals of Glaciology* 3: 125-130

Jacka T H 1983 A computer data base for Antarctic sea ice extent. *ANARE Research Notes* 13

Knapp W W 1966 Formation, persistence and disappearance of open water channels related to the meteorological conditions along the coast of the Antarctic continent. *W.M.O. Technical Note* 87: 89-104

Martin S, Kauffman P, Parkinson C 1983 The movement and decay of ice edge bands in the winter Bering Sea. *Journal of Geophysical Research* 88(C5): 2803-2812

Maykut G A 1978 Energy exchange over young sea ice in the central Arctic. *Journal of Geophysical Research* 83(C7): 3646-3658

NASA 1984 *Passive microwave remote sensing for sea ice research. Report of NASA Science Working Group for the Special Sensor Microwave Imager (SSM/I).* Washington, DC, University of Washington. Applied Physics Laboratory

Rothrock D A, Thorndike A S 1984 Measuring the sea ice floe size distribution. *Journal of Geophysical Research* 89(C4): 6477-6486

Streten N A, Pike D J 1985 Sea ice observations during ADBEX, 1982. *ANARE Research Notes* 28

Wadhams P, Squire V A 1983 An ice-water vortex at the edge of the East Greenland Current. *Journal of Geophysical Research* 88(C5): 2770-2780

Zwally H J, Comiso J C, Parkinson C L, Campbell W J, Carsey F D, Gloersen P 1983 *Antarctic sea ice, 1973-1976: satellite passive-microwave observations.* Washington, DC, National Aeronautics and Space Administration (SP-459)

Annals of Glaciology 9 1987
© International Glaciological Society

IMAGE-ANALYSIS TECHNIQUES FOR DETERMINATION OF
MORPHOLOGY AND KINEMATICS IN ARCTIC SEA ICE

by

Meemong Lee and Wei-Liang Yang

(Image Processing Laboratory, Jet Propulsion Laboratory, California Institute of Technology,
Pasadena, CA 91109, U.S.A.)

ABSTRACT

The synthetic aperture radar (SAR) data have been used to study sea ice with respect to its motion and formation/deformation. With the prospect of the Alaska SAR Facility development in the near future, there is a great need for robust and efficient sea-ice analysis techniques. This paper presents a sea-ice motion analysis technique that can be used for (1) local motion analysis of a selected ice patch and (2) a global ice motion over the entire image area. Though there are several sea-ice motion tracking techniques, they do not provide the required operational speed or robustness. In order to meet the operational speed requirement (over fifty images per day), we have developed a sea-ice motion analysis technique which requires very little human interaction and much simplified computation. The proposed technique uses a subset of easily distinguishable features to predict global motion characteristics and apply template matching over a predicted search area. We applied the developed technique to two pairs of SEASAT SAR images, one pair with a minor motion of "ice pack" and another with a larger and discontinuous motion of "fast ice". The two major achievements of the new approach are: first, development of a set of computer-aided tools for feature selection and registration and, secondly, implementation of an optimal search strategy for automatic template matching via a motion prediction model.

INTRODUCTION

In this paper, morphological and kinematical analysis of sea ice are divided into local and global structures. The rotational and translational motion of a local structure is analyzed via a feature selection and registration process. The motion of a global structure (an entire image covering 100 km by 100 km) is analyzed via an automatic template process guided by a predicted motion model. Currently, most motion analysis techniques are based on tie points selected either manually or automatically. The selected tie points imply geometrical mapping between two images and the motion is analyzed via the displacement of each tie-point pair under the assumption that the two images represent the same region.

In the case of manual selection (1), the global motion is analyzed by selecting tie points over an entire image area. After the tie points have been selected, each displacement vector is computed by connecting the tie-point pair's locations. The magnitude of the displacement vector indicates the amount of displacement of a given feature between two time frames.

$$V_i = ((x_i\ y_i),(x_i',y_i'))$$ (1)

$$|V_i| = \sqrt{(x_i' - x_i)^2 + (y_i' - y_i)^2}$$

where
V_i is displacement vector of ith tie-point pair
(x_i, y_i) is the location of the ith tie-point in a reference image
(x_i', y_i') is the location of the ith tie-point in a compared image

When the two images do not overlap exactly, the actual displacement can be updated by subtracting the regional displacement vector from each displacement vector. Also, relative motion with respect to a selected land mark can be analyzed in a similar manner by replacing the regional displacement vector with the displacement vector of the land mark. Equation 2 shows the displacement vector update where T is a common displacement between two images.

$$V_i' = V_i - T$$ (2)

$$T = ((x_0\ y_0),\ (x_0',\ y_0'))$$

where
(x_0, y_0) is the origin location of a reference image
(x_0', y_0') is the matching location in a compared image

An image pair covering 100 km by 100 km requires several hundreds of tie-point pairs, which takes more than a day, even for an experienced scientist. The manual approach is not only time-consuming but also very strenuous to human eyes. Besides, it does not provide rotational information of specific sea-ice structures since the structural connectivity is not present in tie points.

In order to resolve the strenuous tie-point selecting process of the manual approach, various tie-point selection techniques have been researched. Major problems which exist in developing automatic tie-point selection techniques on SEASAT SAR images are: first, the sea-ice images lack distinctive features; secondly, the boundary features may vary due to seasonal formation and/or deformation; thirdly, the range of sea-ice motion is enormous; fourthly, there is no solid motion unit.

In the case of automatic selection (2), template matching is used to automate tie-point detection, in order to eliminate the laborious tie-point selection process. A template is an image segment inside an arbitrarily chosen window, where a window is usually defined as a regular grid. A selected template from the reference image. is passed over a search area in the compared image until the best match is found. The center of the reference image template and the center of the best matching template are selected as a tie-point pair. The template matching process is then repeated over the entire area for global motion analysis. The template matching process can be expressed as

$$C_n(k,1) = \sum_{j=1}^{ny} \sum_{i=1}^{nx} [T_n(i,j)\ S_n(k+i,\ 1+j)]\ \text{for}\ k=1,sx\ 1=1,sy$$ (3)

where
C_n is a correlation score array of the nth template
T_n is nth template
S_n is nth template search area
(nx,ny) is the size of the template T_n
(sx,sy) is the size of the search area S_n

The location (k,1) that yields the maximum matching score will be the tie-point location for the template T_n.

The number of multiplication required for each template matching is the number of pixels in a template multiplied by the number of pixels in a search area. In order to reduce the computational burden, the template matching is usually performed in a hierarchical manner using a low-resolution image first to find an approximate match and then increase the resolution step by step to narrow down the search area. However, due to the similarity between sea-ice image pairs, using a low-resolution image can be dangerous. The computation time of this approach fluctuates greatly depending on the motion characteristics of the sea-ice images. Also, this approach cannot be used for local ice patch motion analysis since regular templates do not represent sea-ice structures.

This paper presents an alternative for sea-ice structure and motion analysis with a hybrid approach at local and global scale. Major objectives of this approach are to reduce the laborious tie-point selecting process of the manual approach and to reduce the computation time in the automatic template matching approach. The first objective is achieved by replacing the tie-point selecting process with a computer-assisted feature selection and registration process. The second objective is achieved by improving the blind search of automatic template matching with a guided search from a motion prediction model. The procedure flow of this approach is illustrated in the flow chart of Fig.1. First, a subset of primary features are selected from the reference

"rigid body". Thus, movement of a local ice structure can be realized by the translational as well as rotational deviations of a feature. The movement vector field of the local structures can then be utilized to evaluate the motion of the global ice field.

A set of computer-aided tools was developed to assist the feature selection and registration process. The tools for the feature selection process consist of three commands that can be activated by simply pressing a button on the trackball or on the mouse of a frame buffer. The commands are "move cursor", "draw line", and "delete previous line". With these commands, a user can extract features of interest from a given area displayed on the frame buffer. The feature lines are drawn on the overlay plane of the frame buffer. For the local motion analysis of the selected features, the compared image is displayed on the frame buffer and the selected features are registered individually to the compared image. The tools for the feature registration include four commands: "select a feature", "translate (up, down, left, right)", "rotate (clockwise, counter-clockwise)", and "delete" (the selected feature). The "feature delete" command is essential when the feature can not be found from, or registered to, the compared image. As a feature is registered to the compared image, the translation and rotation information is updated and it is displayed on the terminal screen.

The SEASAT SAR images taken over Banks of Island

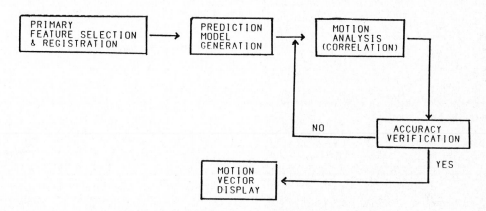

Fig.1.   Procedure flow chart of sea-ice motion anaylsis.

image and they are registered to the compared image. The local motion analysis of specific features can be performed in this procedure. Secondly, a predicted global motion model is generated using the result of the first step. Thirdly, a guided template matching is applied and tie-points are selected. Fourthly, the accuracy of the tie-points is verified. Global ice-floe vector fields are computed when the accuracy test is passed; otherwise, the third step is repeated with larger search areas until the accuracy test is passed. Each step is discussed in detail at the following sections.

LOCAL MOTION ANALYSIS VIA FEATURE SELECTION AND REGISTRATION

Simply stated, motion analysis is performed by evaluation of the "rigid body"'s morphical deviations from an earlier image to a later image (in SEASAT's case, three days apart). The earlier image is used as a reference image and the later image is used as a compared image. This evaluation is performed in part by selecting a set of recognizable features from the reference image and then registering them to the compared image. A feature can be represented by a set of independent points, as in the tie-point methods, or, alternatively, by a set of connected line segments belonging to a single rigid structure. The line representation reduces the redundant tie-point pair selection process since all of the points within a "rigid body" must maintain their geometrical relationships throughout the movements. The relationship of the line segments to a single structure also incorporates the rotational attributes of a

were used to illustrate the local motion analysis. Fig.2a shows selected features from the reference image where five easily recognizable features are drawn with polygons and polylines. Each feature is assigned an identification number so that it can be controlled individually. Fig.2b shows the result of a feature registration process, where the five features are registered on the second image using the translation and rotation commands. Fig.2c shows the local motion vectors of each feature, where the translation of each feature is computed by connecting the rotation center of a feature from the reference image to the compared image. The rotation of each feature is expressed as a set of displacement vectors from every vertex of a feature relative to its rotation center. The rotation center of a feature is defined to be the center of an imaginary box that surrounds the feature. The rotation center can be arbitrarily chosen if desired. This process provides tools to analyze the detailed motion of any local structure very easily without intensive computation or laborious tie-point selection.

MOTION PREDICTION MODEL GENERATION

The feature selection and registration process is also used to assist the global motion analysis so that the automatic template matching technique can be applied without intensive computation on a blind search. The global motion prediction is based on the assumption that sea ice cannot move randomly but in a somewhat correlated manner. Therefore, the correlated motion can be predicted based on the information acquired from the selected features

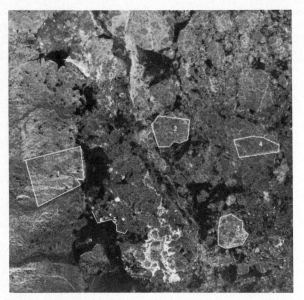

Fig.2a. Reference image features ("fast ice" in Banks of Island area).

Fig.2b. Result of feature registration on the compared image (three days later).

Fig.2c. Local motion analysis of the selected features.

using proper interpolation and extrapolation techniques. The predicted motion model is used to determine the search area for the template matching instead of the hierarchical correlation technique discussed earlier. The generation of prediction model takes insignificant computation time compared to the correlation process. The model is more accurate than other types of prediction model since it is based on manually selected and registered features (i.e. the surface fitting uses the tie points extracted from the vertices of the selected features and the registered features).

Two types of surface fitting methods were considered, the triangulation method for discontinuous surface fit and the least-squares surface fit method for continuous surface fit. The triangulation method connects all the tie points to form the shortest distance triangles and the points inside the triangle are interpolated based on the relation between matching triangles. The mapping of the triangulation method is shown in Equation (4), where three vertices of a pair of mapping triangles are used to compute the coefficients of bi-linear equation.

$$\begin{bmatrix} x_1' & x_2' & x_3' \\ y_1' & y_2' & y_3' \end{bmatrix} = A \begin{bmatrix} x_1 & x_2 & x_3 \\ y_1 & y_2 & y_3 \end{bmatrix} + C \qquad (4)$$

where

$$A = \begin{bmatrix} a_{11} & a_{12} \\ a_{21} & a_{22} \end{bmatrix}$$

$$C = \begin{bmatrix} c_1 & c_1 & c_1 \\ c_2 & c_2 & c_2 \end{bmatrix}$$

where

$(x_i \ y_i)$ is a vertex of a triangle in the reference image
$(x_i' y_i')$ is a vertex of a triangle in the compared image

After the Equation (4) is solved for every triangle in the image plane, a motion prediction model is constructed by computing predicted locations of regular grid points. Each grid point location is analyzed to find the triangle in which the grid point resides; then, the proper set of coefficients to the matching triangle is applied to determine the predicted location of every grid point. This method cannot be used for extrapolation directly. It has to employ the least-squares surface fit method to produce boundary points; then new triangles are formed to the boundary points for extrapolation.

The least-squares surface fit method solves a polynomial equation to find a smooth function that describes the relation between two image surfaces. The least-square surface fit method is described in Equation (5), where any subset of the polynomial equation can be used to estimate the surface fitting. The first-order surface fit requires up to (a_2, b_2), the second-order surface fit requires up to (a_5, b_5), the third-order surface fit requires up to (a_9, b_9), and etc. The order of surface fit may be determined by the characteristics of motion.

$$x' = a_0 + a_1 x + a_2 y + a_3 xy + a_4 x^2 + a_5 y^2 + a_6 x^3 +$$
$$a_7 x^2 y + a_8 xy^2 + a_9 y^3 \ ... \qquad (5)$$

$$y' = b_0 + b_1 x + b_2 y + b_3 xy + b_4 x^2 + b_5 y^2 + b_6 x^3 +$$
$$b_7 x^2 y + b_8 xy^2 + b_9 y^3 + ...$$

where

(x, y) is a location in a reference image
(x', y') is a matching location in a compared image

The Equation (5) can be written as
$$T' = A \ T \qquad (6)$$

where

$$T' = [x', y']^T$$

$$T = [1, x, y, xy, x^2, y^2, ...]^T$$

$$A = \begin{bmatrix} a_0 & a_1 & a_2 & a_3 & a_4 ..... \\ b_0 & b_1 & b_2 & b_3 & b_4 .. \end{bmatrix}$$

The matrix A can be solved by

$$A = \frac{T'T^T}{T\ T^T} \qquad (7)$$

After the least-squares surface fit equation is solved, the predicted motion model is generated by computing the predicted location of the regular grid points that will be used for template matching. Fig.3 shows a predicted motion model using a regular grid of 16 pixels by 16 pixels over a

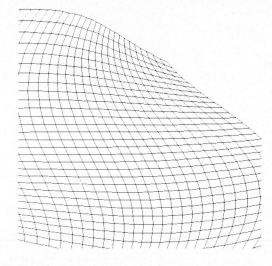

Fig.3. Predicted motion model for "fast ice" case (Fig.2a and 2b).

512 pixel by 512 pixel image area. The sheared grid illustrates a predicted shearing of the compared image. Each grid point is a predicted location of the corresponding regular grid point, which also implies that it is a predicted tie-point.

GLOBAL MOTION ANALYSIS

After a set of predicted tie points is obtained, the final step is to apply automatic template matching to correct the error of the predicted tie points. The prediction error

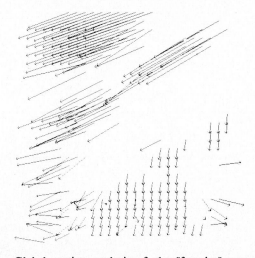

Fig.4. Global motion analysis of the "fast ice" case (Fig.2a and 2b).

can be corrected by applying automatic template matching described earlier around the predicted area. The global motion result displayed in Fig.4 was obtained using a 16 pixel by 16 pixel template to search over a 48 pixel by 48 pixel area, where each pixel represents a 200 m by 200 m area. The search area allows up to a 16 pixel prediction error in four directions (left, right, up, down). In order to analyze motion of the images shown in Fig.2 without motion prediction, the search area should be greater than 200 pixels by 200 pixels, which implies at least 16 times more multiplication for each template and correspondingly more CPU time. For a image of 512 by 512 pixels, the proposed approach takes about 1.5 VAX 11/780 CPU minutes for template matching.

The areas with no motion vector in Fig.4 indicate that the template matching process has failed to find matching templates. The reasons for failure are twofold: (1) features in the area were distorted beyond recognition, and (2) the search area was not large enough to correct the prediction error. The first case cannot be resolved by template matching and can only be interpolated from the surrounding motion. The second case can easily be resolved by iterating the template matching process to a larger search area as shown in Fig.1. The iteration may continue as more templates find their matching templates; otherwise it terminates.

An image pair, collected over the Beaufort Sea three days apart on orbits 1439 and 1482, is analyzed for comparison with existing methods since it is frequently analyzed by other techniques. Fig.5a displays five selected features, Fig.5b shows the result of feature registration,

Fig.5a. Reference image features ("ice pack" in Beaufort Sea).

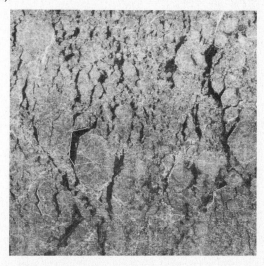

Fig.5b. Result of feature registration on the compared image (three days later).

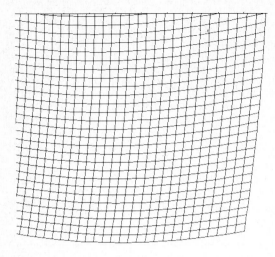

Fig.5c. Predicted motion model.

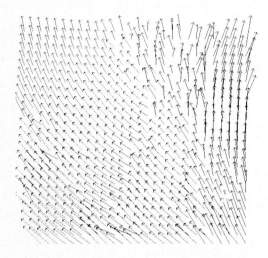

Fig.5d. Global motion analysis.

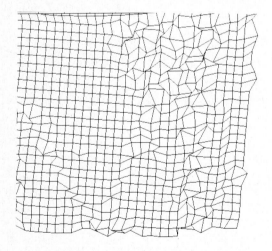

Fig.5e. Surface deformation.

Fig.5c shows the prediction model using the third-order, least-squares surface fit, Fig.5d shows the results of automatic template matching which shows displacement vectors of regularly distributed tie-points and Fig.5e shows the distortion surface characteristics of 1482 due to sea-ice motion. The manual approach took approximately 8 hours to detect 800 tie-points in the central area and 5 hours to detect 200 tie-points in the marginal area. The proposed approach took 80 seconds of VAX 11/780 CPU time and a few minutes of user time to select and register the features. As for the existing automatic approach, timing was not

published by the authors. This particular image pair is an easier case to solve since ice pack does not have the large motion range of fast ice as in the case shown in Fig.2. However, the search area for the template matching process has to be larger without a prediction model, which implies longer processing time.

SUMMARY AND CONCLUSIONS

The entire procedures, starting from the feature selection and registration to the display of final motion vectors, has been implemented using VAX 11/780 and RASTER TECHNOLOGY frame buffer model 125. Without intensive computation or laborious user interaction, this approach not only allows a detailed motion analysis of an individual sea-ice morphism but also analyzes sutructure and motion of global ice fields with high accuracy. Major future research areas include calculation of stress and speed motion map generation using a motion mosaic of multiple image pairs, multi-factor prediction model generation which utilizes climatological information as well as motion history of a given area, intelligent template matching where template matching is regressed from a highly confident area to a less confident area based on neighboring template matching results, and, finally, feature-based motion analysis instead of a template-based motion analysis via automatic feature selection and registration.

ACKNOWLEDGEMENT

The work described in this paper was carried out by the Jet Propulsion Laboratory, California Institute of Technology, under contract with the National Aeronautics and Space Administration.

REFERENCES

Curlander J C, Holt B, Hussey K 1985 Determination of sea-ice motion using digital SAR imagery. *IEEE Journal of Oceanic Engineering* OE 10(4): 358-367

Fily M, Rothrock D A 1986 Extracting sea ice data from satellite SAR imagery. *IEEE Transactions on Geoscience and Remote Sensing* GE-24(6): 849-854

Annals of Glaciology 9 1987
© International Glaciological Society

SNOW MAPPING AND CLASSIFICATION FROM LANDSAT THEMATIC MAPPER DATA

by

Jeff Dozier and Danny Marks

(Center for Remote Sensing and Environmental Optics, University of California, Santa Barbara, CA 93106, U.S.A.)

ABSTRACT

Use of satellite multi-spectral remote-sensing data to map snow and estimate snow characteristics over remote and inaccessible areas requires that we distinguish snow from other surface cover and from clouds, and compensate for the effects of the atmosphere and rugged terrain. Because our space-borne radiometers typically measure reflectance in a few wavelength bands, for climate modeling we must use inferences of snow grain-size and contaminant amount to estimate snow albedo throughout the solar spectrum. Although digital elevation data may be used to simulate typical conditions for a satellite image, precise registration of an elevation data set with satellite data is usually impossible. Instead, an atmospheric model simulates combinations of Thematic Mapper (TM) band radiances for snow of various grain-sizes and contaminant amounts. These can be recognized in TM images and snow can automatically be distinguished from other surfaces and classified into clean new snow, older metamorphosed snow, or snow mixed with vegetation.

INTRODUCTION

In attempting to use satellite multi-spectral remote-sensing data to map snow and estimate snow characteristics over remote and inaccessible areas, we are faced with several problems: (1) We must distinguish snow from other surface cover and from clouds; (2) We must compensate for the effects of the atmosphere and rugged terrain; (3) Our space-borne radiometers typically measure reflectance in a few wavelength bands, but for climate modeling we are interested in snow's reflectance throughout the solar spectrum. Thus, our objective is to map snow and classify it into albedo categories from an existing satellite, the Landsat Thematic Mapper (TM). We also wish to minimize the use of ancillary data. Although digital elevation data may be used to simulate typical conditions for a satellite image, precise registration of an elevation data set with satellite data is often impossible.

The term "mapping" means distinguishing snow from other surfaces or clouds. "Classification" means approximation

of snow grain-size and contaminant amount, such that snow spectral albedo can be calculated with a radiative-transfer model.

SPECTRAL SIGNATURE OF SNOW IN THEMATIC MAPPER BANDS

Table I, which uses data from Markham and Barker (1986), specifies the wavelength bands and saturation radiances for the Landsat-4 and Landsat-5 Thematic Mappers. For the solar part of the electromagnetic spectrum, the values of the exo-atmospheric solar irradiance at the mean Earth–Sun distance integrated over the wavelength bands are also given. The solar-irradiance data used are from Neckel and Labs (1984) and Iqbal (1983). In the last column of the table, the sensor saturation radiance is expressed as a percentage of the exo-atmospheric solar irradiance. If the product of the planetary reflectance and the cosine of the solar zenith angle exceeds this value, the sensor will saturate in this band.

The reflectance of snow can be modeled as a multiple-scattering radiative transfer problem, where the scattering properties of the grains are mimicked by some sort of "equivalent sphere" and near-field effects are assumed unimportant. Warren (1982) has discussed these issues in his review of the optical properties of snow. In the visible wavelengths (TM bands 1 and 2 especially), ice is highly transparent and snow reflectance is insensitive to grain-size but sensitive to modest amounts of absorbing impurities. In the near infra-red (TM4), ice is moderately absorptive and snow reflectance is insensitive to absorbing impurities but sensitive to grain-size. In TM5, ice is more strongly absorptive than in the shorter wavelengths and is more absorptive than water. Snow reflectance in this band is sensitive to grain-size only for very small radii; for most snow, reflectance will be near zero. Both ice and water clouds, however, will be appreciably brighter than snow, allowing for snow/cloud discrimination. Table II shows reflectance for pure snow in the TM bands, and Table III shows reflectances for ice and water clouds for comparison. These were computed by a delta-Eddington approximation

TABLE I. THEMATIC MAPPER RADIOMETRIC CHARACTERISTICS

Band	Wavelength range μm		Radiances W m^{-2} μm^{-1} sr^{-1} Landsat-4 L_{max}	L_{solar}	%	Landsat-5 L_{max}	L_{solar}	%
TM1	0.45	0.52	158.4	623.3	25.4	152.1	622.9	24.4
TM2	0.53	0.61	308.2	581.9	53.0	296.8	582.2	51.0
TM3	0.62	0.69	234.6	496.2	47.3	204.3	495.6	41.2
TM4	0.78	0.90	224.3	332.6	67.4	206.2	333.3	61.9
TM5	1.57	1.78	32.42	69.74	46.4	27.19	69.81	39.0
TM6	10.4	12.5	15.64	(thermal)		15.60	(thermal)	
TM7	2.10	2.35	17.00	23.74	71.6	14.38	23.72	60.6

TABLE II. TM INTEGRATED REFLECTANCE FOR
SNOW
PURE SEMI-INFINITE SNOW, $\theta_0 = 60°$
Optical grain radius
μm

Band	50	100	200	500	1000
TM1	0.992	0.988	0.983	0.974	0.963
TM2	0.988	0.983	0.977	0.964	0.949
TM3	0.978	0.969	0.957	0.932	0.906
TM4	0.934	0.909	0.873	0.809	0.741
TM5	0.223	0.130	0.067	0.024	0.011
TM7	0.197	0.106	0.056	0.019	0.010

TABLE III. TM INTEGRATED REFLECTANCE FOR
WATER AND ICE CLOUD, $\theta_0 = 60°$

Water cloud, 1 mm water
Optical droplet radius
μm

Band	1	2	5	10	20
TM5	0.891	0.866	0.769	0.661	0.547
TM7	0.784	0.750	0.650	0.481	0.345

Ice cloud, 1 mm water equivalent
Optical droplet radius
μm

Band	1	2	5	10	20
TM5	0.817	0.780	0.665	0.513	0.383
TM7	0.765	0.730	0.642	0.478	0.341

(Wiscombe and Warren 1980), which is one of the class of two-stream equations discussed in a later section. It is particularly useful for forward-scattering media, such as ice grains. The refractive-index data are Hale and Querry's (1973) for water and Warren's (1984) for ice, and the Mie-scattering approximation is from Nussenzveig and Wiscombe (1980). When absorbing impurities, dust or soot, are present in the snow, reflectances are reduced most in TM band 1, moderately in TM2, and less so in TM3 (Warren and Wiscombe 1980, Grenfell and others 1981). The effect of moderate concentration of impurities in TM bands 4, 5, and 7 is negligible. Unfortunately, when impurities are present, the reflectance in the visible bands is also sensitive to grain-size. We did not calculate Mie-theory adjustments for contamination to the pure snow reflectances, because of the uncertainties about how to treat impurities that are embedded in the ice grains (Bohren 1986). Instead, we note that the measurements of Grenfell and others (1981) show that albedo in the visible-wavelength TM bands is reduced by moderate contamination about 0.05 in TM1, 0.03 in TM2, and 0.02 in TM3.

The snow-albedo model calculates reflectances only for monochromatic wavelengths, and thus cannot be directly used to obtain band-integrated reflectances for the TM, because both the refractive index of ice and the spectral distribution of the incoming irradiance vary over the bands. For the present investigation, however, we assume we can effectively mimic band-averaged values by monochromatic optical properties.

The parameters that depend directly on wavelength and grain-size are the snow's single-scattering albedo $\tilde{\omega}_s$ and asymmetry parameter g_s; for finite-depth snow, the extinction efficiency Q_{ext} is also needed. These dependencies can be approximated with empirical equations as functions of the grain radius r.

$$\log(1 - \tilde{\omega}_s) = a_0 + a_{1/2}\sqrt{r} + a_1 r \qquad (1a)$$

$$g_s = b_0 + b_{1/2}\sqrt{r} + b_1 r \qquad (1b)$$

$$\log(Q_{ext} - 2) = c_0 + c_{1/2}\sqrt{r} + c_1 r. \qquad (1c)$$

Table IV shows the values of the coefficients for the seven reflective TM bands. Any coefficients not significant at the 10^{-6} confidence level are set to zero.

TABLE IV. COEFFICIENTS FOR SNOW OPTICAL
PROPERTIES IN TM BANDS

Coefficients in Equation (1a): $\log(1 - \tilde{\omega}_s) = f(r)$

Band	a_0	$a_{1/2}$	a_1 $\times 10^{-3}$
TM1	−14.3553	0.217190	−2.65574
TM2	−13.7736	0.221165	−2.71336
TM3	−12.5462	0.218364	−2.66035
TM4	−10.2352	0.217197	−2.70149
TM5	−3.72685	0.183880	−2.89506
TM7	−3.53802	0.178353	−2.86933

Coefficients in Equation (1b): $g_s = f(r)$

Band	b_0	$b_{1/2}$ $\times 10^{-3}$	b_1 $\times 10^{-5}$
TM1	0.885513	0.400541	−0.706325
TM2	0.885480	0.500842	−0.899701
TM3	0.885405	0.561945	−0.998832
TM4	0.885095	0.675243	−1.16128
TM5	0.866603	5.33367	−6.83010
TM7	0.874771	5.50804	−7.50705

Coefficients in Equation (1c): $\log(Q_{ext} - 2) = f(r)$

Band	c_0	$c_{1/2}$	c_1 $\times 10^{-3}$
TM1	−2.75928	−0.149413	1.83038
TM2	−2.62550	−0.152518	1.90884
TM3	−2.58394	−0.148083	1.81885
TM4	−2.40677	−0.149206	1.83269
TM5	−2.06955	−0.137662	1.60004
TM7	−2.57538	−0.0655741	0.0

USE OF DIGITAL ELEVATION DATA IN RADIATION CALCULATIONS

Most radiation calculations over terrain are made with the aid of digital elevation grids, whereby elevation data are represented by a matrix. In the USA, these are available as "Digital Elevation Models" from the US Geological Survey (Elassal and Caruso 1983). The 1 : 250 000 scale $1° \times 2°$ quadrangles are available for the entire USA at 63.5m grid resolution (0.01 in at map scale), and the 1 : 24 000 scale 7.5 min quadrangles are available at 30 m resolution for parts of the country.

From the DEMs, the slope angle S and exposure azimuth E from south can be calculated. We consider a right-handed coordinate system with x increasing toward south and y increasing toward east. The partial derivatives of elevation in the x and y directions are computed from finite differences, and the equations for S and E are given below. Note that the signs of the numerator and denominator allow E to be uniquely determined over $[-\pi, \pi]$.

$$\tan^2 S = (\partial z/\partial x)^2 + (\partial z/\partial y)^2 \qquad (2a)$$

$$\tan E = \frac{-\partial z/\partial y}{-\partial z/\partial x}. \qquad (2b)$$

The most important variable controlling the incident radiation on a slope in mountainous terrain is the local solar illumination angle θ_s, which is determined from the slope orientation, the solar zenith angle on a horizontal surface θ_0, and the Sun's azimuth ϕ_0.

$$\cos\theta_s = \cos\theta_0 \cos S + \sin\theta_0 \sin S \cos(\phi_0 - E). \qquad (3)$$

Dozier and others (1981) described a fast method to determine which points in a DEM are hidden from the Sun by a local horizon for a given solar azimuth. For each point, we calculate the horizon angle $H_{(\phi)}$ in any desired direction ϕ. The horizon can result either from "self-shadowing" or from adjacent ridges. For an unobstructed horizontal surface $H_{(\phi)} = \pi/2$.

For diffuse irradiance, only a part of the overlying hemisphere is visible. The "sky-view factor" V_d is the ratio of the diffuse-sky irradiance at a point to that on an unobstructed horizontal surface, i.e. $0 < V_d \leqslant 1$. We define an anisotropy factor η_d such that $\eta(\theta,\phi)L^{\downarrow} = L(\theta,\phi)$, where $L^{\downarrow}$ is the average down-welling radiance on a horizontal surface. Therefore η_d is normalized such that its hemispheric integral projected on to a horizontal surface is π, so for isotropic diffuse irradiance $\eta_d = 1$. V_d on a slope S with exposure E is given by

$$V_d = \frac{1}{\pi} \int_0^{2\pi} \int_0^{H_{(\phi)}} \eta_d(\theta,\phi)\sin\theta[\cos\theta\cos S + \sin\theta\sin S \cos(\phi - E)]d\theta d\phi.$$

(4)

If diffuse irradiance from the sky is isotropic, the inner integral above can be evaluated analytically. We therefore often use the approximation

$$V_d \approx \frac{1}{2\pi} \int_0^{2\pi} [\cos S \sin^2 H_{(\phi)} - \sin S \cos(\phi-E)(\sin H_{(\phi)}\cos H_{(\phi)} - H_{(\phi)})]d\phi.$$

(5)

By a similar "terrain view factor", it is also possible to account for reflected radiation from the surrounding terrain, but the formulation is more complicated because the isotropic assumption is almost always invalid (Arnfield 1982). Anisotropy results from differing illumination on the surrounding terrain and from geometric effects between the point and the surrounding terrain, even if the surfaces are Lambertian

$$V_t = \frac{1}{\pi} \int_0^{2\pi} \int_{H_{(\phi)}}^{\psi_{(\phi)}} \eta_t(\theta,\phi) \sin\theta[\cos\theta \cos S + \sin\theta\sin S \cos(\phi - E)]d\theta d\phi.$$

(6)

η_t accounts for the anistropy of the reflected or emitted radiation. The limits of integration for the inner integral are from the horizon downward to where a ray is parallel to the slope:

$$\psi_{(\phi)} = \arctan\left[\frac{-1}{\tan S \cos(\phi - E)}\right].$$

(7)

In the up-slope direction, $\cos(\phi - E)$ is negative, so $\psi_{(\phi)} < \pi/2$. In the down-slope direction, $\cos(\phi - E)$ is positive, so $\psi_{(\phi)} > \pi/2$. Across the slope, $\psi_{(\phi)} = \pi/2$. Rigorous calculation of V_t is difficult because it is necessary to consider every terrain facet that is visible from a point in order to calculate η_t. No-one has yet done it. We therefore note that V_d for an infinitely long slope is $(1 + \cos S)/2$ and use the approximation

$$V_t \approx \frac{1 + \cos S}{2} - V_d.$$

(8)

These terrain specifications allow us to specify boundary conditions for an atmospheric radiation calculation over mountainous terrain. However, any mapping or classification algorithm that requires that satellite data be precisely registered to accurate digital elevation data is severely constrained and probably doomed to failure. Digital elevation data in mountainous areas are often of poor quality, with considerable noise from the digitization process, and the differencing operations needed to calculate slope and exposure amplify this noise (See Figs 1 and 2). Thus DEMs can be used to simulate the effects of the combination of

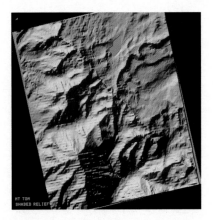

Fig.1. Shaded relief image of the Mount Tom area in the southern Sierra Nevada made from a DEM. The striping in the shaded relief image is from noise created when the DEM is made. Compare this image with Fig.2.

atmosphere, terrain, and surface reflectance, and they can be used for radiation models, where the results are integrated over a drainage basin. Moreover, registration between a satellite image and a DEM can be achieved closely enough so that the elevation of a pixel in the image can be known, but an algorithm that needs the slope and azimuth of a given pixel in order to interpret its multi-spectral radiometric signal imposes an impossible requirement, given the poor quality of available DEM data.

RADIANCE ABOVE THE ATMOSPHERE

The satellite measures upwelling radiance $L^{\uparrow}$ above the Earth's atmosphere, i.e. at optical depth $\tau = 0$. Interpretation of this signal depends on the interaction between the surface reflectance properties, the terrain, and the atmosphere. Because of the difficulties in dealing with the mountainous terrain, it makes little sense to use a sophisticated atmospheric model. Instead, we use a simplified "two-stream" model (Meador and Weaver 1980), which is defined by the following pair of differential equations for a homogeneous plane-parallel atmospheric layer:

$$\frac{dL^{\uparrow}}{d\tau} = \gamma_1 L^{\uparrow} - \gamma_2 L^{\downarrow} - \tilde{\omega}_a S_0 \gamma_3 e^{-\tau/\mu_0}$$

(9a)

$$\frac{dL^{\downarrow}}{d\tau} = \gamma_2 L^{\uparrow} - \gamma_1 L^{\downarrow} + \tilde{\omega}_a S_0 \gamma_4 e^{-\tau/\mu_0}$$

(9b)

where πS_0 is the exo-atmospheric solar irradiance incident at angle θ_0; $\mu_0 = \cos\theta_0$; and $\tilde{\omega}_a$ is the atmosphere's single-scattering albedo. Since the atmospheric layer is homogeneous, the γ values are by definition independent of optical depth. The γ values depend on the single-scattering albedo $\tilde{\omega}_a$, the scattering-asymmetry parameter g_a, and μ_0. How they are calculated is determined by the particular approximation to the scattering-phase function and the radiation-intensity distribution. Meador and Weaver (1980) gave expressions for seven different two-stream approximations.

For a single-layer atmosphere, the solution to Equation (9) depends on the boundary conditions. The usual top-boundary condition is that there is no diffuse irradiance at the top, i.e. $L^{\downarrow}_{(0)} = 0$. At the bottom (optical depth τ_0), however, the terrain effects must be included in the boundary condition, and no-one has yet found a rigorous solution to this problem. Here we resort to an *ad hoc* method. We assume that the atmospheric transmission and scattering properties are the same as over a flat surface with the same mean regional albedo ρ_a, so the lower-

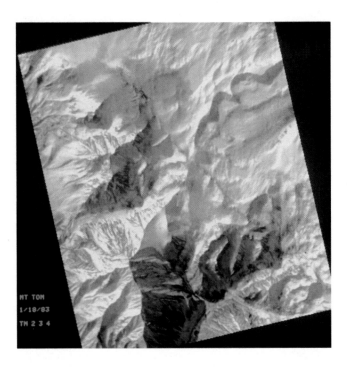

Fig.2. Thematic Mapper image registered to the DEM in Fig.1. Blue = TM2, green = TM3, and red = TM4. The topographic details are much sharper than in the image made from the DEM in Fig.1. The dimensions of this scene are 15 km × 15 km.

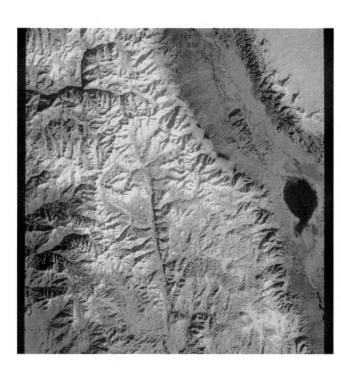

Fig.4. Same scene as in Fig.3, but blue = TM5, green = TM2, and red = TM4.

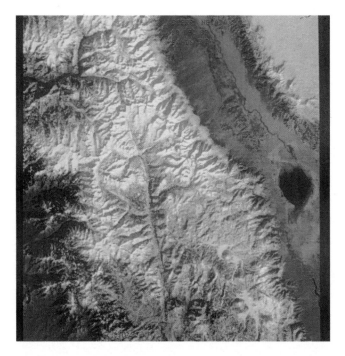

Fig.3. Thematic Mapper image of the southern Sierra Nevada. Blue = TM2, green = TM4, and red = TM3. The dimensions of the scene are 85 km × 80 km. It is sub-sampled such that only every sixth pixel is printed, because the size of our image display screen is 512 × 512.

Fig.6. Automatic mapping of snow-covered area in the southern Sierra Nevada scene using the clusters in Fig.5.

boundary condition for Equations (9a) and (9b) is

$$L^{\uparrow}(\tau_0) = \rho_g(\mu_0 S_0 + L^{\downarrow}(\tau_0)).\qquad(10)$$

The solutions for the upward radiance at the top $L^{\uparrow}(0)$ and the downward radiance at the bottom $L^{\downarrow}(\tau_0)$ are:

$$\frac{L^{\uparrow}(0)}{\mu_0 S_0} = \frac{e^{\xi\tau_0}X^+ + e^{-\xi\tau_0}X^- - 2\xi e^{-\tau_0/\mu_0}[\rho_g(P-1)-Q]}{e^{\xi\tau_0}(\xi - \gamma_2\rho_g + \gamma_1) + e^{-\xi\tau_0}(\xi + \gamma_2\rho_g - \gamma_1)},$$

$$(11a)$$

$$\frac{L^{\downarrow}(\tau_0)}{\mu_0 S_0} = \frac{2P\xi - e^{-\tau_0/\mu_0}(e^{\xi\tau_0}Y^+ + e^{-\xi\tau_0}Y^-)}{e^{\xi\tau_0}(\xi - \gamma_2\rho_g + \gamma_1) + e^{-\xi\tau_0}(\xi + \gamma_2\rho_g - \gamma_1)}.$$

$$(11b)$$

These equations define reflectance and transmittance for the layer. The other symbols are:

$$P = k(\alpha_1\mu_0 + \gamma_4),$$

$$Q = k(\alpha_2\mu_0 - \gamma_3),$$

$$\alpha_1 = \gamma_1\gamma_4 + \gamma_2\gamma_3,$$

$$\alpha_2 = \gamma_1\gamma_3 + \gamma_2\gamma_4,$$

$$\xi^2 = \gamma_1^2 - \gamma_2^2,$$

$$k = \frac{\tilde{\omega}_a}{1 - \xi^2\mu_0^2},$$

$$X^{\pm} = P[\rho_g(\xi \mp \gamma_1) \pm \gamma_2] - Q(\xi \mp \gamma_2\rho_g \pm \gamma_1),$$

$$Y^{\pm} = P(\xi \mp \gamma_2\rho_g \pm \gamma_1) \pm \gamma_2[\rho_g(P-1) - Q].$$

Because of scattering, the relationship between $L^{\uparrow}(0)$ and $L^{\uparrow}(\tau_0)$ depends on ρ_g. However, for a thin atmosphere, one can assume it is linear, and we find the relationship by evaluating Equations (11a) and (11b) for $\rho_g = 0$ and $\rho_g = 1$.

$L^{\uparrow}(\tau_0)$ is given by Equation (11b). For simplification, we assume that $\rho_g = \rho_{s(\mu_0)}$. This is reasonable for thin atmospheres, because direct irradiance is the largest component of the illumination.

ANALYSIS OF THEMATIC MAPPER SIGNALS FROM SNOW

The atmosphere/terrain radiation model described in the previous sections is combined with calculations of the spectral reflectance of snow to simulate radiance at the top of the atmosphere for a range of snow grain-sizes, contamination amounts, and terrain conditions. The atmosphere used in the simulations is the US Standard with a rural background aerosol, 23 km surface visibility, and 50% relative humidity. The optical depths are adjusted to a surface pressure of 650–700 mbar. Table V lists the atmospheric optical properties in the reflective TM bands.

TABLE V. ATMOSPHERIC PROPERTIES USED FOR CALCULATION

Band	τ_0	$\tilde{\omega}_a$	g_a
TM1	0.4–0.6	0.93–0.97	0.44–0.50
TM2	0.25–0.35	0.87–0.92	0.50–0.55
TM3	0.2–0.3	0.87–0.92	0.55–0.58
TM4	0.15–0.25	0.75–0.85	0.61–0.63
TM5	0.1–0.15	0.55–0.65	0.66–0.68
TM7	0.05–0.1	0.4–0.6	0.66–0.68

τ_0 = optical depth.

$\tilde{\omega}_a$ = single-scattering albedo.

g_a = asymmetry parameter.

Figs 3 and 4 show a TM image of the southern Sierra Nevada on 10 December 1982. The solar-zenith angle θ_0 is 64.6° and the azimuth ϕ_0 is 31.9°. For these values we calculated top-of-atmosphere radiance for 500 randomly chosen locations in the digital elevation grid, covering the range of terrain characteristics listed in Table VI, each for a random grain radius between 50 and 1500 μm. For bands

$$L^{\uparrow}(0) \approx \frac{\mu_0 S_0\{2Q\xi e^{-\tau_0/\mu_0} + e^{\xi\tau_0}[\gamma_2 P - Q(\xi + \gamma_1)] - e^{-\xi\tau_0}[\gamma_2 P + Q(\xi - \gamma_1)]\}}{e^{\xi\tau_0}(\xi + \gamma_1) + e^{-\xi\tau_0}(\xi - \gamma_1)} +$$

$$+ \frac{2\xi L^{\uparrow}(\tau_0)[e^{\xi\tau_0}(\xi - \gamma_2 + \gamma_1) + e^{-\xi\tau_0}(\xi + \gamma_2 - \gamma_1)]}{e^{2\xi\tau_0}(\xi + \gamma_1)(\xi - \gamma_2 + \gamma_1) + e^{-2\xi\tau_0}(\xi - \gamma_1)(\xi + \gamma_2 - \gamma_1) + 2\gamma_2(\gamma_1 - \gamma_2)}.$$

$$(12)$$

Now, to find $L^{\uparrow}(\tau_0)$ for a surface in rugged terrain, we assume that the same transmission relation holds. The upwelling radiance for a surface with direct albedo $\rho_{s(\mu_s)}$ at illumination angle θ_s, diffuse albedo ρ_d, and albedo of the "surrounding" terrain ρ_g, is the sum of several terms: reflected direct irradiance, reflected diffuse irradiance, and irradiance reflected from adjacent slopes toward the point. The sum of these must be multiplied by $\cos S$ to find the mean upwelling radiance projected on a horizontal plane. Therefore, using the simplification in Equation (8) and letting $\mu_s = \cos\phi_s$, we find the necessary value to insert $L^{\uparrow}(\tau_0)$ in Equation (12) above

TM1, TM2, and TM3, the simulations included clean, moderately contaminated, and dirty snow.

TABLE VI. TERRAIN CHARACTERISTICS OF SIERRA NEVADA IMAGE SAMPLE

Variable	Mean	Std dev	Min	Max
Elevation (m)	3306	337	2182	4063
Slope (°)	30	15	0.2	68
$\cos\theta_s$	0.41	0.29	0.0	0.98
V_d	0.85	0.10	0.35	1.00
V_τ	0.07	0.06	0.00	0.34

$$L^{\uparrow}(\tau_0) \approx \cos S\{S_0 e^{-\tau_0/\mu_0}[\mu_s\rho_{s(\mu_s)} + \mu_0\rho_g\rho_d(1 + \cos S - V_d)/2] + L^{\downarrow}(\tau_0)\rho_d[V_a + \rho_g(1 + \cos S - V_d)/2]\}$$

$$(13)$$

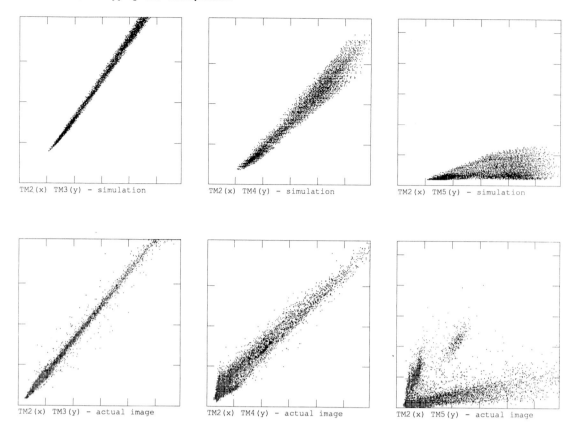

Fig.5. (Top) Theoretical cluster plots of pair-wise combinations of Thematic Mapper top-of-atmosphere radiances for snow surface corresponding to the scene and the topography in Figs 3 and 4. The axes limits are the saturation radiances for the sensor given in Table I. (Bottom) Actual cluster plots of pair-wise combinations of Thematic Mapper top-of-atmosphere radiances for the southern Sierra Nevada, corresponding to the scene in Figs 3 and 4.

The top row in Fig.5 shows theoretical cluster plots of the radiance values from this simulation, paired by some combinations of TM bands. The axes are scaled to correspond to the band-saturation values listed in Table I. From these graphs, it is apparent that the top-of-atmosphere radiance over a snow layer falls into some convenient envelopes for a range of grain-sizes and contamination amounts. The most useful clusters are TM1 and TM4 or TM5 for the shadowed areas, where a threshold brightness distinguishes snow in shadow from other surfaces, and TM2 and TM5, where a threshold brightness distinguishes clouds, vegetation, soil, or rocks from snow in the sunlit areas. Note that most of the values in TM1 saturate, so only a small part of the cluster is on the graph. TM2 and TM3 are redundant. The bottom row in Fig.2 shows actual cluster plots from the TM scene in Figs 4 and 5. These can be used to map snow automatically from the corresponding TM image, without registering the image to a digital elevation grid. A threshold value is chosen from the image for the lowest TM1 radiance for snow in the shadows ($\sim$70 W m^{-2} μm^{-1} sr^{-1} for this mid-winter scene). Then TM2 and TM5 are used to distinguish snow from clouds or other surfaces in the sunlight. In this midwinter scene snow is characterized by:

$$L_{TM5} \leqslant L_{TM2}/25. \qquad (14)$$

Fig.6 shows a map of the snow-covered area made by these criteria. The colours in the snow-covered area are stretched, so that the brightest whites represent new, fine-grained snow, the brownish tints represent older meta-morphosed snow, and the purple areas represent vegetation growing above the snow cover.

Because of the low saturation value in band TM1, it is not feasible to determine snow-contamination amount. The differences between dirty snow and clean snow are not great enough to overcome the other sources of signal variation in mountainous terrain.

CONCLUSIONS

The multi-spectral signature from the Landsat Thematic Mapper can be used effectively to map snow and classify it into albedo categories without requiring that the satellite image be registered to digital elevation data. Thus it is possible to use such data for snow-surface energy-balance models in alpine areas.

ACKNOWLEDGEMENTS

The research was supported by the National Aeronautics and Space Administration (Grant NAS5-28770) and the California Air Resources Board (Grant A3-106-32).

REFERENCES

Arnfield A J 1982 Estimation of diffuse irradiance on sloping, obstructed surfaces: an error analysis. *Archives for Meteorology, Geophysics and Bioclimatology* B-30: 303-320

Bohren C F 1986 Applicability of effective-medium theories to problems of scattering and absorption by nonhomogeneous atmospheric particles. *Journal of the Atmospheric Sciences* 43(5): 468-475

Dozier J, Bruno J, Downey P 1981 A faster solution to the horizon problem. *Computers and Geosciences* 7(2): 145-151

Elassal A A, Caruso V M 1983 *Digital elevation models.* Alexandria, VA, US Geological Survey (Circular 895-B)

Grenfell T C, Perovich D K, Ogren J A 1981 Spectral albedos of an alpine snowpack. *Cold Regions Science and Technology* 4(2): 121-127

Hale G M, Querry M R 1973 Optical constants of water in the 200-nm to 200μm wavelength region. *Applied Optics* 12: 555-563

Iqbal M 1983 *An introduction to solar radiation.* Toronto, Academic Press

Markham B L, Barker J L 1986 Landsat MSS and TM post-calibration dynamic ranges, exoatmospheric reflectances and at-satellite temperatures. *EOSAT Landsat Technical Notes* 1: 2–8

Meador W E, Weaver W R 1980 Two-stream approximations to radiative transfer in planetary atmospheres: a unified description of existing methods and a new improvement. *Journal of the Atmospheric Sciences* 37(3): 630-643

Neckel H, Labs D 1984 The solar radiation between 3300Å and 12500Å. *Solar Physics* 90: 205-258

Nussenzveig H M, Wiscombe W J 1980 Efficiency factors in Mie scattering. *Physical Review Letters* 45(18): 1490-1494.

Warren S G 1982 Optical properties of snow. *Reviews of Geophysics and Space Physics* 20(1): 67-89

Warren S G 1984 Optical constants of ice from the ultraviolet to the microwave. *Applied Optics* 23(8): 1206-1225

Warren S G, Wiscombe W J 1980 A model for the spectral albedo of snow. II: snow containing atmospheric aerosols. *Journal of the Atmospheric Sciences* 37(12): 2734-2745

Wiscombe W J, Warren S G 1980 A model for the spectral albedo of snow. I: pure snow. *Journal of the Atmospheric Sciences* 37(12): 2712-2733

Annals of Glaciology 9 1987
© International Glaciological Society

CHARACTERIZATION OF SNOW AND ICE REFLECTANCE ZONES ON GLACIERS USING LANDSAT THEMATIC MAPPER DATA

by

D.K. Hall and J.P. Ormsby

(Hydrological Sciences Branch, Code 624, NASA/Goddard Space Flight Center, Greenbelt, MD 20771, U.S.A.)

with

R.A. Bindschadler

(Oceans and Ice Branch, Code 671, NASA/Goddard Space Flight Center, Greenbelt, MD 20771, U.S.A.)

and

H. Siddalingaiah

(SEA Corporation, 14700 Mentmore Place, Gaithersburg, MD 20878, U.S.A.)

ABSTRACT

Landsat Thematic Mapper (TM) data have been analyzed to study the reflectivity characteristics of three glaciers: the Grossglockner mountain group of glaciers in Austria and the McCall and Meares Glaciers in Alaska, USA. The ratio of TM band 4 (0.76–0.90 μm) to TM band 5 (1.55–1.75 μm) was found to be useful for enhancing reflectivity differences on the glaciers. Using this ratio, distinct zones of similar reflectivity were noted on the Grossglockner mountain group of glaciers and on the Meares Glacier; no distinct zones were observed on the McCall Glacier. On the TM subscene containing the Grossglockner mountain group of glaciers, 28.2% of the glacierized area was determined to be in the zone corresponding most closely to the ablation area, and 71.8% with the location of the accumulation area. Using these measurements, the glacier system has an accumulation area ratio (AAR) of approximately 0.72. Within the accumulation area, two zones of different reflectivity were delineated. Radiometric surface temperatures were measured using TM band 6 (10.4–12.5 μm) on the Grossglockner mountain group of glaciers and on the Meares Glacier. The average radiometric surface temperature of the Grossglockner mountain group of glaciers decreased from 0.9 ± 0.34 °C in the ablation area, to −0.9 ± 0.83 °C in the accumulation area.

INTRODUCTION

Glaciers and ice sheets are comprised of an ablation and an accumulation area. Within these areas, several facies are present. Facies display a distinctive group of characteristics that reflect the environment under which the snow or ice was formed. The ablation area consists of exposed ice during the summer and contains the ice facies. The accumulation area can be sub-divided into the wet-snow facies, the percolation facies and the dry-snow facies (C.S. Benson, personal communication). Development of the discrete facies is directly related to the temperature regime and mass balance of a glacier or ice sheet. Net loss by melting occurs in the ice facies. In the wet-snow facies all snow deposited since the end of the previous summer is raised to 0 °C and wetted by the end of the melt season. The superimposed ice zone consists of a mass of ice which can overlap both the ice facies and the wet-snow facies. The annual increment of new snow is not completely wetted nor does its temperature reach the melting point in the percolation facies. Negligible melting occurs in the dry-snow facies (Benson 1962; Benson and Motyka 1978). At least

some of these facies can be detected using Landsat Multispectral Scanner (MSS) and Thematic Mapper (TM) data. In this paper, the use of Landsat TM data for detecting glacier surface conditions and for relating these conditions to the glacier facies is studied through analysis of a glacier group in Austria and two glaciers in Alaska, USA.

BACKGROUND

The Landsat TM sensor acquires data in seven spectral bands. TM bands 1 through 5 and 7 are in the visible, near-infrared and middle infrared wavelength regions and have a spatial resolution of each picture element (pixel) of approximately 30 m. TM band 6, a thermal infrared band, is sensitive to infrared surface temperature and has a resolution of 120 m. Spectral reflectivity of snow as determined from the visible, near- and middle-infrared bands is dependent on snow parameters such as grain size and impurity content of the surface layers of the snow (Dozier 1984).

Williams (1983[a], [b], 1987) found that computer-enhanced Landsat MSS images were useful for analysis of ice and snow reflectivity differences present on Vatnajökull, an ice cap in Iceland, especially when Landsat MSS data acquired at the end of the summer melt-season were custom processed. In addition, Crabtree (1976) found that Landsat MSS imagery of an outlet glacier, Merkurjökull, of the Mýrdalsjökull ice cap, Iceland, showed a reflectivity boundary that was attributed to differences in glacier surface conditions. There has also been considerable evidence that Landsat MSS data can show the location of the equilibrium line (Hall and Ormsby 1983).

The MSS band 7 (0.8 – 1.1 μm) is located in a wavelength region which is close to the TM band 4 (0.76 – 0.90 μm) region and has been used for detection of surface water on snow and ice (Holmgren and others 1975; Rango and others 1975). TM band 4 has been found by Dozier (1984) to be sensitive to snow grain size and TM band 2 (0.53 – 0.61 μm) to be sensitive to contamination.

STUDY AREAS

TM digital data of the Grossglockner mountain group of glaciers in the eastern Austrian Alps, the Meares Glacier in the Chugach Mountains in southern Alaska, and the McCall Glacier in the Brooks Range of Alaska, have been analyzed.

The Grossglockner mountain group of glaciers is located

at approximately 47°10'N, 12°45'E in the Noric Alps of Austria. The TM scene (50155-09272) of the Grossglockner group was acquired on 3 August 1984.

The TM scene (50518-20372) of the Meares Glacier in the eastern part of the Chugach Mountains was acquired on 1 August 1985 and is centered at approximately 61°30'N. 148°30'W.

McCall Glacier is located in the Romanzoff Mountains of the Brooks Range at 68°19'N, 143°48'W in northern Alaska. The McCall Glacier TM scene (50196-20474) was acquired on 13 September 1984.

RESULTS

TM band 4 (0.76–0.90 μm) was found to show the greatest variability of the 6 reflective bands, in spectral response in the glacierized areas. Much of the TM band 4 variability in spectral response is caused by snow grain size difference in the accumulation area of the glaciers, and melting or refrozen, previously melted snow. The spectral response pattern of TM band 2 (0.52–0.60 μm) generally follows that of TM band 4 but detector saturation is more common in band 2 over snow-covered areas. TM band 5 (1.55–1.75 μm) is quite useful for distinguishing between clouds and snow, and also shows subtle surface reflectivity differences on the glaciers. TM band 6 (10.4–12.5 μm), the thermal band, is useful for measuring radiometric surface temperature and detecting high cirrus clouds over snow and ice.

Using the computer compatible tapes (CCTs), the contrast between imaged features can often be enhanced by band ratioing (Moik 1980). This technique is particularly useful in eliminating the intensity variations caused by shadows. The ratio of TM band 4 to TM band 5 produces an image product that enhances snow and ice features because of the large difference in spectral response in snow and ice features between band 4 where high digital numbers (DNs, a measure of spectral reflectance) are common and band 5 where low DNs characterize snow and ice. Contrast enhancement is especially evident in the accumulation area of the glaciers, where the difference between the TM band 4 and 5 spectral response is the greatest.

Grossglockner mountain group of glaciers

Observations of TM imagery and transects across the Grossglockner mountain group of glaciers using TM bands 2, 4 and 5 digital data from CCTs reveal that there are three separate zones in which spectral reflectivity is distinctive. These zones relate to differences in snow and ice surface conditions, e.g. presence of surface water and differences in snow grain size. Fig.1 is an image processed by employing the TM band 4/5 ratio and assigning colors according to ranges of DNs as seen in Table I. Zone I is within the ablation area or ice facies and may be underestimated due to the similarity in DN between the debris-covered margin of the ice facies and the background. The snow line delineates the ablation area from Zone II

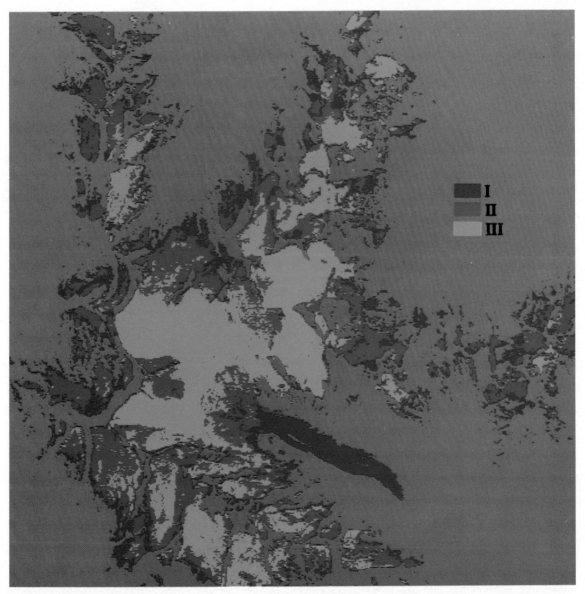

Fig.1. Image obtained from ratioing TM bands 4 and 5 (4/5) and assigning colors to reflectance zones of the Grossglockner mountain group of glaciers; Landsat TM image (50155-09272) was acquired on 3 August 1984. Zone I is believed to correspond to the ablation area and Zones II and III are within the accumulation area.

TABLE I.   STATISTICS FOR THE GROSSGLOCKNER MOUNTAIN GROUP OF GLACIERS USING THE TM BAND 4/5 RATIO

Zone	Number of pixels	Mean DN	Range of DN	Standard deviation	Area in km²	Percent area
I	24 270	106.4	73-144	21.02	21.84	28.18
II	35 595	190.3	145-226	23.61	32.04	41.34
III	26 239	244.8	227-255	9.31	23.61	30.48
TOTALS	86 104				77.49	100.00

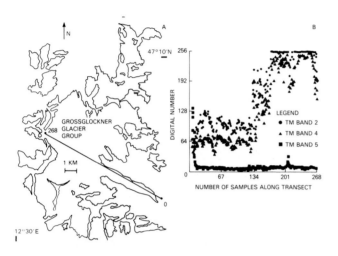

Fig.2.   The location of the transect is shown in A. The DNs which correspond to spectral reflectance are shown in B for the Grossglockner mountain group of glaciers; Landsat TM image (50155-09272) was acquired on 3 August 1984.

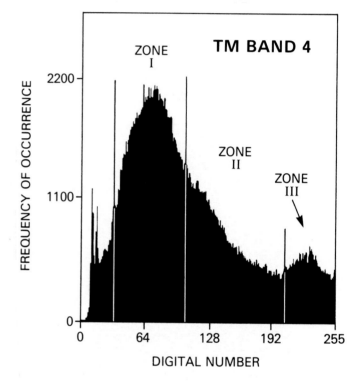

Fig.3.   Histogram showing the frequency of occurrence of each DN for TM band 4 of the Grossglockner mountain group of glaciers; Landsat TM image (50155-09272) was acquired on 3 August 1984.

which is believed to represent some or all of the wet-snow facies. This is quite possibly an area of slush. Zone III may be an area of fresh snow as there was a late-season snowfall in the higher elevations of these mountains (H. Rott, personal communication). The small crystals characterizing fresh snow are highly reflective in TM band 4.

A transect across the Grossglockner mountain group of glaciers using TM bands 2, 4 and 5 is shown in Fig.2. Note that some detector saturation occurs in TM bands 2 and 4. In Fig.2B the relatively low DNs for TM band 4 in the ablation area are due to actively melting ice on the glacier tongue, while the highest DNs correspond to that portion of the accumulation area in which the snow that overlies the glacier ice was apparently new at the time of the overpass. A small increase in reflectivity upglacier can be seen in TM band 5 data with DNs being the highest toward the end of the transect near the highest portions of the glacier system.

Histograms showing the frequency of occurrence of each DN for TM bands 4 and 5, for the Grossglockner mountain group of glaciers are shown in Figs 3 and 4, respectively. In Fig.3, Zones I and III are clearly distinguishable from other features. Zone II appears more as a transition region with no obvious central peak. On the TM band 5 histogram (Fig.4) the snow and ice are shown to have very low reflectivities (DN ≤ 30) and low variability in reflectivity.

The number of pixels comprising each zone as shown in Fig.1 was determined (Table I). Note that the largest surface area in the Grossglockner mountain group of glaciers is comprised of Zone II. It is important to note that the range of DN selected to delineate each zone governs the boundaries and thus the percentage area in each zone.

The accumulation area ratio (AAR), the accumulation area divided by the area of the entire glacier, is 0.72 as calculated roughly for the Grossglockner mountain group of glaciers from the data given in Table I. An AAR of about 0.70 corresponds to a net mass balance of zero for temperate mountain glaciers (Paterson 1981). (The fact that the TM data were acquired on 3 August 1984 instead of at the end of the melt season makes the calculation of AAR less precise.)

Meares Glacier

Using the 1 August 1985 TM data, Meares Glacier shows two distinct areas of similar reflectivity which appear to correspond to the ablation area (ice facies) and a portion of the wet-snow facies. Fig.5 shows a transect across the Meares Glacier. The variability in spectral reflectance in TM bands 2 and 4 in Fig.5B in the central part of the transect results from an icefall in Meares Glacier in which crevassing is present. On this subscene there is apparently no snow present that could be considered to be new.

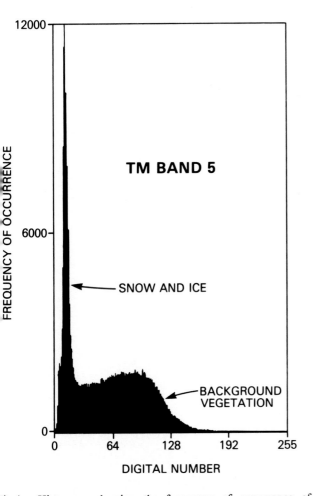

Fig.4. Histogram showing the frequency of occurrence of each DN for TM band 5 of the Grossglockner mountain group of glaciers; Landsat TM image (50155-09272) was acquired on 3 August 1984.

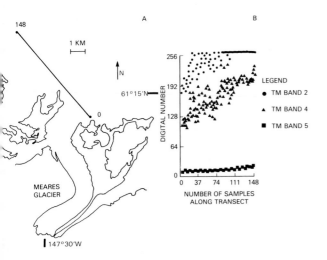

Fig.5. The location of the transect is shown in A. The DNs which correspond to spectral reflectance are shown in B for Meares Glacier; Landsat TM image (50518-20372) was acquired on 1 August 1985.

McCall Glacier

Transects across McCall Glacier using the 13 September 1984 TM digital data reveal that the reflectivity variations in TM band 4 are generally quite small. Wendler and others (1975) reported that the entire annual increment of snow on McCall Glacier reached $0\,^{\circ}C$ by the end of the melt season in 1969. Similar conditions have been observed in other years on McCall Glacier (C.S. Benson, personal communication) and this probably occurred during the summer of 1984. At the glacier terminus a large decrease in reflectivity is observed. This drop in reflectivity is probably due to the presence of surficial morainic debris on that part of the glacier.

RADIOMETRIC SURFACE TEMPERATURES

The TM-derived radiometric surface temperature can be calculated from the "at-satellite" infrared temperature recorded by the TM band 6 detectors by using a two-step process in which both the absolute calibration of the band 6 detectors and atmospheric conditions are considered (Schott and Volchok 1985). A clear, dry atmosphere is assumed for the Grossglockner and Meares scenes and has been assumed in the calculation used to derive the radiometric surface temperature from the "at-satellite" temperature; an emissivity value of 1.0 is assumed for the glacier surface (J. Barker, personal communication). The TM band 6-derived radiometric surface temperatures can be expressed in $^{\circ}C$ and are calculated from the DNs by using the following formula:

$$T = 1260.56 \left[\ln\left[\frac{49.836}{0.0338 + 0.005626\,Q_{cal}} + 1\right]\right]^{-1} - 273.15 \tag{1}$$

where Q_{cal} is the digital number after calibration (J. Barker, personal communication).

The average radiometric surface temperatures determined from TM band 6 for each zone and associated standard deviations are shown in Table II for zones I, II, and III for the Grossglockner scene, and zones I and II for the Meares scene. The radiometric surface temperature differences are small with approximately $1\,^{\circ}C$ difference between the ablation and accumulation areas on the Grossglockner group. Because atmospheric conditions are different in Austria from those in Alaska, the radiometric surface temperatures between scenes may not be comparable. However, relative differences in within-scene radiometric surface temperatures are considered to be meaningful (J. Barker, personal communication). Small errors in the radiometric surface temperatures may result from the assumption of unity for the emissivity of the ice and snow surfaces. In fact, the grain size variability that is believed to be present between facies on a glacier will result in different emissivities. However, Dozier and Warren (1982) report snow emissivity variability to be small, from 0.985 to 0.990 for all snow grain sizes, as determined from model calculations.

CONCLUSIONS

Landsat TM data have been analyzed to study the reflectivity variations on glaciers in order to relate TM spectral signatures to glacier facies. Results have shown that the TM band 4/5 ratio permits a good delineation to be made between areas having different surface melt histories and grain sizes on the Grossglockner mountain group of glaciers in Austria and Meares Glacier in southern Alaska.

TABLE II. AVERAGE RADIOMETRIC SURFACE TEMPERATURES ($^{\circ}C$) AND STANDARD DEVIATIONS AS DETERMINED FROM TM BAND 6 (n = 15)

	Zone I	Zone II	Zone III
Grossglockner	0.9 ± 0.34	-0.4 ± 0.64	-0.9 ± 0.83
Meares Glacier	0.1 ± 0.54	-1.9 ± 0.79	(outside image area)

The reflectivity differences between the ablation and accumulation areas are pronounced. Within the accumulation area, two zones can be delineated spectrally on the Grossglockner mountain group of glaciers. Extensive surface melt on the McCall Glacier may have caused the reflectance to be similar over much of the glacier surface. In addition, glacier radiometric surface temperatures which are pertinent to the development and distribution of the facies have been calculated using TM band 6 data for the Grossglockner mountain group of glaciers and the Meares Glacier.

ACKNOWLEDGEMENTS

The authors would like to thank Dr Carl S. Benson of the Geophysical Institute of the University of Alaska, and Dr Richard S. Williams, Jr of the US Geological Survey in Reston, VA, for their reviews of this paper; Dr John Barker of NASA/GSFC for his comments on the calculation of surface temperature using TM data; and Dr Helmut Rott of the Institut für Meteorologie und Geophysik, Innsbruck, Austria for his comments on conditions of the Grossglockner mountain group of glaciers during the summer of 1984.

REFERENCES

Benson C S 1962 Stratigraphic studies in the snow and firn of the Greenland ice sheet. *US Army Snow, Ice and Permafrost Research Establishment. Research Report* 70

Benson C S, Motyka R J 1978 *Glacier-volcano interactions on Mt. Wrangell, Alaska.* Fairbanks, AK, University of Alaska. Geophysical Institute (Annual Report 1977-78)

Crabtree R D 1976 Changes in the Mýrdalsjökull ice cap, south Iceland: possible uses of satellite imagery. *Polar Record* 18(112): 73-76

Dozier J 1984 Snow reflectance from Landsat 4 thematic mapper. *IEEE Transactions on Geoscience and Remote Sensing* GE-22: 323-328

Dozier J, Warren S G 1982 Effect of viewing angle on the infrared brightness temperature of snow. *Water Resources Research* 18(5): 1424-1434

Hall D K, Ormsby J P 1983 Use of SEASAT synthetic aperture radar and LANDSAT multispectral scanner subsystem data for Alaskan glaciology studies. *Journal of Geophysical Research* 88(C3): 1597-1607

Holmgren B, Benson C, Weller G 1975 A study of the breakup on the Arctic slope of Alaska by ground, air and satellite observations. *In* Weller G, Bowling S A (eds) *Climate of the Arctic. Proceedings of the Twenty-Fourth Annual Alaska Science Conference, August 15-17, 1973.* Fairbanks, AK, University of Alaska: 358-366

Moik J G 1980 *Digital processing of remotely sensed images.* Washington, DC, NASA Scientific and Technical Information Branch (NASA SP-431)

Paterson W S B 1981 *The physics of glaciers. Second edition.* Oxford etc, Pergamon Press

Rango A, Salomonson V V, Foster J L 1975 Employment of satellite snowcover observations for improving seasonal runoff estimates. *In* Rango A (ed) *Operational applications of satellite snowcover observations.* Washington, DC, National Aeronautics and Space Administration: 157-174

Schott J R, Volchok W J 1985 Thematic mapper thermal infrared calibration. *Photogrammetric Engineering and Remote Sensing* 51(9): 1351-1357

Wendler G, Benson C, Fahl C, Ishikawa N, Trabant D, Weller G 1975 Glacio-meteorological studies of McCall Glacier. *In* Weller G, Bowling S A (eds) *Climate of the Arctic. Proceedings of the Twenty-Fourth Annual Alaska Science Conference, August 15-17, 1973.* Fairbanks, AK, University of Alaska: 334-338

Williams R S Jr 1983[a] Remote sensing of glaciers. *In Manual of remote sensing. Second edition.* Falls Church, VA, American Society for Photogrammetry and Remote Sensing: 1852-1868

Williams R S Jr 1983[b] Satellite glaciology of Iceland. *Jökull* 33, 1982: 3-12

Williams R S Jr 1987 Satellite remote sensing of Vatnajökull, Iceland: a review. *Annals of Glaciology* 9: 127-135

Annals of Glaciology 9 1987
International Glaciological Society

SNOW AND ICE STUDIES BY THEMATIC MAPPER AND MULTISPECTRAL SCANNER LANDSAT IMAGES

by

Olav Orheim

(Norsk Polarinstitutt, P.O. Box 158, 1330 Oslo Lufthavn, Norway)

and

Baerbel K. Lucchitta

(U.S. Geological Survey, 2255 North Gemini Drive, Flagstaff, AZ 86001, U.S.A.)

ABSTRACT

Digitally enhanced Landsat Thematic Mapper (TM) images of Antarctica reveal snow and ice features to a detail never seen before in satellite images. The six TM reflective spectral bands have a nominal spatial resolution of 30 m, compared to 80 m for the Multispectral Scanner (MSS). TM bands 2–4 are similar to the MSS bands. TM infra-red bands 5 and 7 discriminate better between clouds and snow than MSS or the lower TM bands. They also reveal snow features related to grain-size and possibly other snow properties. These features are not observed in the visible wavelengths. Large features such as flow lines show best in the MSS and lower TM bands. Their visibility is due to photometric effects on slopes. TM thermal band 6 has a resolution of 120 m. It shows ground radiation temperatures and may serve to detect liquid water and to discriminate between features having similar reflectivities in the other bands, such as blue ice.

Repeated Landsat images can be used for sophisticated glaciological studies. By comparing images from 1975 and 1985, flow rates averaging 0.72 km a^{-1}, and mean longitudinal and transverse strains of respectively 1.3 × 10^{-4} a^{-1} and 130 × 10^{-4} a^{-1} have been measured for Jutulstraumen, Dronning Maud Land.

INTRODUCTION

The launch of the Landsat satellite series, beginning in 1972, provided a new instrument for glaciological studies which has proved especially valuable in remote areas. Thus Landsat 1, 2, and 3 provided MSS (Multispectral Scanner) coverage of Antarctica (north of the satellite limit at lat. 81°S.). These images have four spectral band passes with a nominal spatial resolution (pixel size) of about 80 m. Because they have various internal distortions caused by roll, pitch, and yaw of the spacecraft during picture acquisition, it is difficult to produce controlled rectified maps even from the digital data unless a large number of fixed ground-control points are available for geometric corrections (Ødegaard and Welle 1982).

Further images of Antarctica were obtained by Landsat 4 and 5. These include both MSS and TM (Thematic Mapper) images. The latter have additional spectral bands and a spatial resolution of about 30 m; they also have a thermal band which has a resolution of 120 m. Landsat 4 and 5 images are of high quality geometrically and can be placed into map projections with fewer ground-control points.

The first glaciological studies of MSS data tended to work with standard photographic products. Digital analyses, including enhancement techniques, have become more common in recent years and are used for this report.

This paper presents, we believe, the first analyses of TM data of Antarctica (or of any other large ice mass), and discusses features displayed in different TM bands. The data from TM are much more useful than those from MSS, not only because of the three times higher resolution, but also because the expanded spectral range gives additional information. Because the Landsat systems may be unfamiliar to some readers we preface our discussion with some general comments. The paper further compares overlapping images acquired in 1975 and 1985, notes various transient and stationary features of the Jutulstraumen area, and presents values of flow rates and longitudinal and transverse strains (Fig.1).

THE LANDSAT DATA SETS

Hundreds of images of Antarctica were obtained by Landsat 1, 2, and 3. A nearly complete coverage of the northern part of Dronning [Queen] Maud Land was obtained as a result of NASA project 28550 (Orheim, principal investigator). This project generated 67 scenes, which were mostly collected in the 1975–76 season, and which have relatively low cloud cover. Various unenhanced photographic prints of those scenes were investigated (Orheim 1978). Recently, the US Geological Survey acquired about 170 computer-compatible tapes (CCTs) of the early Landsat MSS images, including some of Dronning Maud Land (Lucchitta and others 1985). One of these scenes, Landsat 2 No. 2281-07424 recorded on 30 October 1975, is discussed here.

Fifty-seven images from central and western Dronning Maud Land were acquired in early 1985 by Landsat 5. Most of these were recorded both by the MSS and TM systems, and most have low cloud cover. These scenes were requested by Norsk Polarinstitutt, which carried out a simultaneous dedicated ground programme in connection with the Norwegian Antarctic Research Expedition (NARE) 1984–85.

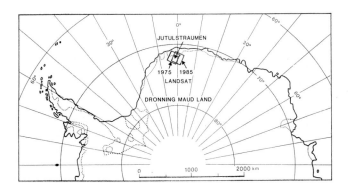

Fig.1. Index map, showing the location of the study area and images.

A survey team determined geographical coordinates of points − such as small isolated nunataks − that were especially selected to be readily identified on Landsat images. Some of the CCTs of these images have been bought by Norsk Polarinstitutt and are being used in the production of new maps. One scene, No. 5034407520, recorded on 8 February 1985, is discussed in this article.

DATA PROCESSING AND DIGITAL ENHANCEMENT

The present study was conducted with digitally enhanced images. The TM data set for one scene comprises 12 CCTs, compared to two CCTs for the MSS set. For this reason, the TM data are divided into four quadrants, numbered 1, 2, 4, and 3 reading clockwise from the upper left quadrant. Image processing for the Landsat 2 MSS scene included routine procedures such as radiometric corrections, noise and striping removal, and standard geometric corrections. Further processing included haze removal and linear stretching. For Landsat 5, some routine procedures were already incorporated in purchased CCTs.

Linear stretches were applied to individual TM and MSS bands after inspection of grey-value histograms. Special stretches, and in places filters, were applied to bring out detail in snow and ice. Stretches in TM bands 5 and 7 may be severe because most of the data are concentrated at the lower end of the grey scale. Therefore, TM bands 5 and 7 images appear to have degraded resolution when compared to images of TM bands 1−4 (compare Figs 3 and 6). Grey values and colours may not match across adjacent quadrants within one scene, because different stretches were applied.

Most TM bands have a better dynamic range than those of MSS, so that saturation over snow-covered scenes is less severe. We find that saturation chiefly affects TM band 1, which agrees with the observations of Dozier (1985). Because of the limited information provided by TM band 1, it is not discussed in this article.

IMAGE INTERPRETATION
General remarks

As is shown in Table I, the MSS and TM reflective bands have partly overlapping wavelengths. TM bands 2, 3, and 4 have approximately the same wavelengths as MSS bands 4, 5, and 7 (1, 2, and 4 of Landsat 4 and 5).

The basic physics of remote sensing of snow and ice in the MSS and TM bands are well established. The spectral reflectivity of snow depends on grain-size and shape, near-surface liquid water content, depth and surface roughness, impurities, and angle of solar incidence. Various models have been developed to calculate changes in snow reflectivity (e.g. Colwell 1983, chapter 3), and it was found that the reflectivity generally decreases with increasing wavelength in the visual and near infra-red, and that the variations with wavelength of albedo of snow, firn, and ice are broadly parallel (Qunzhu and others 1984).

The snow albedo decreases with increasing grain-size at wavelengths above 0.7 μm (TM bands 4, 5, and 7), but it is little affected by grain-size in shorter wavelengths (TM bands 1−3) (Choudhury and Chang 1981). Thus grain-size variations do not show in the lower MSS or TM bands. The

MSS bands also do not discriminate well between snow and clouds, because snow and clouds both reflect highly at MSS wavelengths. TM bands 5 and 7, on the other hand discriminate well between snow, ice clouds, and water clouds, which have increasingly higher albedos, in that order (Dozier 1985). Therefore, snow will appear darker than clouds in the reflective infra-red bands.

The total emitted thermal radiation is given by: R $\epsilon \sigma T^4$, where ϵ is the emissivity, σ the Stefan−Boltzmann constant, and T the absolute temperature. The thermal radiation of snow is affected by temperature, crystal size and liquid-water content (Hall and Martinec 1985). The radiation at terrestrial temperatures is strongest between and 14 μm, with a maximum around 11 μm at temperatures in the −25° to 0°C range. This maximum is covered by TM band 6, whose spectral band path also lies in an atmospheric window of transparency to water vapour, so that the band records the ground temperatures. No radiation occurs at the above temperatures in the range of the TM reflective bands.

The following presents some examples of the information available from the TM data. In some cases the interpretations seem well established. In others the full understanding of the phenomena shown by the TM image requires ground confirmation that is not presently available. We find that understanding is increased by studying the separate images of individual bands in addition to the colour composites.

TM bands 2−4

These bands are most easily interpreted, as they are similar to the familiar MSS bands, for which interpretations were verified on the ground. Comparison of TM and corresponding MSS images shows that these are indeed similar, except for the higher resolution of TM. The similarity is illustrated by Figs 2 and 3, which show composites of MSS bands 1, 2, and 4 (4, 5, and 7) and TM bands 2, 3, and 4 of the same scene. It is well established that among MSS bands, band 4 (7) gives the greatest information for snow and ice, both because band 4 (7) lacks saturation, and because it shows good contrast in snow and ice. The same applies to TM band 4.

The three times higher resolution of TM compared to MSS represents a major advance in our ability to detect small snow and ice features. Enhanced TM images can be produced at a scale of 1:100 000 without unacceptable degradation caused by the pixel size. Typically, a scale 1:250 000 is the maximum acceptable magnification for MSS images. A picture comparison is given in Fig.4, which shows the shear rifts along the east side of Jutulstraumen. Fig.4a and b show that an enlargement to 1:100 000 results in unacceptable quality for the MSS image but good quality for the TM image. Fig.4c shows, for comparison, the most recent oblique aerial photograph of the area available to us. The activity in this fractured area is so intense that no common features are visible in the TM image and the photograph taken 26 years earlier.

We have determined the depths of various rifts in the TM image from trigonometric relations between shadow lengths and Sun angle. Most shadows measure 1−3 pixels, corresponding to depths of 10−30 m; the floors of the

TABLE I

MSS band*	Wavelength μm	Region	TM band	Wavelength μm	Region
4 or 1	0.5−0.6	green-yellow	1	0.45−0.52	blue-green
5 or 2	0.6−0.7	red	2	0.52−0.60	green-yellow
6 or 3	0.7−0.8	"near" infra-red	3	0.63−0.69	red
7 or 4	0.8−1.1	"near" infra-red	4	0.76−0.90	"near" infra-red
			5	1.55−1.75	"middle" infra-red
			6	10.40−12.50	thermal
			7	2.08−2.35	"middle" infra-red

*The MSS bands were numbered 4−7 for Landsat 1−3, and 1−4 for Landsat 4 and 5. In this article we use 4−7 for the Landsat 2 image, and 1−4 for the Landsat 5 image. (Band labels for the respective alternate system are shown in parentheses for clarity.)

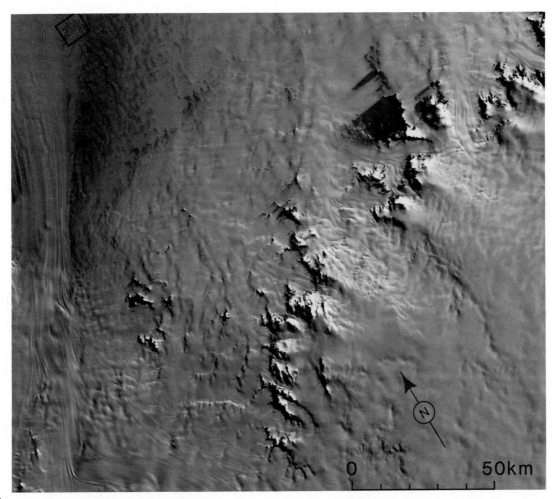

↑ Fig.2. Composite image of MSS bands 1, 2, and 4 of scene 5034407520, recorded by Landsat 5 on 8 February 1985. Sun angle 21°. Location of Fig.4a and b is shown in upper left.

↓ Fig.4. Different representations of rift area on eastern side of Jutulstraumen. North is upwards, parallel with sides, of Fig.4a and b.

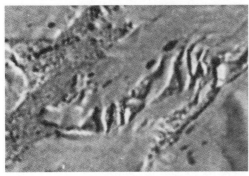

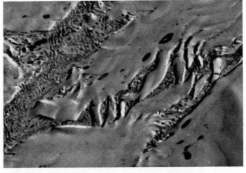

(a) Enlargement to 1 : 100 000 of MSS band 4 of part of scene 5034407520. Note poor resolution compared with 4b.

(b) Enlargement to 1 : 100 000 of TM band 4 of the same scene. Note excellent detail, including lakes, and features within the rifts.

(c) Oblique aerial photograph Nr. 1653 of the same rift area, taken by Norsk Polarinstitutt at 17.40 GMT on 6 January 1959, from 1800 m altitude, looking south-south-east. We see the rifts on the eastern margin of Jutulstraumen and the mountains behind on both sides of the ice stream.

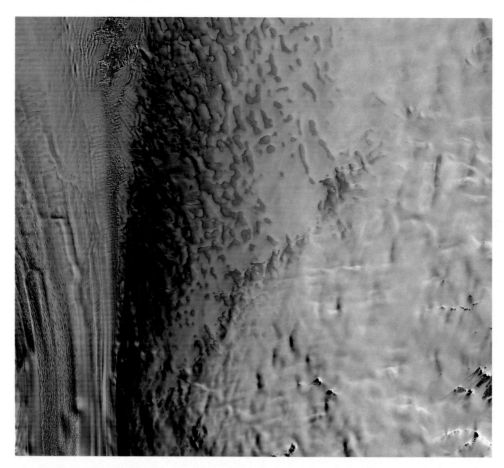

Fig.3. Multi-spectral composite of TM bands 2, 3, and 4 of quadrant 1 of scene 5034407520. Note diagonal blue band across image, reflecting step in bedrock topography. Area to north of step is lower and melting may have occurred. Area to south of line is higher and shows bare blue-ice areas (blue) down-wind of nunataks.

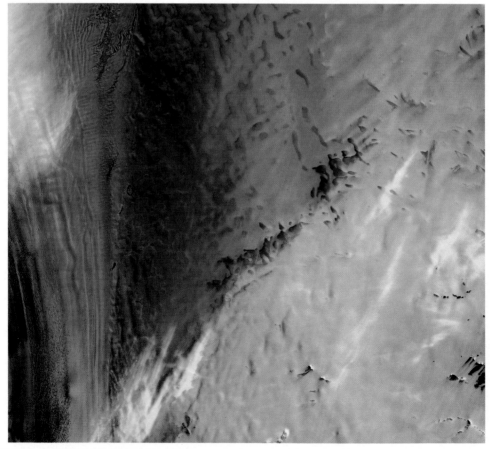

Fig.6. Multi-spectral composite of TM bands 4, 5, and 7 of quadrant 1 of scene 5034407520. Note clouds of high reflectivity (yellow) in the upper left corner, left of lower centre, and as streaks across right part of image. These clouds are transparent in the visible spectral bands (Figs 2 and 3). Note also dendritic pattern (faint yellow) running down-slope to Jutulstraumen in the left centre of image.

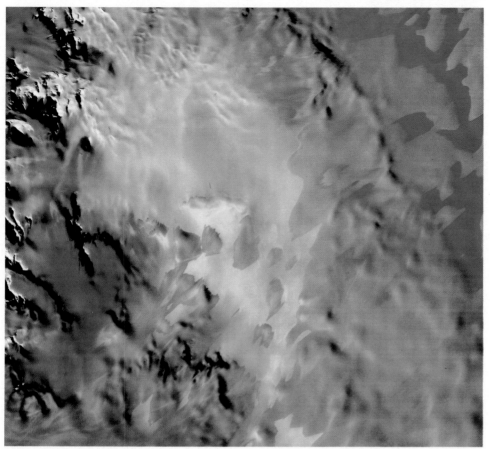

Fig.8. Multi-spectral composite of TM bands 4, 5, and 7 of quadrant 4 of scene 5034407520. Note the strong patterns in the snow in top right (blue) and centre (yellow) of image. Both show as vague patterns in MSS bands.

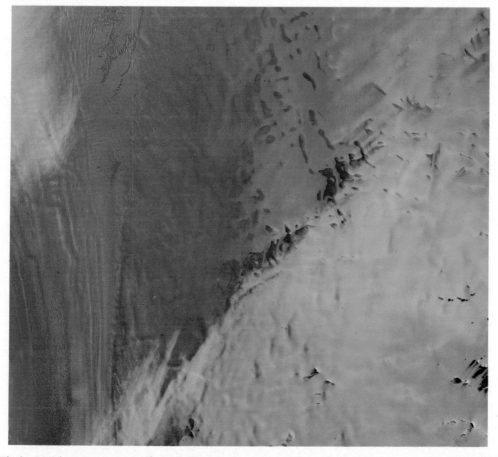

Fig.9. Multi-spectral composite of TM bands 4, 5, and 6 of quadrant 1 of scene 5034407520. Note especially warm areas at lower elevations (red areas in left of picture), and in lakes at the eastern side of Jutulstraumen. Also note high temperatures associated with nunataks in bottom right of picture (orange-yellow), and band of higher temperatures associated with possible "blue-ice" areas crossing picture diagonally.

fractured areas shown in Fig.4 lie 20–30 m deeper than the adjoining broken-off segments of the ice stream. The resolution of MSS images is not adequate for such measurements.

It is well established that the Landsat images also display large-scale snow and ice features that extend over many pixels, such as (paleo)flow lines and sub-surface (bedrock) topography. Such features are shown in Figs 2, 3, and 5. Note especially the continuous flow lines of Jutulstraumen and of smaller glaciers. Note also the east-north-east to west-south-west trending "steps" in the subglacial bedrock to the south-east of Jutulstraumen. MSS and TM images depict large features in a similar fashion even though TM images have greater acuity. The reason is that large but subtle features, both on MSS and TM pictures, are enhanced by photometric shading caused by the low Sun angle prevalent in Antarctica.

Our analyses agree with those of Allison and others (1985) in recognizing different types of dark snow and ice signatures in the visible bands. The enhanced composite images (Fig.3) and the individual bands (including the thermal band; see below) show different types of low-albedo features. One of these crosses the image diagonally and is inferred to consist of bare blue ice, which is also warmer than most of the other areas of low albedo (see below). The area north of this diagonal feature is mostly at lower elevations, between 100 and 600 m where melting may occur, and also shows dark patterns surrounded by snow of lighter signature. The southern area is generally at elevations from 600 to 1200 m. The snow here is lighter, and all low-albedo areas within the snow and ice are inferred to be bare blue ice, because they generally lie directly down-wind from nunataks and receive less drifting snow. The differ-

ences between bare blue ice and other areas of low-albedo ice or snow are better seen in false-colour composites of MSS bands 4, 5, and 7, or TM bands 2, 3, and 4, than in the individual bands.

We finally note that the high resolution of TM makes it very attractive for geological and biological studies (e.g. we recognize bird colonies), and for planning field operations to avoid crevassed areas.

TM bands 5 and 7

(a) Cloud discrimination

As stated earlier, the images of reflected infra-red light (TM bands 5 and 7) discriminate well between snow and clouds. The differences are illustrated by Figs 2, 3, 5, and 6. We recognize two types of cloud.

Clouds of the first type are familiar. They are small, dense and low, and cast shadows in the images. They show as a group in the upper central section of Fig.5, and as isolated "dots" to the west of the larger group, and as a group in the lower central section of Fig.6. They seem to be alto-cumulus. Their shadows cause these clouds to be recognized easily, and they show clearly in all spectral bands.

The second type is more enigmatic. These clouds show strongly in TM bands 5 and 7 (Fig.6), but are not seen at all in TM bands 2–4 (or the MSS bands) (Figs 2 and 3). These clouds have similar reflectivities in TM bands 5 and 7, which suggests that they are ice clouds, because the data of Dozier (1985) show that reflectivity in bands 5 and 7 is high and similar for ice clouds of small crystal radius. That the clouds consist of small ice crystals is also supported by the observation that they are apparently transparent in the

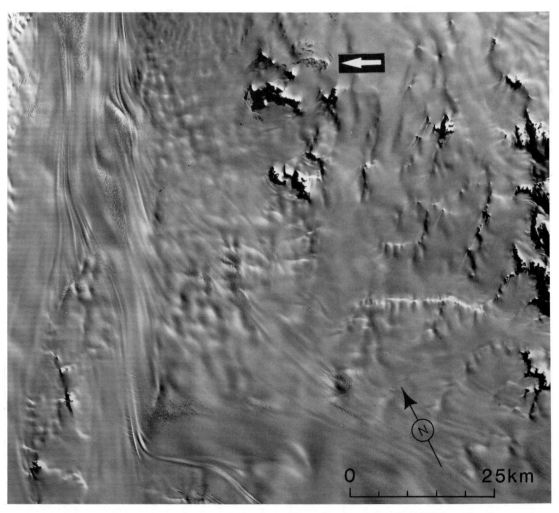

Fig.5. Composite of TM bands 2, 3, and 4 of quadrant 3 of scene 5034407520. The image abuts Fig.3 on the south and shows the southern continuation from Fig.3 of Jutulstraumen. Note flow lines entering into the ice stream. Alto-cumulus (?) clouds are seen in the upper centre of the image (arrow).

visible wavelengths, and that they do not cast shadows. We conclude that these "clouds" are thin (cirrus?), consisting of ice crystals with optical radius <1 μm, and that they are probably located at high altitudes.

The lower section of Fig.5 shows additionally a third phenomenon, which may be low clouds or drifts near the surface, because they seem to be affected by topography and surface obstacles in the flow path.

b) Snow phenomena

The theoretical considerations referred to earlier indicate that TM bands 5 and 7 should be more sensitive to grain-size variations than the lower bands, with high reflectivities corresponding to snow of small grain-sizes. We believe the images confirm this.

The composite of infra-red reflected bands (Fig.6) supports the conclusion that different types of snow and ice, looking similar in the visual bands, show as different in infra-red reflective bands. The low-albedo feature crossing the image diagonally has similar signature in TM bands 5 and 7 to the blue ice near nunataks, whereas the northern area of low albedo in the visible bands has a different signature of high reflectivity in TM bands 5 and 7.

TM bands 5 and 7 show several snow features not seen in the visible wavelengths. One of these is only seen up-hill of, and near, the melt-water lakes described below. It consists of faint, thin, dendritic patterns running down-hill toward the water (Fig.6); we suggest that these are water courses, which have changed the snow properties. The channels were apparently dry when the image was recorded, because, if wet, they should have shown more conspicuously in the lower bands.

Different and much larger features are seen in the inland snow areas, generally above 2000 m elevation (Figs 7 and 8). Whereas only faint patterns are visible in TM bands 2–4 (Fig.7), well-developed snow patterns of both relatively high and low albedo are shown in TM bands 5 and 7 (yellow and blue) in Fig.8. The patterns do not show in the thermal band. Examination of a partly overlapping image, recorded 19 days later, shows that the features are persistent. Fig.8 also shows that the albedo patterns are partly aligned with the prevailing wind direction, which can be inferred from snowdrifts and margins of blue-ice areas down-wind of nunataks (Fig.2). The up-wind side of the patterns appears to have a sharp edge, perhaps of some relief, because it appears to trap material of different reflectivity. The patterns are probably formed by wind and reflect different types of snow.

We conclude that the different reflectivities are most likely related to variations in grain-size, and that the lightest areas show very fine-grained, perhaps new snow, whereas the patterns of lower albedo are caused by coarser snow or possibly glazed surfaces.

The large flow lines on glaciers that appear so clearly in MSS bands and TM bands 2–4 show less clearly in bands 5 and 7. Nor do bands 5 and 7 show bedrock-related topographic features as well as the lower bands. Altogether, the observations support the theoretical principles which suggest that TM bands 2–4 are best for the detection of topographic features owing to photometrically accentuated slopes, whereas TM bands 5 and 7 are best for the detection of clouds and grain-size variations, or other surface properties of snow.

TM band 6

The thermal band is a useful supplement to the reflective bands. It provides ground-temperature information

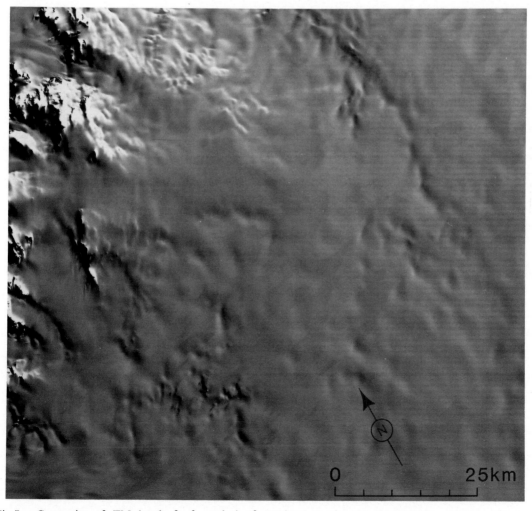

Fig.7. Composite of TM bands 2, 3, and 4 of quadrant 4 of scene 5034407520. Patterns in snow show faintly.

and, in particular, it shows a strong signal from "warm" bodies within regions of snow and ice. The thermal data can also be used to discriminate between some features that are similar in the reflective bands.

Fig.9 shows a multi-spectral composite including the thermal band (red signature) of quadrant 1. Generally, the temperatures decrease with increasing elevation and are higher on surfaces of low albedo. Jutulstraumen, even though of low elevation, shows relatively low surface temperatures, presumably because cold air follows the topography and flows down along the ice stream. The rifts near the edge of Jutulstraumen show considerably higher temperatures than the surrounding ice, even though the elevation differences are only 20–30 m. This could be caused by increased absorbed solar radiation resulting from multiple reflections within the rifts, and at some locations from the presence of small amounts of water.

Lake-like features are observed both within the rifts,

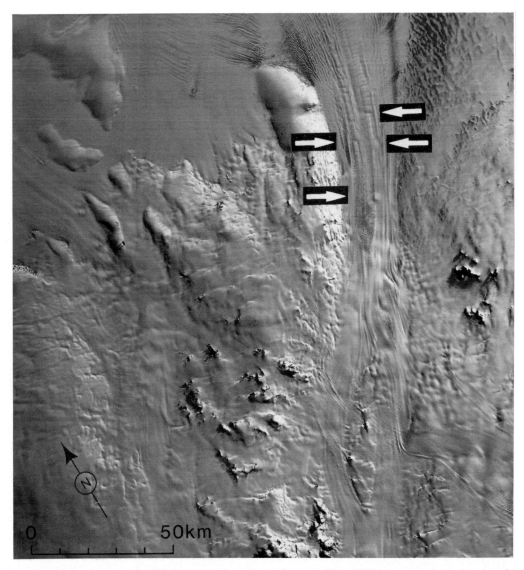

Fig.10. Composite images of MSS bands 4, 5, and 7 of scene 2281-97424, recorded by Landsat 2 on 30 October 1975. Sun angle 22°. Comparison of this image with that from 1985 (Fig.2) reveals a number of both permanent and transitory features. Four stationary ice rumples show in right centre of image, of which the northernmost has a "peak" shape (arrow). These, and the up-stream flow lines, are "permanent", and reflect the bedrock. All rifts and large crevasses are down-stream of these rumples, where the ice is afloat. Note especially the two large crevasses and a deviation in the flow line (arrows). Note also that the low-albedo slopes in the upper part of the comparison images (see Fig.2) cover approximately the same area, and have many identical details, thus demonstrating relationship with topography. The February 1985 image (this figure) shows a more continuous zone of low albedo, presumably because it was recorded at the end of the melt season.

TABLE II

	Mean flow rate	Standard deviation	Strain-rate Longitudinal	Transverse
	m a^{-1}	m a^{-1}	10^{-4} a^{-1}	10^{-4} a^{-1}
West side (8)	698	22	2.9	
East side (14)	730	13	0.7	
All localities (22)	718	22		131 (4 sets)

along the south-eastern margin of Jutulstraumen, and on the north-west-facing slopes leading down to the ice stream. All of these features have ≈0 albedo in the six reflective bands, indicating the presence of water. The thermal signal shows very high temperatures, probably around $0°C$. We conclude that these are "open" water bodies. The largest of the lakes is seen left of centre in Figs 3 and 9; it is elliptically shaped and $1.5 km × 2 km$ in size. Both this and other lakes have a central area covered by snow or ice thick enough to give snow or ice signatures. We do not know from the signatures whether the so-called "open" water is perhaps covered by a very thin layer of ice.

Fig.9 shows relatively high temperatures on blue-ice areas near nunataks and on areas located along a band crossing the image diagonally and located at around 500 m elevation (compare with Fig.3). Most of the north-west-facing slopes of lower elevation leading down to Jutulstraumen show lower temperatures. This anomaly can be explained if the higher-lying band of higher reflectivity also contains bare blue ice. Thus, the thermal band "flags" the areas of bare blue ice and it helps in discriminating between these and visually similar-looking sites.

Concluding remarks

The above examples have illustrated the potential of the TM data. Whereas the MSS bands and low TM bands are most useful in recognizing topographic features, enhanced by shading due to low Sun angles, the upper TM bands are best suited to discriminating between properties in snow and ice. Because of limited space, we have discussed only some of the more obvious features. Many more are visible in the TM images but will have to await future discussions.

COMPARISON OF IMAGES OF JUTULSTRAUMEN TAKEN IN 1975 AND 1985

More than a decade has now passed since the first Landsat data were collected from Antarctica. These early images form an important basic data set, which can be especially valuable because of the remoteness of the Antarctic continent, the difficulties of access, and the expenses of field work. We therefore examined whether repeated Landsat images can be used to provide quantitative glaciological data. A similar study was conducted by Lucchitta and Ferguson (1986).

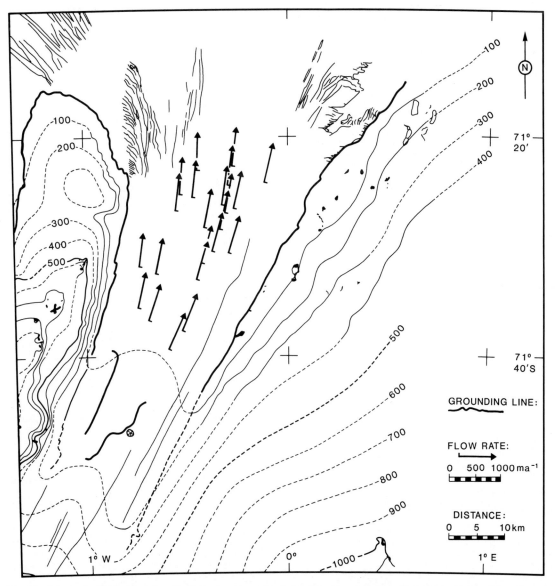

Fig.11. Map of displacements of features recognized in the two images of Jutulstraumen (Figs 2 and 10) recorded 9.3 years apart. The arrows show flow rates determined from 22 points recognized in both images. The elevation contours are based on Norsk Polarinstitutt surveys and are shown dashed where they are approximate. Lakes are shown as solid black. Zones of definite grounding within the ice stream and near the 100 m contour are also shown. The flotation line is probably located down-stream of this contour.

Fig.10 shows the enhanced MSS image recorded on 30 October 1975. Altogether 22 point features, moving with the glacier and mostly located at the edges of large crevasses, have been identified in this image. These features are also visible in the image of 8 February 1985 (Fig.2), and remain recognizable, even though subdued by snow accumulation, because they are located where crevasses are few and distinct. Fig.11 shows the displacements and corresponding flow rates that could be measured from the distinctive features on the two images, acquired at an interval of 9.3 years.

The flow rates are surprisingly similar. Table II shows the results for all points, divided into two groups. Of the two groups, the eight points on the west side of Jutulstraumen are closer to the edge and have generally lower velocities. The point closest to the edge shows the lowest flow rate at 649 m a^{-1}. The 14 points on the east side have generally higher flow rates, with a maximum of 744 m a^{-1}.

A flow rate of 0.72 km a^{-1} agrees well with maximum flow rates of 0.39 km a^{-1}, measured $\approx$90 km up-stream at 900 m elevation (Gjessing 1972), and a rate of $\approx$2 km a^{-1} at the ice front that we have determined from repeated surveys.

The main errors in the flow-rate determinations result from poor geometric accuracy in the 1975 image, and from imprecise positioning of points caused by uncertainties in identifying identical details of the features in the two images. Geometric inaccuracies were partly removed by manually adjusting the 1975 and 1985 images to fixed points, but some errors remain. We had expected the positioning of the points to be more difficult and to lead to larger scatter in the results than observed. The relatively good agreement in the measurements indicates that the technique works well in spite of potential difficulties. We believe that the standard deviations (Table II) give a good estimate of the likely error in the mean flow rates.

The mean flow rates increase from the edge to the centre, and also down-stream, and are thus consistent with glaciological principles. However, comparison of individual points shows that the down-stream changes in flow rates are too small and irregular to allow a rigorous strain analysis involving all individual points. Instead, Table II shows the mean strains along, and transverse to, the flow.

The longitudinal surface strain is calculated from the mean change in flow rate between the up-stream and the down-stream group of points on each side of the centre line. The change in flow rates is only 8 m a^{-1} on the west side and 2 m a^{-1} on the east side, over distances >20 km, corresponding to only a small acceleration along this section of the ice stream.

The transverse strains have been determined at 90° to the flow direction, by measuring the change in distance between pairs of points moving coherently in longitudinal direction. It varies for four sets between 113 and 162 × $10^{-4}a^{-1}$. Thus the transverse strain is two orders of magnitude larger than the strain along the flow direction. This large difference results from the widening of the ice stream over this section.

Figs 2 and 10 show the fractures resulting from the high transverse strains, and indicate in addition that larger longitudinal strains occur further down-stream.

It was not possible to define the flotation (grounding) line within Jutulstraumen with confidence even on enhanced false-colour composite TM image at maximum enlargement. Fig.11 shows locations of grounded ice, which are identified either by bedrock topography, which shows as a stationary feature, or by stationary crevasse patterns. The actual grounding line is an unknown distance down-stream from these places. Also shown is the 100 m contour line from Norsk Polarinstitutt 1 : 250 000 map sheet G5, Ahlmannryggen. This contour is approximate. It seems likely, from comparison with similar ice streams, that the flotation line should be close to, and down-stream of, the 100 m elevation.

There is an intriguing aspect of the crevasses which may be related to the position of the grounding line. Fig. 10 shows two large crevasses, in the western and central part of Jutulstraumen, located about 20 and 30 km respectively from a stationary "ice peak" farther south, and

a deviation in the flow line at 20 km. Fig.2 shows that these three features have moved about 7 km down-stream, and that no new large crevasses have formed in the former positions. This suggests that these features were caused by singular events, rather than by a permanent aspect of the strain regime such as would be related to topography. One such event that could have caused the flow deviation would have been a small surge from a tributary glacier. However, we consider it more likely that these features are related to a change in flow regime brought about by a major calving on the ice-shelf front.

The distance from these features to the seaward edge is presently nearly 200 km. The ice shelf was, however, formerly even larger, and extended as a 50 km wide, 100 km long ice tongue, named Trolltunga. This tongue calved off in July 1967 (Vinje 1977). Calving could have changed the stress regime, causing an acceleration of the inland ice shelf and the two large crevasses.

No field data are available to evaluate how the calving may have changed the flow regime of the ice shelf. But a simple calculation can test the explanation. In the October 1975 image, the deviation in flow line and the large crevasses are respectively 7.4, 7.5, and 16.5 km down-stream from the 100 m contour, which translates into a time span of 10.1, 10.5, and 22.2 years considering the flow rates at each locality. The image was recorded 8.3 years after Trolltunga calved. Thus the intervals match well for two of the large features if the flotation line was near, and down-stream of, the 100 m contour at the time of calving. If this explanation is correct, then this also implies that the stress and the acceleration must have propagated rapidly across the ice shelf.

ACKNOWLEDGEMENTS
We gratefully acknowledge much help from colleagues at Norsk Polarinstitutt and at US Geological Survey, Flagstaff. Image 5034407520 was reproduced by permission of EOSAT.

REFERENCES

Allison I, Young N W, Medhurst T 1985 Correspondence. On re-assessment of the mass balance of the Lambert Glacier drainage basin, Antarctica. *Journal of Glaciology* 31(109): 378-381

Choudhury B J, Chang A T C 1981 On the angular variation of solar reflectance of snow. *Journal of Geophysical Research* 86(C1): 465-472

Colwell R N (ed) 1983 *Manual of remote sensing. Second edition.* Falls Church, VA, American Society of Photogrammetry

Dozier J 1985 Snow reflectance from thematic mapper. *Landsat-4 science characterization early results. Vol IV. Applications.* Washington, DC, National Aeronautics and Space Administration: 349-357 (NASA Conference Publication 2355)

Gjessing Y T 1972 Mass transport of Jutulstraumen ice stream in Dronning Maud Land. *Norsk Polarinstitutt. Årbok* 1970: 227-232

Hall D K, Martinec J 1985 *Remote sensing of ice and snow.* London and New York, Chapman and Hall

Lucchitta B K, Ferguson H M 1986 Antarctica: measuring glacier velocity from satellite images. *Science* 234(4780): 1105-1108

Lucchitta B K, Eliason E M, Southworth S 1985 Multispectral digital mapping of Antarctica with Landsat images. *Antarctic Journal of the United States* 19(5): 249-250

Ødegaard H, Helle S G 1982 *Polar mapping using Landsat data. Final report.* Oslo, IBM and Norsk Polarinstitutt

Orheim O 1978 Glaciological studies by Landsat imagery of perimeter of Dronning Maud Land, Antarctica. *Norsk Polarinstitutt. Skrifter* 169: 69-80

Qunzhu Z, Meisheng C, Xuezhi F, Fengxian L, Xianzhang C, Wenkun S 1984 Study on spectral reflection characteristics of snow, ice and water of northwest China. *Scientia Sinica* Ser B 27(6): 647-656

Vinje T E 1977 Drift av Trolltunga i Weddellhavet. *Norsk Polarinstitutt. Årbok* 1975: 213

Annals of Glaciology 9   1987
© International Glaciological Society

SEASONAL AND REGIONAL VARIATIONS OF NORTHERN HEMISPHERE SEA ICE AS ILLUSTRATED WITH SATELLITE PASSIVE-MICROWAVE DATA FOR 1974

by

C.L. Parkinson, J.C. Comiso, H.J. Zwally, D.J. Cavalieri, P. Gloersen

(Laboratory for Oceans, NASA/Goddard Space Flight Center, Greenbelt, MD 20771, U.S.A.)

and

W.J. Campbell

(U.S. Geological Survey, University of Puget Sound, Tacoma, WA 98416, U.S.A.)

ABSTRACT

A detailed description of the seasonal cycle of Northern Hemisphere sea ice for 1974 is provided by the passive microwave data from the Nimbus 5 Electrically Scanning Microwave Radiometer (ESMR). Sea ice extent has been mapped and analyzed in eight regions of the Arctic and marginal seas. In the seasonal sea ice areas, the ice concentration is also mapped, whereas in areas of first-year and multiyear ice mixtures, the corresponding mapping is of a parameter representing a combination of ice concentration and multiyear ice fraction. The total monthly ice extent increased from a sharp minimum of 7.6×10^6 km^2 in September, when the ice pack was mostly confined to the central Arctic Ocean and portions of the Greenland Sea, Kara Sea, and Canadian Archipelago, to a broad maximum of 14.4×10^6 km^2 in March, when the ice cover was nearly complete in the Arctic Ocean, Hudson Bay, Kara Sea, and Canadian Archipelago and was extensive for large portions of the other peripheral seas and bays. In the areas of seasonal sea ice coverage, the average ice concentration was approximately 75% in winter, which is close to the values observed in the Southern Ocean and significantly less than the greater-than-95% concentrations observed in the central Arctic Ocean and Hudson Bay, where the ice packs are constrained by land boundaries. Midwinter decreases in ice extent for 1–2 months are noted in the regions of the Greenland Sea and the Kara and Barents Seas.

INTRODUCTION

Data collected by the Electrically Scanning Microwave Radiometer (ESMR) on board the Nimbus 5 satellite from its launch in December 1972 through most of the next four years provide the earliest all-weather, all-season imagery of global sea ice. For 39 months of the four-year period, good quality Northern and Southern Hemisphere data were obtained, at a wavelength of 1.55 cm and a resolution of about 30 km, allowing a detailed description of the seasonal cycle of sea ice conditions and the interannual variations within that cycle over the period 1973–1976. The Southern Hemisphere data were compiled into an Antarctic sea ice atlas by Zwally and others (1983), and the Northern Hemisphere data have been compiled into an Arctic sea ice atlas by Parkinson and others (1987). This paper describes some of the results obtained for the Northern Hemisphere for the year 1974, which is the year with the most complete ESMR data set.

ESMR data have been used in several previous studies to analyze ice conditions in the Northern Hemisphere, including work by Crane and others (1982), Carsey (1982), and Campbell and others (1984). Interested readers are referred to those articles and the Arctic atlas itself for an extensive bibliography.

The value of the ESMR data for sea ice studies derives from the large contrast in microwave emissivities between sea ice and open water. At the 1.55 cm wavelength of the ESMR instrument, open water has an emissivity of approximately 0.44, whereas sea ice has an emissivity of 0.80 to 0.97, depending upon the composition and surface characteristics of the ice. Although the large range in sea ice emissivities introduces an ambiguity in sea ice concentration calculations from the ESMR brightness temperature data, the large contrast between the 0.44 emissivity of water and the 0.80–0.97 range for sea ice enables a ready determination of the location of the sea ice edge from the ESMR data.

The most important contrast in sea ice emissivities for this study is the contrast between the emissivity of first-year ice and the emissivity of multiyear ice, the latter being defined as ice having survived a summer melt period. In winter, first-year ice has an emissivity of approximately 0.92, whereas multiyear ice has an emissivity of approximately 0.84. The lowered emissivity of multiyear ice derives in large part from the desalination produced by brine drainage. During summer, surface characteristics tend to dominate the microwave emission, often making the two ice types indistinguishable. This would assist the determination of ice concentrations from the microwave data except for the fact that the varied surface characteristics of the ice can produce a wide range in microwave emissivities. For instance, large-scale melt ponding reduces the emissivity whereas wet snow increases the emissivity. Nonetheless, the summertime sea ice emissivities remain predominantly in the range 0.80 to 0.97 (Mätzler and others 1983; Parkinson and others 1987).

SEASONAL SEA ICE CYCLE

Fig.1 shows monthly averaged ESMR brightness temperature data mapped and color-coded onto polar stereographic projections for March, June, September, and December, 1974, revealing the seasonal cycle of Northern Hemisphere sea ice coverage. The low microwave emissivity of open water leads to brightness temperatures over open water predominantly in the range 130–150 K, while the higher microwave emissivity of sea ice leads to brightness temperatures exceeding 200 K. Intermediate values within the sea ice region indicate mixtures of sea ice and open water. The sea ice edge is readily apparent on these images at the occurrence in ocean regions of the sharp color gradient from the grays of open water to the oranges, browns, and reds of the sea ice region. Ice sheets, such as the Greenland ice sheet in the lower center of the images of Fig.1, have microwave brightness temperatures predominantly in the range 150–200 K, whereas other land surfaces have microwave brightness temperatures predominantly exceeding 230 K.

The large contrast in microwave emissivities between sea ice and open water enables a conversion of the ESMR brightness temperatures to sea ice concentrations (percentages of the ocean area covered by sea ice) as long as all sea ice

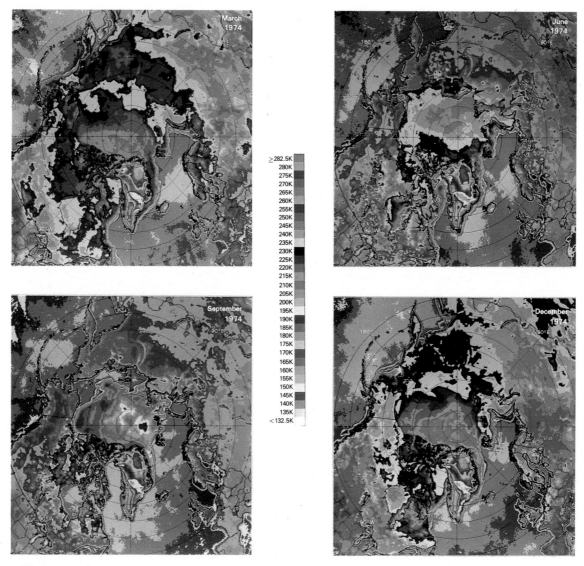

Fig.1. Mean monthly brightness temperatures from the Nimbus 5 ESMR for March, June, September, and December 1974

in the field of view has approximately the same emissivity. This is the case throughout most of the Southern Ocean and in many of the seas and bays peripheral to the Arctic Ocean, where the ice is predominantly first-year sea ice, with an emissivity near 0.92. In such a case, the sea ice concentration, C, can be calculated as:

$$C = \frac{T_B - T_o}{\epsilon_I T_{eff} - T_o} \qquad (1)$$

where T_B is the microwave brightness temperature, ϵ_I is the emissivity of ice (taken as 0.92 for first-year ice), T_{eff} is the effective radiating temperature of the ice, and T_o is the average observed brightness temperature over open ocean. For the calculations here, a constant value of 138.3 K has been used for T_o. This value was determined from the ESMR data as an average open water value in the north polar region over the 1973–1976 period. (In the Southern Hemisphere a somewhat lower value of 135 K was used, again determined from the ESMR data. In both cases, these observed brightness temperatures include an atmospheric component.) T_{eff} is calculated as follows from mean monthly climatological air temperatures, T_{air}:

$$T_{eff} = T_{air} + f (T_F - T_{air}) \qquad (2)$$

where $T_F = 271.16$ K is the freezing point of sea water

and f = 0.25 is a parameter determined empirically from observed data. The use of mean climatological data in Equation 2 could result in ice concentration errors as large as 20%. Further details can be found in Parkinson and others (1987).

In the Arctic Ocean itself, and some of the immediately adjacent waters, there is a large multiyear ice component, with an emissivity near 0.84, in addition to first-year ice, with an emissivity near 0.92. In these areas, sea ice concentrations cannot be calculated directly with Equation 1, although they can be calculated in a similar fashion from the brightness temperatures as a function of the fraction, F_{MY}, of the ice cover which is multiyear ice:

$$C = \frac{T_B - T_o}{F_{MY} \, \epsilon_{IM} \, (T_{eff})_M + (1 - F_{MY}) \, \epsilon_I \, (T_{eff})_F - T_o} \qquad (3)$$

where ϵ_{IM} is the emissivity of multiyear ice, and $(T_{eff})_M$ and $(T_{eff})_F$ are the effective radiating temperatures of the multiyear ice and first-year ice respectively.

The brightness temperature data of Fig.1 have been converted to sea ice concentrations, using Equations 1–3. The resulting sea ice concentrations are presented in color-coded images in Fig.2. Equation 3 is depicted on this figure by a nomogram relating the color scale, multiyear ice fraction, and sea ice concentration.

Before describing the sea ice cycle revealed in Figs 1 and 2, one further set of images will be presented. These

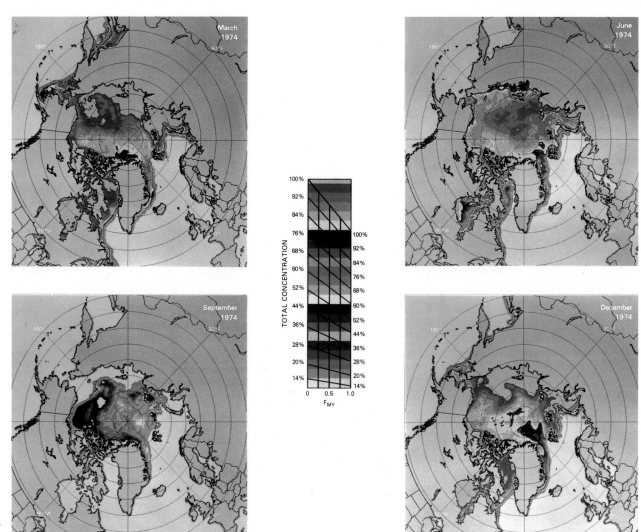

Fig.2. Mean monthly sea ice concentrations, calculated from Nimbus 5 ESMR brightness temperatures and mean monthly climatological atmospheric temperatures, for March, June, September, and December 1974. The nomogram in the center relates the color scale to the total ice concentration and the fraction, F_{MY}, of multiyear ice.

re monthly differences, calculated from the monthly averaged ice concentrations by subtracting the concentrations in one month from those in the subsequent month. By olor coding positive values as blues and greens and egative values as reds, purples, and oranges, the growth nd decay regions become immediately identifiable on these monthly difference maps. The full 1974 annual cycle of monthly differences is presented in Figs 3 and 4. The redominance of blues and greens in the images from eptember–October through January–February and the redominance of reds, purples, and oranges from April–May hrough July–August clearly distinguish the growth and lecay seasons and identify August and September as the months of minimum ice cover and February, March, and April as the months of maximum ice cover.

Figs 1–4 together provide a detailed depiction of the ull annual cycle of sea ice coverage in the Northern Iemisphere for 1974. In March, at the time of maximum ce extent, the ice cover was nearly complete in the Arctic)cean, Hudson Bay, the Kara Sea, and the Canadian Archipelago, and was extensive for large portions of the ther peripheral seas and bays (Figs 1 and 2). The pringtime retreat of the ice edge in 1974 began first in he Sea of Okhotsk and the southern Greenland Sea and ater in the Bering Sea, Baffin Bay, Hudson Bay, the orthern Greenland Sea, and the Barents Sea (Fig.3). In eptember, at the time of minimum ice extent, the ice pack vas mostly confined to the central Arctic Ocean and

portions of the Greenland Sea, the Kara Sea, and the Canadian Archipelago. Essentially no ice remained in the Bering Sea, Hudson Bay, the Sea of Okhotsk, or Baffin Bay/Davis Strait (Figs 1 and 2). Noticeable autumn ice-edge advance began first in the Greenland Sea, between August and September, then became apparent throughout the remainder of the ice-covered region between September and October (Fig.4). These basic features of the annual cycle of sea ice are fairly consistent over the four years of ESMR data, 1973–1976, although with differences in specifics, as detailed in Parkinson and others (1987).

The monthly averaged ice concentration data have been used to quantify the annual cycle of ice extents and of ice within various ice concentration classes. Specifically, the data have been spatially integrated over the Arctic Ocean and many peripheral seas and bays (see Fig.5) to determine the area of the ice cover as a function of time (upper left diagram of Fig.6). For each month, the area of ice-covered ocean (termed the total ice extent) is calculated by summing the areas of all map elements having at least 15% ice coverage. The area of an individual map element is approximately 30 × 30 km. Labelling as "pseudo ice concentrations" the ice concentrations calculated from Equation 1 with an emissivity ϵ_I of 0.92, the areas of the ocean covered by ice of at least 35%, 50%, 65%, and 85% pseudo ice concentration are determined by summing the areas of all map elements having at least those respective percentages of pseudo sea ice coverage. Because of the sharpness of the

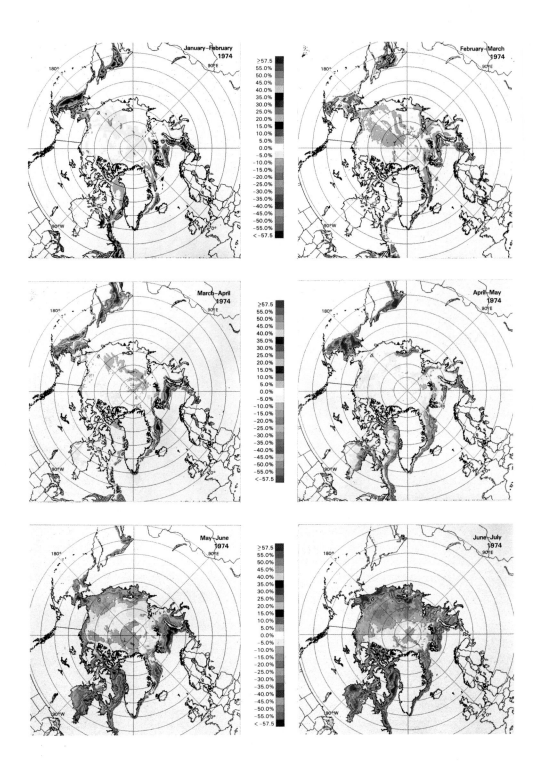

Fig.3. Sea ice concentration differences between consecutive months, from January—February 1974 through June—July 1974. Differences are calculated such that concentrations for the first month are subtracted from those for the subsequent month.

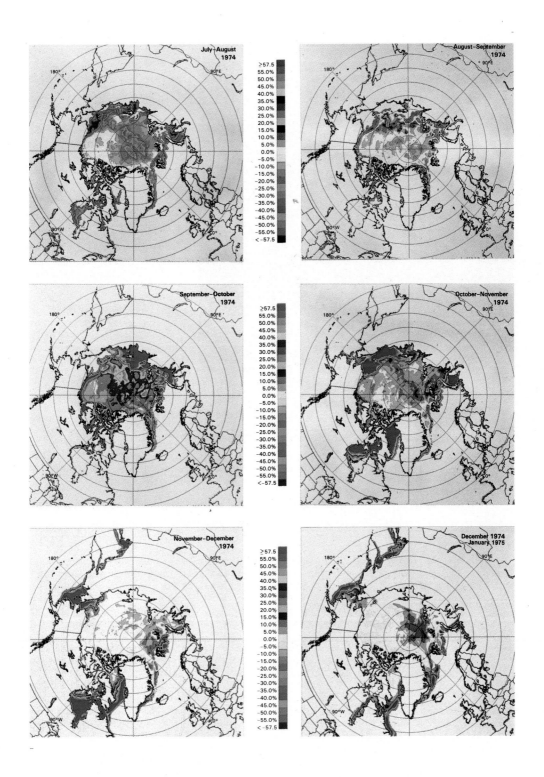

Fig.4. Sea ice concentration differences between consecutive months, from July—August 1974 through December 1974 to January 1975. Differences are calculated such that concentrations for the first month are subtracted from those for the subsequent month.

ARCTIC SEA ICE REGIONS

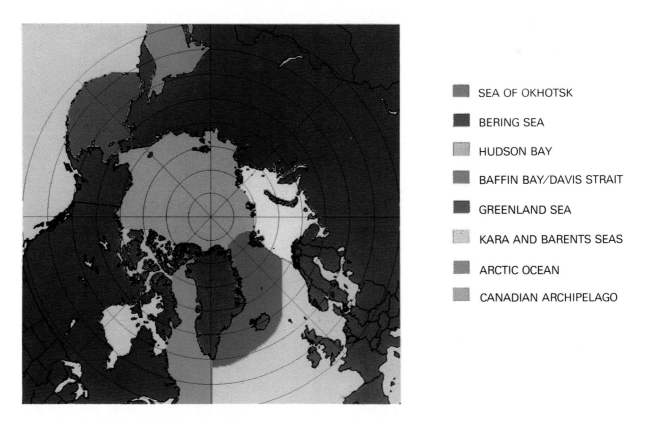

SEA OF OKHOTSK

BERING SEA

HUDSON BAY

BAFFIN BAY/DAVIS STRAIT

GREENLAND SEA

KARA AND BARENTS SEAS

ARCTIC OCEAN

CANADIAN ARCHIPELAGO

Fig.5. Regions identified for analysis and for calculation of areal ice coverages.

ice edge and its weak sensitivity to the ice emissivity, it was not felt necessary to attach the adjective "pseudo" in the case of the 15% curve. The upper left diagram of Fig.6 also includes the area obtained by summing the products of the pseudo ice concentration and the geographic area of each map element. These are termed "pseudo actual ice areas."

Monthly average ice extents in 1974 increased from a minimum of 7.6×10^6 km^2 in September to a maximum of 14.4×10^6 km^2 in March (Fig.6). The maximum is broad, with February, March, and April all having ice extents between 14.3 and 14.4×10^6 km^2 and with December, January, and May also having extents greater than 12.8×10^6 km^2. The minimum is much sharper, with the extents in August and September, 8.1 and 7.6×10^6 km^2 respectively, being significantly less than the extent in any other month. The annual cycles for the other ice concentration classes are similar to the cycle for the ice extent, with the various curves running almost parallel for much of the year (Fig.6). In August, less than 1000 km^2 contain ice with pseudo ice concentration exceeding 85%.

REGIONAL DIFFERENCES

Eight regions covering most of the Northern Hemisphere sea ice area have been selected for individual analysis. These are the Arctic Ocean, the Sea of Okhotsk, the Bering Sea, Hudson Bay, Baffin Bay/Davis Strait, the Greenland Sea, the Kara and Barents Seas, and the Canadian Archipelago (Fig.5). The same set of calculations used to create the upper left diagram of Fig.6 have been performed on the data for each of the eight regions, with the results also presented in Fig.6. In the cases of regions with almost exclusively first-year ice, the qualifier "pseudo" is not felt necessary and so has been omitted from the plots and the discussion.

The regions where the ice cover essentially goes to zero in summer are the Sea of Okhotsk, the Bering Sea, Hudson

Bay, and Baffin Bay/Davis Strait. The seasonal and inter-annual variabilities of the ice covers in these seasonal sea ice zones reflect the variabilities of the regional meteorology and oceanography, making the wintertime ice covers of the Sea of Okhotsk and the Bering Sea, for instance, far more variable than the ice cover of Hudson Bay. The Bering Sea ice area increases sharply from January to February, increases only slightly from February to March (the ice extent, or ≥ 15% ice concentration class, actually decreases slightly), and, except for the ≥ 15% and ≥ 35% ice concentration classes, decreases from March to April (Fig.6). This contrasts with the rapid growth of the Sea of Okhotsk ice cover from January to March and the subsequent rapid decay from March until the end of the season (Fig.6). Such regional differences, as well as interannual differences, between the ice covers of the Bering Sea and the Sea of Okhotsk are the subject of recent air-sea-ice interaction studies (e.g., Cavalieri and Parkinson in press; Parkinson and Gratz 1983).

Hudson Bay and Baffin Bay/Davis Strait show less wintertime variability than the Bering Sea, partly because of being largely surrounded by land and therefore not being as susceptible to the influences of the open ocean. The ice covers of both regions grow largely from the northwest toward the southeast, with the most rapid growth occurring during October/November (Figs 4 and 6). By December, the entire Hudson Bay region fills with ice, resulting in the uniform ice area calculated for the region throughout the winter months (Figs 2 and 6). Baffin Bay/Davis Strait has a broad opening to the south and is subject to some variability as indicated in Figure 6. The sea ice covers of both these regions survive much later in the season than do the ice covers of the Bering and Okhotsk Seas (Figs 2 and 6).

The remaining four regions each have large multiyear ice components, with considerable ice amounts remaining during the summer months. The Greenland Sea has a large influx of multiyear ice from the Arctic Ocean due to the

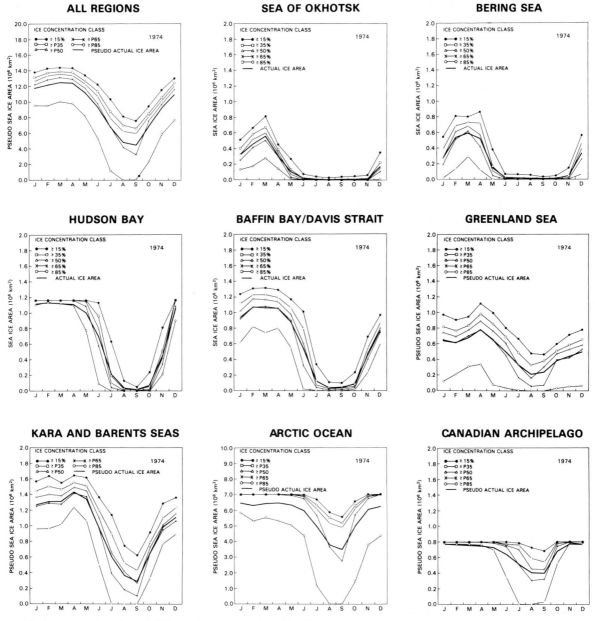

Fig.6. Annual cycles of (1) sea ice extents (≥ 15%), (2) areas with pseudo ice concentrations exceeding 35%, 50%, 65%, and 85% (labelled ≥ P35, ≥ P50, ≥ P65, and ≥ P85, respectively, for regions with multiyear ice), and (3) pseudo actual ice areas, for each of the regions identified in Fig.5 and for the sum of the regions. In regions with no or almost no multiyear ice (the Sea of Okhotsk, Bering Sea, Hudson Bay, and Baffin Bay/Davis Strait), the "pseudo" qualifier is unnecessary and is therefore omitted.

Transpolar Drift Stream, which proceeds across the Arctic Basin from near the Siberian coast and out of the Arctic through the Fram Strait passage between Greenland and Svalbard. Both the multiyear ice transported from the Arctic and the first-year ice formed in the Greenland Sea are carried south along the east coast of Greenland by the East Greenland Current, accounting for the prominent southward extension of the ice cover in this region throughout the year (Fig.2). The annual cycle of ice in the Greenland Sea (Fig.6) is unusual in that there is a late but strong peaking of the ice cover in April and an earlier midwinter decrease of ice from January to February with only a slight rebounding from February to March.

The Kara and Barents Seas also experienced a midwinter decrease in ice extent, from February to March, in 1974 although the peak afterwards was not as prominent as the April peak for the Greenland Sea (Fig.6). The Kara and Barents Seas provide an interesting contrast in that the Barents Sea is subject to the warm water inflow from the northward flowing Norwegian Current, whereas the Kara Sea is effectively blocked from that inflow by the island of Novaya Zemlya. The result is an almost total absence of ice in most of the Barents Sea throughout the year, in spite of the high latitudes and in spite of the very heavy ice cover in the adjacent Kara Sea for all except the late summer months (Fig.2).

Both the Arctic Ocean and the Canadian Archipelago are almost fully ice covered from November to May (Figs 2 and 6). The September minimum ice extent in each of these two regions exceeds 80% of the maximum ice extent, indicating a much smaller seasonal cycle of percentage ice cover than in any of the other regions. This also suggests a greater percentage of multiyear ice and the accompanying complications in interpreting the microwave brightness temperatures. From Figs 1 and 2, the retreat of the Arctic ice from the coasts in summer is clearly greater in the eastern hemisphere than in the western hemisphere, and the retreat of the ice in the Archipelago is clearly greatest in the southern passageways, with the exception of the southern Gulf of Boothia at about 86°W, 69°N.

DISCUSSION AND SUMMARY

The Nimbus 5 ESMR data provide a detailed depiction of the annual cycle of global sea ice coverage. In this paper we present this cycle for the north polar region for 1974. The ESMR data reveal that the sea ice extent in the eight regions of the Northern Hemisphere identified in Fig.5 fluctuated in 1974 from a minimum of 7.6×10^6 km^2 in September to a maximum of 14.4×10^6 km^2 in March. The data also reveal that the sea ice cycle includes considerable interannual variability and interregional contrasts.

Many of the large-scale latitudinal asymmetries in the extent of the ice visible in Figs 1 and 2 are consistent from year to year (Parkinson and others 1987), and the consistencies are in many cases readily explained by ocean currents and bathymetry. In particular, currents with major impacts include the warm, north-flowing Norwegian, West Greenland, and West Kamchatka currents, which prevent or delay ice formation in the Barents Sea, immediately southwest of Greenland, and along the west coast of the Kamchatka Peninsula, respectively, and the cold south-flowing East Greenland and Labrador currents, which transport ice far to the south along the east coasts of Greenland and Canada.

There are other features of the ice cover which are not consistent from year to year. For instance, the February ice extent in the Sea of Okhotsk, 0.66×10^6 km^2 in 1974 (Fig.6), was much higher in 1973, then being 1.09×10^6 km^2. This and many of the other regional interannual contrasts in the ice can be explained by interannual differences in the atmospheric pressure and wind fields. In particular, for the case mentioned, the Siberian High pressure system was a more dominating influence over the Sea of Okhotsk in 1973 than in 1974, bringing in very cold air from northern Siberia.

Although the ESMR data set is too short to identify any long-term climate trends, the fact that the microwave data enable a very clear depiction throughout the year of global sea ice coverage establishes their usefulness for the future for identifying and confirming both major and smaller-scale changes in the climate system. The ESMR data reveal no systematic trend in the overall area of Northern Hemisphere ice coverage over the four years 1973–1976, but the extended data set possible from continued collection of global microwave data will contribute to the accumulating information on the Earth's climatic components and their seasonal and interannual variabilities.

REFERENCES

Campbell W J, Gloersen P, Zwally H J 1984 Aspects of Arctic sea ice observable by sequential passive microwave observations from the Nimbus-5 satellite. *In* Dyer I, Chryssostomidis C (eds) *Arctic technology and policy.* Washington, DC, etc, Hemisphere Publishing/McGraw-Hill International: 197-222

Carsey F D 1982 Arctic sea ice distribution at end of summer 1973-1976 from satellite microwave data. *Journal of Geophysical Research* 87(C8): 5809-5835

Cavalieri D J, Parkinson C L In press On the relationship between atmospheric circulation and the fluctuations in the sea ice extents of the Bering and Okhotsk seas. *Journal of Geophysical Research*

Crane R G, Barry R G, Zwally H J 1982 Analysis of atmosphere-sea ice interactions in the Arctic basin using ESMR microwave data. *International Journal of Remote Sensing* 3(3): 259-276

Mätzler C, Olaussen T, Svendsen E 1983 *Microwave and surface observations of water and ice carried out from R/V Polarstern in the marginal ice zone north and west of Svalbard.* Bergen, University of Bergen. Geophysical Institute

Parkinson C L, Gratz A J 1983 On the seasonal sea ice cover of the Sea of Okhotsk. *Journal of Geophysical Research* 88(C5): 2793-2802

Parkinson C L, Comiso J C, Zwally H J, Cavalieri D J, Gloersen P, Campbell W J 1987 *Arctic sea ice, 1973-1976: satellite passive-microwave observations.* Washington, DC, National Aeronautics and Space Administration (NASA SP-489)

Zwally H J, Comiso J C, Parkinson C L, Campbell W J, Carsey F D, Gloersen P 1983 *Antarctic sea ice, 1973-1976: satellite passive-microwave observations.* Washington, DC, National Aeronautics and Space Administration (NASA SP-459)

Annals of Glaciology 9 1987
International Glaciological Society

SATELLITE REMOTE SENSING OF VATNAJÖKULL, ICELAND

by

Richard S. Williams Jr

(U.S. Geological Survey, 927 National Center, Reston, VA 22092, U.S.A.)

ABSTRACT

Iceland's largest ice cap, Vatnajökull, has been the test site for a series of airborne and satellite remote-sensing studies since 1966. Various types of image data acquired by the Landsat Multispectral Scanner (MSS) and the Seasat Synthetic Aperture Radar (SAR) are assessed for their value to glaciological studies of Vatnajökull. A low Sun angle winter 1973 MSS band 7 Landsat image of Vatnajökull provides information about the distribution and size of subglacial volcanic calderas, cauldron subsidence caused by subglacial geothermal and (or) intrusive volcanic activity, and delineation of the probable position of surface ice divides. Two types of multi-spectral digital enhancements were applied to a late summer 1973 MSS image of Vatnajökull. The first type was used to prepare a planimetric base map showing the location of the principal surface features and an inventory of 38 named outlet glaciers, one internal ice cap (Öraefajökull), and two detached glaciers which comprise this complex ice cap, and to measure its area (8300 km²). The second type provides information about the position of the snow line at the approximate end of the 1973 melt season, the areas encompassed by the ice facies of the ablation area and the slush zone and wet-snow facies/percolation facies of the accumulation area. More information about the surface morphology of Vatnajökull was available from the low Sun angle winter and the digitally enhanced summer Landsat image of the ice cap than from the Seasat SAR image.

INTRODUCTION

Iceland's largest ice cap, Vatnajökull (Figs 1 and 2A), has been the focal point for a series of volcanological and glaciological airborne and satellite remote-sensing studies during the past 20 years. In the late 1960s and early 1970s three aerial thermographic surveys of known or suggested geothermal and volcanic areas within and peripheral to the ice cap were conducted (Friedman and others 1969, 1972), including the high-temperature geothermal area (Hveradalur) and volcanic features associated with the calderas at Kverkfjöll on the north-central margin of Vatnajökull. The knowledge gained from these surveys provided the scientific impetus for a succession of satellite remote-sensing experiments on Vatnajökull, beginning in 1973 with the first usable Landsat Multispectral Scanner (MSS) images. Research with Landsat data was initially directed at achieving a better understanding of the subglacial geomorphology and regional tectonic setting of the ice cap (Thorarinsson and others 1974); subsequent research has focused on glaciological studies (Williams 1983[a], 1986). Only limited analysis has been done with Seasat Synthetic Aperture Radar (SAR) images of Vatnajökull.

Vatnajökull is situated in south-eastern Iceland. It has an area of 8300 km², according to measurements made from the 22 September 1973 Landsat image (Williams 1983[a]), or 8% of the area of Iceland (73.7% of the area covered by

---

Author's note: The Icelandic letters, thorn (þ), eth (ð), and (æ) are used on all figures and in Table I. In the text, however, thorn, eth, and æ are transliterated to th, d, and ae, respectively. Diacritics over Icelandic vowels are retained in both figures and text.

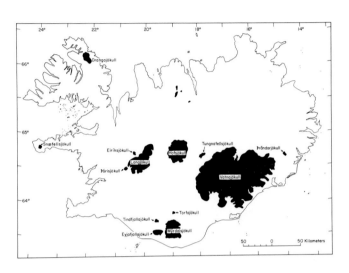

Fig.1. Index map (after Williams 1983[a]) of the 13 principal ice caps of Iceland. Base map modified from Ríkisútgáfa Námsbóka, Reykjavík (no date).

glaciers (Björnsson 1980[a])). It has an estimated volume of 3520 km³ (Bauer 1955, Sigbjarnarson 1971). Although Vatnajökull appears to be simply a single large ice cap, in reality it is dynamically complex, the result of the coalescence of several independent foci of ice accumulation centered on subglacial volcanic highs. Emanating from this 8300 km² composite ice cap are 38 named outlet glaciers (Table I; Fig.2A, B), each with its own physical and flow characteristics. There may be at least as many additional unnamed outlet glaciers, if all the various protruberances around its margin are included (Figs 2A, B, and 3). West of a line between the western edge of Kverkfjöll and the eastern margin of Síðujökull, Vatnajökull straddles the northern part of the Eastern and the southern part of the Northern Volcanic Zones of active volcanism, including the dormant subglacial volcano at Thórdarhyrna which last erupted in 1903 (Thorarinsson and Saemundsson 1980), and intrusive and extrusive volcanic and geothermal activity associated with the calderas at Grímsvötn and Kverkfjöll. Grímsvötn has had a long history of volcanic activity (sporadic documentation since the sixteenth century), with two eruptions in this century (1934 and 1983); there have been seven (or eight?) known volcanic eruptions from the Kverkfjöll area during the past 500 years (Thorarinsson and Saemundsson 1980). The south-eastern part of Vatnajökull is dominated by the ice center and large composite volcano, Öraefajökull, which has the highest elevation in Iceland at 2119 m; two historic eruptions took place in 1362 and 1727, respectively (Iceland Geodetic Survey 1979, Thorarinsson and Saemundsson 1980). Volcanic eruptions from Kverkfjöll and Öraefajökull cause jökulhlaups, or glacier outburst floods, down the valley of Jökulsá á Fjöllum to the north and across the eastern part of the outwash plain of Skeidarársandur on the south, respectively. Grímsvötn is also the source of more periodic jökulhlaups (Rist 1955), following penetration of the ice dam on the east side of the

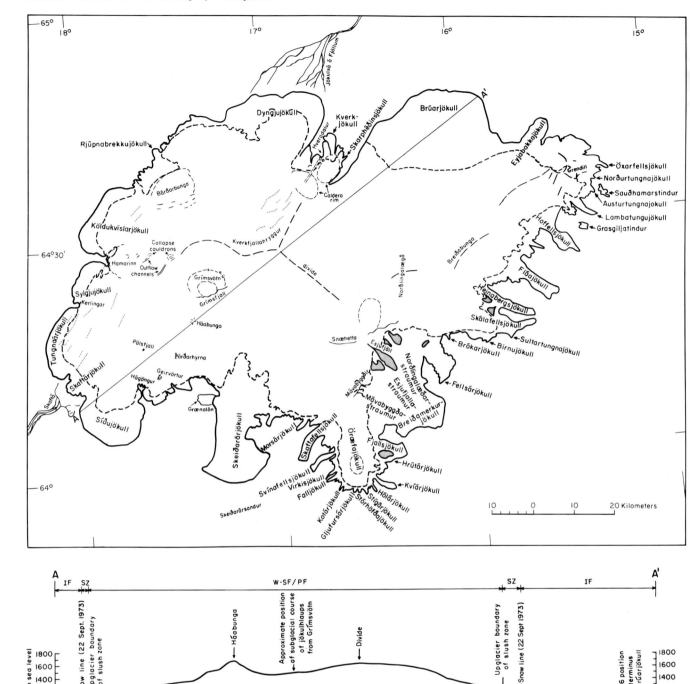

Fig.2A. Map of Vatnajökull, Iceland, showing the 38 named outlet glaciers, one interior ice cap (Öraefajökull), and two detached glaciers, and principal morphological features of the ice-cap surface, such as calderas (short-dashed lines), collapse cauldrons, surface lineations (dotted lines), and the probable positions of surface ice divides (long-dashed lines) derived from Figs 3, 4, and 5. Geographic place-names are from published maps and articles (see Table I). The transient snow line on 22 September 1973 is shown by a heavy dashed line. The stippled areas are nunataks. The portrayal of the glacial headwaters of Jökulsá á Fjöllum and Skaftá are derived from the third edition of the 1 : 500 000 scale *Touring Map of Iceland* published by the Iceland Geodetic Survey (1984).

B. Cross-section (12.5× vertical exaggeration) along line A–A' north-east across Vatnajökull from the terminus of Síðujökull to the terminus of Brúarjökull (123.5 km) showing the altitudinal positions of the transient snow line and the up-glacier boundary of the slush zone on 22 September 1973 (derived from Fig.5). The ice facies (IF) of the ablation area, and the slush zone (SZ) and wet-snow facies (W-SF)/percolation facies (PF) of the accumulation area are also shown. The position of the surface ice divide is derived from Figs 3 and 4. The probable subglacial course of flood water from the March 1972 jökulhlaup from Grímsvötn (Thorarinsson and others 1974) is derived from Fig.3 (series of shallow depressions forming a chain east to south-south-east of Grímsvötn). Tómasson (1975) and Björnsson (1975) also showed the subglacial course of the 1972 jökulhlaup. Topographic elevations along A–A' derived from 1 : 250 000 scale Adalkort (General) maps of Iceland, Iceland Geodetic Survey: Sheet 5 (1978), Sheet 6 (1973), Sheet 8 (1976), and Sheet 9 (1971).

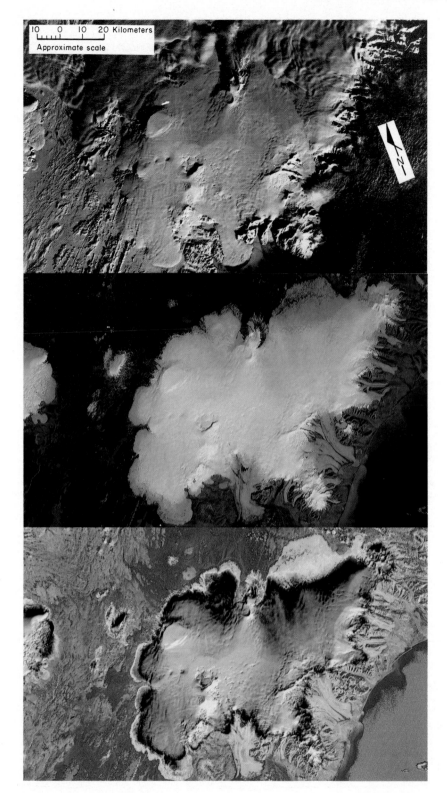

Fig.3. Conventionally processed, low Sun angle (7°) Landsat 1 MSS image (1192-12084, band 7; 31 January 1973), showing detailed surface morphology on Vatnajökull (see Fig.2). (Image by courtesy of EROS Data Center, US Geological Survey.)

Fig.4. Digitally enhanced Landsat 1 MSS false-color composite image (1426-12070, bands 4, 5, and 7; 22 September 1973), showing the principal morphological features on the surface of Vatnajökull (see Fig.2). (Custom digital image-processing by courtesy of Lincoln Perry, EROS Data Center, US Geological Survey.)

Fig.5. Digitally enhanced Landsat 1 MSS false-color composite image (1426-12070), bands 4, 5, and 7; 22 September 1973), showing differences in spectral reflectance of the various facies in the ablation and accumulation areas on the surface of Vatnajökull at the approximate end of the 1973 melt season. (Digital image processing by courtesy of Michael J. Abrams, Jet Propulsion Laboratory, California Institute of Technology.)

TABLE I. NAMED OUTLET AND DETACHED GLACIERS OF VATNAJÖKULL.[1] SEE FIG.2 FOR
GEOGRAPHIC LOCATIONS OF EACH GLACIER. ALTERNATIVE NAMES WHICH APPEAR ON
SOME MAPS ARE GIVEN IN THE FOOTNOTES

Number	Name of outlet glacier	Number	Name of outlet glacier
1	Skeiðarárjökull[2]	23	Brókarjökull
2	Síðujökull[3]	24	Fellsárjökull[7]
3	Skaftárjökull[3]	25	Breiðamerkurjökull[8]
4	Tungnaárjökull[3]	26	Fjallsjökull
5	Sylgjujökull	27	Hrútárjökull
6	Köldukvíslarjökull	28	Kvíárjökull
7	Rjúpnabrekkujökull	39	Hólárjökull
8	Dyngjujökull	30	Stígárjökull
9	Kverkjökull	31	Stórhöfðajökull
10	Skarphéðinsjökull	32	Gljúfursárjökull
11	Brúarjökull	33	Kotárjökull
12	Eyjabakkajökull	34	Falljökull
13	Öxarfellsjökull[4]	35	Virkisjökull
14	Norðurtungnajökull	36	Svínafellsjökull
15	Austurtungnajökull	37	Skaftafellsjökull
16	Lambatungujökull	38	Morsárjökull
17	Hoffellsjökull[5]	39	Öræfajökull[9]
18	Fláajökull		
19	Heinabergsjökull		Name of detached glacier
20	Skálafellsjökull		
21	Sultartungnajökull[6]	40	Sauðhamarstindur
22	Birnujökull	41	Grasgiljatindur[10]

[1]Pálsson (unpublished) used the name Klofajökull in preference to Vatnajökull, although the latter is the accepted name today. Other place-names of outlet glaciers shown on his 1794 map of Klofajökull were as follows: Skeiðarárjökull, Síðujökull, Skaptárjökull (entire western margin), Dýngjujökull, Brúarjökull, Fljótsdalsjökull (north-eastern margin), Lónjökull (north-eastern part of south-eastern margin), Hornafjarðarjöklar (middle part of south-eastern margin), and Breiðamerkrjökull (south-western part of south-eastern margin), and Øræfajökull. Some of these place-names have survived to the present; others have become obsolete.

[2]Some twentieth century maps show the name Súlujökull for the south-western part of Skeiðarárjökull. Súlujökull has now been dropped as a place-name (personal communication from S. Rist, 1986).

[3]Thoroddsen (1892) applied the name Síðujökull collectively to the following three outlet glaciers in south-western Vatnajökull: Síðujökull, Skaftárjökull, and Tungnaárjökull.

[4]Axarjökull.

[5]Some twentieth century maps use the name Svínafellsjökull for the south-western part of Hoffellsjökull. Because of confusion with an outlet glacier of the same name from the western side of Öræfajökull, it has been replaced by the name Hoffellsjökull vestri (personal communication from S. Rist, 1986).

[6]Sultartungujökull; Eyvindstungnakollur.

[7]Fellsjökull; Birnujökull.

[8]Sigbjarnarson (1971) subdivided Breiðamerkurjökull into three parts: Mávabyggðajökull on the west, Esjufjallajökull in the center, and Norðlingalæðarjökull on the east. Rist (personal communication, 1986) stated that these separate parts of Breiðamerkurjökull do not really exist as independent glaciers; he prefers that they be referred to as ice "currents": Mávabyggðastraumur, Esjufjallastraumur, and Norðlingalæðarstraumur.

[9]Öræfajökull is not a true outlet glacier. It is a separate ice cap within the southern margin of Vatnajökull, from which 11 outlet glaciers (26–36) originate.

[10]Hoffellsdalsjökull.

caldera (Björnsson 1975, Tómasson 1975) or, less frequently, from volcanic activity (Björnsson and Kristmannsdóttir 1984, Grönvold and Jóhannesson 1984). The surface manifestation on Vatnajökull of subglacial geothermal and (or) intrusive volcanic activity is often in the form of collapse cauldrons (cone-shaped, concentric fault-bounded depressions) in the ice-cap surface, of which the two collapse cauldrons north-west of Grímsvötn are the most prominent at present. Both of these depressions are the source of periodic jökulhlaups on the south-western margin of Vatnajökull, down the valley of Skaftá (Thorarinsson and Rist 1955, Björnsson 1983). According to Thorarinsson and others (1974), the easternmost cauldron first appeared in 1955; according to Björnsson (1983), the westernmost cauldron was first noted in 1971.

LANDSAT
Landsat images, especially those acquired during 1973, have proven to be the most valuable of all the Landsat images acquired of Iceland between 1972 and 1986, in terms of providing new information about the regional volcanic geomorphology and tectonics of the terrane beneath Vatnajökull (Figs 3 and 4) (Thorarinsson and others 1974). Although the basic image initially used for the geomorphic analysis of the surface of Vatnajökull was partially cloud-covered and acquired at a solar elevation angle of only 7° on 31 January (1192-12084) (Fig.3) (Thorarinsson and others 1974), special digital processing of the 22 September 1973 image (1426-12070) (Fig.4) provides essentially the same information for the entire ice cap. On the 1973 winter image, the northern margin and some of the eastern and

south-eastern parts of Vatnajökull are obscured by clouds; the 1973 late summer image is cloud-free. Image maps at a scale of 1 : 500 000 have been published for both the winter image (US Geological Survey 1977[a]) and the summer image (US Geological Survey 1976) of Vatnajökull.

Although the low Sun angle 1973 winter image provided the best information about the surface morphology of Vatnajökull, the digitally enhanced, cloud-free late summer image was more valuable for glaciological studies, because it was fortuitously acquired at the approximate end of the 1973 melt season. Delineation of the margin of the ice cap, surface ice divides, and the transient snow line (Fig.2A, B) can be easily accomplished from the image (Fig.4). Special digital image processing of the late summer image also permits delineation of the ice facies of the ablation area and the slush zone and wet-snow facies/percolation facies of the accumulation area of Vatnajökull (Figs 2B and 5). The terminology associated with the concept of glacier facies will be defined and discussed in a later section of this paper. Münzer and Bodechtel (1980) also experimented with various digital image-processing techniques on the fall and winter 1973 images of Vatnajökull to analyze the surface expression of subglacial topography and to plot lineaments.

Digital image-enhancement methods

Fig.3 is a standard black-and-white print · of Vatnajökull produced from a multi-spectral scanner (MSS) band 7, 70 mm negative processed by an electron-beam image recorder (EBIR). The Landsat 1 image was acquired at a very low solar elevation angle on 31 January 1973. Michael J. Abrams (Jet Propulsion Laboratory, California Institute of Technology) prepared a series of digitally enhanced images from computer-compatible tape (CCT) data, including linear stretches of the optimum range of digital numbers (DNs) of each of the four MSS bands and bi-band ratios of several band combinations. None of these digitally processed data, however, added any new information about the surface morphology other than that which was already evident on the conventionally processed MSS band 7 EBIR image.

Low Sun angle images of snow-covered surfaces, such as Fig.3, enable subtle morphological features to be recorded by the sensors of the Landsat MSS system. This is the result of differential illumination produced by slight changes in slope orientation, especially those at right-angles to the solar azimuth of 162°. At higher Sun angles, morphological and spectral information is often lost on snow-covered glaciers because of saturation of the MSS detectors (Ferrigno and Williams 1983, Dowdeswell and McIntyre 1986). An analysis of the 31 January 1973 image of Vatnajökull (Fig.3) yielded considerable information about the regional structural, volcanic, and tectonic setting of the terrane under the ice cap (Thorarinsson and others 1974, Williams and Thorarinsson 1974). Fig.3 was especially useful in plotting the probable location of surface ice divides (see Fig.2A and B), calderas, collapse cauldrons, and north-east-trending surface features in the western part of Vatnajökull.

Fig.4 is a digitally enhanced Landsat 1 MSS false-color composite, cloud-free image (bands 4, 5, and 7) of Vatnajökull acquired on 22 September 1973. It was custom processed by Lincoln Perry at the EROS Data Center of the US Geological Survey. A series of standard digital image-processing techniques was applied to all four MSS bands: radiometric restoration (also called destriping), edge enhancement, synthetic line generation, sampling geometric restoration, Earth rotation correction, and detector misregistration correction (US Geological Survey, 1977[b]). Perry also employed a triple piecewise linear-stretch (contrast-enhancement) technique on the three groups of brightness ranges on the Vatnajökull image which were identified from computer-generated histograms of the DNs of each of the three MSS bands (4, 5, and 7): highly reflective snow, medium-reflective vegetation, and low-reflective basalt flows, deep water, and outwash plains (sandar). Each of these three groups of reflectance was independently stretched linearly over a specific DN interval to capture the full brightness range (Table II). Each group in each of the three bands was re-combined to create three new digitally enhanced MSS bands. These new digital data

TABLE II. BRIGHTNESS RANGES (RANGE OF DIGITAL NUMBERS (DNs)) BEFORE AND AFTER ADJUSTMENT FOR THE LINEAR CONTRAST STRETCH OF THE THREE REFLECTANCE GROUPS ON THE 22 SEPTEMBER 1973 LANDSAT 1 IMAGE OF VATNAJÖKULL

Reflectance group	DN range before contrast stretch	Adjusted DN range used for the linear contrast stretch
Basalt flows, deep water, outwash plains (sandar)	0–64	32–100
Vegetation, sediment-laden water, bare glacier ice	65–144	100–140
Snow-covered glaciers, snow-pack	145–255	140–255

were then converted to analog form (film transparency) by a laser beam image recorder (LBIR) and composited on to color film by projecting MSS bands 4, 5, and 7 through yellow, red, and blue filters, respectively, thereby producing the image shown in Fig.4. From a glaciological viewpoint, Perry's achievement was to bring out morphological detail on the highly reflective snow-covered surface of Vatnajökull while still retaining detail in the low-reflectance areas of the ice cap, such as the bare glacier ice and debris-covered termini, and at the same time preserving the standard false-color composite image colors. By comparison, virtually all of the morphological information in the snow-covered areas is lost in the EBIR-generated image. The digitally enhanced image (Fig.4) also very nearly recaptures the morphological detail of the ice-cap surface shown on Fig.3, but at a solar elevation angle of 25°, not 7°, thus indicating that this method can be used on scenes acquired under a wide range of solar elevations, as long as a particular MSS band is not saturated.

Glacio-volcanic geomorphology

The morphological information shown on Fig.4 was used to prepare Fig.2A. Four well-defined, elliptically shaped calderas can be delineated (Thorarinsson and others 1974): one at Grímsvötn, two at Kverkfjöll, and one at Öraefajökull. The 6 km × 4 km outer rim and the inner 3 km × 2 km perimeter of the snow- and ice-covered lake on the caldera floor can be delineated at Grímsvötn. The caldera is the source of periodic jökulhlaups which exit to the east before emerging to the south under the terminus of the Skeidarárjökull outlet glacier (Thorarinsson 1953, 1974, Rist 1955, Björnsson 1975, Tómasson 1975, Björnsson and Kristmannsdóttir 1984). Two well-defined calderas both measuring 5 km × 3 km can be identified in the Kverkfjöll area, either or both being the likely source for jökulhlaups flowing to the north down the valley of the Jökulsá á Fjöllum (Thorarinsson 1950). A linear depression in the ice-cap surface extends 7 km to the south-west from the Hveradalur geothermal area into the westernmost caldera, about doubling the previous known linear extent of the geothermal area (Friedman and others 1972). The total thermal yield from Hveradalur and a subglacial geothermal area located approximately 2 km up-glacier from the terminus of the Kverkfjöll outlet glacier (Kverkjökull) was estimated at $300–540 × 10^6$ cal s^{-1} (by Friedman and others 1972) from calculations of areas of thermal activity interpreted from aerial thermographs and photographs, energy estimates of thermal surface drainage from these areas, and energy estimates from temperature and volumetric discharge measurements at the thermal stream emanating from the terminus of Kverkjökull. The extension under the ice cap to the south could raise this total to as much as $1 × 10^7$ cal s^{-1}, if the thermal output is commensurate with the areal increase. Indistinct linear features (dotted lines on

Fig.2A) on the ice-cap surface extend another 14 km to the south-west from the westernmost caldera towards the Grímsvötn caldera.

Approximately 10 and 14 km north-west of the western margin of the floor of the caldera in Grímsvötn, respectively, are large (2 km diameter) and small (1 km diameter) cauldrons, the result of subglacial geothermal and (or) intrusive volcanic activity at the base of approximately 500 m of ice (Eythórsson 1951, Björnsson 1986[a], [b]). From the larger cauldron a 5 km long sinuous depression extends to the south-west; a 2 km long depression also extends to the south-west from the smaller cauldron. The channel-like depressions are the probable morphological expressions on the ice-cap surface of the initial part of the sinuous sub-glacial cavities of water courses which periodically conduct melt water from the geothermal and (or) intrusive volcanic activity beneath the cauldrons to the beginning of the subaerial part of Skaftá river valley at the terminus of Skaftárjökull (Thorarinsson and Rist 1955). Rist (Thorarinsson and Rist 1955) calculated the total volume of water discharged from the September 1955 Skaftárhlaup at $226 \times 10^6 \, m^3 \pm 45 \times 10^6 \, m^3$, approximately equivalent in volume to the largest cauldron. The dotted lines on Fig. 2A show linear features on the surface of the ice cap which are the subdued surface expression of subglacial north-east-trending móberg (palagonite) ridges, so strongly expressed south-west of the Vatnajökull margin on Fig.3. Melt water that produces jökulhlaups on the Skaftá (Björnsson 1978[a]) are constrained to flow in a south-westerly direction because of the subglacial orientation of the móberg ridges (Thorarinsson and Rist 1955, Björnsson 1986[b]). Details of the subglacial morphology in this part of Vatnajökull have been well defined by analysis of data from radio echo-sounding surveys (Björnsson 1978[b], 1986[a], [b]).

A well-defined 3 km × 2 km caldera can also be delineated in the summit area of Öraefajökull. A subglacial caldera, with dimensions similar to those at Kverkfjöll, is concealed at Bárdarbunga. Another subglacial, extinct(?) volcano is partially exposed at Esjufjöll with an indistinct 5 km × 3 km caldera visible north-east of Snaehetta. The eastern part of Vatnajökull apparently conceals another extinct(?) volcano at Grendill, with a possible eroded summit caldera having dimensions similar to the one at Öraefajökull.

Glaciology

Surface morphology

On Fig.2A, the probable locations of surface ice divides are delineated north of Grímsvötn, including one branch towards Kverkfjöll (Kverkfjallahryggur) and another towards Esjufjöll. Other probable divides are mapped at Snaehetta, transverse to Nordlingalaegd, at Breidabunga, and another extends south-west from Grendill. Dowdeswell (unpublished) delineated surface ice divides on the Nordaustlandet ice caps of Svalbard from Landsat images. No attempt was made to delineate the approximate boundaries between outlet glaciers on the basis on the surface ice divides, such as was done by Dowdeswell and Drewry (1985) in Svalbard, because of the complex character of Vatnajökull with its 38 named outlet glaciers and its variable subglacial volcanic morphology. Comprehensive radio echo-sounding surveys, such as those of Björnsson (1986[a], [b]) will be a necessary part of delineating the drainage-basin boundaries for each outlet glacier. The margin of Vatnajökull, shown on Fig. 2A, which was determined from Fig.4, was used to compile an inventory of the 38 outlet glaciers, one intra-ice-cap ice cap (Öraefajökull), and two detached glaciers (Table I), and to calculate the area of the ice cap (8300 km²). As was already established in the work on Langjökull (Fig.1), Landsat is an excellent source for compiling inventories of large ice masses, especially ice caps and outlet glaciers (Williams 1986). The geographic place-names of the outlet and detached glaciers of Vatnajökull are derived from maps published by the Iceland Geodetic Survey and from maps included in articles published in *Jökull* and other scientific publications with corrections furnished by Sigurjón Rist (personal communication 1986).

Glacier facies

The concept of glacier facies was developed by Benson (1959, 1961, 1962) and Müller (1962). Benson's original concept was modified somewhat in subsequent papers (Benson 1967, Benson and Motyka [1979]). In the discussion of glacier facies which follows, however, the latest published work by Benson (Benson and Motyka [1979]) and the work by Müller (1962) are used.

Benson and Motyka ([1979]) divided a glacier into an *ablation area* and an *accumulation area*; the dividing line between the two areas represents the *equilibrium line*. The equilibrium line also represents the down-glacier boundary of the *superimposed ice zone*. Up-glacier from the equilibrium line is the *snow line*. The bare glacier ice of the ablation area and the exposed part of the superimposed ice zone between the equilibrium line and the snow line are collectively referred to as the *ice facies*. Up-glacier from the snow line are the *wet-snow facies*, the *percolation facies*, and the *dry-snow facies*, separated by the *wet-snow line* and the *dry-snow line*, respectively. Müller (1962) developed a similar scheme but he subdivided the wet-snow facies and the percolation facies of Benson into three zones up-glacier from the annual snow line: *slush zone*, *percolation zone B*, and *percolation zone A*, with a *slush limit* defining the up-glacier boundary of the slush zone and percolation zone B.

By special digital image-processing techniques, it appears possible to spectrally delineate some aspects of glacier facies on Landsat images of glaciers acquired during the ablation season (in particular the ice facies, transient snow line, slush zone, wet-snow facies/percolation facies, and possibly the dry-snow facies, although the latter remains to be established). Fig.5 is a digitally enhanced Landsat 1 MSS false-color composite, cloud-free image (bands 4, 5, and 7) of Vatnajökull acquired on 22 September 1973 (processed by Michael J. Abrams from the same CCTs used to produce Fig.4 (Soha and others 1976, Williams and others 1977)). The image was separated into three reflectance zones: dark (deep water), medium (land including vegetation), and bright (snow), the three zones being originally defined by histograms. Each of these zones was linearly stretched in each of the four MSS bands over the *full* dynamic range for each band, 0–255 DNs for MSS bands 4, 5, and 6, 0–127 DNs for MSS band 7. Unlike Fig. 4, in which the resulting image retained normal false-color image colors, the projection of the fully stretched MSS bands 4, 5, and 7 through yellow, red, and blue filters, respectively, yielded markedly different colors (Soha and others 1976, personal communication from Michael J. Abrams).

Fig.5 shows the position of the transient snow line on Vatnajökull at 12.07 h GMT (local Icelandic time) on 22 September 1973. A review of all Landsat images of Vatnajökull acquired in late August or September indicates that the transient snow line on the 22 September 1973 image is probably at its highest altitude. The elevation of the snow line varies around Vatnajökull (Fig.2B), being higher in the north because of less precipitation (Eythórsson 1960, Björnsson 1980[b]). On 22 September 1973, the "snow line" generally paralleled the 1000 m contour and often conformed to it on several outlet glaciers; on the southern margins, the "snow line" is generally lower than 1000 m, descending to about 700 m on the terminus of Skálafellsjökull; on the northern margin, the snow line is above 1000 m, reaching 1300 m or more on the north-west slope of Bárdarbunga, around Kverkfjöll, and on the north slope of the area around Grendill.

Fig.2B, a cross-section along A–A' on Fig.2A, a distance of 123.5 km from the terminus of Sídujökull in south-western Vatnajökull to the terminus of Brúarjökull on the northern part of the ice cap, shows the snow line at about 900 m on the former and about 1150 m on the latter. The planimetric map position of the "snow line" on 22 September 1973, derived from the Landsat imagery, for the entire ice cap is shown as a heavy dashed line on Fig.2A. Sídujökull and Brúarjökull are classified as surging glaciers, and both surged in 1963–64, 0.5 km and 8.0 km, respectively (Thorarinsson 1969). The position of the terminus of Brúarjökull in 1936 is shown by an arrow, approximately 6 km further south than its position on 22 September 1973.

Also on Fig.2B, I have shown by arrows the position of the up-glacier edge of the slush zone on Síðujökull and Brúarjökull at 1000 m and 1200 m, respectively. The positions of these arrows on the cross-section were determined by analysis of the color patterns on Fig.5.

The type of computer-enhanced color image shown in Fig.5 appears to provide a good correlation of spectral variations on the surface of Vatnajökull, with the ice facies of the ablation area and some of the facies of the accumulation area of a temperate glacier. On Vatnajökull, the colors apparently correlate as follows: ice facies (ablation area and superimposed ice zone of the accumulation area): light blue, bare glacier ice; orange, concentration of supra-glacial debris (on the surface of the glacier) or englacial material (within the glacier; for example, "dirty" ice versus clean ice). According to Benson and Motyka ([1979]), the ice facies includes the entire ablation area *and* extends into the accumulation area as a superimposed ice zone (if one is present), between the equilibrium line (boundary between the ablation area and the accumulation area) and the snow line, *and* also extends beneath the wet-snow facies. It is probably not possible to differentiate between the spectral signatures of bare glacier ice of the ablation area and the exposed part of the superimposed ice zone of the accumulation area, which is also bare glacier ice. On Fig. 2B all bare glacier ice is shown as ice facies. It is not absolutely certain what the orange color represents. It may be a concentration of dirt cones, other surficial debris, or englacial material in the ice facies; field spectra and observations are obviously needed to confirm or reject these conclusions. Laboratory studies of ice suggest that small amounts (~1% or less) or particulate matter can reduce the reflectance at 0.55 μm (Landsat MSS band 4 is 0.50–0.60 μm) by 50% or more (Clark 1982). Perhaps the presence of englacial or supraglacial particulate matter is the explanation for the orange color within some parts of the ice facies. Accumulation area: black, slush zone; dark to light gray, wet-snow facies/percolation facies. The black color does apparently correspond with the slush zone. It is a relatively narrow zone on the steeper outlet glaciers (a few kilometers wide) but much wider on gently sloping glaciers, such as on Brúarjökull and Dyngjujökull, where the slush zone can extend for more than 10 km. Above the up-glacier boundary of the slush zone (Fig.2B) is the wet-snow facies which grades further up-glacier into the percolation facies.

It is doubtful whether a dry-snow facies of the accumulation area exists on Vatnajökull, except perhaps at the two highest elevations at Bárdarbunga (2000 m) and Öraefajökull (2119 m). According to Einarsson (1976), during July, the warmest month in Iceland, only two small areas around the two highest parts of Vatnajökull lie within the 0 °C isotherm of mean monthly temperature. Because of the absence of permanent year-round weather stations on Vatnajökull, the isotherms on the ice cap are calculated from the average adiabatic lapse-rate in Iceland, 0.067 deg km^{-1}, using interior highland weather stations, such as that at the Mödrudalur farm about 65 km north of Brúarjökull, as the basis for extrapolation (personal communication from Trausti Jónsson, 1986). Because of slope orientation away from the Sun (solar azimuth of 164°), the lighter gray areas on the ice cap are probably the result of this aspect difference in snow reflectance rather than the snow being dry. In all likelihood, the entire ice cap above the up-glacier edge of the slush zone is in the wet-snow facies/percolation facies (Figs 2B and 5), although it is probably not possible to differentiate between the wet-snow facies and the percolation facies solely on the basis of spectral information from the snow-covered surface. The two facies are therefore shown together.

Østrem (1975) showed how the specific mass balance for several Scandinavian glaciers can be determined remotely (on satellite images or aerial photographs) by determining the elevation of the snow line (if it conforms to the equilibrium line) at the end of the melt season. From more than 20 years of mass-balance studies on Nigardsbreen, Norway, Østrem and Haakensen (in press) found a linear relationship between the elevation of the equilibrium line at the end of the ablation season and the specific net mass balance of a glacier. Once the curve is established for a glacier, remote determination of the equilibrium line will

give the specific net mass balance for that year. An alternative, although less precise technique is to use the accumulation-area ratio (AAR) to infer whether a given glacier has a positive or negative mass balance (Krimmel and Meier 1975). On the 22 September 1973 image of Vatnajökull (Figs 2B, 4, and 5), the estimated minimum AAR for Vatnajökull was 0.70, if the position of the snow line on that date was approximately the same as the position of the equilibrium line. Most glaciologists consider a range of 0.5–0.8 in the AAR for b = 0 to be reasonable for temperate mountain glaciers.

SEASAT

Seasat Synthetic Aperture Radar (SAR) images of all but the extreme western and north-eastern parts of Iceland were acquired during late summer and early fall of 1978. Bodechtel and others (1979) were the first to compare Seasat SAR and Landsat MSS images of Iceland to delineate morphologic and tectonic features. Ford and others (1980) published a cursory analysis of a Seasat SAR image (revolution 719; 16 August 1978) and a companion Landsat image (2494-11503; 30 May 1976) of north-central Iceland which included parts of the Dyngjujökull outlet glacier and the Kverkfjöll area. Hunting Geology and Geophysics Ltd (undated) published a 1 : 500 000 scale Seasat SAR image mosaic of most of Iceland, including all of its glaciers.

An analysis of the Hunting Geology and Geophysics Ltd mosaic does not provide any new or improved information about the morphology of the surface of Vatnajökull when compared with either the January 1973 or digitally processed September 1973 Landsat images. Whether or not morphological features are imaged is strongly influenced by the look-angle of the Seasat SAR with respect to the orientation of the landform during the orbital passes (north-west to south-east or vice versa across Iceland). The two calderas at Kverkfjöll and the linear extension of the Hveradalur geothermal area are prominently imaged. The Grímsvötn caldera is only faintly visible; the caldera on Öraefajökull was not even recorded on Seasat SAR images. Medial moraines are faintly visible as are the two collapse cauldrons and associated sinuous depressions north-west of Grímsvötn. No surface ice divides can be delineated; the pro-glacial lake Graenalón is difficult to delineate, and the various facies of the ablation and accumulation areas were not recorded. The transient snow line is also difficult to delineate over most of the ice cap. Rott (1984[a], [b]) briefly discussed his analysis of a 24 August 1978 Seasat SAR image of Hofsjökull (Fig.1), a small ice cap (915 km^2) which lies about 30 km north-west of Vatnajökull. He found a low radar return from the wet-snow facies at the higher elevations on this ice cap. Radar back-scatter increased down-glacier until it was difficult to discriminate between bare glacier ice and snow in the marginal part of the wet-snow facies. If the back-scatter intensity were identical from bare glacier ice and the slush zone at L-band frequencies, this would explain the difficulty in determining the position of the transient snow line on the L-band SAR image of Vatnajökull.

Neither the Seasat SAR image mosaic of Vatnajökull nor the individual image strips correctly portray the ice cap geometrically; the image cannot be used to produce an accurate planimetric map. In mountainous areas (south-eastern part of Vatnajökull) it is difficult to delineate the termini of the outlet glaciers. Quite the opposite is the case with the lobate outlet glaciers of most of the rest of Vatnajökull. The termini of these outlet glaciers, such as Síðujökull, Dyngjujökull, and Brúarjökull, and associated terminal moraines are clearly shown; their delineation on the Seasat SAR image is superior to the digitally enhanced Landsat MSS image and about equal to a Landsat 3 RBV image (Williams 1979). The combination of improved spatial resolution and the response of two different parts of the electromagnetic spectrum is considered to be the key.

CONCLUSIONS

Landsat MSS images to date have proven to be the best source of satellite-image data for studying surface morphology and glaciological phenomena, such as glacier

facies, on the large and complex Vatnajökull ice cap, and in preparing a preliminary inventory of its outlet glaciers (Fig.2A). The successful use of Landsat images to prepare preliminary inventories of some of Iceland's glaciers (Williams 1983[a], [b], 1986) was also used by the Temporary Technical Secretariat for the World Glacier Inventory as the basis for preparing new guidelines for preliminary glacier inventories in polar areas (Scherler 1983). It seems likely that satellite-imaging technology will be increasingly used for some types of glaciological studies, especially the preparation of preliminary inventories and dynamic changes in glacier facies of large ice masses, such as the polar ice sheets and ice caps (Swithinbank 1984, Haeberli 1985, Hall and Martinec 1985, Williams 1985).

Future work on Vatnajökull should include analysis of Système Probatoire d'Observation de la Terre (SPOT) and Landsat Thematic Mapper (TM) images, especially the thermal infra-red and mid-infra-red bands of the latter to determine their applicability to glaciological studies. To correlate better the different spectral reflectance of snow, ice, and morainic debris on the surface of Vatnajökull to the various facies of the ablation and accumulation areas, snow and (or) ice observations, field spectra, and ambient-temperature data need to be collected during traverses of outlet glaciers of the ice cap during the summer melt season.

REFERENCES

Bauer A 1955 Contribution à la connaissance du Vatnajökull-Islande. *Jökull* 5: 11-22

Benson C S 1959 Physical investigations on the snow and firn of northwest Greenland 1952, 1953, and 1954. *US Army Snow, Ice, and Permafrost Research Establishment. Research Report* 26

Benson C S 1961 Stratigraphic studies in the snow and firn of the Greenland ice sheet. *Folia Geographica Danica* 9: 13-37

Benson C S 1962 Stratigraphic studies in the snow and firn of the Greenland ice sheet. *US Army Snow, Ice, and Permafrost Research Establishment. Research Report* 70

Benson C S 1967 Polar regions snow cover. *In* Oura H (ed) *Physics of Snow and Ice. International Conference on Low Temperature Science ... 1966 ... Proceedings.* Vol 1, Pt 2. [Sapporo], Hokkaido University. Institute of Low Temperature Science: 1039-1063

Benson C S, Motyka R J 1979 Glacier-volcano interactions on Mt. Wrangell, Alaska. *University of Alaska. Geophysical Institute. Annual Report,* 1977-78: 1-25

Björnsson H 1975 Explanation of jökulhlaups from Grímsvötn, Vatnajökull, Iceland. *Jökull* 24, 1974: 1-26

Björnsson H 1978[a] Könnun á jöklum með rafsegulbylgjum. *Náttúrufrædingurinn* 47(3-4): 184-194

Björnsson H 1978[b] The cause of jökulhlaups in the Skaftá river, Vatnajökull. *Jökull* 27, 1977: 71-78

Björnsson H 1980[a] Glaciers in Iceland. *Jökull* 29, 1979: 74-80

Björnsson H 1980[b] The surface area of glaciers in Iceland. *Jökull* 28, 1978: 31

Björnsson H 1983 A natural calorimeter at Grímsvötn; an indicator of geothermal and volcanic activity. *Jökull* 33: 13-18

Björnsson H 1986[a] Delineation of glacier drainage basins on western Vatnajökull. *Annals of Glaciology* 8: 19-21

Björnsson H 1986[b] Surface and bedrock topography of ice caps in Iceland, mapped by radio echo-sounding. *Annals of Glaciology* 8: 11-18

Björnsson H, Kristmannsdóttir H 1984 The Grímsvötn geothermal area, Vatnajökull, Iceland. *Jökull* 34: 24-50

Bodechtel J, Hiller K, Münzer U 1979 Comparison of Seasat and Landsat data of Iceland for qualitative geologic applications. *In Seasat-SAR Processor Workshop, Frascati. Proceedings.* Paris, European Space Agency: 61-67

Clark R N 1982 Implications of using broadband photometry for compositional remote sensing of icy objects. *Icarus* 49(2): 244-257

Dowdeswell J A Unpublished Remote sensing studies of Svalbard glaciers. (PhD thesis, University of Cambridge, 1984)

Dowdeswell J A, Drewry D J 1985 Place names on the Nordaustlandet ice caps, Svalbard. *Polar Record* 22(140): 519-523

Dowdeswell J A, McIntyre N F 1986 The saturation of LANDSAT MSS detectors over large ice masses. *International Journal of Remote Sensing* 7(1): 151-164

Einarsson M A 1976 *Vedurfar á Íslandi.* Reykjavík, Idunn

Eythórsson J 1945 Jöklaritid. *In* Eythórsson J (ed) *Ferdabók Sveins Pálssonar.* Reykjavík, Prentsmidjan Oddi: 423-552

Eythórsson J 1951 Thykkt Vatnajökuls. *Jökull* 1: 1-6

Eythórsson J 1960 *Vatnajökull.* Reykjavík, Almenna Bókafélagid

Ferrigno J G, Williams R S Jr 1983 Limitations in the use of Landsat images for mapping and other purposes in snow- and ice-covered regions: Antarctica, Iceland, and Cape Cod, Massachusetts. *Proceedings of the Seventeenth International Symposium on Remote Sensing of Environment.* Ann Arbor, Environmental Research Institute of Michigan: 335-355

Ford J P and 6 others 1980 *Seasat views North America, the Caribbean, and western Europe with imaging radar.* Pasadena, CA, Jet Propulsion Laboratory (JPL Publication 80-67)

Friedman J D, Williams R S Jr, Pálmason G, Miller C D 1969 Infrared surveys in Iceland - preliminary report. *US Geological Survey. Professional Paper* 650-C: C89-C105

Friedman J D, Williams R S Jr, Thorarinsson S, Pálmason G 1972 Infrared emission from Kverkfjöll subglacial volcanic and geothermal area, Iceland. *Jökull* 22: 27-43

Grönvold K, Jóhannesson H 1984 Eruption in Grímsvötn 1983; course of events and chemical studies of the tephra. *Jökull* 34: 1-11

Hall D K, Martinec J 1985 *Remote sensing of ice and snow.* London, Chapman and Hall

Haeberli W 1985 Global land-ice monitoring: present status and future perspectives. *In Glaciers, Ice Sheets, and Sea Level: Effect of a CO_2-induced Climatic Change. Report of a workshop held in Seattle, Washington September 13-15, 1984.* Washington, DC, United States Department of Energy: 216-231

Helland A 1882 Islaendingen Sveinn Pálssons beskrivelser af islandske vulkaner og braeer. *Den Norske Turistforenings Árbok* 1882: 19-79

Hunting Geology and Geophysics Undated *Seasat-1 radar mosaic (of)' Iceland. 1 : 500 000-scale mosaic.* Borehamwood, England, Hunting Geology and Geophysics Ltd

Iceland Geodetic Survey 1979 *Ísland. 1 : 750 000-scale map.* Reykjavík, Uppdráttur Ferdafélags Íslands

Krimmel R M, Meier M F 1975 Glacier applications of ERTS images. *Journal of Glaciology* 15(73): 391-402

Müller F 1962 Zonation in the accumulation area of the glaciers of Axel Heiberg Island, N.W.T., Canada. *Journal of Glaciology* 4(33): 302-311

Münzer U, Bodechtel J 1980 Digitale Verarbeitung von Landsat-Daten in den Eis- und Schneegebieten des Vatnajökulls (Island). *Bildmessung und Luftbildwesen* 48: 21-28

Østrem G 1975 ERTS data in glaciology - an effort to monitor glacier mass balance from satellite imagery. *Journal of Glaciology* 15(73): 403-415

Østrem G, Haakensen N In press Glaciers of Norway. *US Geological Survey. Professional Paper*

Pálsson S Unpublished Forsög til en physisk, geographisk og historisk Beskrivelse over de islandske Isbjerge i Anledning af en Reise til de fornemste deraf i Aarene 1792-1794 med 4 Situations- og Prospect-Tegninger. "Om isbjerge 1792-1794." Reykjavík, Landsbókasafn Íslands JS 26 Fol (Sveinn Pálsson's handwritten manuscript, which was completed in 1795, was translated from the original Danish to Icelandic by Jón Eythórsson and published by him in 1945. Amund Helland (1882) published (set in type) excerpts from Sveinn Pálsson's original text in Danish, but Helland did not include any maps or drawings included in the original manuscript. Helland also did not publish Part I, Sections 1-10 or Part III, Sections 23-28)

Rist S 1955 Skeidarárhlaup 1954. *Jökull* 5: 30-36
Rott H 1984[a] Synthetic aperture radar capabilities for snow and glacier monitoring. *Advances in Space Research* 4(11): 241-246
Rott H 1984[b] The analysis of backscattering properties from SAR data of mountain regions. *IEEE Journal of Oceanic Engineering* OE-9(5): 347-355
Scherler K E 1983 *Guidelines for preliminary glacier inventories.* Zürich, Swiss Federal Institute of Technology. Temporary Technical Secretariat for the World Glacier Inventory
Sigbjarnarson G 1971 On the recession of Vatnajökull. *Jökull* 20,1970: 50-61
Soha J M, Gillespie A R, Abrams M J, Madura D P 1976 Computer techniques for geological applications. *Proceedings of the Caltech/JPL Conference on Image Processing Technology, Data Sources and Software for Commercial and Scientific Applications.* Pasadena, CA, Jet Propulsion Laboratory: 4-1-4-21
Swithinbank C 1984 A distant look at the cryosphere. *Advances in Space Research* 5(6): 263-274
Thorarinsson S 1950 Jökulhlaup og eldgos á jökulvatnasvædi Jökulsár á Fjöllum. *Náttúrufrædingurinn* 20(3): 113-133
Thorarinsson S 1953 Some new aspects of the Grímsvötn problem. *Journal of Glaciology* 2(14): 267-275
Thorarinsson S 1969 Glacier surges in Iceland, with special reference to the surges of Brúarjökull. *Canadian Journal of Earth Sciences* 6(4, Pt2): 875-882
Thorarinsson S 1974 *Vötnin strid. Saga Skeidarárhlaupa og Grímsvatnagosa.* Reykjavík, Bókaútgáfa Menningarsjóds
Thorarinsson S, Rist S 1955 Skaftárhlaup í september 1955. *Jökull* 5: 37-40
Thorarinsson S, Sæmundsson K 1980 Volcanic activity in historical time. *Jökull* 29, 1979: 29-32
Thorarinsson S, Sæmundsson K, Williams R S Jr 1974 ERTS-1 image of Vatnajökull: analysis of glaciological, structural, and volcanic features. *Jökull* 23, 1973: 7-17
Thoroddsen Th 1892 Islands Jøkler i Fortid og Nutid. *Geografisk Tidskrift* 11(5-6), 1891–1892: 111-146
Tómasson H 1975 Grímsvatnahlaup 1972, mechanism and sediment discharge. *Jökull* 24, 1974: 27-39
US Geological Survey 1976 *Vatnajökull, Iceland (fall scene). 1 : 500 000 scale.* Reston, VA, US Geological Survey (Landsat Image Format Series N6359WO1723. Experimental printing)
US Geological Survey 1977[a] *EROS digital image enhancement system (EDIES) fact sheet.* Sioux Falls, SD, EROS Data Center
US Geological Survey 1977[b] *Vatnajökull, Iceland (winter scene). 1 : 500 000 scale.* Reston, VA, US Geological Survey (Landsat Image Format Series N6359WO1723. Experimental printing)
Williams R S Jr 1979 Regional geologic mapping using Landsat 3 return beam vidicon images: examples from Iceland and Cape Cod, Massachusetts. *Geological Society of America. Abstracts with Programs* 11(7): 541
Williams R S Jr 1983[a] Remote sensing of glaciers. *In* Colwell R N (ed) *Manual of remote sensing. Geological applications. Second edition.* Falls Church, VA, American Society for Photogrammetry and Remote Sensing: 1852–1866
Williams R S Jr 1983[b] Satellite glaciology of Iceland. *Jökull* 33: 3-12
Williams R S Jr 1985 Monitoring the area and volume of ice caps and ice sheets: present and future opportunities using satellite remote-sensing technology. *In Glaciers, Ice Sheets, and Sea Level: Effects of a CO_2-induced Climatic Change. Report of a workshop held in Seattle, Washington September 13–15, 1984.* Washington, DC, United States Department of Energy: 232-240
Williams R S Jr 1986 Glacier inventories of Iceland: evaluation and use of sources of data. *Annals of Glaciology* 8: 184-191
Williams R S Jr, Thorarinsson S 1974 ERTS-1 image of the Vatnajökull area: general comments. *Jökull* 23, 1973: 1-6
Williams R S Jr, Mecklenburg T N, Abrams M J, Gudmundsson B 1977 Conventional vs. computer-enhanced Landsat image maps of Vatnajökull, Iceland. *Geological Society of America. Abstracts with Programs* 9(7): 1228-1229

Annals of Glaciology 9 1987
© International Glaciological Society

COMPARISON OF OBSERVED AND MODELED ICE MOTION IN THE ARCTIC OCEAN

by

H. Jay Zwally

(Laboratory for Oceans, NASA/Goddard Space Flight Center, Greenbelt, MD 20771, U.S.A.)

and

John E. Walsh

(Department of Atmospheric Sciences, University of Illinois, Urbana, IL 61801, U.S.A.)

ABSTRACT

Daily maps of multiyear ice concentration, derived from Nimbus-7 SMMR passive microwave data, are analyzed to obtain the displacement of the multiyear ice edge and information on the convergence/divergence within the pack. The dynamic–thermodynamic sea-ice model of Hibler (1979) is run with daily time steps and with forcing by the interannually varying fields of geostrophic wind and temperature-derived thermodynamic fluxes. Model-data comparisons are made for the net drift during the months of November through January of the 1978–79, 1979–80, and 1980–81 seasons, and for the shorter-term drift during a 52 day period. Both the model and the data-based drifts for the 25 November 1978 to 28 January 1979 period differ from the classical Beaufort-gyre pattern exhibited in the other two winters. For the 52 day period of November–December 1978, both the model and the data show an eastward drift followed by a westward drift of the ice edges in the Laptev Sea, and for the 25 November 1978 to 28 January 1979 period, a net westward drift of about 250 km. Overall, the model and the data exhibit the same patterns of ice movement with marked month-to-month and large interannual variations in the drift. Good agreement is found in most regions of the central Arctic, but pronounced discrepancies occur near the edge of the total ice pack in the East Greenland Sea. During a short period of large changes in multiyear ice concentration in the central Arctic around 2 December 1980, the divergence implied by the changes in multiyear concentration is qualitatively compared with the divergence computed from the modeled velocity fields. Both the microwave data and the model results indicate similar temporal characteristics of pack-ice response during this major deformation event.

1. INTRODUCTION

Large-scale geophysical modeling requires effective use of commensurate data fields for initialization and verification. In the case of large-scale sea-ice modeling (e.g. Hibler 1979, 1980; Parkinson and Washington 1979), climatological data fields have been mostly used with few exceptions. This situation exists largely because only a few sea-ice parameters are observable on global scales with sufficient frequency. Furthermore, the observable parameters are mostly not the same as the ice parameters on which the models have been structured. The exception is sea-ice extent, which is usually defined as the 10–15% total ice-concentration boundary. Sea-ice extent has been well measured by satellites since 1973, and has been used for comparison with sea-ice models by several investigators (Hibler and Ackley 1983; Parkinson and Bindschadler 1984). Limited use has also been made of sea-ice concentration from passive microwave imaging for comparison with a similar model parameter, compactness (Hibler 1979; Parkinson 1983). Recently, observed atmospheric data over 25 years have been used to model the interannual

fluctuations of the Arctic ice pack with particular attention to the ice-velocity fields (Hibler and Walsh 1982; Walsh and others 1985).

The modeled ice-velocity fields are highly variable on the short time-scales characteristic of synoptic weather systems and on interannual time-scales. Consequently, large-scale observational data sets on daily to interannual time-scales are essential for effective comparisons with the modeled parameters. Sea-ice thickness is a fundamental parameter describing the ice pack and has been a basic parameter in the formulation of sea-ice models (Thorndike and others 1975). However, there is no known technique for large-scale measurement of sea-ice thickness. Inferences on ice thicknesses from ice-type distributions have some potential utility, but it should be emphasized that direct large-scale measurement of sea-ice thickness is not a realistic possibility. Consequently, it is essential to examine the relationships among the parameters that are observable on large scales, which implies measurement by satellite remote sensing, and the parameters that are presently used (or could be used) in ice models.

Recent analysis of time series of multiyear ice-concentration fields derived from satellite passive microwave data has shown that the winter-time drift of the multiyear ice pack in the central Arctic and information on the convergence and divergence can be obtained. The principle, on which the derivation of drift and convergence/divergence is based, is the approximate conservation of multiyear ice area during winter. Consequently, changes in multiyear concentration can be related to ice advection in places of large concentration gradients and to convergence and divergence within the ice pack. (Multiyear ice is defined in this study as ice that has survived one summer's melt season and includes second-year and older ice types.)

As discussed in the following sections, multiyear concentration is not directly a model parameter, but ice-velocity fields and, consequently, convergence and divergence fields are basic model outputs. The purpose of this paper is to examine the relationships between the modeled ice velocity and the drift and convergence/divergence derived from the multiyear concentration maps. In particular, the compatibility of the fluctuations of the quantities deduced from the SMMR imagery and from a dynamic–thermodynamic sea-ice model is addressed. The emphasis is on time-scales ranging from several days to several months.

2. SEA-ICE MODEL SUMMARY

The ice model used here is based on the formulation of Hibler (1979). Modifications to the model thermodynamics have been described by Walsh and others (1985), who also described the domain, initialization, and other aspects of the multi-decadal simulations performed with the model. The essential features of the model are: (1) a momentum balance based on geostrophically derived air

and water stresses, Coriolis force, ocean tilt, internal ice stress, and inertial terms; (2) an ice rheology based on a viscous–plastic constitutive law and an ice-strength parameter, P*; (3) an ice-thickness distribution characterized by the compactness and the mean ice thickness averaged over an entire grid cell; and (4) a thermodynamic code in which vertical growth rates are estimated from heat-budget computations at the top and bottom surfaces of the ice and from heat stored in a motionless oceanic boundary layer. In order to incorporate the strong thickness dependence of ice-growth rates, the thermodynamic computations utilize a seven-level distribution of thicknessess equally spaced between 0 and $2h_m$, where h_m is the mean thickness of the ice in a grid cell. Each of these thicknesses is assumed to represent one-seventh of the ice-covered area of each grid cell.

The thickness distribution described above represents merely a first-order attempt to include a range of thicknessess in the model thermodynamics. In reality, a considerable part of the central Arctic contains deformed ice, some of which will exceed $2h_m$ in thickness. Perhaps more importantly, the mix of first-year and multiyear ice may favor a tendency toward non-linear thickness distributions with peaks corresponding to new, second-year, and/or older multiyear ice (e.g. Williams and others 1975, Fig.9). The distinction between first-year and multiyear ice, while readily apparent in the SMMR imagery, is lost in the model formulation because of the oversimplification of the thickness distribution.

The mean thickness, h_m, and the compactness, A, of the ice in a grid cell change over time through the processes of advection, convergence/divergence, and growth/melt (Hibler 1979):

$$\frac{\partial h_m}{\partial t} = -\frac{\partial(uh_m)}{\partial x} - \frac{\partial(vh_m)}{\partial y} + S_h + \text{diffusion} \quad (1)$$

$$\frac{\partial A}{\partial t} = \frac{\partial(uA)}{\partial x} - \frac{\partial(vA)}{\partial y} + S_A + \text{diffusion} \quad (2)$$

where u and v are the ice-velocity components in a Cartesian (x,y) coordinate system. The diffusion terms in Equations (1) and (2) are only included for numerical stability and are small. S_h and S_A represent the changes due to growth and melt:

$$S_h = \sum_{i=1}^{7} \frac{A}{7} f(h_i) + (1-A)f(0) \quad (3)$$

$$S_A = \left\{ \begin{matrix} (1-A)\dfrac{f(0)}{h_0} & \text{if } f(0) > 0 \\ 0 & \text{if } f(0) < 0 \end{matrix} \right\} + \left\{ \begin{matrix} 0 & \text{if } S_h > 0 \\ \dfrac{A}{2h_m}S_h & \text{if } S_h < 0 \end{matrix} \right\} \quad (4)$$

where $f(h_i)$ is the growth rate of ice of thickness h_i and $h_0 = 0.5$ m is a prescribed upper bound on the thickness of "thin ice". The term S_h in Equation (3) indicates simply that the net growth rate is a concentration-weighted mean of the growth rates of the ice of different thicknesses (including growth over open water). The above equations are subject to the constraint that $A \leqslant 1$, which effectively introduces a "sink term" when $A = 1$ under convergent ice motion. This sink term allows h_m to increase while A remains equal to 1.0, thus permitting some thickness build-up under condition of 100% ice cover. The S_A term in Equation (4) allows for the rapid decay of open water (1 - A) under freezing conditions, as well as for the decrease of A under melting conditions ($S_A \leqslant 0$). Under the assumption of a linear distribution of ice thickness, the ice of thickness less than $S_h \Delta t$ will melt and form open water over time Δt; this ice covers an areal fraction equal to $S_h \Delta t A / 2h$, which leads to the second term in Equation (4). This term does not represent lateral melt but rather the melt of the thinnest ice within an assumed thickness distribution.

The growth rates $f(h_i)$ are evaluated from an energy-balance computation including the major thermodynamic fluxes at the top and bottom ice surfaces (e.g. radiative fluxes, sensible and latent heat transfer,

conduction). The formulation of these terms is essentially the same as that used by Parkinson and Washington (1979), although the treatment here partitions the snow cover as well as the sea ice into seven thickness categories (Walsh and others 1985).

As noted earlier, the linear distribution assumed for the ice thickness is somewhat unrealistic. An associated problem in the model formulation is the temporal dependence of the thickness-category delimiters. Because these delimiters are equally spaced between 0 and $2h_m$, the delimiters change as the ice thickness changes. It is therefore not possible in the model results to distinguish multiyear and new ice by "tagging" segments of ice cover through periods of growth or melt. Because the simplified treatment of ice thickness in the model thus precludes a direct comparison with the SMMR imagery, the fields of model ice drift and deformation are used for comparison with the SMMR data in the following sections.

3. MULTIYEAR CONCENTRATON DATA

Since 1978, 6 day average maps of multiyear concentration, C_m, have been made by the Nimbus project from the multi-frequency and dual-polarization passive microwave imaging data obtained by the SMMR (Scanning Multichannel Microwave Radiometer) on the Nimbus-7 satellite (e.g. Cavalieri and others 1984). Using the hypothesis that multiyear ice area should be approximately conserved during winter, except for small reductions due to ridging of multiyear ice, Zwally and Cavalieri (paper in preparation) showed that the total area of the observed multiyear ice, (integral $C_m(x,y,t)$), over regional-scale boxes (~1500 km × 1500 km) is approximately conserved from November through April, with short-term variations of about 8% (one sigma) due to advection and undetermined measurement errors.

As discussed by Zwally and Cavalieri (paper in preparation), $C_m(x,y,t)$ follows the continuity equation,

$$\frac{\partial C_m}{\partial t} = -\text{div.}(C_m V) + S_{cm} \quad (5)$$

where V is the ice volocity and S_{cm} is the multiyear source term that is approximately zero during winter. Equation (5) formally relates changes in C_m to divergence of the velocity field and advection. Therefore, the time dependence of C_m can be used to study deformation of the ice pack. In contrast to $C_m(x,y,t)$, the total concentration $C_t(x,y,t)$ also decreases during divergence, for example, but under conditions of ice growth in open leads and polynyas C_t quickly increases as new ice is formed. During convergence, it is presumed that most of the ridging occurs in the weaker new and first-year ice.

Locations of large gradients of C_m, such as near the edge of the multiyear ice pack, provide markers for studying the large-scale drift of the ice pack, using the following relation between the velocity, W_x, of a constant C_m line in the x-direction, for example, and the actual ice velocity u:

$$W_x = \frac{-\partial C_m/\partial t}{\partial C_m/\partial x} = u + \frac{C_m \text{div.}V}{\partial C_m/\partial x} . \quad (6)$$

The second term in Equation (6) tends to be negligible at the edge of the multiyear ice pack, as shown by Zwally and Cavalieri (paper in preparation), where the edge is defined by the $C_m = 20\%$ line. Together, Equations (5) and (6) provide quantitative relations between changes in C_m and drift, and convergence/divergence. A descriptive relationship between motion of the observed multiyear ice pack and surface winds deduced from surface-pressure maps has been illustrated by Zwally and Cavalieri. Here, the sea-ice model is used to relate the observed atmospheric forcing to the observed changes in sea-ice distribution.

4. COMPARISON OF SIMULATED AND SMMR-DERIVED ICE DRIFT

The sea-ice drift obtained from the model results and the SMMR imagery are compared over three time-scales. First, the interannual variability of seasonal means is

NOV. 25   JAN. 30

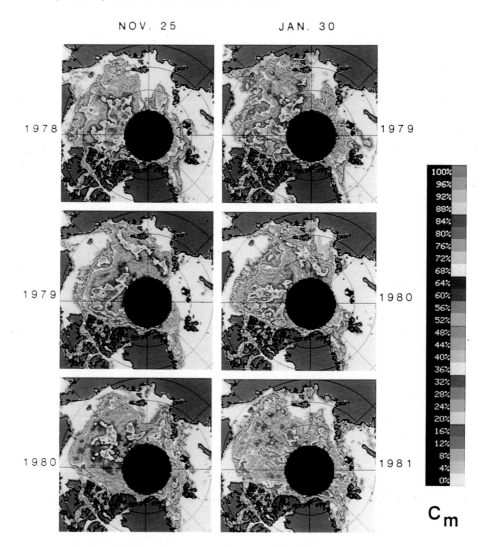

Fig.1. SMMR images of multiyear sea-ice concentration for 25 November and 30 January of 1978–79, 1979–80, and 1980–81.

examined for a set of three successive winters, 1978–79 through 1980–81. Secondly, the two sets of drift information are used in a comparison of intra-seasonal variations during a period characterized by two distinctive regimes of ice drift. Finally, the comparison focuses on the short-term fluctuation associated with an extreme synoptic event that appears to have produced a relatively long-lasting signal in the SMMR imagery for the winter of 1980–81. Comparisons with fluctuations of the analyzed pressure fields for these periods are also noted in this section. In the following section, the results of the December 1980 event are analyzed in terms of a derived quantity, the velocity divergence.

a. Interannual variability of seasonal mean drift

As an illustration of the interannual variability of multiyear sea-ice coverage in the Arctic, Fig.1 shows the multiyear ice concentrations in late November and late January of the winters of 1978–79 through 1980–81. Large interannual fluctuations of the ice edge are apparent in the Laptev, East Siberian, and Chukchi Seas. The multiyear concentrations within the pack also vary considerably from year to year (e.g. January 1979 vs January 1981). A tendency for the concentration anomalies to be of opposite sign in different sectors of the Arctic is apparent. For example, the multiyear ice concentration is noticeably lighter near long. 120–150°E. and heavier near long. 150–180°E. in 1981 than in 1980. Seasonal shifts in the position of the multiyear ice edge are also apparent in Fig.1, as illustrated by the westward shift of the multiyear ice edge in the Laptev Sea between late November 1978 and late January 1979. The seasonal advances and retreats of the multiyear edge are summarized for each year in Fig.2. Displacements

of 100–200 km during the 10 week period are not uncommon, and the tendency for corresponding displacements of opposite sign in adjacent sectors is again apparent.

Fig.3 shows the model-simulated drift vectors for the periods corresponding to the SMMR-deduced drift in Fig.2. While an anticyclonic gyre dominates the North American sector of the central Arctic in 1979–80 and 1980–81, the trans-polar drift stream dominates the entire Arctic Basin in 1978–79. The contrasting flow patterns result in a flux of ice toward the Laptev Sea in 1978–79, modest outflow from the Laptev Sea in 1979–80, and strong outflow in 1980–81. These year-to-year differences are apparent in the areas of multiyear advance and retreat in Fig.2, which also supports the interannual variability of the model's meridional flow component superimposed on the westward drift north of Alaska and the Bering Strait (e.g. retreat in 1978–79, advance in 1980–81, and the gyre-induced juxtaposition of advance and retreat in 1979–80). The largest discrepancy between the model- and SMMR-derived drift is in the East Greenland Sea, especially in 1980–81, when the model indicates outflow rather than the retreat implied by the SMMR data. The model results have been shown elsewhere to be deficient in the North Atlantic waters because of the absence of oceanic coupling (Walsh and others 1985). Since ambiguities in the microwave signature are also known to occur in this region, there is little reason to expect close agreement there.

b. Intra-seasonal variability, late 1978, Siberian sector

By the procedure used to deduce the ice-edge motion depicted in Fig.2, the SMMR-derived multiyear ice concentrations were used to deduce the short-term

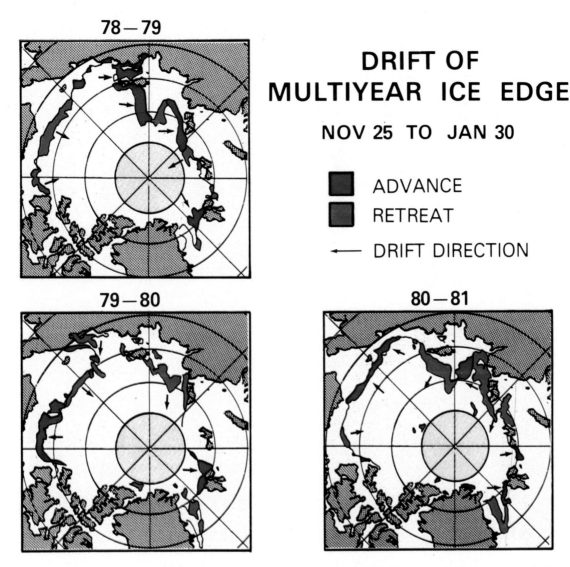

Fig.2. Motion of the multiyear ice edge derived from the SMMR images of Fig.1. Areas of ice advance and retreat are indicated by red and blue, respectively.

movement of the ice edge near long. 140°E. (shown in Fig.1, 25 November 1978) during the period 1 November to 23 December 1978. Inspection of the images had shown that the first part of this period was characterized by generally eastward motion of the ice edge, while the second part was characterized by generally westward motion. Fig.4 shows clearly the contrast in the longitudinal component of the drift of this ice edge prior and subsequent to 23 November. The model-simulated drift for the two sub-periods is shown in Fig.5. The flux of ice across the pole during the first sub-period contrasts with the absence of trans-polar drift in the second sub-period. The dramatic reversal of the simulated drift is especially apparent in the Laptev and East Siberian Seas. The simulated ice thickness averaged over the Laptev Sea is 0.8 m at the end of October, 0.9 m at the end of November, and 1.3 m at the end of December. The more rapid increase in simulated thickness during December is caused by the large ice motion toward the coast.

Support for the hypothesis that wind forcing is primarily responsible for the reversal is provided by the corresponding fields of sea-level pressure (Fig.6). The circulation during the first sub-period is dominated by south-westward flow around a strong Barents Sea cyclone, which is essentially absent in the second sub-period. South-eastward flow during the latter sub-period occurs between a weak ridge near the North Pole and a Bering Sea cyclone displaced about 500 km north-west of its normal position. The larger anomalies relative to the climatological mean circulation occurred during the first sub-period, when departures from normal pressure were as large as 20 mbar in the Barents Sea.

c. Decrease of multiyear coverage, December 1980

The multiyear ice-concentration fields for 30 November (day 335) and 2 December (day 337) 1980 (Fig.7a and b) indicate a rapid reduction of multiyear ice concentration in the pack ice north of Alaska. Decreases of as much as 30–40% occurred during this period in the region between the Chukchi Sea and the North Pole (Fig.7c). Multiyear concentrations of less than 40% then persisted throughout much of the following 2 months, as indicated by the concentration field for 31 January 1981 (see Fig.1). These concentration changes correspond to reductions in total multiyear ice area of more than 50% over scales of several hundred kilometers. During this interval, the observed total multiyear ice area (integral of C_m) over a regional-scale box enclosing the low-concentration area showed a temporary decrease of about 15%, which is typical of the maximum variations of this quantity about a mean value during the winter season and considered to be at least partly due to ice advection through the box boundaries. Specifically, the multiyear ice area in the box (lat. 67.6°N., long. 152.5°W.; lat. 81.1°N., long. 185.2°W.; lat. 67.8°N., long. 118.9°W.; lat. 81.5°N., long. 87.3°W.) enclosing the Beaufort Sea and northward varied from 8.5 to 6.9 to 7.4 to 8.8 × 10⁵ km² on days 335, 337, 339, and 365. The corresponding values of total multiyear ice area in the Chukchi and East Siberian Seas box (lat. 61.9°N., long. 176.0°W.; lat. 70.6°N., long. 208.1°W.; lat. 67.6°N., long. 152.5°W.; lat. 81.1°N., long. 185.2°W.) are 4.6, 4.5, 4.4 to 4.8 × 10⁵ km² on days 335, 337, 339, and 365.

The low concentrations around lat. 80°N. and long.

165°W. appear to be consequences of the strong deformation associated with a major cyclonic system that moved northward into the Arctic on 1–2 December. The intensity of the cyclonic system is apparent in Fig.8, which is an analysis of sea-level pressure observations from the network of Arctic drifting buoys (Thorndike and Colony 1981), one of which reported a pressure of 984 mbar near the analyzed cyclone center in Fig.8. The gradient of sea-level pressure in the Alaskan Arctic is stronger during this event than in any other December analysis of the 5 year record of buoy-derived analyses. While the cyclonic system weakened and migrated northward after 2 December, it remained the

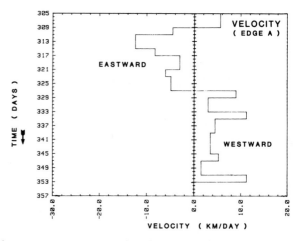

Fig.4. East–west component of motion of multiyear ice edge between lat. 77° and 82°N. at approximately long. 140°E. in the Laptev Sea from 1 November to 23 December 1978, as derived from displacement of the SMMR 20% multiyear concentration line.

1978-1979 (NOV.25-JAN.28)

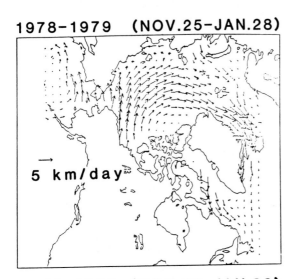

1979-1980 (NOV.25-JAN.28)

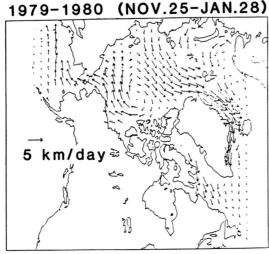

1980-1981 (NOV.25-JAN.28)

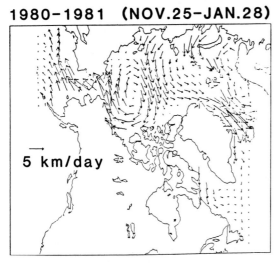

SIMULATED DRIFT VECTORS

Fig.3. Simulated drift vectors for 25 November–30 January of 1978–79, 1979–80, and 1980–81.

OCT. 29 - NOV. 25, 1978

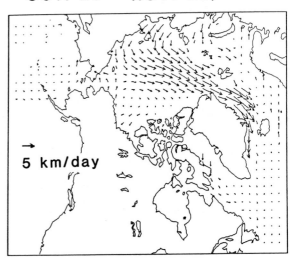

NOV. 26 - DEC. 23, 1978

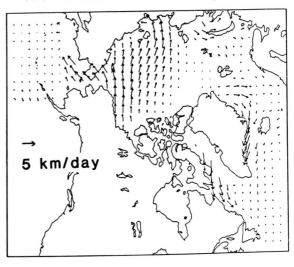

SIMULATED DRIFT VECTORS

Fig.5. Simulated drift vectors for 29 October–25 November 1978 (upper) and 26 November–23 December 1978 (lower) showing correlation with observed motion (Fig. 4).

dominant feature of the Arctic circulation for the subsequent 3–4 days.

Fig.9 shows the simulated drift vectors for the period of 2–8 December 1980. Strong cyclonic drift opposite to the climatological mean drift for early December dominates the Arctic Basin. The apparent divergence of the drift vectors near the cyclone center is consistent with the view that sea ice generally drifts at a slight angle to the right of the surface wind (Zubov 1945). The angular deviation of 5–6° found by Thorndike and Colony (1982) for the winter season is quite similar to the corresponding deflection from the geostrophic wind in the model drift (Walsh and others 1985). The apparent divergence in Fig.9 is clearly consistent with the decrease of ice concentration in the SMMR-derived concentrations of Fig.8. The fact that the reduced concentrations persist through January suggests that major synpotic events can have long-lived effects on the characteristics of sea ice in the central Arctic. This "irreversibility" of the consequences of deformation events requires further investigation in other years and seasons in order to determine the generality of such impacts.

5. DIVERGENCE COMPUTATIONS

While velocity fields such as those discussed in section 4 provide information pertinent to large-scale transport, the spatial gradients of the velocities are more relevant to some fundamental aspects of sea-ice behavior. As components of the strain-rate tensor, the velocity gradients are measures of the deformation that dictates changes in quantities of major importance to the dynamics and thermodynamics of sea ice: the areal fraction of open water, the internal ice stress, and the ice-thickness distribution. In this section, some of the temporal and spatial aspects of the divergence associated with the December 1980 event discussed in section 4 are evaluated.

Fig.10 shows the time series of the daily model divergence averaged over the nine grid cells centered on lat. 80°N., long. 165°W. Fluctuations characteristic of the passage of transient synoptic systems occur throughout much of November and December 1980. The largest value of either sign occurs on 2 December, when the cyclonic system discussed earlier migrated over the nine-point area. Interestingly, the effect of this system appears as a "spike" confined to about 3 days. The short-term nature of the divergence event is supported by the SMMR-derived changes of multiyear ice concentration along a transect from the north-western Canadian coast (lat. 79.4°N, long. 104.3°W.) to the New Siberian Islands (lat. 76.1°N., long. 142.3°E.). Fig.11 shows that the decreases of concentration between 30 November and 2 December are large (20–40% in the central Arctic), but that the antecedent and subsequent 2 day periods show little or no evidence of such a decrease. These results support the contention made earlier that the persistent area of first-year ice during the 1980–81 winter may be attributed largely to a strong but short-lived deformation event.

The short-term forcing of this event must also be viewed in the context of the large-scale distribution of multiyear ice. The SMMR images of Fig.1 show that 1980 is characterized by generally lower concentrations of multiyear ice in the Alaskan sector during late November 1980 than at the corresponding time in 1978 and 1979. The contrast is especially apparent near the Canadian Archipelago, where the multiyear concentrations are 70–80% in the 1978 and 1979 images, but only 40–60% in the 1980 image. Moreover, the total ice concentrations of ~80% depicted for this region during September 1980 in the weekly ice analyses of the US Navy/NOAA Joint Ice Center were less than normal, while the model-derived concentrations of 80–90% for this period were also less than the 30 year September mean of the model results (although several other years had similar concentrations in September). The pack ice in the Alaskan part of the Arctic Basin may thus have been predisposed to a major deformation in response to an intense synoptic system. It is therefore quite possible that the pre-existing state of the pack, as well as the intensity of a particular synoptic system, contributed to the large and rapid decrease of multiyear concentration in December 1980.

OCT. 29 – NOV. 25, 1978

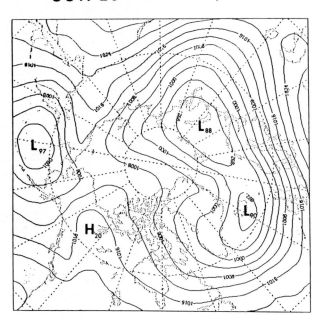

NOV. 26 – DEC. 23, 1978

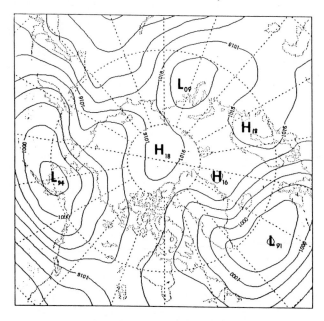

SURFACE PRESSURE

Fig.6. Fields of sea-level pressure averaged over the periods of ice drift shown in Fig.5.

The model-generated fields of divergence during the critical period of the deformation event are also examined on a daily basis for the first 2 weeks of December 1980. In agreement with the inference made visually from Fig.10, the simulated divergence is strongest in the immediate vicinity of the low-pressure center. The area of strongest divergence indeed migrated with the pressure center in the model simulation. It should be noted, however, that the model simulation was forced by gridded pressure analyses produced by the US Navy. The central pressure of the Arctic cyclone in the US Navy analysis for 2 December was 992 mbar (see Fig.8). The cyclonic circulation and associated ice divergence are thus likely to have been weaker in the present model simulation than if the same quantities had been computed from a more accurate analysis of sea-level pressure.

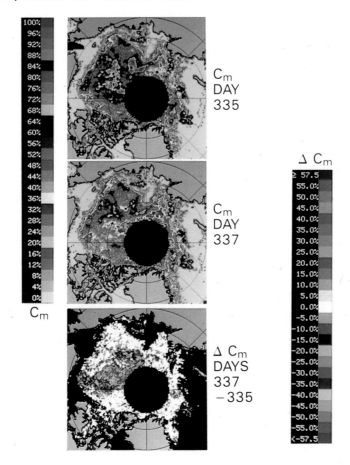

SHORT-TERM MULTIYEAR CONCENTRATION
CHANGES

Fig.7. SMMR images of multiyear sea-ice concentration for 30 November (day 335, top) and 2 December (day 337, center) 1980. Field of change of multiyear concentration between 30 November and 2 December is shown at bottom.

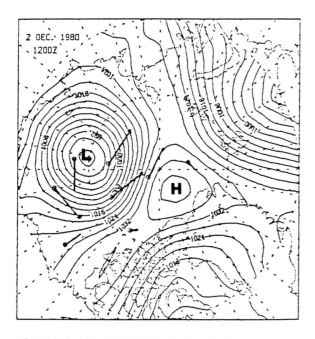

SURFACE PRESSURE/BUOY DRIFT

Fig.8. Sea-level pressure analysis for 12.00Z 2 December 1980. Positions of Arctic buoys are shown by open circles, direction of buoy drift by linear segments (from Thorndike and Colony 1981).

6. CONCLUSIONS

The comparisons presented here show that the sea-ice velocity fluctuations deduced from SMMR data and from a dynamic–thermodynamic sea-ice model are generally compatible over the daily to seasonal time-scales. Several conclusions are drawn from the results in sections 4 and 5.

(1) Daily, monthly, and even seasonally averaged fields of sea-ice motion are highly variable, and departures from the climatological "normal" field can dominate the mean patterns for periods of several days to several months. Thus, the field of motion for a particular month or season cannot be assumed to be representative of the long-term mean for the corresponding month or season.

(2) Because the variable air stress is the primary determinant of the model's drift fluctuations, which are generally consistent with the data-derived drift fluctuations, the variability examined here is attributed primarily to the fluctuations of the geostrophic wind or sea-level pressure. This conclusion applies to the daily to seasonal time-scales, and is not incompatible with the notion that oceanic variability may account for larger parts of the drift variability on time-scales longer than the monthly or seasonal (Thorndike and Colony 1982).

(3) Major synoptic events can have long-lived impacts on the concentration of multiyear ice in the central Arctic. The model-derived velocity divergence supports the inference from the SMMR data that the large decrease of multiyear ice concentration in December 1980 was limited to a period

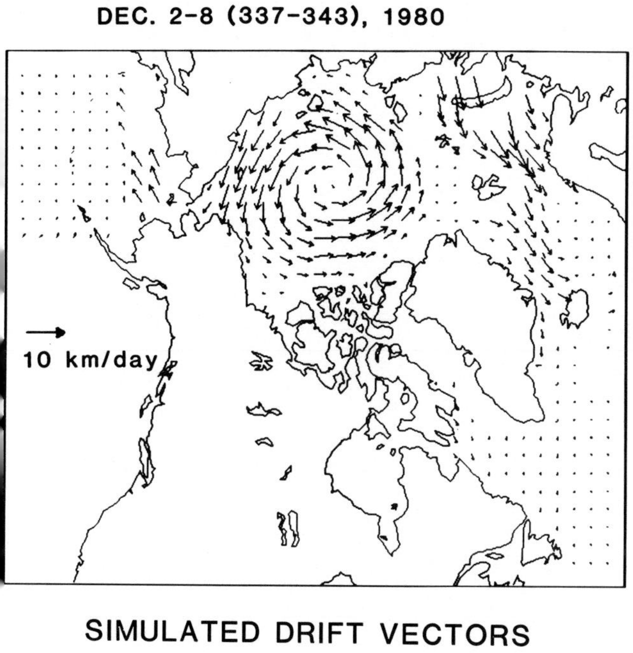
DEC. 2-8 (337-343), 1980
10 km/day
SIMULATED DRIFT VECTORS

'ig.9. Simulated drift vectors for 2–8 December 1980.

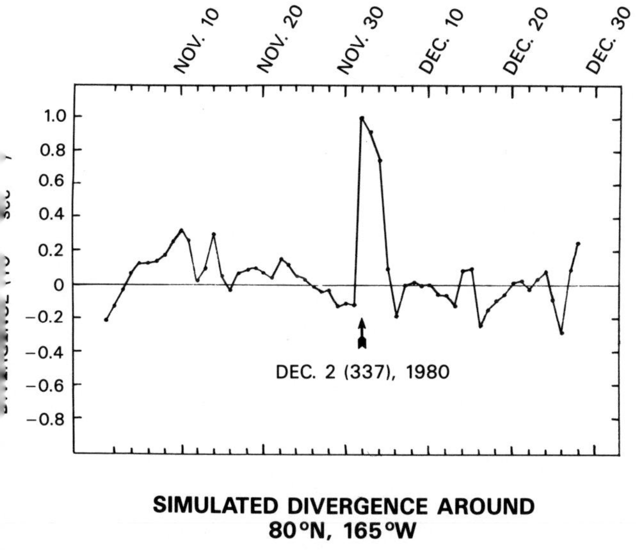
NOV. 10
NOV. 20
NOV. 30
DEC. 10
DEC. 20
DEC. 30
1.0
0.8
0.6
0.4
0.2
0
−0.2
−0.4
−0.6
−0.8
DEC. 2 (337), 1980
SIMULATED DIVERGENCE AROUND
80°N, 165°W

'ig.10. Daily values of divergence of ice-model drift vectors for 1 November–30 December 1980. Divergences are averages over 666 km square (nine grid cells) centered at lat. 80°N., long. 165°W.

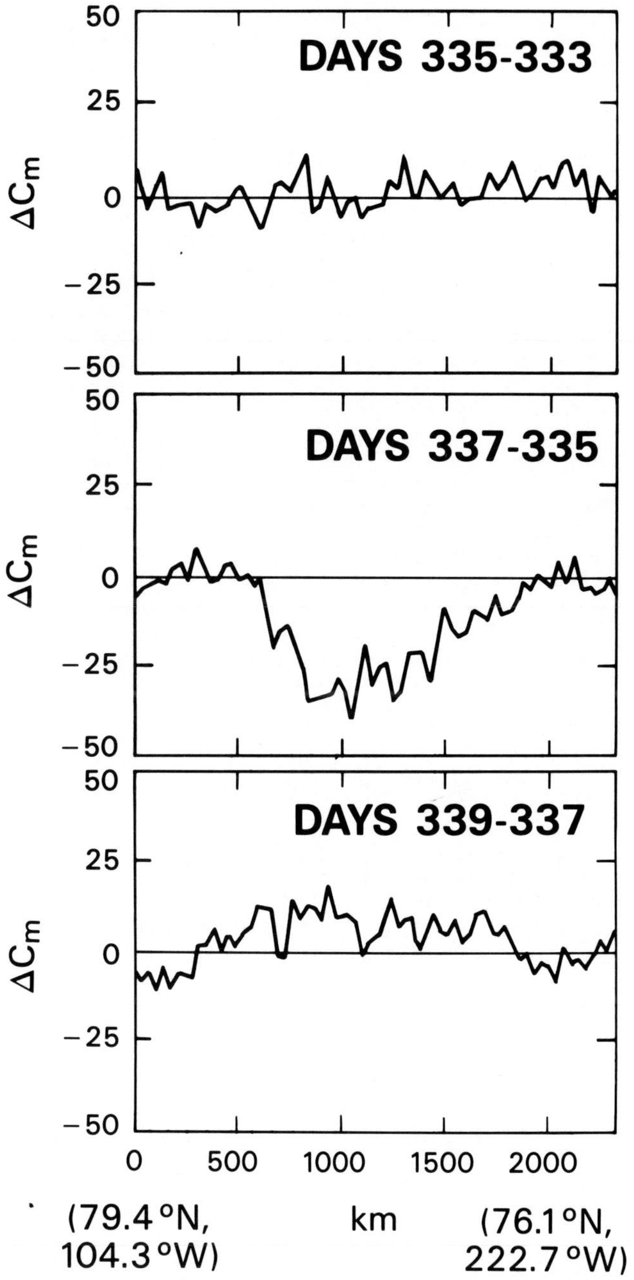
DAYS 335-333
DAYS 337-335
DAYS 339-337
ΔCm
50
25
0
−25
−50
0
500
1000
1500
2000
(79.4°N,
104.3°W)
km
(76.1°N,
222.7°W)

Fig.11. Changes of SMMR multiyear ice concentration along transect from lat. 79.4°N., long. 104.3°W. to lat. 76.1°N., long. 142.3°E. for three 2 day periods of 1980: 28–30 November (top), 30 November–2 December (middle), and 2–4 December (bottom).

: 1–3 days, although the resulting low concentrations were ɔparent in the SMMR imagery for the next several months. he role of the large-scale state of the ice prior to the eformation event remains to be clarified.

) In view of the compatibility of the SMMR- and odel-derived fields of ice divergence, it appears that the MMR data represent an excellent tool for studies of the ass balance of Arctic pack ice. Work described by Zwally nd Cavalieri (paper in preparation) has shown that the total eal coverage of multiyear ice is quasi-conservative during e winter months, implying that estimates of regional vergence of the multiyear ice can provide a measure of e area of new ice formation and growth within the pack uring winter. Because the vast majority of ice growth kes place in areas of open water or young ice, the SMMR ata may provide valuable input to assessments of a crucial omponent of the mass balance. Estimates of the seasonal d interannual variability of the areas susceptible to new e growth can be used in assessing the validity of large-ale ice models, as well as in assessing the variability of the large-scale ice mass balance and the associated thermal and salinity fluxes to the atmosphere and ocean.

ACKNOWLEDGEMENTS

Support for this study was provided by the National Science Foundation, Division of Polar Programs, through grant DPP-8511443 for author J.E.W. and by NASA's Oceanic Processes Program for author H.J.Z. Computer-programming assistance was provided by Becky Ross, Steve Fiegles, and Scott Bringen.

REFERENCES

Cavalieri D J, Gloersen P, Campbell W J 1984 Determination of sea ice parameters with the Nimbus 7 SMMR. *Journal of Geophysical Research* 89(D4): 5355–5369

Hibler W D III 1979 A dynamic thermodynamic sea ice model. *Journal of Physical Oceanography* 9(4): 815–846

Hibler W D III 1980 Modeling a variable thickness sea ice cover. *Monthly Weather Review* 108(12): 1943–1973

Hibler W D III, Ackley S F 1983 Numerical simulation of the Weddell Sea pack ice. *Journal of Geophysical Research* 88(C5): 2873–2887

Hibler W D III, Walsh J E 1982 On modeling seasonal and interannual fluctuations of Arctic sea ice. *Journal of Physical Oceanography* 12(12): 1514–1523

Parkinson C L 1983 On the development and cause of the Weddell polynya in a sea ice simulation. *Journal of Physical Oceanography* 13(3): 501–511

Parkinson C L, Bindschadler R A 1984 Response of Antarctic sea ice to uniform atmospheric temperature increases. *In* Hansen J E, Takahashi T (*eds*) *Climate processes and climate sensitivities.* Washington, DC, American Geophysical Union: 254–264 (Geophysical Monograph 29; Maurice Ewing Volume 5)

Parkinson C L, Washington W M 1979 A large-scale numerical model of sea ice. *Journal of Geophysical Research* 84(C1): 311–337

Thorndike A S, Colony R 1981 *Arctic Ocean buoy program, 1 January 1980–31 December 1980.* Seattle, University of Washington. Polar Science Center

Thorndike A S, Colony R 1982 Sea ice motion in response to geostrophic winds. *Journal of Geophysical Research* 87(C8): 5845–5852

Thorndike A S, Rothrock D A, Maykut G A, Colony R 1975 The thickness distribution of sea ice. *Journal of Geophysical Research* 80(33): 4501–4513

Walsh J E, Hibler W D III, Ross B 1985 Numerical simulation of northern hemisphere sea ice variability, 1951–1980. *Journal of Geophysical Research* 90(C3): 4847–4865

Williams E, Swithinbank C, Robin G de Q 1975 A submarine sonar study of Arctic pack ice. *Journal of Glaciology* 15(73): 349–362

Zubov N N 1945 *L'dy Arktiki,* Moscow, Izdatel'stvo Glavsevmorputi [English translation: *Arctic ice.* San Diego, US Navy Electronics Laboratory, 1963]

Annals of Glaciology 9 1987
© International Glaciological Society

REMOTE SENSING OF SEA-ICE GROWTH AND MELT-POOL EVOLUTION, MILNE ICE SHELF, ELLESMERE ISLAND, CANADA

by

Martin O. Jeffries and William M. Sackinger

(Geophysical Institute, University of Alaska, Fairbanks, AK 99775-0800, U.S.A.)

and

Harold V. Serson

(997 Stellycross Road, Brentwood Bay, British Columbia V0F 1A0, Canada)

ABSTRACT

Periodically since 1950, air photographs and SLAR images have been taken of the Arctic ice shelves. The study of air photographs and SLAR images of the outer part of Milne Ice Shelf had three aims: (1) to map losses and ice re-growth at the shelf front, (2) to map the evolution of melt pools on shelf ice and multi-year land-fast sea ice, and (3) to assess the usefulness of air photographs and SLAR for these purposes. For mapping of ice calvings and subsequent sea-ice growth, both air photographs and radar images have been used sucessfully. However, air photographs are better than radar for mapping ice-surface features. The ridge-and-trough systems that characterize the surface of the ice shelf and old sea ice are clearly visible on each type of imagery but, because of their larger scale, air photographs proved to be most useful for a study of melt-pool evolution. The orientation of the melt pools is parallel to the prevailing winds which drive water along the troughs. The drainage system evolves by a process of elongation and coalesence.

INTRODUCTION

The present Arctic ice shelves off the north coast of Ellesmere Island (Fig.1a) have a total area of about 1350 km² and are the remnants of the once-extensive Ellesmere Ice Shelf. The latter most likely existed as a continuous fringe of thick ice from Point Moss to Nansen Sound (Koenig and others 1952) (Fig.1a). During the last 100 years, the disintegration of the Ellesmere Ice Shelf has created ice islands that drift in the Arctic Ocean (Koenig and others 1952).

The present ice-shelf extent is of concern with regard to possible locations of future ice-island calvings and areas of ice-shelf re-growth. Commonly, the ice shelves increase their lateral extent by the seaward addition of pack ice, although horizontal growth rates are not constant due to interruption and reversal by calving events (Lyons and Ragle 1962). Jeffries and Serson (1986) mapped the extent of multi-year land-fast sea ice between Yelverton Bay and Ward Hunt Ice Shelf in 1984.

The link between ice shelves and ice islands was established largely due to the similarity of their striking surface topography of parallel ridges and troughs (Koenig and others 1952). Similar, but smaller undulations occur on multi-year land-fast sea ice (Hattersley-Smith 1957; Ragle and others 1964; Jeffries and Serson 1986). Hattersley-Smith (1957) reviewed possible origins of the undulations or "rolls", but there have been no systematic studies of this phenomenon.

Ice shelves, multi-year land-fast sea ice, and their surface features are visible on air photographs and SLAR (side-looking airborne radar) images of the north coast of Ellesmere Island. The remote-sensing record for Milne Ice Shelf (Fig.1a) is particularly good and has been chosen for

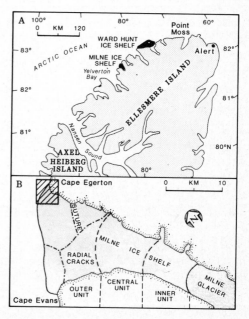

Fig.1a. Location map of Milne Ice Shelf on the north coast of Ellesmere Island. b. Map of Milne Ice Shelf showing ice-shelf units, radial cracks, and a suture. The cross-hatched area at Cape Egerton corresponds to the area mapped in Fig.4.

the study reported here. The purpose of the study was to use air photographs, SLAR imagery, and ground observations, and to asssess the usefulness of each, for detailed mapping of ice losses and ice gains, and mapping of ice-surface features.

REMOTE SENSING AND GROUND OBSERVATIONS OF ARCTIC ICE SHELVES
Background

Air photographs have proven to be useful in establishing the location of ice-island calvings. Koenig and others (1952) compared early photographs of ice islands with 1950 trimetrogon air photography of the ice shelves and established that the ice islands had broken away from the ice shelves. The massive calving of ice from Ward Hunt Ice Shelf was first observed during RCAF reconnaissance flights in early 1962 and photographed in June 1962 (Hattersley-Smith 1963). A comparative study of 1959 and 1974 photography established that Ayles Ice Shelf had moved 5 km out of Ayles Fiord (Jeffries 1986[a]). Jeffries also showed that 33 km² of ice calved from Milne Ice Shelf prior to 1974.

In recent years there has been renewed interest in ice-shelf remote sensing in view of potential ice-island calvings that could pose a subsequent threat to offshore operations in the southern Beaufort Sea. Imagery obtained in the 1980s includes air photographs, SLAR, and SAR (synthetic aperture radar). The SAR imagery is proprietary and, therefore, unavailable. Also, since 1982 field work in this region has included observations of coastal ice conditions, particularly between Yelverton Bay and Ward Hunt Ice Shelf (Fig.1a; Jeffries and Serson 1986).

For the purposes of visual and photographic observations, the surface topography of the ice shelves appears best on nearly cloud-free summer days. However, there are few days on which conditions are suitable for high-level air photography. Thus, summer and non-summer opportunities for successful visual and photographic reconnaissance are limited. The SLAR technique has the advantage of being able to operate year-round under almost any weather conditions. SLAR images of the ice shelves have been obtained from aircraft flying parallel to the coastline.

Outer Milne Ice Shelf

Milne Ice Shelf is the second largest remaining ice shelf and has a total area of about 290 km². On the basis of ice thickness, surface morphology, and surface features, the ice shelf has been subdivided into three distinct units (Fig.1b) (Jeffries 1986[b]). The outer unit, with an area of 140 km² and a maximum thickness of 90 m (Prager unpublished), is the subject of this paper. The purpose of this section is to provide a background description of the surface morphology and features of shelf ice and multi-year land-fast sea ice according to their appearance on air photographs and SLAR imagery.

On outer Milne Ice Shelf the authors have measured the troughs to be as much as 7.5 m deep (ridge top to trough bottom), but most are about 5 m deep. Beginning in July, melt water accumulates in the bottom of the troughs and creates elongated melt pools. On air photographs the melt water creates a dark tone that contrasts with the lighter tone of ice exposed on the ridges (Fig.2a). As at other ice shelves, the rolls on the outer unit are aligned approximately parallel to the coastline. Details of melt-pool orientation will be discussed in a later section. The "look-direction" of the SLAR is essentially perpendicular to the rolls and the troughs are picked out by the radar shadows cast by adjacent ridges (Fig.2b). The ridges appear as a lighter grey tone and suggest greater relative relief or a steeper slope angle, giving rise to the brighter radar reflection.

Cutting across the general orientation of the rolls are three radial cracks, one of which shows as a very dark tone as a result of being water-filled (Figs 1b and 2a). Only one of these cracks remained in 1983 (for reasons to be discussed) and it shows clearly in Fig.2b. The crack itself is a dark tone or shadow cast by the north slope, while the south slope (north-facing) is a lighter grey reflection facing the incoming radar signal. Similar shadows and bright edges also emphasize a second linear, crack-like feature (or suture) at the north-east side of the outer unit (Figs 1b and 2b). This feature is less obvious on air photographs because it does not contain any water (Fig.2a). This was also true in 1984 air photographs.

The fractures cut across the general trend of the rolls, but ground observations show that their morphology does not markedly differ from that of typical troughs. The fractures are similar in width to troughs, and their smooth slopes, as opposed to sharp edges, suggest that the fractures are very old features that have been modified by melt-water processes. The erosion of sharp edges smooths the slopes of the fractures and hence they do not have a very bright radar signature (Fig.2b).

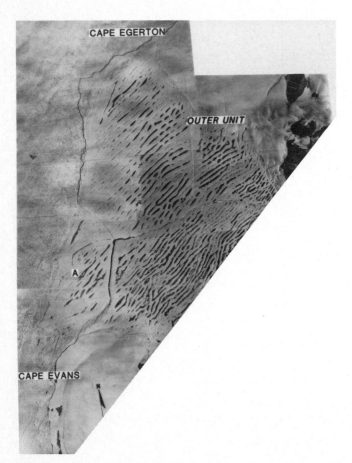

Fig.2a. Left. Vertical air-photo mosaic of outer Milne Ice Shelf taken in July 1959. The ice island that calved in 1964–67 is marked A. Photographs available from the National Air Photographic Library, Ottawa, Ontario, Canada. b. Right. X-band SLAR image of outer Milne Ice Shelf taken in May 1983. The dry snow-pack had a mean depth of about 60 cm. Imagery available from Atmospheric Environment Service, Downsview, Ontario, Canada.

Two of the radial cracks noted in the previous paragraph formed the boundaries of an ice-island calving that occurred at some time during the interval 1959–74 (Jeffries 1986[a]). The calving includes the loss of shelf ice and old land-fast sea ice at Cape Evans. While the sea ice at Cape Evans was lost with the ice island, sea ice at Cape Egerton remained. On 1983 SLAR imagery, this long narrow strip of sea ice appears as a very dark grey tone at the northern edge of the ice shelf (Figs 2b and 4). After the ice-island calving at Cape Evans, the lost shelf ice was replaced by thick sea ice now known as the Milne Re-entrant. In 1983, this extensive area of ice at the south-west shelf front appeared as a featureless, mottled, dark grey tone (Fig.2b). Milne Re-entrant, however, has evolving features as evidenced by the small-scale ridge-and-trough systems shown in Fig.5. During surface traverses in spring 1985, the troughs were found to be up to 1 m deep and, therefore, were probably too small to be resolved by the SLAR system. On the other hand, pressure ridges up to 5 m high, separating the re-entrant from the pack ice, created a clear bright reflection on SLAR (Fig.2b). Likewise, there is a bright linear reflection at Cape Egerton that suggests a pressure ridge in the fast ice. The presence of many flat facets, oriented in many directions, in such ridges means that some of them will provide specular reflection of the incident SLAR radiation and produce bright returns on the images.

ICE CALVING AND RE-GROWTH

Since the advent of aerial and ground survey of the ice shelves after 1950, it has been possible to document ice-shelf losses and ice re-growth. In this section, the available information is used to map recent changes at the front of Milne Ice Shelf. It is noted that there is no evidence to suggest that the ice shelf is moving forward, rather the ice front is quasi-stationary.

Sequences of ice loss and re-growth are shown in Fig.3. There are two notable features of the ice front. First, there is the persistence of an area of very old sea ice at Cape Egerton. The ice was first photographed in 1950 and has, therefore, remained in place for at least 36 years. Secondly, there is the ice-island calving that occurred in 1959–74 near Cape Evans (Fig.3c).

The cause of the calving is unknown, but it occurred along two of the radial cracks which were clearly weaknesses in the ice. Likewise, the exact date of calving is unknown, but by 1974 the lost shelf ice had been replaced entirely by the sea ice of Milne Re-entrant. Ground observations and salinity and isotope analysis of ice cores indicate that the sea ice is 19–22 years old and, therefore, that the calving occurred that long ago (Serson 1984; Jeffries unpublished).

Since 1974, the area of sea ice at the ice-shelf front has remained fairly constant, maintaining a balance between ice loss and ice re-growth. In the previous section, a pressure ridge was noted in the sea ice east of Cape Egerton (Fig.2b). In spring 1985 it was observed that some of this ice had broken away along the line of the pressure ridge, reducing the sea-ice area to about 40 km².

Variations in the extent and location of sea-ice accretion are probably related in some degree to pack-ice movement across the mouth of Milne Fiord. The dominant sea-ice motion is episodic and to the west along the coast. It is reasonable, therefore, to expect that pack ice will pile up against prominent coastal features. This occurred at Markham Bay Re-entrant, for example, where pack-ice interaction with Ward Hunt Ice Shelf caused rafting and hummocking in the sea ice that subsequently became fast and an integral part of the ice shelf (Ragle and others 1964). At Cape Egerton, Milne Ice Shelf protrudes slightly into the pack ice, acting as a thick ice obstacle against which some pile-up will occur. Some of the very old sea-ice strip might have formed in this way. Ground observations in 1983–85 show no evidence of old, weathered pressure ridges on Milne Re-entrant. Lying in the shelter of the main ice-shelf mass and less subject to pack-ice motion, except at the outer margin, the sea ice in Milne Re-entrant has entirely replaced the former ice shelf.

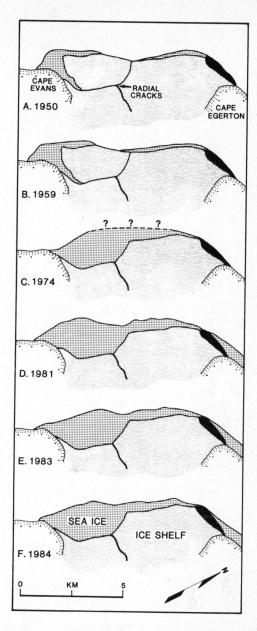

Fig.3. Sequence of sea-ice loss and re-growth at the front of Milne Ice Shelf mapped from air photographs (A, B, C, and F) and SLAR imagery (D and E). The dark shading at Cape Egerton denotes very old sea ice. The area of sea ice in each year is as follows: 1950, 4.5 km²; 1959, 5.0 km²; 1959, 5.0 km²; 1974, 40.0 km²; 1981, 1983, 1984, 42.0 km².

MELT POOLS

The ridge-and-trough systems on the ice shelves and land-fast sea ice are best observed and photographed in summer when the troughs contain water in the form of elongated melt pools. Hattersley-Smith (1957) and Crary (1960) concluded that the ridge-and-trough systems are the result of strong summer winds in association with annual melt-water drainage. But, it has also been suggested that pack-ice pressure and ridge formation will influence the original siting of the rolls (Hattersley-Smith 1957). With the completion of air photography in 1984 there was available a high-quality and long-term photographic record of the ice shelf and sea ice of outer Milne Ice Shelf. Using this series of photographs, the melt-pool patterns on the ice near Cape Egerton (Fig.1a) have been mapped (Fig.4). The purpose of this section is to examine melt-pool sequences and evolution near Cape Egerton (Fig.4a) and on Milne Re-entrant (Fig.5).

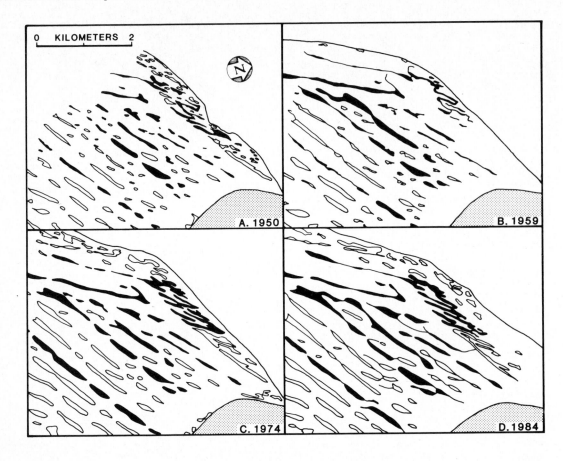

Fig.4. Sequential maps of melt-pool development and orientation on ice shelf and sea ice at Cape Egerton mapped from air photographs. The shaded melt pools represent particular sequences of elongation and coalescence. Fig.4a shows all the information available from the 1950 photography.

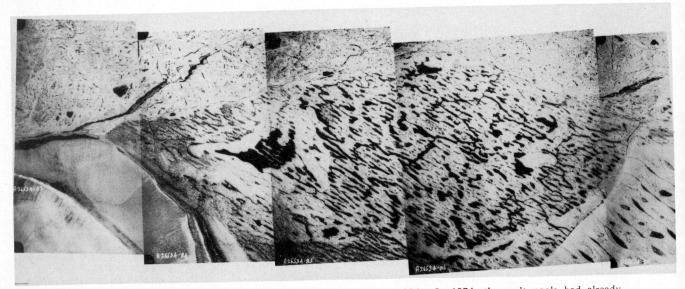

Fig.5. Vertical air-photo mosaic of Milne Re-entrant, July 1984. In 1974, the melt pools had already developed an elongated pattern which was even better defined by 1984. The better definition might be a function of the amount of melt water present at the ice surface, but is more likely to be the result of melt-pool evolution by elongation and coalescence.

Wavelength, elongation, and coalescence

On Milne Re-entrant and the old sea ice at Cape Egerton, the wavelength of the rolls varies from 60 to 100 m with a mean of 90 m (Figs 4 and 5). On the ice shelf the wavelength varies from 135 to 450 m with a mean spacing of almost 330 m (Figs 2 and 4). It is also noted that sea-ice melt pools are neither as long nor as wide as those on the ice shelf (Fig.4). Furthermore, on shelf or sea ice, the width and length of melt pools increases as the age of the ice increases (Figs 4 and 5). For example, in 1950, the melt pools on the old sea ice at Cape Egerton showed a quite high degree of elongation (Fig.4a)

and, therefore, had probably been in place for a number of years. Since that time, many of the smaller and more randomly scattered pools have coalesced until the entire melt-pool system has evolved into one of fewer, but better defined, elongate pools (Fig.4c and d).

The amount of water in the troughs determines, to some extent, the appearance of the melt pools. However, not only the amount of melt water must be considered, but also lateral convective processes of the melt water. Langleben (1972) has described the processes that tend to create a series of randomly scattered pools on freely floating sea ice. In a discussion of the ice-shelf rolls, Crary (1960) noted that fresh water on ice is a very efficient medium for melting ice because of its lower albedo and convection processes below 4°C. High surface winds pile up water down-wind and at the same time increase the convection currents. On fast ice, therefore, any series of randomly scattered ponds should in time form elongated lake systems (Fig.4) under the influence of a system of prevailing winds. Fast ice remaining in a fixed position throughout the summer melt season will have melt pools subject to the same dominant wind direction. Therefore, melt-pool elongation will tend to occur as the heat-transfer processes are concentrated at the down-wind ends of melt pools. As melt pools lengthen and drainage from the adjacent ridges proceeds, the hills separating melt pools will narrow and eventually be completely removed. Some melt pools appear to be connected by narrow streams (Fig.4d) that are evidence of the removal of such hills. As a consequence, melt pools will coalesce and lengthen further. The processes of elongation and coalescence are suggestive of stream-drainage network evolution and give rise to the characteristic linear surface drainage patterns on the ice in the presence of a system of prevailing winds.

Orientation

There is no wind-direction data available for outer Milne Ice Shelf. However, if it is the case that the melt pools elongate and coalesce in the direction of the prevailing winds, the dominant wind directions can be ascertained from the orientation of the ridges and troughs. Roll orientations have been mapped from air photographs.

The majority of the large or primary rolls on the ice shelf, together with the small undulations on Milne Re-entrant, are oriented approximately 030–210° with respect to true north. A second orientation of about 050–230° corresponds to the rolls on the old sea ice at Cape Egerton and two melt-water lakes (one with a distinct hook; Fig.4) or secondary rolls, at the front of the ice shelf. The primary and secondary roll orientations also correspond respectively to the two coastal trends at this location: (1) a general coastal trend approximated by the long fetch of Yelverton Bay (Fig.1a), and (2) the mouth of Milne Fiord between Cape Evans and Cape Egerton.

The data indicate two principal wind directions that essentially blow along the coast. A similar situation was recorded at Ward Hunt Ice Shelf where roll orientations were well correlated with prevailing east–west, summer winds blowing parallel to the coastline (Crary 1960). At Milne Ice Shelf the melt-pool evidence suggests that south-westerly winds blow from across Yelverton Bay while north-easterlies follow the Cape Evans–Cape Egerton trend. Moreover, since the winds have their greatest effect during summer and the greater proportion of the rolls are oriented parallel with westerly winds, in summer the latter are probably more dominant than easterly winds.

Easterly winds appear to cause some secondary coalescence and elongation at the expense of the primary melt pools. This process is especially evident in the region of the hooked melt pool where the primary pools have gradually changed orientation since 1950 (Fig.4). This might be evidence of a shift in the prevailing wind directions, but it seems unlikely in view of the fact that melt-pool orientations on the re-entrant have not developed an easterly trend.

CONCLUSION

Ice shelves and land-fast sea ice are clearly visible on air photographs and SLAR imagery. Both SLAR and air photographs show extensive sea-ice growth at the front of Milne Ice Shelf since 1950 alone. For the purpose of mapping ice loss and re-growth, SLAR imagery is preferable to air photographs, since with SLAR the entire area of concern is shown in one image, with sufficient detail, whereas air photographs require the composition of air-photo mosaics and subsequent reduction. SLAR covers a wider area in a shorter time than air photographs and is, therefore, more cost effective. Conversely, air photographs are more effective than SLAR for detailed study of ice-surface features. The undulating surface topography of the ice shelf and sea ice is delineated by melt pools in summer air photographs. The large-scale ice-shelf undulations are clearly visible on SLAR, but smaller-scale sea-ice undulations can not be resolved. Furthermore, SLAR images are of insufficient scale to show the development of individual ridges and troughs. Using sequential air photographs, melt pools can be mapped. It is apparent that the rolls evolve over a long period of time as a result of melt-pool coalescence and elongation. This process tends to occur in the direction of prevailing summer winds. Summer westerlies account for the primary melt-pool orientation, although in some instances this has been modified by summer easterlies. The modification of the melt-pool orientations is probably related more to local effects (topography and ice-shelf extent) rather than a shift in prevailing wind direction.

ACKNOWLEDGEMENTS

Work on Arctic ice shelves was initiated while Martin O Jeffries was a graduate student at the University of Calgary. Field work along the north coast of Ellesmere Island was made possible through the logistic support of the Polar Continental Shelf Project (PCSP, Mr G D Hobson, Director). Financial support came from the Arctic Institute of North America, Defence Research Establishment Pacific, Gulf Canada Resources Inc., Dome Petroleum Ltd, Petro-Canada, and the University of Calgary (Alberta Research Council Scholarship). PCSP also provided the Twin Otter aircraft for the 1984 air photography mission and we thank David Terroux for operating the aerial camera. Work at the Geophysical Institute, University of Alaska, is funded by the US Department of Energy, Morgantown Energy Technology Center, Morgantown, West Virginia.

REFERENCES

Crary A P 1960 Arctic ice island and ice shelf studies. Part II. *Arctic* 13(1): 32–50
Hattersley-Smith G 1957 The rolls on the Ellesmere Ice Shelf. *Arctic* 10(1): 32–44
Hattersley-Smith G 1963 The Ward Hunt Ice Shelf: recent changes of the ice front. *Journal of Glaciology* 4(34): 415–424
Hattersley-Smith G 1967 Note on ice shelves off the north coast of Ellesmere Island. *Arctic Circular* 17(1): 13–14
Jeffries M O 1986[a] Glaciers and the morphology and structure of Milne Ice Shelf, Ellesmere Island, N.W.T., Canada. *Arctic and Alpine Research* 18(4): 397–405
Jeffries M O 1986[b] Ice island calvings and ice shelf changes, Milne Ice Shelf and Ayles Ice Shelf, Ellesmere Island, N.W.T. *Arctic* 39(1): 15–19
Jeffries M O Unpublished Physical, chemical and isotopic investigations of Ward Hunt Ice Shelf and Milne Ice Shelf, Ellesmere Island, N.W.T. (PhD thesis, University of Calgary, 1985)
Jeffries M O, Serson H V 1986 Survey and mapping of recent ice shelf changes and landfast sea ice growth along the north coast of Ellesmere Island, NWT, Canada. *Annals of Glaciology* 8: 96–99
Koenig L S, Greenaway K R, Dunbar M, Hattersley-Smith G 1952 Arctic ice islands. *Arctic* 5(2): 67–103
Langleben M P 1972 The decay of an annual cover of sea ice. *Journal of Glaciology* 11(63): 337–344

Lyons J B, Ragle R H 1962 Thermal history and growth of the Ward Hunt Shelf. *International Association of Scientific Hydrology Publication* 58 (Colloque d'Obergurgl 10-9–18-9 1962 – *Variations of the Regime of Existing Glaciers*): 88–97

Prager B T Unpublished Digital signal processing of UHF radio-echo sounding data from northern Ellesmere Island. (MSc thesis, University of British Columbia, 1983)

Ragle R H, Blair R G, Persson L E 1964 Ice core studies of Ward Hunt Ice Shelf, 1960. *Journal of Glaciology* 5(37): 39–59

Serson H V 1984 Ice conditions off the north coast of Ellesmere Island, Nansen Sound, Sverdrup and Peary channels, 1963 to 1980. Appendix A. (*In*) Sackinger W M, Stringer W J, Serson H V *Arctic ice island and sea ice movements and mechanical properties. Second quarterly report.* Submitted to US Department of Energy, Morgantown, WV

Annals of Glaciology 9 1987
© International Glaciological Society

RADIO ECHO-SOUNDING OF SUB-POLAR GLACIERS IN SVALBARD:
SOME PROBLEMS AND RESULTS OF SOVIET STUDIES

by

V.M. Kotlyakov and Yu. Ya. Macheret

(Institute of Geography, U.S.S.R. Academy of Sciences, Moscow 109017, U.S.S.R.)

ABSTRACT

The paper discusses data analysed from airborne radio echo-sounding of Svalbard glaciers at frequencies of 440 and 620 MHz. Bottom returns from depths greater than 200 m are recorded with fewer gaps if the more powerful 620 MHz radar is used, and if measurements are carried out in the spring before intensive melt on glaciers. For all relatively thin glaciers and some glaciers up to 320–625 m thick, the track with bed returns is still rather common, apparently caused by their colder temperature regime. However, because of severe scattering of radio waves, this procedure still does not solve the problems of the echo-sounding of accumulation areas of many of the larger glaciers, the ice plateau, and heavily crevassed parts of glaciers.

For considerable areas of those Spitsbergen glaciers which have a thickness greater than 200 m, internal radar reflections (IRR) were registered as a single isolated layer from depths usually ranging from ¼ to ½ of their thickness. Studies of two deep bore holes on Fridtjovbreen have demonstrated that such IRR are related to a boundary between cold ice and water-bearing ice near the melting point. These IRR can be interpreted as indicators of a special class of two-layered or transitional glacier, and of the location within them of the ice-melt isotherm.

INTRODUCTION

Sub-polar and temperate glaciers are more complicated than cold ice sheets for radio echo-sounding (RES) surveys, because they have higher ice temperatures (up to the melting point on so-called "warm" glaciers) and a more inhomogeneous internal structure. In summer, a surface layer of water or soaked firn (slush) is formed on their surface. These water-soaked facies result in higher absorption, attenuation, and scattering of radio waves. In some cases this hampers or makes ineffective the use of radars operating in the ultra-high frequency (UHF) and very high frequency (VHF) range, which have been used successfully for sounding of cold ice sheets in Antarctica, Greenland, and the High Arctic.

Results of radio echo-sounding of Svalbard glaciers with different morphology, types of accumulation, range of zones of ice formation and temperature regime, which were obtained by Soviet expeditions between 1974 and 1984 using radars operating in the VHF and high frequency (HF) range, are illustrated in the following sections. Special attention is paid to the influence of melting on results of sounding, to the problem of sounding of accumulation areas of glaciers, and to the investigation of the nature of internal reflections.

UHF AIRBORNE RADIO ECHO-SOUNDING OF SVALBARD GLACIERS

Soaked snow (slush) and water on the surface of glaciers are transparent to radio waves in the UHF range. However, they do attenuate radar signals and also increase their reflection from the upper boundary. Melt water accumulating in the snow–firn facies, in streams, and in crevasses also increases radio-wave scattering. These factors reduce the depth of possible sounding and make the identification of bed returns difficult. Therefore, the cold season, particularly the spring months of April and May, are most favourable for measurements on glaciers which experience intense summer melting. However, because of logistical constraints, we managed to undertake most survey flights over Svalbard in the period June–August.

In summer field seasons during 1974–75 and 1978–79 a large number of valley glaciers, ice fields, and ice caps on Spitsbergen and Nordaustlandet (Fig.1) were sounded employing Mikoyan MI-4 and MI-8 helicopters. In 1984, measurements were carried out in some other areas of the archipelago, in particular, on Austfonna in Nordaustlandet (Fig.2), and in the central and southern parts of Spitsbergen. On Spitsbergen, the flights were begun in the middle of May, prior to intense glacier melting.

The flights were generally carried out along longitudinal profiles at a flying height of about 300 m above the glacier surface. Navigation along flight lines plotted on 1 : 100 000 and 1 : 500 000 scale maps was carried out by visual sighting to known points. Norsk

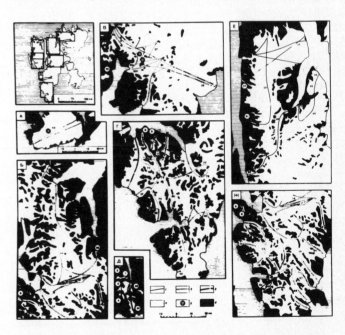

Fig.1. Index maps showing airborne radio echo-sounding (RES; 440 and 620 MHz) of Svalbard glaciers during the summer field seasons of 1974–75 and 1978–79. Delineation of flight lines of the RES surveys and coding of the types of reflections are shown as follows: 1, from bed; 2, from internal layer; 3, simultaneously from bed and from internal layer; 4, traverses where no reflections from bed and internal layer were registered; 5, location of bore holes; and 6, unglaciated terrain. For glaciers 1–7 data were obtained at a frequency of 440 MHz; for the remainder of the glaciers a frequency of 620 MHz was used. The numbers on the index maps denote glaciers shown in Tables I and II, and on Fig.3.

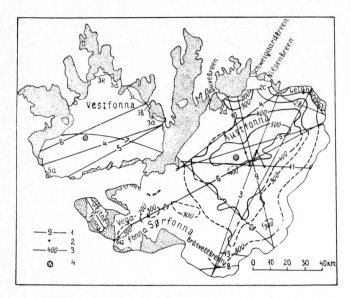

Fig.2. Index map showing flight lines followed during airborne radio echo-sounding (620 MHz) of glaciers on Nordaustlandet during the summer of 1984, and contours of ice thickness of Austfonna and Vestfonna. Legend: 1, flight lines and flight-line numbers of airborne measurements; 2, field calibration sites on the surface of the ice caps; 3, ice-thickness contours derived from airborne echo-sounding data; and 4, positions of bore holes.

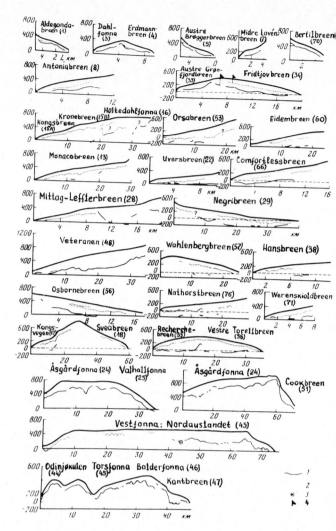

Fig.3. Profiles of Svalbard glaciers derived from airborne echo-sounding data acquired during the summer field seasons of 1974–75 and 1978–79 at 440 MHz and 620 MHz, respectively. Legend: 1 and 2, bottom and internal reflecting layer, respectively, according to airborne echo-sounding (RES) data; 3, location of bore holes. For glaciers 1, 3, 4, 5, and 7 RES data were obtained at a frequency of 440 MHz; for the rest of the glaciers at a frequency of 620 MHz. The figures in parentheses correspond to the glacier numbers shown on Fig.1 and in Tables I and II.

Polarinstitutt maps based on air-photogrammetric surveys carried out in 1936, 1938, and 1966 were used. In 1984, navigation along flight lines over Nordaustlandet used helicopter navigational equipment to control distances between the initial and final points of the RES profiles.

During the first two field seasons an aircraft-mounted RV-17 radio-altimeter was used. It operated at a frequency of 440 MHz, had a total system performance of 130 dB, and employed a wide-beam antenna of about 100°. The altimeter was successfully used for echo-sounding of small mountain glaciers not more than 150 m in thickness (Macheret and Zhuravlev 1980). However, for one larger glacier, Fridtjovbreen, internal reflections and bed returns were obtained from ice thicknesses as great as 250 m. In later field seasons, measurements were carried out with an RLS-620 operating at a frequency of 620 MHz. Its total system performance was 185 dB and the radar had a narrow antenna beam width of 18° in E- and H-planes (Kotlyakov and others 1980). Radio echo-sounding measurements were carried out with the RLS-620 over most of the glaciers studied earlier.

Summer radio echo-sounding surveys at the 620 MHz frequency

Analysis of data from radio echo-sounding surveys at the 620 MHz frequency during the summers of 1978–79 are shown in a series of 27 profiles (Fig.3). For several glaciers the interpretations differ from those published earlier by Macheret (1981) and Macheret and Zhuravlev (1982). The variation is thought to be the result of erroneous interpretation of internal reflections as actual bottom returns (Dowdeswell and others 1984[a]; Macheret and others 1984[b]). This revised interpretation is supported by data from airborne RES surveys carried out by British and Norwegian researchers during the spring of 1980 and 1983 with a system which operated at 60 MHz (Dowdeswell and others 1984[a], [b]; Dowdeswell unpublished).

At the 620 MHz frequency, bottom returns throughout most echo-sounding profiles were obtained principally for comparatively small glaciers and ice caps (see Fig.1) not greater than 150–200 m thick and for only a few outlet glaciers not greater than 320–430 m thick. For other relatively large glaciers not greater than 200 m thick, bottom returns were registered with many breaks or gaps in profiles

(see Figs 1 and 3) or were not identified at all, especially in accumulation areas and in the heavily crevassed tongues of tide-water glaciers. Examples of such records obtained for Eidembreen and Borebreen are shown in Fig.4a–d. Internal reflections (R), two- and three-fold multiple reflections from glacier surface (S_2 and S_3), and complicated multiples from the ice surface and the internal boundary (R + S) are also shown.

At Austfonna and Sørfonna in Nordaustlandet, the maximum measured ice thickness was 556 m, and bottom returns were obtained for about 70% of the length of radio echo-sounding profiles. This compares well with 60 MHz RES on Austfonna in 1983 by Dowdeswell and others (1986), where bed echoes were recorded over more than 91% of the traverses flown. Three profiles are shown in Fig.5. Bottom returns are generally absent in the heavily crevassed zones on outlet glaciers and in summit parts of the ice caps, where ice thickness exceeds 400–550 m (see Fig.2). On Vestfonna, tracks with bed returns (TBR) are fewer, about 50%. This compares with 52% bed returns observed by Dowdeswell and others (1986). Ice thicknesses up to 400 m were recorded in these areas. At Veteranen glacier, returns from as much as 500 m of ice were

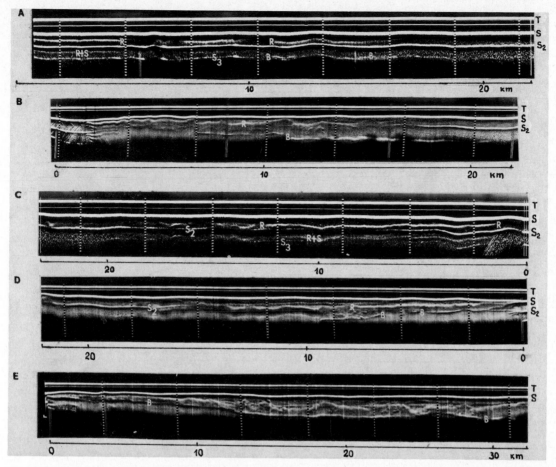

Fig.4. Examples of 620 MHz frequency records of radar reflections registered at Eidembreen (a and b), Borebreen (c and d), and Doctorbreen (e) in the summer of 1979 (a and c) and spring of 1984 (b, d, and e). T, echo-sounding pulses; S_2 and S_3, two- and three-fold multiple reflections from the glacier surface; R, reflection from internal layer; B, bottom return; and R + S, multiple reflections from surface and internal boundary.

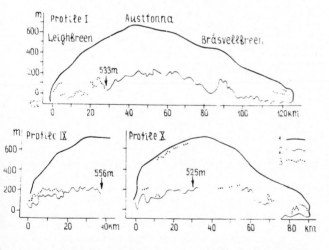

Fig.5. Thickness of glacier ice on Austfonna–Sørfonna in Nordaustlandet according to RES surveys carried out during the summer of 1984 at a frequency of 620 MHz. Legend: 1, glacier surface; 2 and 3, bottom and internal reflecting layers, respectively, according to analysis of airborne radio echo-sounding data. The location of survey flight lines for the profiles is shown on Fig.2.

recorded in 1979 (see Fig.3) and, in 1984, even deeper bottom returns were observed to as much as 625 m.

Spring radio echo-sounding surveys at the 620 MHz frequency

During aerial RES surveys in May 1984 several glaciers were surveyed (Table I). The same glaciers were surveyed during the summers of 1978 and 1979 at a frequency of 620 MHz (Fig.1) and in the spring of 1980 at 60 MHz (Dowdeswell and others 1984[b]). Examples of 620 MHz frequency RES records for Eidembreen, Borebreen, and Doctorbreen are shown in Fig.4.

Figs 1, 3, and 4 indicate that on Eidembreen and Borebreen the TBR increased markedly when sounded at 620 MHz in the spring rather than in the summer, from 17 to 75% and from 0 to 30%, respectively. In addition, the pattern of two- and three-fold reflections, as well as of multiple reflections (R + S) have also changed. In particular, the S_3 reflections have practically disappeared from the records, and S_2 and R + S reflections were observed on only a few plots. Additional information in the form of internal reflections with a complicated form was also present.

The TBR at 620 MHz also increased on other glaciers, many of which were studied repeatedly during the spring of 1984 (Table I). On Amundsenisen and Mühlbacherbreen, bottom returns were not recorded in 1984. Nor were they observed during ground measurements by the RLS-620 radar at Amundsenisen in June 1984, when melting was still insignificant.

Influence of melting on data from high-frequency radio echo-sounding

Our experiments have shown (Macheret and Zhuravlev 1980) that in accumulation areas of small Spitsbergen glaciers attenuation of radio signals at 440 MHz is 17–34 dB higher during the melt season, and 10 dB higher in the ablation areas. In accumulation areas the amount of additional attenuation depends on many factors: thickness and structure of the ice–firn layer, its temperature regime, intensity of melting and precipitation, depth of soaking, and spatial distribution and thickness of water horizons.

TABLE I. TRACK WITH BOTTOM RETURNS (TBR, %) OF SELECTED SPITSBERGEN GLACIERS FROM SUMMER AND SPRING AIRBORNE RADIO ECHO-SOUNDING SURVEYS AT FREQUENCIES OF 620 MHz (AUTHORS' WORK) AND 60 MHz (DOWDESWELL AND OTHERS 1984[b])

| Name of glacier (number in parentheses refers to glacier number on Fig.1) | 620 MHz | | | | | | 60 MHz | | |
| | Summer | | | Spring | | | Spring | | |
	L	TBR	h_{max}	L	TBR	h_{max}	L	TBR	h_{max}
Wahlenbergbreen (57)	22	6	430	18	56	325			
Borebreen (58)	23	0	–	22	30	330			
Nansenbreen (59)	16	0	–	16	13	295			
Eidembreen (60)	22	17	380	22	75	400			
Austre Grønfjordbreen (33)	7	80	170	7	84	170	7	100	170
Fridtjovbreen (34)	12	77	320	12	90	320	11	100	320
Amundsenisen (73)	30	0	–	30	0	–			
Mühlbacherbreen (80)	10	0	–	10	0	–			
Finsterwalderbreen (12)	15	3	160	13	35	260	12	100	260
Storbreen (82)				10	40	220	7[1]	21	140
Nathorstbreen (76)	28	10	400	27	70	380	28	61	420
Liestølbreen [3]				9[2]	55	100	3[2]	0	–
Doctorbreen [3]				21	82	315	28	61	340
Paulabreen [3]				12	76	260	18	72	240
Hayesbreen [3]				25	64	200	3[1]	34	135
Von Postbreen [3]				18	40	260	29	52	300
Total:	185	12.5		272	47		146	65	

Notes: L, length of echo-sounding profiles in kilometres; h_{max}, maximum measured thickness of ice in metres; (1) measurements on cross-profile, (2) measurements on tongue of glacier only; (3) measurements on longitudinal profile.

Theoretical assessments of signal attenuation at 440 MHz by a rain-soaked firn layer with a density of 0.5 Mg/m³ and 10–20 m thick gave a range of figures from 8 to 16 dB (Dowdeswell and others 1984[b]), and a water layer 5 mm thick spread over the ice surface, about 10 dB (Smith and Evans 1972). At the 60 MHz frequency these figures are several times less.

According to our calculations when using the 440 MHz altimeter, an additional 20–30 dB attenuation of signal at a flying height of 300 m is equivalent to a reduction in depth of echo-sounding of the glacier bottom which has a bulk ice temperature from $0°$ to $-3°C$, by approximately 300–350 to 50–150 m when compared to the situation of a dry ice surface and absence of water in the snow–firn layers. At 620 MHz the attenuation may be still greater, and can reach up to 30–40 dB in glacier accumulation areas which have thick snow–firn layers and severe melting. For the RLS-620 radar, this will be equivalent to a decrease in depth of echo-sounding of the bed by approximately 650–850 to 250–400 m. As for the 440 MHz equipment during the period of melting, this should result in a significant reduction of TBR on relatively thick glaciers which have low equilibrium-line altitudes (ELAs). Experimental data on summer and spring measurements at high frequencies support this conclusion (see Figs 1 and 3, and Table I).

The reflection coefficient of the melting glacier surface increases up to values which makes it possible to use the more powerful 620 MHz radar to record n-fold S_2 and S_3 reflections (see Fig.4a and c). However, in some cases, RES at this frequency may mask bottom returns and reflections from internal layers which makes interpretation of Z-records more difficult. Spring measurements give a different pattern (see Fig.4b, d, and e). This allows us to interpret n-fold reflections as an indicator of surface melting and spatial–temporal changes of snow cover on glaciers. However, this pattern may be complicated by water accumulating in the snow–firn layer at its contact with solid ice. As revealed by the drilling of several shallow bore holes on Austfonna in 1985, these water horizons and lenses may be preserved until the following spring.

COMPARISON OF AIRBORNE RADIO ECHO-SOUNDING DATA AT FREQUENCIES OF 620 AND 60 MHz

The above data on the thickness of Spitsbergen glaciers obtained from 620 and 60 MHz RES agree rather well, despite differences in flight lines and navigational errors; as a rule, the difference is within 10–15%. Valuable information is also obtained by comparison of data on the TBR of glaciers in the course of spring and summer echo-sounding at various frequencies (Tables I and II).

Table I demonstrates that during spring measurements

TABLE II. TRACK WITH BOTTOM RETURNS (TBR, %) OF SELECTED SPITSBERGEN AND NORDAUSTLANDET GLACIERS FROM SUMMER AND SPRING AIRBORNE RADIO ECHO-SOUNDING SURVEYS AT FREQUENCIES OF 620 MHz (AUTHORS' WORK) AND 60 MHz, (DOWDESWELL AND OTHERS 1984[b]; DOWDESWELL AND OTHERS 1986), RESPECTIVELY

Name of glacier (number in parentheses refers to glacier number on Fig.1)	620 MHz (summer)			60 MHz (spring)		
	L	TBR	h_{max}	L	TBR	h_{max}
Spitsbergen						
Isachsenfonna–Kongsbreen– –Kronebreen–Holtedahlfonna (14, 15a, 15b, 16)	124	10	585	125	13	530
Kongsvegen (17)	29	7	320	27 33.5 [1]	33 46	390
Sveabreen (18)	31	0	–	34	0	–
Tunabreen (18)	21	0	–	33	18	250
Negribreen (29)	39	25	400	62[2]	84	430
Penckbreen (10)	12	2	160	13	96	235
Vestre Torellbreen (36)	21	58	430	34[2]	93	310
Austre Torellbreen (37)	11	5	380	8	0	–
Hansbreen (38)	12	20	330	14	29	310
Werenskioldbreen (71)	8	30	320	7	71	240
Total:	308	14	–	390.5	39	–
Nordaustlandet						
Vestfonna (43)	300	50	400		52	300
Austfonna–Sørfonna	950	70	556		91	583

Notes: L, length of echo-sounding profiles in kilometres; h_{max}, maximum measured thickness of ice in metres; measurements made on cross-section profiles [1] and in lower part of glacier [2].

the TBR on the same glaciers at 620 MHz is somewhat lower than data recorded at 60 MHz (51 and 64% respectively), although the total system performance of the 60 MHz Scott Polar Research Institute (SPRI) Mk. IV radar is lower, 144 dB. This is apparently caused by the greater attenuation of radio waves at higher frequencies. The greatest difference was registered at Finsterwalderbreen, 35 and 100%, respectively. But for Doctorbreen, Paulabreen, Nathorstbreen, and Von Postbreen, bottom returns at 620 MHz were measured at a greater distance from the end of the glacier tongue than at 60 MHz, and on Doctorbreen (Fig.6) and Paulabreen they were observed up to the ice divide. On the whole, during spring survey flights in 1984 and 1980 the TBR was about the same: at 620 MHz it was equal to 47%, and at 60 MHz equal to 49% (Dowdeswell and others 1984[b]) (length of the profiles of the radio echo-sounding was about 270 and 740 km, respectively).

But in the summer period at 620 MHz, the TBR of most glaciers was much lower than during spring measurements at 60 MHz. Both the summer and spring measurements at 620 MHz have not recorded bottom returns, mostly in the upper parts of large glaciers and in their lower heavily crevassed parts. The same situation occurred during spring measurements at 60 MHz (Dowdeswell and others 1984[a], [b]).

Therefore, carrying out the surveys in the spring before melting begins, and reducing the radar frequency from 620 to 60 MHz will not solve the problem of echo-sounding in the accumulation area of many Spitsbergen glaciers (Smith and Evans 1972; Dowdeswell and others

1984[b]). Simple increases in the total system performance of a radar are not able to solve the problem either. The main reason is that, because of melting and refreezing of water, many sub-polar glaciers have higher temperatures of firn and ice in their accumulation areas than ice temperatures in the ablation areas, and some glaciers may have temperatures equal to the pressure melting-point (Troitskiy and others 1975; Baranowski 1977). In addition, the snow–firn layer in many sub-polar glaciers has an inhomogeneous structure because of abundant interbeds and lenses of ice and water inclusions. This results in heavier absorption and attenuation of radio waves of UHF and VHF ranges as well as heavy scattering and masking of bottom returns (Smith and Evans 1972). This is seen on the records discussed previously (Fig.4). Therefore, heavy scattering may be interpreted to be a qualitative indicator of warmer firn and ice in the accumulation areas of Spitsbergen glaciers (Dowdeswell and others 1984[b]). An effective way to reduce scattering is to decrease the radar frequency to 10 MHz or lower (Watts and England 1976).

INVESTIGATION OF THE NATURE OF INTERNAL REFLECTIONS

Airborne RES over selected glaciers on Spitsbergen has shown that internal radar reflections (IRR) are present (cf. Macheret and others 1984[b]; Dowdeswell and others 1984[a]). In most cases they could be followed within a depth range from 70 to 200 m, which represents from ¼ to ½ of the glacier thickness, and were, for the most part,

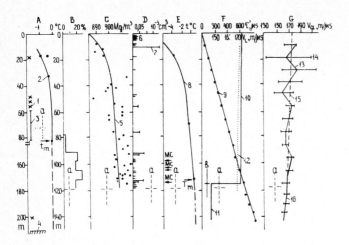

Fig.6. Data from bore-hole site 1 in 1975 (A) and from site 2 in 1979 (B–G) on Fridtjovbreen. A: 1, interbeds of dirty ice layers; 2, ice temperature from 1979 measurements; 3, depth of bore hole beyond which the water level in the bore hole, after water addition and withdrawal at the beginning and at the end of drilling periods, attained equilibrium at a permanent depth of 27 m; t_m isotherm of ice melting, and 4, glacier bed. B: concentration of ice layers with sub-horizontal orientation of air bubbles. C (density of ice from point measurements): 5, approximation by exponential function (I). D (insoluble mineral admixtures in ice); 6, layers of dirty ice; 7, weight concentration of mineral admixtures in selected layers; E (ice temperature): 8, 1981 measurements; T_m, calculated pressure melting-point of ice; MC, sub-vertical micro-channels 3–5 mm² found in ice core. F and G (vertical velocity profiles of the glacier according to data from radar logging of bore hole at a frequency of 620 MHz, according to the following calculations): 9, curve $\tau'(h)$; 10 and 11, curves of layer velocities V_{L1} and V_{L2}, respectively; 12, graph of v_h velocity calculated on the basis of empirical Equations (1) and (2) (see text). G (curves of the average velocity of radio-wave travel in the glacier $v_a(h)$: 13, after two series of measurements; 14, range of v_a in each series of measurements; 15 and 16, approximation of curve 13 by two segments of a straight line; a and c, respectively, depths and errors of definition of the internal reflecting boundary and turning point of the curve $\tau'(h)$.

parallel to the glacier surface. On Z-records they were usually observed as an isolated single-layer echo (Fig.4). An explanation for such IRR may be either: (1) an extended layer with dielectric contrast, or a series of such layers, with a thickness more or less comparable with the transmitted pulse length in the ice (about 30 m for the RLS-620 radar), or (2) a separation of two "semi-infinite" media which differ in their dielectric properties (Clough 1977). Extended IRRs within a depth range of 70 to 190 m were also registered by airborne radio echo-sounding at 60 MHz (Dowdeswell and others 1984[a], [b]). A specific feature of the IRR is that their thickness did not appreciably change with time, within the accuracy of measurements and navigation, and that they were recorded on only relatively thick glaciers (more than 200–250 m thick).

The probable reason for the extended IRR may be ice layers, differring in density, structure, concentration of mineral inclusions, and chemical impurities from the rest of the ice mass, and water lenses within the glacier (e.g. Millar 1982; Bogorodskiy and others 1983). These reflections may also be caused by different concentrations of water in cold ice and in ice at the pressure melting-point (Luchininov and Macheret 1971).

For glaciological interpretations the physical nature of IRR is important; evidence comes from a comparison of radio echo-sounding data with data from deep bore holes (Kotlyakov and others 1980). Such data are available, for example, for Fridtjovbreen, a typical sub-polar glacier

(Macheret and others 1985). Studies of Fridtjovbreen included ground and airborne RES, recovery of a deep ice core, and thermal drilling at two sites where IRR were recorded. A description and instrumental investigation of the ice core was carried out in addition to measurement of the temperature of the ice in bore holes and recording of the water level in one of them. Radar echo-sounding surveys were carried out at frequencies of 440 and 620 MHz along a network of profiles. Along one of the profiles the following frequencies were also used: 450, 686, 735, 786, and 865 MHz.

Two bore holes exceeding 200 m in depth were made; the data collected are shown on Fig.6. In the bore hole at site 1 ice temperatures reached the melting point at a depth of about 80 m (Fig.6a); in the bore hole at site 2 the melting point was reached at about 113 m (Fig.6e). By reference to indirect evidence, the ice temperature at considerable depths is probably close to the pressure melting-point (for example, constant speed of thermal drilling, no sticking of boring tools resulting from icing of the bore hole even if drilling is carried out without antifreeze, absence of sludge, etc.). At depths of less than 115 m the average ice temperature in the bore hole at site 2 equals −1.7°C. The mean integral ice density ρ within the range of depths less than 119 m is 904 kg m⁻³, and gradually changes with depth h from 888 to 917 kg m⁻³ according to the equation:

$$\rho(h) = 879.5h^{0.0072}.$$

Volume concentrations of mineral impurities in dirty ice layers generally do not exceed 0.04% (Fig.6d).

A comparison of data from ground and airborne RES obtained during different seasons, at different radar frequencies, and under different flight altitudes has shown that at the ice divide the depth, character, and patterns of the internal reflecting boundary (IRB) are not subject to significant changes over time, taking into account the range of accuracy of measurements and navigational precision. The bore hole at site 1 has a depth of 72 ± 5 m, the one at site 2, 120 ± 10 m. This denotes, first of all, the absence of internal reflecting layers which are "transparent" or "not transparent" for some frequencies; and secondly, that the character of IRB is basically determined by the nature of the IRB and not by equipment parameters or conditions of measurements. Taking this into account, one can, as a first approximation, interpret the IRB at the drilling sites as a comparatively thick homogeneous dielectric layer which is stable with depth.

Measurements of reflected signals from the IRB at bore-hole site 2 provided data of an attenuation of 107 dB at a frequency of 620 MHz. Taking into account the errors of measurements of reflected signals, errors in depth to the IRB, and accuracy of parameters assumed for the calculated model, the reflection coefficient in power from IRB R_R = −35 ± 10 dB. At the same time, changes in density, structure, mineral admixtures, and chemical impurities in individual layers or in a series of layers found in the ice core (Fig.6b, c, and d) give, according to our estimates, lower reflection coefficients (from −55 to −100 dB or less), and cannot explain the reported high reflection coefficient from the IRB. The high reflection coefficient from the IRB is, therefore, a result of a dielectrically more contrasting medium, as, for example, from a layer of water-bearing ice or of water.

To investigate the vertical velocity profile of Fridtjovbreen, radar logging of the bore hole at site 2 was carried out (Macheret and others 1984[a]). Measurements down to a depth of 145 m are given in Fig.6f and g at a specified time:

$$\tau' = \tau \frac{h}{\sqrt{(h^2 + d^2)^{\frac{1}{2}}}}$$

and V_a, the average velocity of radio waves in the glacier:

$$V_a = \frac{h}{\tau'} = \frac{\sqrt{(h^2 + d^2)^{\frac{1}{2}}}}{\tau}$$

where τ is the time of travel of a signal from the

transmitting antenna on the glacier surface to the receiving antenna in the bore hole, h is the vertical depth of sinking of the receiving antenna (with correction for bending of the bore hole), d is the distance from the transmitting antenna to the bore-hole outlet. The interval of depths corresponding to every linear interval on the diagram, $\tau'(h)$, is characterized by an approximately constant value of travel time of radio waves and may be interpreted as about homogeneous. In such a layer, the velocity is denoted as layer velocity and is determined as follows:

$$V_L = \frac{\Delta h_i}{\Delta \tau_i'}$$

where Δh_i is the thickness of layer i, $\Delta \tau_i'$ is the time of wave travel in this layer. If a glacier profile consists of two layers with different layer velocities V_{L1} and V_{L2}, then the average velocity $V_{a2}(h)$ in the lower layer will change in a non-linear way with an increase in thickness of the lower layer $h_2(Z)$:

$$V_{a2}(h) = \frac{h_1 + h_2(Z)}{h_1/V_{L1} + h_2(Z)/V_{L2}} \qquad (1)$$

where h_1 is the thickness of the upper layer.

On Fig.6, $\tau'(h)$, despite rather high dispersion of points resulting mainly from errors of measurements τ, one can see an abrupt change at the depth range of 105–130 m. The curve $\tau'(h)$ can be satisfactorily approximated by two linear segments. Therefore, the cross-section of the glacier may be presented as two layers with different layer velocities and, correspondingly, with different dielectric constants. The separation of these layers is at depth $h^* = 117 \pm 12$ m. This is very close to the depth of the IRR for the bore hole at site 2 ($h_R = 120 \pm 10$ m); therefore, it is suggested that this variation is what produces the IRR. Taking into account small losses in the ice, the following values of layer velocities V_L and of dielectric constant $\epsilon = (c/V_L)^2$ are calculated as follows for the upper layer (0–117 m):

$$V_{L1} = 172.2 \text{ m } \mu s^{-1}, \qquad \epsilon_1 = 3.035; \quad \text{for the lower}$$

(117–145 m):

$$V_{L2} = 147.7 \text{ m } \mu s^{-1}, \qquad \epsilon_2 = 4.125.$$

Fig.6f and g show V_{L1} and V_{L2}, and curves $\overline{V}_{a1}(h)$ and $\overline{V}_{a2}(h)$, produced with an approximation of a $V_a(h)$ curve with two linear segments within the same depth interval as the graph $\tau'(h)$. They also demonstrate the correlation of average velocity V_h of radio–wave travel in the glacier with depth calculations taking into account the ratio (1) according to the formula of Robin and others (1969)

$$\sqrt{\epsilon} = 1 + 0.00085\rho, \qquad (2)$$

which describes ϵ of "dry" cold ice as a function of ρ in kg m^{-3}.

From Fig.6f and g, it follows that the value of V_{L1} differs by no more than 3.3% from $\overline{V}_{a1}(h)$ and by only 2% from V_h. This confirms the comparatively high accuracy in the calculations of V_{L1} and in the validity of the chosen two-layer model of the glacier. The comparatively low mean velocity of radio waves $\overline{V} = 161.4$ m μs^{-1}, determined through comparison of radio echo-sounding and drilling data for the bore hole at site 1, is explained by a layer of ice at depths $h > h^*$ with an average radio-wave velocity close to V_{L2}.

The relationship of the IRR with englacial water is indicated by the following points: depth of the IRR at bore-hole site 1 ($h_R = 72 \pm 5$ m) is close to the level where inflow of liquid water from the glacier to the bore hole begins, that is, where englacial water appears ($h_w = 55–81$ m) and the ice temperature reaches the melting point ($h_0 \approx 80$ m). Also, the level of the water in the bore hole was at 27 m, or 12.6% of the thickness of the glacier. This may be explained by the effects of hydrostatic pressure and the hydraulic connection of the bore hole with the englacial water. The depth of the IRR at bore-hole site

2 ($h_R = 120 \pm 10$ m) is close to the level where the ice temperature reaches the melting point ($h_0 \approx 113$ m). Also, in the ice core there were micro-channels, which are specific to "warm" ice (Zagorodnov and Zotikov 1981) ($h_w = 100–116$ m); the calculated coefficient of reflection from the separation of "cold" and "warm" ice with a low concentration of water (several per cent in volume) is −30 dB (Luchininov and Macheret 1971) which is very close to the actual measurements ($R_R = -35 \pm 10$ dB).

Data on the layer velocity V_L allows the content of water in the lower layer to be assessed using the formula of J Paren (Smith and Evans 1972):

$$\epsilon_s = \epsilon_d + \frac{1}{3}\epsilon_w(1 - \upsilon), \qquad (3)$$

showing the function of the dielectric constant ϵ_s of water-saturated ice, in which all pores and cavities are filled with water and have a chaotic spatial pattern, in which water content in per cent $W = (1 - \upsilon)/100\%$, where $\upsilon = \rho_d/\rho_i$, ρ_d and ρ_i are the density of "dry" glacial ice and of compact ice ($\rho_i = 916.8$ kg m^{-3}); ϵ_d and ϵ_w are the dielectric constant of "dry" glacial ice and water ($\epsilon_w = 86$). However, Equation (3) provides only the maximum estimate of water in ice. Taking the data of radar logging of the bore hole at site 2 and assuming $\epsilon_d = \epsilon_1 = 3.035$ and $\epsilon_s = \epsilon_2 = 4.125$, we get $W = 3.8\%$. The value of ϵ_1 may also be estimated according to Equation (2). Taking $\rho = 904$ kg m^{-3}, we get $\epsilon_1 = 3.13$, and $W = 3.4\%$. The value of ϵ_2 can also be assessed by the value of the reflection coefficient R_R from the IRB. If, starting from the above, we assume that the IRB is a division of two "semi-infinite" media (two-layer model), then:

$$\epsilon_2 = \epsilon_1(1 + \kappa)$$

where $\kappa = \dfrac{\epsilon_2 - \epsilon_1}{\epsilon_1} = \left[\dfrac{1 + 10^{0.05R_R}}{1 - 10^{0.05R_R}}\right]^2 - 1.$

If we assume that $R_R = -35 \pm 10$ dB and ϵ_1 equals 3.035 and 3.13, we get ϵ_2 varying from 3.1 to 3.3 and from 3.20 to 3.92, and W changing within 0.23–2.67 and 0.24–2.76%. These estimates are similar to the available data on the water content in warm glaciers of 1–2% (Golubev 1976).

These data and their analysis suggest that the reason for the IRR at Fridtjovbreen bore-hole sites 1 and 2 is the isothermal layer of "warm" wet ice underlying "dry" cold ice. The depth of the IRR practically coincides with the depth of the isotherm of ice melt in the glacier. Indicators of such a two-layer structure are IRBs, which were recorded along a considerable stretch of this glacier, including cross-profiles. Therefore, it is likely that much of Fridtjovbreen has a two-layer structure.

RELATIONSHIP OF INTERNAL REFLECTIONS WITH THE HYDROTHERMAL STATE AND REGIME OF GLACIERS

Similar IRR observed by airborne radio echo-sounding along considerable lengths of many Svalbard glaciers are likely to be of the same nature as those observed at Fridtjovbreen. This allows us to classify as two-layered nearly one-half of the glaciers we studied in Svalbard. However, there is still a chance that in some glaciers, or at least along part of their lengths, IRRs with a similar pattern correspond to reflections from englacial and (or) surface reflectors (for example, water lenses, water streams, cavities, etc.), producing on the RES records a series of hyperbolae encircling their upper parts, or individual branches (Fig.4d and e).

Fig.1 shows that on Spitsbergen the selected glaciers, for which IRR were registered (the supposed indicators of their two-layer structure), occur most often either in the margins of the plateau glaciers or correspond to outlet glaciers. Some of them have (or had in the past) a pattern of warm firn in their accumulation areas and originate in the glacial plateau with warm firn accumulation areas, such as with Fridtjovbreen, the plateau glaciers Amundsenisen and Isachsenfonna, and others. The outlet glaciers descend

from plateau glaciers with cold firn in the accumulation areas and negative ice temperature throughout the entire ice thickness such as with Lomonosovfonna (Troitskiy and others 1975; Kotlyakov 1985). It was noted previously that the characteristic thickness of ice for these glaciers is greater than 200–250 m, and location of the IRR is usually at ¼ to ½ of their depth. At the same time, in glaciers with a smaller ice thickness, IRR were not usually observed.

The preceding data suggest that glaciers without IRR have cold temperature regimes either throughout the whole ice mass or very nearly to the base of the glacier. This does not, however, preclude the presence of a bottom layer of warm, water-soaked ice, its thickness being beyond the radar resolution in delay time (i.e. about 30 m) as well as the occurrence of water in the bottom layer and on the glacier bed. The IRR should be absent in the accumulation areas of glaciers with warm firn zones, where location of the 0 °C isotherm is similar to the thickness of the active layer, that is, in the range of 10–15 m. Therefore, in some cases, extended IRR may not be recorded on both cold and warm glaciers. In addition, there may be other factors causing the absence of such IRR on records, such as their absence in records which record bottom returns.

The existence on Spitsbergen of glaciers with a bottom layer of ice at the melting temperature was earlier suggested on the basis of observations of ice temperature measured in shallow pits and bore holes (Schytt 1969; Baranowski 1977). Hypothetical and theoretical ideas were also put forward, that the availability of small amounts of water in the bottom layer of warm ice may result in its outflow from a glacier, thereby producing a notable influence on the mechanical stability of glaciers (Krass 1983).

Analysis of radio echo-sounding data (except for 1984) has demonstrated that of the total number of 43 Spitsbergen glaciers for which IRR were recorded, 16 glaciers have experienced surge behaviour during the past century (Liestøl 1969; Kotlyakov 1985). This accounts for 40% of the total number of selected glaciers under study; data on winter englacial run-off and ice naleds are available for an additional five glaciers. Of the remaining 22 glaciers, 15 are tidal and probably also have englacial run-off during the winter; five are tributary glaciers. In May 1984, for example, we observed differently coloured streams of water discharging into the sea from Eidembreen.

However, surges and naleds were also observed on glaciers which cannot be classified as two-layered ones (for example, Hessbreen, Midre Lovenbreen). On the other hand, ice naleds were found in the tongue of Austre Grønfjordbreen, where hyperbolic IRR were observed from the bottom strata, denoting, apparently, their relation to englacial water accumulations. Hyperbolic reflections were also recorded from the bottom strata of Eidembreen (Fig.4b), where extended IRR were also recorded. In the past 100 years, surges of Spitsbergen glaciers were more often observed from the middle–end of the nineteenth century to the 1920s and 1930s, when the "Little Ice Age" (c. sixteenth to eighteenth centuries) was replaced by warming of the climate, accompanied by glacier recession and thinning. At the same time, many glaciers of the archipelago have apparently experienced changes in the type of accumulation. For some glaciers this could cause cooling of the ice mass from the surface; for other glaciers, warming from the surface (Kotlyakov 1985; Kotlyakov and Troitskiy 1985). For some glaciers the internal warming is caused by dissipation of energy (Krass 1983; Grigoryan and others 1985), that is, a reconstitution of the glaciers' thermal regime to form a two-layer one. This mechanism of englacial warming provides a theoretical explanation for the two-layer structure of selected Spitsbergen glaciers, for which extended IRR were registered, and for a good correlation of the calculated depth of location of the ice-melting isotherm in those glaciers which have experimental data on the location of the IRR. In particular, a two-layer structure of Veteranen glacier was theoretically predicted and then supported by radio echo-sounding survey data of 1984.

CONCLUSIONS

The RLS-620 radar, which operates at a frequency of 620 MHz, can be successfully employed for radio echo-

sounding of a certain type of sub-polar glacier in both the spring and summer. This type of glacier includes comparatively small glaciers up to 200 m thick, as well as some larger ones and ice caps up to 320–625 m thick with, apparently, a colder temperature regime. However, melting considerably worsens the results of radio echo-sounding in the upper parts of many glaciers which are more than 200–250 m thick. In the accumulation areas of some of these glaciers severe scattering of radio waves caused by englacial inhomogeneities formed by melting and refreezing of water often masks bottom returns even in the case of spring measurements. Studies of Amundsenisen, which has a warm temperature regime and is 700 m thick, have established that this problem may be eliminated by using radar operating in the HF range.

Investigations of the nature of IRR recorded along a considerable length of Fridtjovbreen at a depth range from 70 to 180 m have established that they are caused by a small (of the order of 1–2% in volume) amount of water in the bottom layer and the presence of isothermal ice, which is close to the melting point. Therefore, such IRR may be interpreted as indicators of a special class of two-layered glacier and of the location of the isotherm of melting within them. Many of these glaciers have experienced surge behaviour during the past century and have winter englacial run-off and ice naleds.

REFERENCES

Baranowski S 1977 The subpolar glaciers of Spitsbergen seen against the climate of this region. *Acta Universitatis Wratislaviensis* 410

Bogorodskiy V V, Bentley C P, Gudmandsen P 1983 *Radioglyatsiologiya [Radioglaciology]*. Leningrad, Gidrometeoizdat

Clough J W 1977 Radio-echo sounding: reflections from internal layers in ice sheets. *Journal of Glaciology* 18(78): 3–14

Dowdeswell J A Unpublished Remote sensing studies of Svalbard glaciers. (PhD thesis, University of Cambridge, 1984)

Dowdeswell J A, Drewry D J, Liestøl O, Orheim O 1984[a] Airborne radio echo sounding of sub-polar glaciers in Spitsbergen. *Norsk Polarinstitutt. Skrifter* 182

Dowdeswell J A, Drewry D J, Liestøl O, Orheim O 1984[b] Radio echo-sounding of Spitsbergen glaciers: problems in the interpretation of layer and bottom returns. *Journal of Glaciology* 30(104): 16–21

Dowdeswell J A, Drewry D J, Cooper A P R, Gorman M R, Liestøl O, Orheim O 1986 Digital mapping of the Nordaustlandet ice caps from airborne geophysical investigations. *Annals of Glaciology* 8: 51–58

Golubev G N 1976 *Gidrologiya lednikov [The hydrology of glaciers]*. Leningrad, Gidrometeoizdat

Grigoryan S S, Bozhinskiy A N, Krass M S, Macheret Yu Ya 1985 Yavleniye vnutrennego razogreva "kholodnykh" lednikov i obrazovaniye lednikov perekhodnogo tipa [Phenomenon of internal heating of "cold" glaciers and formation of glaciers of transitional type]. *Materialy Glyatsiologicheskikh Issledovaniy. Khronika. Obsuzhdeniya* 52: 105–110

Kotlyakov V M (ed) 1985 *Glyatsiologiya Shpitsbergena [Glaciology of Spitsbergen]*. Moscow, "Nauka"

Kotlyakov V M, Troitskiy L S 1985 Novyye dannyye ob oledenenii Shpitsbergena [New data on glaciation of Spitsbergen]. *Akademiya Nauk SSSR. Vestnik* 2: 128–136

Kotlyakov V M, Macheret Yu Ya, Gordiyenko F G, Zhuravlev A B 1980 Geofizicheskiye i izotopnyye issledovaniya lednikov Shpitsbergena [Geophysical and isotopic investigations of Spitsbergen glaciers]. *Akademiya Nauk SSSR. Vestnik* 2: 132–138

Krass M S 1983 *Matematicheskaya teoriya glyatsiomekhaniki [Mathematical theory of glaciomechanics]*. Moscow, Vsesoyuznyy Institut Nauchnoy i Tekhnicheskoy Informatsii (Itogi Nauki i Tekhniki. Seriya Glyatsiologiya 3)

Liestøl O 1969 Glacier surges in West Spitsbergen. *Canadian Journal of Earth Sciences* 6(4): 895–897

Luchininov V S, Macheret Yu Ya 1971 Elektromagnitnoye zondirovaniye teplykh gornykh lednikov [Electromagnetic sounding of mountain-type temperate glaciers]. *Zhurnal Tekhnicheskoy Fiziki* 41(6): 1299–1309

Macheret Yu Ya 1981 Forms of glacial relief of Spitsbergen glaciers. *Annals of Glaciology* 2: 45–51

Macheret Yu Ya, Zhuravlev A B 1980 Radiolokatsionnoye zondirovaniye lednikov Shpitsbergena s vertoleta [Radio echo-sounding of Spitsbergen glaciers from helicopters]. *Materialy Glyatsiologicheskikh Issledovaniy. Khronika. Obsuzhdeniya* 37: 109–131

Macheret Yu Ya, Zhuravlev A B 1982 Radio echo-sounding of Svalbard glaciers. *Journal of Glaciology* 28(99): 295–314

Macheret Yu Ya, Vasilenko Ye V, Gromyko A N, Zhuravlev A B 1984 Radiolokatsionnyy karotazh skvazhiny na lednike Frit'of, Shpitsbergen [Radio echo logging of the bore hole on Fridtjovbreen, Spitsbergen]. *Materialy Glyatsiologicheskikh Issledovaniy. Khronika. Obsuzhdeniya* 50: 198–203

Macheret Yu Ya, Zagorodnov V S, Vasilenko Ye V, Gromyko A N, Zhuravlev A B 1985 Issledovaniye prirody vnutrennikh radiolokatsionnykh otrazheniy na subpolyarnom lednike [Study of the nature of internal radio echo reflections in a subpolar glacier]. *Materialy Glyatsiologicheskikh Issledovaniy. Khronika. Obsuzhdeniya* 54: 120–130

Macheret, Yu Ya, Zhuralev A B, Bobrova L I 1984[b] Tolshchina, podlednyy rel'yef i ob"yem lednikov Shpitsbergena po dannym radiozondirovaniya [Thickness, subglacial relief and volume of Svalbard glaciers based on radio echo-sounding data]. *Materialy Glyatsiologicheskikh Issledovaniy. Khronika. Obsuzhdeniya* 51: 49–63

Macheret Yu Ya, Zhuravlev A B, Bobrova L I 1985 Thickness, subglacial relief and volume of Svalbard glaciers based on radio echo-sounding data. *Polar Geography and Geology* 9(3): 224–243

Robin G de Q, Evans S, Bailey J T 1969 Interpretation of radio echo sounding in polar ice sheets. *Philosophical Transactions of the Royal Society of London* 265A(1166): 437–505

Schytt V 1969 Some comments on glacier surges in eastern Svalbard. *Canadian Journal of Earth Sciences* 6(4): 867–873

Smith B M E, Evans S 1972 Radio echo sounding: absorption and scattering by water inclusion and ice lenses. *Journal of Glaciology* 11(61): 133–146

Troitskiy L S, Zinger Ye M, Koryakin V S, Markin V A, Mikhalev V I 1975 *Oledeneniye Shpitsbergena (Svalbarda) [Glaciation of Spitsbergen (Svalbard)]*. Moscow, "Nauka"

Watts R D, England A W 1976 Radio-echo sounding of temperate glaciers: ice properties and sounder design criteria. *Journal of Glaciology* 17(75): 39–48

Zagorodnov V S, Zotikov I A 1981 Vnutrilednikovyye kanaly [Intraglacial channels]. *Materialy Glyatsiologicheskikh Issledovaniy. Khronika. Obsuzhdeniya* 41: 200–202

Annals of Glaciology 9   1987
© International Glaciological Society

AIRBORNE RADIO ECHO-SOUNDING IN SHIRASE GLACIER
DRAINAGE BASIN, ANTARCTICA

by

S. Mae

(Department of Applied Physics, Faculty of Engineering, Hokkaido University, Sapporo 060, Japan)

and

M. Yoshida

(National Institute of Polar Research, Tokyo 173, Japan)

ABSTRACT

Airborne radio echo-sounding was carried out in order to measure the thickness of the ice sheet in the Shirase Glacier drainage basin and map the bedrock topography. It was found that the elevation of bedrock was approximately at sea-level from Shirase Glacier to 100 km up-stream of the glacier and thereafter it was 500–100 m higher. Investigation of the echo intensity reflected from the bedrock indicates that at ice thicknesses less than 1000 m absorption was about 5.2 dB/100 m, but at greater ice thicknesses echo intensity did not depend upon the ice thickness but became approximately constant. Where ice thicknesses were greater than 1000 m in the main flow area of the Shirase Glacier drainage basin, the reflection strengths of about 9 dB were greater than outside the basin. Since the increase in echo intensity was considered to be due to the existence of water, the strong echo observed in the main part of the basin supported an hypothesis that the base of the basin was wet and the ice sheet was sliding on the bedrock.

INTRODUCTION

In 1974 the Japanese Antarctic Research Expedition (JARE) established a 150 km trilateration network, which was tied to Motoi Nunatak in the Yamato Mountains in order to measure the flow of the Shirase Glacier drainage basin (Fig.1). JARE re-surveyed the network and obtained horizontal and vertical velocities (Naruse 1978). Based on these data, Naruse (1978) found that the ice sheet was thinning at a rate of 0.7 m a^{-1} along the network only in the Shirase Glacier drainage basin.

Mae and Naruse (1978) and Mae (1979a, b) calculated the basal sliding velocity, v_b, along the network from the thinning rate, using an equation of continuity, and obtained $v_b \simeq 10$ m/year — about half of the horizontal flow velocity. They considered that such a large basal sliding velocity was caused by the presence of water at the base of ice sheet.   Mae (1979a, b) concluded that the ice sheet in the Shirase Glacier drainage basin was not in a steady state by comparing similar changes of basal shear stress along the flow line in the Shirase Glacier drainage basin with those of surging glaciers. He proposed an hypothesis that the area of the wet base which caused the ice-sheet thinning had been initiated at Shirase Glacier (Fig.1) and had expanded up-stream from Shirase Glacier, reaching the position of the trilateration network a few hundred years ago.

In order to obtain more information on the ice dynamics of the Shirase Glacier drainage basin, JARE planned a 5-year project on glaciology, named Glaciological Programme in East Queen Maud Land, which began in 1982. JARE installed many JMR positioning stations in the basin to measure the velocity of the ice sheet and changes in the ice thickness. The programme of measurement and analysis will be completed in a few years.

Since the dynamics of the ice sheet depends upon ice thickness and the bedrock topography, JARE developed a 179 MHz airborne radio echo-sounder which was installed in a Pilatus Porter PC-6 aircraft. In 1979, the sounder was

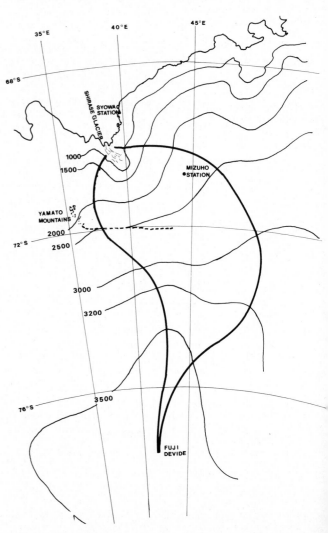

Fig.1. Map of the Shirase Glacier drainage basin (thick solid line). The trilateration network is indicated by dashed line.

used to make a preliminary sounding of Shirase Glacier and the Yamato Mountains. At that time the aircraft possessed poor naviagational instruments and the position of the aircraft was determined only by map reading. The sounding flight passed over as many known sites (e.g. nunataks or exposed rocks) as possible. On the basis of these data, Wada and others (1982) made a bedrock-topography map in the vicinity of the Yamato Mountains, but it proved impossible to map the bedrock elevations near Shirase Glacier due to poor navigation. In 1983, JARE carried out further airborne radio echo-sounding using an Omega navigation system. A preliminary report of the sounding was published by Mae (1986), and in this paper the results obtained will be reported in detail.

INSTRUMENTS AND SOUNDING AREA

Since the 179 MHz sounder was designed to be installed in the small (Pilatus Porter PC-6) aircraft, electric power and aerial dimensions were limited. The peak transmitter power was restricted to about 1 kW and aerials, 1 m in length, were attached to the wings of the aircraft. These limited the depth of penetration to 1500–2000 m. The details of the sounder have been reported by Wada and Mae (1981) and Wada and others (1982), and its specifications are listed in Table I. The previous recording system of A-scopes was changed from a 35 mm camera to a video set.

The radio echo-sounding was carried out in November 1983 along the routes shown in Fig.2. Flight lines 4, 5, 6, and 7 are in the Shirase Glacier drainage basin.

The up-stream area of the Shirase Glacier drainage basin was not surveyed because of logistics difficulties in supporting the aircraft. At present, only oversnow vehicles can survey this up-stream area. In this paper, therefore, only the results of the radio echo-sounding along the routes shown in Fig.2 are discussed.

TABLE I. SPECIFICATIONS OF SOUNDER

Transmitter	Carrier frequency	179 MHz
	Pulse width	0.3 μs
	Rise time	0.15 μs
	Peak power	1 kW
	Pulse-repetition frequency	1 kHz
	Total power consumption	dc 28 V, 2.7 A
	RF gain	39 dB
Receiver	Central frequency	179 MHz
	Band width	5 kHz
	Noise	3 dB
	Receiver sensitivity	-104 dBm
Aerials	3-element Yagi	
	Absolute power gain	8 dB
Monitor and recorder	Oscilloscope (National VP-5260)	
	Rise time: 35 ns	
	Video set	

RESULTS
Bedrock topography

In order to estimate ice thickness, the wave velocity in ice of 169 m μs^{-1} (Robin and others 1969) was used. Surface-elevation data for the Mizuho Plateau area, including the Shirase Glacier drainage basin, were compiled by Moriwaki of the Japanese National Institute of Polar Research and the bedrock topography was obtained by subtraction of the ice thicknesses.

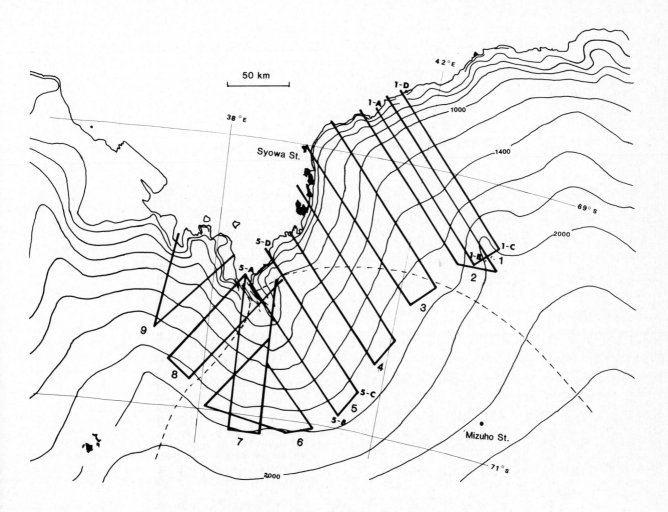

Fig.2. Radio echo-sounding flight lines (thick solid lines).

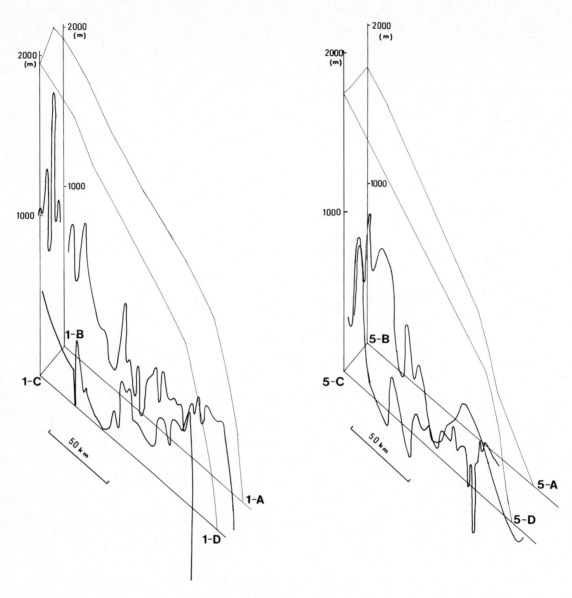

Fig.3a. Bedrock topography (thick lines) and surface topography (thin line) of flight 1. b. Bedrock topography (thick lines) and surface topography (thin line) of flight 5.

The elevation of the ice surface and the bedrock along flights 1 and 5 are shown in Fig.3a and b. The elevations of the bedrock surface are at sea-level or higher. On the basis of these results, a contour map of bedrock elevation was compiled and this is shown in Fig.4.

Fig.3a and Fig.4 indicate that there are subglacial mountains at the up-stream edge of flights 1 and 2 which are about 2000 m above sea-level. The presence of the mountains had been previously assumed from areas of bare-ice field with many crevasses and containing rock debris. In the Shirase Glacier drainage basin there are not such high subglacial mountains. From Fig.4 it can be seen that a shallow and wide valley lies between Shirase Glacier and Mizuho Station. Even in this basin, the elevation of the bedrock on the upstream side of the sounding area is about 500 m above sea-level. This is consistent with the result obtained from a 60 MHz radio echo-sounding survey of an oversnow traverse from Mizuho Station to the Yamato mountains (Shimizu and others 1978; Nishio, private communication). It can be assumed that the bedrock elevation is 500-1000 m above sea-level in the up-stream area of the Shirase Glacier drainage basin (personal communication from M. Yoshida).

Echos reflected from bedrock

The intensity, E, of the reflected echo from the bedrock surface varies with ice thickness. As shown in

Fig.5a and b, E decreases in proportion to the increase of ice thickness, h, up to 1000 m, thereafter E is approximately constant.

When h is less than 1000 m, E (in dB) along flight 1 is given by

$$E = -0.0052h - 26. \qquad (1)$$

and E along flight 5 is given by

$$E = -0.0052h - 25. \qquad (2)$$

Since Equations (1) and (2) are approximately the same, the characteristic feature of the reflection from bedrock shows little difference between the Shirase Glacier drainage basin and the surrounding area. If the reflection coefficient at the bedrock surface is uniform, the absorption within the ice mass is estimated to be 5.2 dB/100 m. This value is larger than 4.5 dB/100 m obtained by Robin and others (1969) in Greenland.

At ice thicknesses greater than 1000 m, E does not depend on h and becomes approximately constant. The value of E in the Shirase Glacier drainage basin, ($\approx$ −78 dB) is larger than in the surrounding area ($\approx$ −87 dB). It is difficult to explain the difference, 9 dB. It is not clear, if water exists at a base, where there are high hydrostatic pressures of about 15−20 MPa and how such water may influence the dielectric properties of ice at radio frequencies.

The spatial distribution of E is illustrated in Fig.6. In the Shirase Glacier drainage basin E at A and C is smaller

162

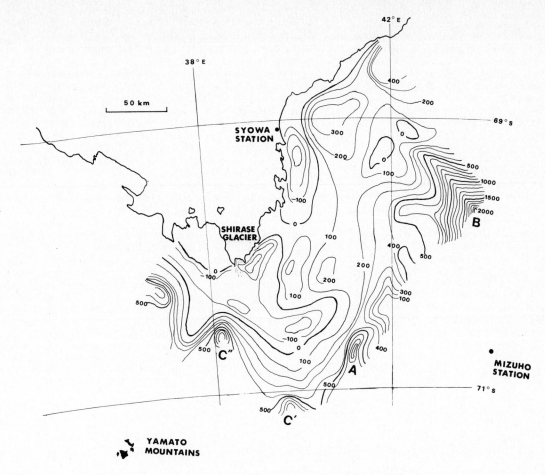

Fig.4.   Map of bedrock topography.

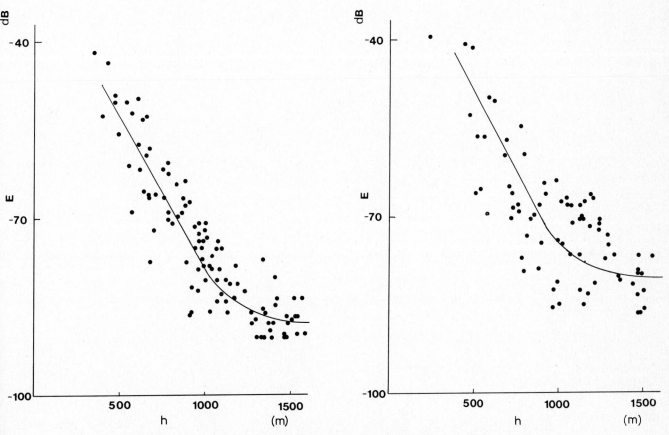

Fig.5a. Relationship between E and h, flight 1. b. Relationship between E and h, flight 5.

than E in the neighbouring area. At A in Fig.6 there is a mountain at A in Fig.4 and at C in Fig.6 there is a ridge from C' to C" in Fig.4. Down-stream of A and C in Fig.6 there are areas, D and F, with high E. However, in the down-stream area, B', in Fig.6 of the highest mountain B in Fig.4, E is low as shown in Fig.6 and there is no high E area such as D and F.

In Fig.6 there is a area of low E values, G, in the main flow area in the Shirase Glacier drainage basin.

DISCUSSION

The reflection of radio-echo energy may be affected by the properties of the bedrock and the roughness of the bedrock surface. It is impossible, however, to investigate these factors under a thick ice sheet. When h is less than 1000 m, the relationship between E and h is given by approximately the same equation as in the sounding area.

Yoshida and others (1987) have found that echo intensity depends upon the angle between the direction of the ice flow and the aerial orientation. When they are parallel, the echo intensity is strong. In this survey, the direction of the aerials is the same as the direction of the aircraft long axis, which varied with wind direction and wind velocity during flight. It is conceivable, therefore, that the influence of the aerial direction is negligibly small in this survey.

If the properties of the bedrock and its surface roughness are uniform, a possible explanation of the differences in E is that the echo strength in the Shirase Glacier drainage basin is affected by the presence of water at the base. Outside of the basin, ice is frozen to the bedrock.

At A in Fig.4 there is a subglacial mountain with an ice thickness of about 1200 m. The average temperature at the ice surface at this point is about −30°C, given that the average temperature gradient with elevation is −1.3°C/100 m at Mizuho Plateau (Satow 1978). The flow velocity has not been measured but is estimated to be 30 m a^{-1}. The accumulation rate is 100−200 mm a^{-1}. Based on these data, temperature calculations carried out by Nishio and Mae (1979), the basal temperature is about -10° − -15°C, and matches with the low value of E.

Since the flow velocity along the ridge from C' to C" is less than the velocity of the main flow area of the Shirase Glacier drainage basin (Naruse 1978), it is reasonable to suppose that the basal sliding velocity is zero and the basal ice is frozen to the bedrock.

E is also low around the mountain B and in the down-stream area of B. Although E becomes higher in the down-stream area of low B in the Shirase Glacier drainage basin, there is no high area such as D and F in Fig.6 in the down-stream area of B. This suggests the basal ice is frozen to bedrock.

As shown in Fig.6, a wide, curved valley extends from Shirase Glacier towards Mizuho Station, where the elevation of the bedrock is approximately 150 m above sea-level. Elsewhere in the sounded area the bedrock elevation is

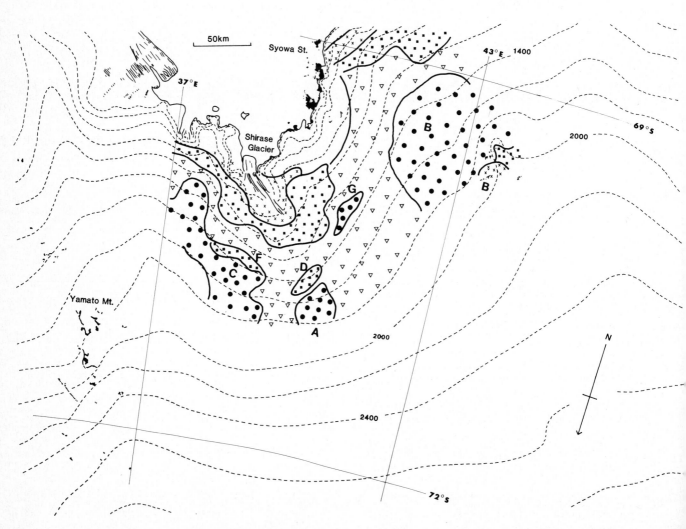

Fig.6. Map of the bedrock reflection coefficient. x, −40 dB > E > −60 dB; ■, −60 dB > E > −70 dB;
△, −70 dB > E > −80 dB; ●, E < −80 dB.

about 500 m above sea-level. According to radio echo-sounding conducted on oversnow traverses (National Institute of Polar Research 1985; Yoshida and others 1987), a high plateau (500–1000 m) extends further inland.

Mae (1979a, b) and Mae and Naruse (1978) proposed that the basal ice is wet along the trilateration network in the Shirase Glacier drainage basin (Fig.1) and such a wet base is related to basal sliding of about 10 m a^{-1} and an ice-sheet thinning rate of 700 m a^{-1}. Mae (1979a, b) and Mae (1982) suggested that the area of wet basal ice (approximately the same area as the region of ice-sheet thinning) expanded in size a few hundred years ago. This idea of disequilibrium is supported by the profile of basal shear stress, τ_b, along the flow line in the Shirase Glacier drainage basin which displays characteristics similar to those of surging glaciers (Mae 1979a, b; Mae 1982). Mae (1979a, b) estimated that τ_b reached a maximum of about 180 kPa in the region between the boundary of this sounding area and the trilateration network.

If such an hypothesis is correct, the ice-sheet thinning must take place in the down-stream area of the Shirase Glacier drainage basin. Although no direct proof of this hypothesis is available, it is reasonable to consider that the wet basal zone expanded from Shirase Glacier to the relatively low area of bedrock surface which is surrounded by the 500 m high plateau. At present, the mechanism of formation of the wet base and its expansion is not clear, but it is important for understanding the historical variations of the ice sheet in this area.

The low values of the reflection coefficient at G in the Shirase Glacier drainage basin are difficult to explain as the bedrock surface is only about 100 m higher than the surrounding area; this is insufficient evidence to suggest that the base is frozen to the bedrock.

ACKNOWLEDGEMENT

This study is partly supported by a Grant-in-Aid to Scientific Research of the Ministry of Education, Science and Culture, Japan. This is a contribution to the Glaciological Research Programme in east Queen Maud Land, Antarctica, by JARE.

REFERENCES

Mae S 1979[a] The basal sliding of a thinning ice sheet, Mizuho Plateau, East Antarctica. *Journal of Glaciology* 24(90): 53–61

Mae S 1979[b] The recent variation of ice sheet in Mizuho Plateau. *Memoirs of National Institute of Polar Research.* Special Issue 14: 1–7

Mae S 1982 Hyo-Sho no dorikigaku [Dynamics of ice sheet]. *In Kori to yuki [Ice and snow].* Tokyo, National Institute of Polar Research: 117–163

Mae S 1986 Radio echo sounding in the Shirase Glacier drainage basin. *Antarctic Record* 30(1): 11–18

Mae S, Naruse R 1978 Possible causes of ice sheet thinning in the Mizuho Plateau. *Nature* 273(5660): 291–292

Naruse R 1978 Surface flow and strain of the ice sheet measured by a triangulation chain in Mizuho Plateau. *Memoirs of National Institute of Polar Research.* Special Issue 7: 198–226

National Institute of Polar Research 1985 *Data of Antarctic research.* Tokyo, National Institute of Polar Research

Nishio F, Mae S 1979 Temperature profile in the bare ice area near the Yamato Mountains, Antarctica. *Memoirs of National Institute of Polar Research.* Special Issue 12: 25–37

Robin G de Q, Evans S, Bailey J T 1969 Interpretation of radio echo sounding in polar ice sheets. *Geophysical Transactions of the Royal Society of London* Ser A 265(1166): 437–505

Satow K 1978 Distribution of 10 m snow temperatures in Mizuho Plateau. *Memoirs of National Institute of Polar Research.* Special Issue 7: 63–71

Wada M, Mae S 1981 Airborne radio echo sounding on the Shirase Glacier and its drainage basin, East Antarctica. *Antarctic Record* 72: 16–25

Wada M, Yamanouchi T, Mae S 1982 Radio echo-sounding of Shirase Glacier and the Yamato Mountains area. *Annals of Glaciology* 3: 312–315

Yoshida M, Yamashita K, Mae S 1987 Bottom topography and internal layers in east Dronning Maud Land, East Antarctica, from 179 MHz radio echo-sounding. *Annals of Glaciology* 9: 221–224

Annals of Glaciology 9 1987
© International Glaciological Society

INTERPRETATION AND UTILIZATION OF AREAL SNOW-COVER DATA FROM SATELLITES

by

J. Martinec

(Swiss Federal Institute for Snow and Avalanche Research, 7260 Weissfluhjoch/Davos, Switzerland)

and

A. Rango

(USDA-ARS Hydrology Laboratory, Beltsville, MD 20705, U.S.A.)

ABSTRACT

Areal snow-cover data provided by remote sensing enable the areal water equivalent at the start of the snow melt season to be evaluated. To this end, the time scale in the graphical representation of the snow coverage curves is replaced by the totalized computed daily melt depths. These refer to the seasonal snow cover at the starting date and disregard subsequent snowfalls.

INTRODUCTION

Since the beginning of Earth satellite observations, snow cover monitoring has been a primary remote sensing objective. In recent years, measurement of the areal extent of snow cover has improved through better spatial resolution and discrimination between snow and clouds. An interpretation of these measurements is required to produce meaningful snow information.

SNOW-COVER MAPPING

Conventional depletion curves of snow-covered areas in mountain basins are useful for those snow-melt run-off models which are designed to use this information. The shape of these curves, varying from year to year, has even been used to estimate the accumulation of snow in terms of the water equivalent (Meier 1973; Ødegaard and Østrem 1977; Rango and others 1977; Moravec and Danielson 1980). Fig.1 shows, however, that substantial errors may be involved. From the depletion curves it would seem that the highest snow accumulation occurred in 1978 and the lowest in 1973, with average values in 1975 and 1972. On the contrary, from direct measurements, the extremes were in 1975 and 1972, with about three times more snow on 1 May 1975 than on 1 May 1972. The course of depletion curves is influenced by weather conditions during the snow-melt season as explained elsewhere (Hall and Martinec 1985).

MODIFIED DEPLETION CURVES OF SNOW COVERAGE

The snow-covered area declines not because time elapses, but because snow is melting. Modified depletion curves can be derived by replacing the time scale by cumulative melt depths from daily computed values. As shown in Fig.2, these curves represent the snow accumulation in the selected years in the right order and even, especially with regard to 1975 and 1972, in the right proportion. Since the water equivalent at the starting date of the snow-melt season is required, the effect of subsequent snowfalls must be eliminated. Therefore, the energy input required to melt the intermittent new snow is disregarded in calculating the cumulative melt depths. As illustrated in Fig.2, irregularities in the shape of some curves occur due to interpolation problems with the original conventional curves.

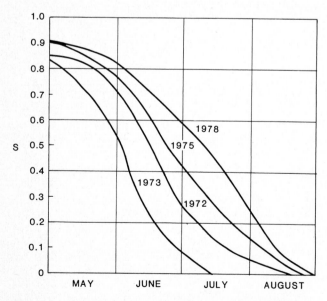

Fig.1. Conventional depletion curves of snow-covered areas in the elevation zone 2100–2600 m a.s.l. (24.5 km²) of the Dischma basin, Swiss Alps. S = proportion of the snow-covered area.

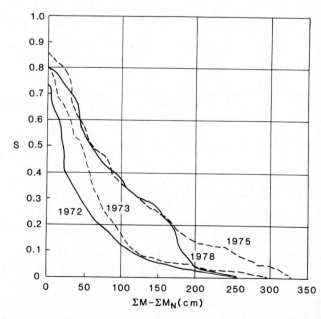

Fig.2. Modified depletion curves of snow-covered areas in the elevation zone 2100–2600 m a.s.l. of the Dischma basin, Swiss Alps. $\Sigma M - \Sigma M_N$ = cumulative snow-melt depth excluding snowfalls after 1 May.

The interpretation of the modified depletion curves in terms of the water volume stored in the snow cover or, in other words, of the areal water equivalent, is illustrated in Fig.3. These values are directly proportional to the area between the x, y axes and the curve. The results in the given example refer to 1 May, but any other date can be selected for evaluation according to the climate in the investigated basin or partial area.

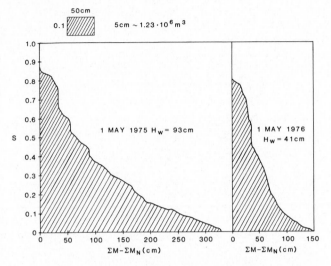

Fig.3. Areal water equivalent from modified depletion curves in two extreme years, zone 2100–2600 m a.s.l. of the Dischma basin.

EVALUATION OF AREAL WATER EQUIVALENT OF SNOW

As a verification, the areal water equivalents derived from the modified depletion curves are compared with the point measurements carried out in the period 1970–1979 on 1 May or close to this date. The melt depths were calculated just from the daily maxima and minima of the air temperature (WMO 1986) in order to test the applicability of the method with normally available data and not just in well-equipped representative basins. For the same reason, snowfall in May–August was distinguished from rainfall by a critical temperature although actual observations were available. Modified depletion curves for the elevation zone 1700-2100 m a.s.l. of the Dischma basin and for the four characteristic years are shown in Figure 4. Results are listed in Table I and plotted in Fig.5.

Details of the point measurements are given in the standard data set of Dischma (WMO 1986). Areal water equivalents for the zone 2100–3100 m a.s.l. are arithmetic

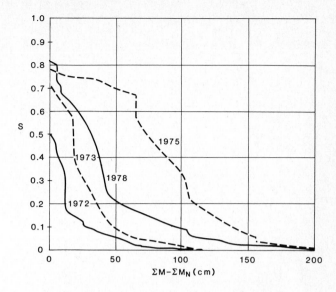

Fig.4. Modified depletion curves of snow-covered area in the elevation zone 1700–2100 m a.s.l. (8.9 km^2) of the Dischma basin.

averages of areal values for the zones 2100–2600 m a.s.l. and 2600–3100 m a.s.l. Weighted averages according to the respective areas in the Dischma basin would be about 5–10% lower.

OPERATIONAL USE OF MODIFIED DEPLETION CURVES

According to Fig.5, the point measurements at Weissfluhjoch are fairly representative for the elevation zone 2100–3100 m a.s.l. and the point measurements at the three elevations indicated in Table I for the elevation zone 1700–2100 m a.s.l. Similarly, the representativeness of selected stations in a snow-gauging network can be evaluated and correction factors derived, if necessary. Thus the operational data measured, for example, for avalanche warning or river flow forecasts can be improved.

Another possible application involves real-time evaluation of the snow accumulation from the modified depletion curves in a current year. However, the processing of satellite images as shown in Fig.6 and completion of a curve requires time. The accumulation of snow can vary over a wide range from year to year and these variations are revealed by the course of the modified depletion curves of the snow coverage. This is illustrated by Fig.7, in which the curves have been derived from Landsat data. A low accumulation of snow is indicated in 1977 and a high

TABLE I. AREAL WATER EQUIVALENTS (H_W) IN THE DISCHMA BASIN COMPARED WITH POINT MEASUREMENTS

Date: 1 May	Zone 1700–2100 m a.s.l.		Zone 2100–3100 m a.s.l.	
	Derived areal value H_W (mm)	Average of point measurements at 1818, 1910, 2007 m a.s.l. H_W (mm)	Derived areal value H_W (mm)	Point measurements at Weissfluhjoch, 2450 m a.s.l. H_W (mm)
1970	552	598	870	1192
1971	129	97	593	621
1972	95	90	484	454
1973	205	263	665	745
1974	304	259	964	987
1975	720	561	1073	1121
1976	169	203	545	493
1977	313	406	841	979
1978	360	385	847	873
1979	478	391	820	814

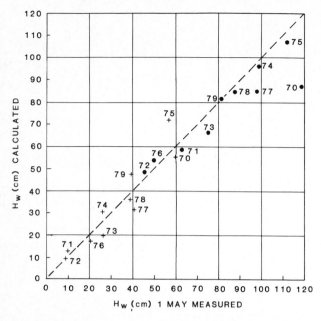

Fig.5. Comparison of areal water equivalents of snow in two elevation zones of the Dischma basin derived from snow-cover mapping and from point measurements at comparable altitudes. + = 1700–2100 m a.s.l., average elevation 1938 m a.s.l., and 8.9 km²; ● = 2100–3100 m a.s.l., average elevation 2560 m a.s.l., and 34.4 km².

accumulation in 1979, while the year 1976 represents an average. These conditions are similar in the remaining elevation zones of the South Fork basin and are confirmed by the subsequent run-off volumes. In real time of a future year the problem is how to identify the proper modified curve of the snow coverage as soon as possible after the start of the snow-melt season.

If the snow-covered areas in a basin are available for

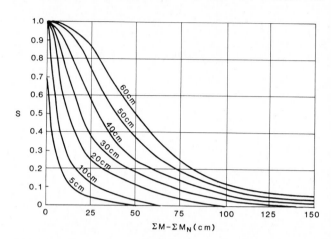

Fig.8. Nomograph for estimating the areal water equivalent of the snow cover on 1 April in the elevation zone 2900–3350 m a.s.l. of the South Fork basin from real-time periodic measurements of the snow coverage and from calculated totalized daily melt depths.

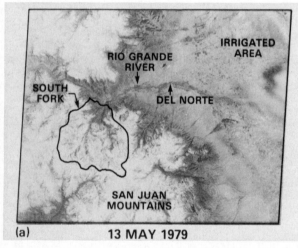

Fig.6. Landsat 0.6–0.7 μm images of the South Fork basin in the Rocky Mountains showing the decrease of the snow-covered area from May to June 1979.

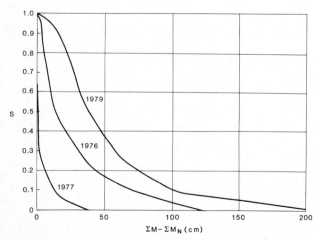

Fig.7. Modified depletion curves of snow-covered area in the elevation zone 2900–3350 m a.s.l. (269 km²) of the South Fork basin in three characteristic years, starting on 1 April.

several snow-melt seasons, preferably including some extreme years, it is possible to draw a set of typical curves, such as the nomograph for the South Fork basin shown in Fig.8. In a future year, snow–water equivalent data such as those available from the SNOTEL system (Rallison 1981) can be used to select an initial curve of the nomograph to use at the beginning of the snow-melt season. Then the snow-covered areas monitored by satellites and the corresponding snow-melt depths, calculated and totalized daily, either confirm this selection or signal a necessary correction. If no measurements of the water equivalent of snow are available, the proper curve may still be recognized in a month or two. For example, if the snow coverage measured on 1 May (1 month after the starting date) is 50% and the total melt depth, excluding new snow, to date is 25 cm, the point falls between the 30 cm and 40 cm curves of the nomograph, indicating an initial water equivalent of snow of 35 cm as an average for the respective area. However, in a year with an unusual spatial distribution of snow at the end of the accumulation period, deviations from the typified shape of the curves must be expected. If such conditions are indicated, for example, by an unusual prevailing wind direction in the winter months,

the identification of the proper curve is only provisional and must be verified or updated by subsequent measurements of the snow-covered area.

CONCLUSION

The modified depletion curves of snow coverage can be derived in basins or elevation zones in which the areal extent of the snow cover is periodically measured by satellites or aircraft during the snow-melt season. Further requirements are temperature data which must be extrapolated from a nearby station and precipitation data. The calculation of snow-melt depths, normally based only on temperature, can be refined if pertinent data are available.

Curves derived for past years from historical data can be used to assess the representativeness of point measurements in snow-gauging networks. This serves, for example, to determine the regional anomalies of the maximum expected snow loads on structures. With a timely evaluation of the satellite imagery, the snow accumulation in the given area at the beginning of a current snow-melt season can be evaluated with a certain delay and then updated after each satellite overflight.

REFERENCES

Hall D K, Martinec J 1985 *Remote sensing of ice and snow*. London, Chapman and Hall

Meier M F 1973 Evaluation of ERTS imagery for mapping of changes of snow cover on land and on glaciers. *Symposium on Significant Results Obtained from the Earth Resources Technology Satellite-1*. New Carrollton, MD, NASA: 863-875

Moravec G F, Danielson J A 1980 A graphical method of stream runoff prediction from Landsat derived snowcover data for watersheds in the upper Rio Grande basin of Colorado. *In* Rango A, Peterson R (eds) *Operational Applications of Satellite Snowcover Observations. Proceedings of a final workshop ... at Sparks, Nevada, April 16-17, 1979*. Greenbelt, MD, NASA: 171-183 (NASA Conference Publication 2116)

Ødegaard H A, Østrem G 1977 *Application of Landsat imagery for snow mapping in Norway. Final report*. Oslo, Norwegian Water Resources and Electricity Board

Rallison R 1981 Automated system for collecting snow and related hydrological data in mountains of the western United States. *Hydrological Sciences Bulletin* 26(1): 83-89

Rango A, Salomonson V V, Foster J L 1977 Seasonal streamflow estimation in the Himalayan region employing meteorological satellite snow cover observations. *Water Resources Research* 13(1): 109-112

WMO 1986 *Intercomparison of models of snowmelt runoff*. Geneva, World Meteorological Organization (Operational Hydrology Report 23; WMO Publication 646)

Annals of Glaciology 9 1987
© International Glaciological Society

IMAGING SUBGLACIAL TOPOGRAPHY BY A SYNTHETIC APERTURE RADAR TECHNIQUE

by

G.J. Musil* and C.S.M. Doake

(British Antarctic Survey, Natural Environment Research Council, High Cross, Madingley Road, Cambridge CB3 0ET, U.K.)

ABSTRACT

A synthetic aperture radar (SAR) technique has been used to image part of the grounding-line region of Bach Ice Shelf in the Antarctic Peninsula. The radar was sledge-mounted and operated in a pulsed mode with a carrier frequency of 120 MHz. The coherently detected output was recorded photographically as in-phase and quadrature components. Because the system was essentially stationary for each measurement, there was no doppler information about the reflecting points as in the more commonly used airborne and satellite-based SARs. Instead, the phase history was used directly to identify point targets by a correlation method.

Three sounding runs were carried out over the grounding line to give views of the area from separate directions. An aperture length of 104 m was necessary to achieve 8 m resolution in the along-track direction for an ice thickness of 290 m. The mapped swath was 88 m wide. Corrections to the data were made to allow for density variations and absorption in the ice. The back-scatter coefficient showed greater variations in echo strength over grounded ice compared with floating ice and texture analysis of the radar image revealed a statistically significant difference between these two regimes.

1. INTRODUCTION

Conventional radio echo-sounding through ice sheets measures range to a reflector, leaving an ambiguity in position. When mounted on a moving platform, such as a sledge or aircraft, a two-dimensional representation of reflector range against along-track distance is obtained. If the reflecting surfaces are relatively smooth, their range will be a close approximation to nadir distance, and errors due to contributions from side echoes will be negligible. For steeper slopes, strong reflecting points give characteristic hyperbolic echoes of range against distance that allow these points to be identified and the reflecting surface to be more accurately reconstructed (Harrison 1971). There still remains the uncertainty about topography to the sides of the track and the effect of cross-track slopes. However, statistical analysis of the variation of echo strength with range and position can give an indication of the type of reflecting surface being observed and can allow different types to be discriminated (Berry 1975; Oswald 1975; Neal 1979). Despite the promise of these methods in helping to unravel sub-ice conditions, in practice little has been achieved in applying them to problems such as identifying geological boundaries between nunataks. There are several reasons for this lack of progress, among them being uncertainty about the relationship between the radio-echo reflecting surface and the bed lithology. The presence of water and till at or near the base of the ice sheet could provide complicated reflecting surfaces that would not necessarily be characteristic of the underlying bed.

In order to study some aspects of the basal reflecting surface in more detail, a synthetic aperture radar (SAR) was developed. The advantages of this technique are that a high resolution can be achieved and that a two-dimensional image of the reflecting surface is obtained (Kovaly 1976). In the usual echo-sounding equipment that is used to penetrate thick polar ice the resolution is limited by the transmitted pulse length. For pulsed radio-frequency radars operating at 60 MHz, the pulse length is typically around 0.3 μs, equivalent to 50 m in ice. For impulse radars the equivalent pulse length can be an order of magnitude less (e.g. 5 m for a single cycle at 35 MHz), but there is often a penalty of reduced power and system performance that limits the depth that can be penetrated. The theoretical limit to the resolution of a SAR depends on the radar frequency and is proportional to the physical size of the antenna (Kovaly 1976). At 120 MHz (the carrier frequency used in the equipment described here) the resolution would be about 1 m. However, the achievable resolution is usually degraded by other equipment parameters as well as by the requirements of data processing. It is worth noting that with a SAR the resolution can be made independent of range by varying the length of the synthetic aperture, in contrast to the range-dependent resolution of sideways-looking radars that use real aperture techniques.

Intensity variations across two-dimensional SAR images are caused by changes in the back-scatter coefficient of a rough surface. Because the back-scatter coefficient of most surfaces is a strong function of viewing angle, the resulting SAR image can often be interpreted by the eye in the same way as an optical image. However, if there are any significant topographic features within the images, then their position will be displaced because the radar can only measure range. This "layover" effect can give significant distortion in mountainous regions. Airborne and satellite-based SAR images of the Earth's surface are often used as an alternative to aerial photographs or high-resolution optical images because active radars such as SARs can operate at night time and through clouds (Elachi 1980).

In order to achieve a given resolution, data must be collected along a track (or aperture) preserving both in-phase and quadrature components of the signal (unlike conventional radars which preserve only amplitude). For SARs mounted on moving platforms these requirements can lead to very high data rates and sophisticated processing equipment. Photographic techniques can be used to record data as a kind of hologram but complicated optical processors are required to decode the images (Cutrona and others 1966). For this initial study of the application of SAR to sounding ice sheets it was decided that a ground-based system with a simple method for recording and analysing wave forms was appropriate. The major question was whether a realistic image of the glacier bed could be obtained through the non-homogeneous ice sheet. Corrections had to be made when processing the signals to allow for ice density, anisotropy, and absorption affecting the ray path and signal strength. The successful conclusion to this feasibility study means that further development of recording techniques is worthwhile.

---

* Present address: Antarctic Division, Melbourne, Australia.

2. PRINCIPLES OF GROUND-BASED SAR

High-resolution images can be recovered from coherent radar data by analysing the complex echo signatures from individual ground cells. In airborne or satellite systems, the platform velocity generates a doppler frequency for each reflecting point which can be isolated and identified if data are collected over a long enough aperture (Harger 1970). However, a ground-based system (which is essentially static when each radar pulse is transmitted and received) must instead rely on the phase history to identify point targets. In both cases, to avoid ambiguity of reflecting points at equal ranges on either side of the ground track, the radar is arranged to look to one side only. Consider the geometry shown in Fig.1. The ground swath illuminated as the radar moves can be considered to consist of a number of pixels,

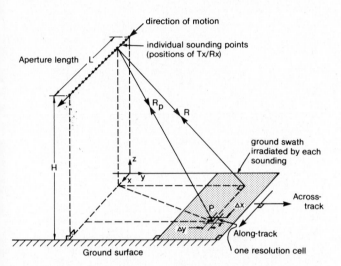

Fig.1. Geometry of SAR sounding (without an ice sheet). A combined transmitter/receiver (Tx/Rx) aerial moves along a path at height H above the reflecting surface, at range R from pixels of dimension Δx by Δy.

size Δx by Δy. The cross-track resolution, Δy, is calculated from the requirement that two reflecting targets with a separation Δy satisfies the relationship

$$\Delta y \geqslant \frac{v\tau}{2} \frac{1}{\sin \Theta}$$

where v is the velocity of radio waves, τ is the pulse duration, and Θ is the angle of incidence (Fig.2). The cross-track resolution therefore depends not only on the pulse length but also on the angle of incidence. The location of reflecting points in the y-coordinate is derived from the echo delay time.

To locate reflectors in the x-direction, an appropriate reference function is defined as the complex conjugate of the expected return from a point target in a given pixel. The recorded echo signals are then cross-correlated with the reference function centred on each pixel in turn and the relative strengths of the reflected signals evaluated for each cell (Wu 1978). A two-dimensional map of echo strength is built up, with cells of size Δx by Δy. The along-track resolution Δx can be deduced from the amplitude spectrum of the cross-correlation result, which can be written as

$$\left|V_0(x_0)\right| = LE_p^2 \left|\text{sinc} \frac{\omega_0 x_0 L}{v|R|}\right|$$

where $|V_0|$ is a scalar quantity which represents the magnitude of the echo received from a point P with range R from the transmitter/receiver after it has been cross-correlated with the reference function located at an offset x_0 from P. The length of the aperture is L, ω_0 is the angular frequency of the radar wave, and E_p is the

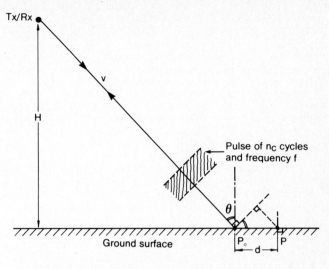

Fig.2. Resolution in the across-track direction is determined by the separation d between adjacent reflecting points P_0 and P. The transmitter/receiver aerial (Tx/Rx) broadcasts and receives a pulse of radio waves of duration $\tau = n_c f$ which travels with velocity v at angle of incidence θ with the ground surface.

amplitude of the echo. The function $|V_0(x_0)|$ is plotted in Fig.3. It is centred about a value x corresponding to the location of the reflector and its amplitude is proportional to the strength of the signal reflected from the target. The width of the sinc envelope therefore defines the along-track resolution, Δx. Taking the nulls as defining the half-width, Δx is given by

$$\Delta x = \frac{v|R|}{Lf} .$$

It can be seen that the resolution can be made independent of target range R by adjusting the length of the synthetic aperture L.

A similar expression can be derived for a doppler-orientated SAR, showing that the resolving capabilities are the same as for the phase-sensitive SAR. However, by cross-correlating the doppler frequencies as well as the phases of the echo signals, the along-track resolution can be improved by a factor of two.

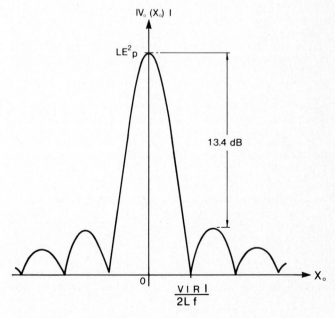

Fig.3. Amplitude spectrum of cross-correlation result. The symbols are defined in the text.

171

3. SAR SOUNDING THROUGH ICE: GENERAL PRINCIPLES

The overall effect of an ice mass on a SAR is to move the ground swath closer to the radar than it would be in free space. But the variable nature of the physical properties of an ice sheet also have to be taken into account if a worthwhile resolution is to be achieved (i.e. better than 20 m). The increase in permittivity with depth further refracts the radar signals and alters the apparent position of the echo source. The speed of the signals also changes with depth and has to be included when calculating the delay time of an echo signal.

The errors introduced in estimating the horizontal range of a reflector by assuming the ice to have constant permittivity (i.e. constant density) instead of a variation with depth, depend on the angle of incidence and ice thickness. The range would typically be underestimated by 9 m for an angle of incidence on the ice surface of 70° in ice more than about 300 m thick. Similarly, the two-way echo-time delay would be about 0.3 μs longer for constant-density ice. These errors, which are mainly introduced within the firn layer, are virtually constant for ice depths more than about 250 m.

A significant restriction on the size of the ground swath that can be imaged through an ice sheet arises from the requirement that echoes from the surface do not obscure echoes from the bed. This requirement places an upper limit on the angle of incidence on the ice surface and is a function of radar height above the snow surface, of ice thickness and of density. A typical value for the maximum angle would be about 70°. The lower limit is set by the cross-track resolution. Once the greatest angle of incidence has been calculated, the length of the aperture (L) required to synthesize a ground swath to a given along-track resolution (Δx) is provided by the equation

$$ L = \frac{R\lambda_i}{\Delta x} $$

where R is the greatest slant range of a pixel and λ_i is the wavelength of the radar waves in ice.

Corrections must be made to the echo amplitudes to allow for absorption, scattering, partial reflection at the air–ice boundary, geometrical losses, and the antenna-gain pattern. Although the losses due to geometrical effects are the greatest, it is the antenna-gain correction which is most sensitive to the relative position of the radar and the target. Fig.4 shows that signals that originated from the centre of the ground swath could appear to be 14 dB stronger than those from the outer corner, mainly because of the angular dependence of the antenna pattern.

The rate of change of echo phase along the aperture determines how often successive soundings must be made. In most practical cases this separation between soundings reduces to $\lambda_i/2$, but shorter distances are preferable to avoid approaching the Nyquist sampling limit too closely. Any variation in the radar frequency will introduce errors, so limits can be put on allowable oscillator jitter and drift by considering the phase history of reflecting points. Typical figures would place jitter at less than 2 kHz, and drift at less than 2 Hz s^{-1} for a radar operating at 120 MHz. These restrictions can be met by placing a crystal oscillator in a temperature-controlled environment.

4. IMAGING A GROUNDING LINE: FIELD RESULTS

A coherent radar working at 120 MHz with a half-wave dipole mounted in a 90° corner reflector was used to survey a grounding-line region on Bach Ice Shelf, Alexander Island, in the Antarctic Peninsula (Fig.5). The snow surface was smooth and flat to within a small fraction of a wavelength. The in-phase and quadrature components of the receiver output were recorded on photographic film and later digitized.

Fig.6 shows the ground swaths mapped by the SAR during three sounding runs. The position of the grounding line was originally estimated by using a 60 MHz radar and

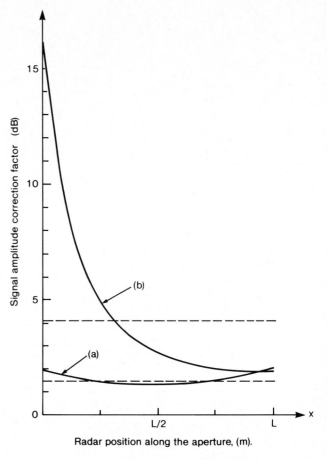

Fig.4. Amplitude correction profiles for echo signals originating at the centre of the ground swath (a) and at an outer corner (b).

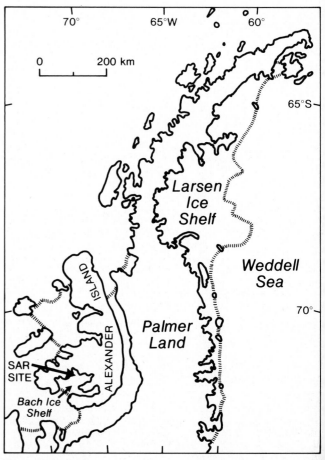

Fig.5. Location map of area of SAR survey.

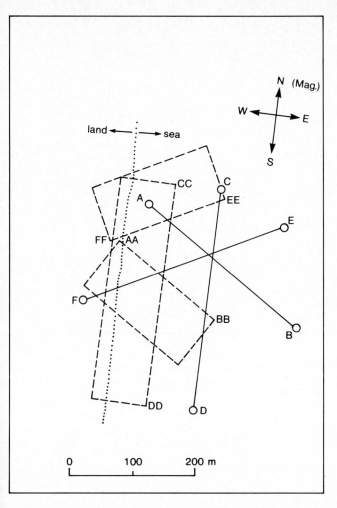

Fig.6. Plan view of sounding runs (full lines) and ground swaths (dashed lines). Grounding line is shown by dotted line.

by observing the presence of strand cracks, a diagnostic feature of tidal bending at a hinge zone (Swithinbank 1957). The ice thickness was around 290 m and the length of the sounding runs varied from about 300 m to 350 m. Records were obtained every 10 cm of aerial movement long the runs.

The digitized echo data were processed to give a chosen ground resolution of 8 m in both the along- and cross-track directions. The width of the ground swath that is mapped when sounding through 290 m of ice is 88 m and the length of the synthetic aperture needed to produce 8 m resolution in the along-track direction is 104 m. Thus, images containing 13 × 11 pixels are produced. Processing for higher resolution would give a narrower swath that also had proportionally fewer pixels. The values used here were chosen as a compromise between a high resolution with a narrow swath and a lower resolution that also gave a reasonable areal coverage. Because square pixels make subsequent analysis of the images easier to interpret, the pixel size in the along- and cross-track directions was chosen to be the same. Because the cross-correlation method does not consider the length of the transmitter pulse, the data can be processed to any desired resolution, but the measured pulse length of the transmitter, 140 ns, would give a nominal resolution across the track of about 30 m, thus blurring the image in that direction. Although the blurring is significant, it does not obscure the major details of the images.

The two images in Fig.7 were independently processed from data collected along the single traverse AB (Fig.6). Each image was derived from an aperture of length 104 m, but for the second image the aperture was displaced along track by 40 m (i.e. five pixels) with respect to the first. Although nearly 40% of the raw data are different in the two images, where they overlap, similar features can be identified. Ridges of relatively high signal back-scatter (R_1 to R_3), as well as an isolated peak P, are identifiable in both images. These features are displaced by 40 m, as

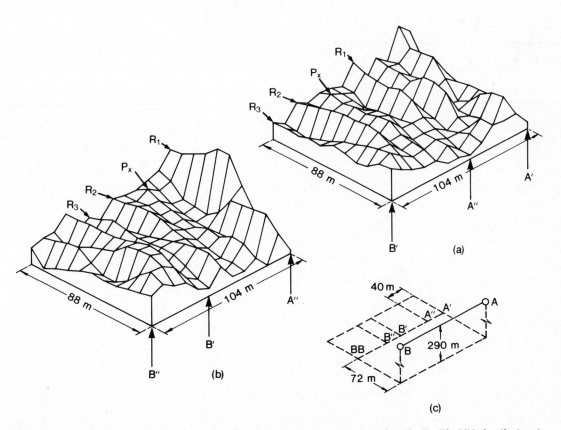

Fig.7. Two images of echo strength taken along the same traverse (AB in Fig.6). Fig.7(b) is displaced by 40 m (five pixels) along track compared to Fig.7(a), as shown by sounding geometry in Fig.7(c).

173

expected if the processing were correctly computing aperture-averaged amplitudes of a spatially stable surface. Maximum amplitudes of ridges R_1 to R_3 in image (b) are 210%, 75%, and 76%, respectively, of the amplitudes image (a). This is the behaviour expected from linear reflectors such as ridges or troughs trending almost perpendicular to the traverse. As the viewing angle approaches the normal to the reflecting surface, a higher back-scatter will be produced. Thus, in Fig.7, ridges of high back-scatter increase in amplitude towards the ends of the aperture but are relatively low near the centre. (Note that the images are of signal back-scatter, not of topography.)

The three images of the ground swaths in Fig.6 are shown in Fig.8. Each image was built up of smaller overlapping images, derived by processing successive apertures of length 104 m shifted 8 m (one pixel cell) along track each time. In the final (composite) image, each pixel value is the average of the individual image values for that particular pixel. This procedure reduces noise (often known as speckle) in the final image, although results from individual apertures seem to contain little noise anyway.

Quantitative texture analysis was used to classify and separate automatically the different types of surface. The procedure was based on second-order grey-level statistics (Haralick and others 1973) in which a set of grey-tone spatial-dependent matrices is computed from the original images (Fig.8). Four measures of texture defined from values of the elements in the matrices provide a set of numbers which describe and identify an image. These numbers, known as "identification vectors", can be used to compare textures of SAR images. Because of the relatively small number of pixels in our SAR images, the "Min–Max" decision rule was used to classify the various textures

(Haralick and others 1973): accuracies exceeding 80% have been reported for this method. The images from traverses AB and EF were analysed by splitting both into three equal parts (i.e. six portions each 8 × 11 pixels in size). The Min–Max decision rule automatically classified the two image portions closest to A and to F as one category and the four remaining portions into a different category. We suggest that these categories correspond to the division between grounded and floating ice, respectively.

A consistent interpretation of the three images can be given by assuming that a boundary between grounded and floating ice crosses each image as shown in Fig.6. Traverse CD (Fig.6) runs approximately parallel to the grounding line so that the strong echoes in the upper half of the image originate from grounded ice and the weaker echoes result from increased specular reflection from floating ice. Traverses AB and EF are orientated at angles of about $57°$ and $117°$, respectively, to the grounding line. Both of these images show a significant amount of radar back-scatter from beneath floating ice, in contrast to the image from traverse CD. This implies that the subglacial surface of the floating ice scatters back virtually no signal to the receiver when the radar is looking directly towards the grounding line, but scatters back significant energy when the radar views the grounding line obliquely. A plausible explanation is that as the ice flows into the sea it retains an imprint of the rock surface where it was last grounded. Grooves running in the direction of flow would return comparatively little incident radar signal when viewed along the ice-flow direction but would give stronger returns when viewed across the line of flow. The elongation across track of strong reflecting points is due to the length of the transmitter pulse.

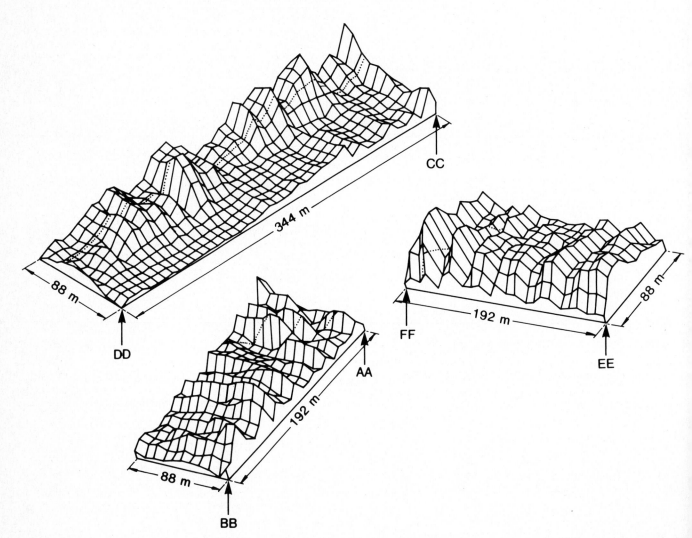

Fig.8. Images of echo strength for the three ground swaths shown in Fig.6.

5. CONCLUSIONS

A method of imaging subglacial topography by using phase-sensitive radio echo-sounding gives realistic results across a grounding-line area. The technique of looking to one side of a sounding run is similar to SARs used in aircraft and space satellites. Because the equipment was used on the ground in essentially a static mode, no doppler information was available as is the case when sounding from moving platforms. Instead, the phase history was used in an analogous cross-correlation procedure to reconstruct a two-dimensional image of the target back-scatter coefficient. Textural analysis of the resulting images automatically classifies the surfaces into categories that correspond to those derived from visual interpretation.

An alternative method of mapping bedrock characteristics by sounding over a square grid on the surface has been described by Walford and Harper (1981). Both the amplitude and phase information are used to reconstruct the glacier-bed geometry and reflection coefficient. However, a large area would be easier to survey with a side-looking radar. Both methods suffer from not considering polarization effects, although it is recognized that the polarization behaviour of the radar waves contains information not only on the birefringent properties of the ice through which the signals are propagating but also on the nature of the reflecting surface and englacial scatters (Doake 1982, Walford and others 1986).

ACKNOWLEDGEMENTS

We wish to thank Howard Thompson for his help in constructing the equipment and Ron Ferrari for his valuable advice and support.

REFERENCES

Berry M V 1975 Theory of radio echoes from glacier beds. *Journal of Glaciology* 15(73): 65-74

Cutrona L J, Leith E N, Porcello L J, Vivian W E 1966 On the application of coherent optical processing techniques to synthetic-aperture radar. *Proceedings of the IEEE* 54: 1026-1032

Doake C S M 1982 Polarization of radio waves propagated through George VI Ice Shelf. *British Antarctic Survey Bulletin* 56: 1-6

Elachi C 1980 Spaceborne imaging radar: geologic and oceanographic applications. *Science* 209(4461): 1073-1082

Haralick R M, Shanmugan K, Dinsteiri I 1973 Textural features for image classification. *IEEE Transactions on Systems Man, and Cybernetics* SMC-3(6): 610-621

Harger R O 1970 *Synthetic aperture radar systems, theory and design.* New York, Academic Press

Harrison C H 1971 Radio-echo sounding: focusing effects in wavy strata. *Geophysical Journal of the Royal Astronomical Society* 24(4): 383-400

Kovaly J J 1976 *Synthetic aperture radar.* Dedham, MA, Artech House

Neal C S 1979 The dynamics of the Ross Ice Shelf revealed by radio echo-sounding. *Journal of Glaciology* 24(90): 295-307

Oswald G K A 1975 Investigation of sub-ice bedrock characteristics by radio echo-sounding. *Journal of Glaciology* 15(73): 75-87

Swithinbank C W M 1957 Glaciology I.A. The morphology of the ice shelves of western Dronning Maud Land. *Norwegian-British-Swedish Antarctic Expedition 1949-52. Scientific Results* III

Walford M E R, Harper M F L 1981 The detailed study of glacier beds using radio-echo techniques. *Geophysical Journal of the Royal Astronomical Society* 67(3): 487-514

Walford M E R, Kennett M I, Holmlund P 1986 Interpretation of radio echoes from Storglaciären, northern Sweden. *Journal of Glaciology* 32(110): 39-49

Wu C 1978 *Optimal sampling and quantization of synthetic aperture radar signals.* Jet Propulsion Laboratory (JPL Publication 78–41)

Annals of Glaciology 9   1987
© International Glaciological Society

EVOLUTION OF UNDER-WATER SIDES OF ICE SHELVES AND ICEBERGS

by

Olav Orheim

(Norsk Polarinstitutt, P.O. Box 158, 1330 Oslo Lufthavn, Norway)

ABSTRACT

A systematic programme of side-scan sonar and plumb-line soundings was carried out in the Weddell Sea area in 1985 to measure the under-water sides of ice shelves and icebergs. From these observations the following model is suggested for the evolution of the ice front:

(1) Initial stage: fracturing of the ice shelves takes place along smooth, curvi-linear segments with vertical faces.

(2) Formative stage: the freshly formed vertical face is eroded both by wave and swell action around the water line, by small calvings from the undercut, overhanging subaerial face, and by submarine melting. The melting has a minimum at 50–100 m depth, and increases with depth to a rate of around $10 \, \text{m a}^{-1}$ at 200 m. This is about twice the rate of erosion at the water line. The variation in melting with depth results from a combination of summer melting by near-surface water, and year-round melting by water masses that are increasingly warmer than the pressure melting-point with depth.

(3) Mature stage: this stage is reached after a few years of exposure. The backward erosion of the face leads to a shape with a prominent under-water "nose" with a maximum projection to more than 50 m at 50–100 m depth. The ramp above this slopes upwards to meet the vertical wall about 5 m below the water line. The ice below the nose is melted back beyond the above-water face. There is no net buoyancy and ice shelves at this mature stage are generally *not* up-warped at the front.

INTRODUCTION

This paper discusses the evolution of the front of ice shelves and the sides of icebergs, with main emphasis on changes below the water line.

Ice shelves float, and are modified in contact with sea-water. A freshly calved tabular iceberg is a sample of the ice shelf, with the experimental advantage of including sides that have not been affected by the sea, and at least one side that has been exposed for some time.

The shapes and the physical dimensions of ice fronts tend to be similar around Antarctica. This, and the opportunity of recognizing ice fronts and icebergs recently calved, gives good conditions for systematic studies of the evolution of the free faces. Furthermore, ice fronts can sometimes be observed to have moved steadily outwards without undergoing large-scale calving. Such locations are especially attractive for investigation of changes with time.

A typical ice-shelf front is around 200 m high, of which 30–35 m is above water. The freeboard is greater than for pure ice because the ice shelves consist of firn in their upper layers. The elevation of the ice front generally varies little over short distances, except where grounding has taken place. Our typical measurements show a height of freeboard within 28 ± 4 m over a distance of several kilometres.

The map shape shown by the ice front is partly a question of scale. It will generally appear straight or slightly curved on a scale of a kilometre or less. The curved segments have the concave side to the sea, i.e. the fracture lines meet in cusps pointing away from the ice

shelf. For present purposes, the ice front is considered constant in thickness and linear in map shape. Thus, we are only concerned with the changes in the profile at right-angles to the ice front.

This paper presents data on profiles at various stages. These are used to develop a model for the evolution of the free face, and determine which processes are most effective in forming the face. Practically all field data were collected on the Norwegian Antarctic Research Expedition (NARE) 1984–85, where part of the ship time was used for a dedicated study of the under-water ice. To my knowledge, no previous systematic studies have been conducted to determine the under-water shapes of ice shelves and tabular icebergs, apart from preliminary studies on NARE 1978–79 (Klepsvik and Fossum 1980).

OBSERVATION TECHNIQUES

(a) Side-scan sonar

The main part of the under-water data is based on sonographs obtained by a Klein model 400 side-scan sonar (SSS). The SSS system consisted of three units: an under-water unit (the "fish"), a combined data, power, and tow cable, and a dual-channel graphic printer. The fish is a streamlined, hydrodynamically balanced body containing two sets of transducers. These transmit a 0.1 ms ultrasonic sinusoidal pulse at 100 kHz and then change to a receiving mode for a time interval determined by the range.

Most of the pulse energy is concentrated in two beams with 40° openings. For these experiments, the fish was arranged so that both beams faced the ice wall, looking up and down from the horizontal. The beam width is 1° in the direction of travel. The resolution is determined both by graphic capability and range. It was typically 2–4 m in these studies.

The plotted image has a linear time-scale with zero at the centre, and each returning signal is plotted on the paper according to the time it is received after the outgoing pulse. As the fish moves through the water, the adjacent scans form an acoustic image with one scale determined by the speed of the fish. Strength of the return signal determines darkness of the plotted image. A strong signal appears black. For angles of incidence ≠ 90°, the strength of the return signal depends upon the roughness of the ice at scales comparable with the wavelength of the acoustic pulse (0.015 m) (Klepsvik and Fossum 1980).

The present system was not equipped with a correcting device, the scales along the two axes of the sonograph were different, and the return times gave a non-linear representation of the object. The sonograph therefore needs careful analysis to give a correct image. The sonographs may also be distorted by phenomena such as: (a) cross-talk of various kinds between the channels, (b) deviating fish motions, (c) multiple reflections, and (d) focusing of the sound beams.

Further descriptions of the SSS can be found in Leenhardt (1974) and Flemming (1976). Information on this kind of SSS investigation of ice shapes has been given by Klepsvik and Fossum (1980). Their studies on NARE 1978–79 led to the present programme.

Two profiling techniques were used in the present study. Most of the data, altogether 130 line km, were

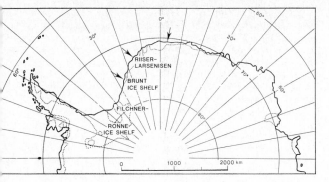

Fig.1. Index map showing location of side-scan studies.

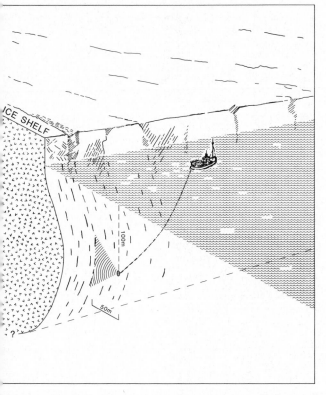

Fig.2. Principle of horizontal, under-water sounding. The figure does not show the whole sector covered by the sound beam.

causing damage, which reflects on the smoothness of the under-water ice.) The increased offset, in combination with the back-sloping ice, unfortunately meant poor records from the greatest depths.

The fish was also used in a vertical mode, being lowered and raised by the cable. This technique had the advantage that it provided a direct visual representation of the face of the ice, which was readily interpreted. A disadvantage with our arrangement was that the fish hung freely. Thus, there could be gaps in the vertical records because the fish rotated around the cable so that both transducers faced away from the ice (Fig.3).

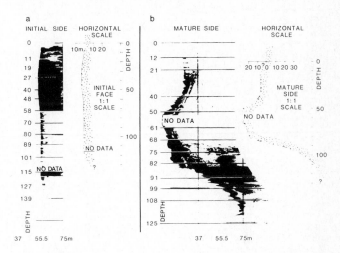

Fig.3a and b. Results of vertical sounding of two sides of iceberg Billie (B).
(a) Original record of initial (newly fractured) face, and rectified to a 1 : 1 scale.
(b) Original record of mature face, and rectified to a 1 : 1 scale. The spikes are not from the ice, but are probably caused by ship motion. Note vertical grooves at the side of the face, with the signal alternating between reflection and shadow. These are most noticeable between 21 m and 50 m depth.
Fig.3a and b show data from one channel. The fish rotated while being lowered and raised so that this channel did not always face the ice. For those sections where the other channel then faced the wall this recorded profile is shown. There are also gaps in the data where neither channel faced the ice.

collected by towing the fish behind the vessel, usually at around 50 m from the ice face (Figs 1 and 2). The best results were obtained by close profiling. However, navigating the vessel at a constant short distance from the ice face was not feasible because of reduced manoeuvrability at the profiling speed of 2–3 knots. We found it possible to profile as close as 30 m along straight sections.

The fish was towed at several depths at most locations. A major difference between the present study and that of Klepsvik and Fossum (1980) was that the present fish was now fitted with a depth sensor. This was read on board, and the depth controlled by adjusting cable length or vessel speed. More importantly, it became possible to combine different profiles to build three-dimensional images of the ice face. Such reconstructions are unfortunately labour intensive, because variations in scales caused by changes in speed and range meant that the assembly had to be done manually.

The offset of the fish causes problems both in field observations and later reconstructions. The fish trails behind the ship at cable lengths typically three times the depth, and will therefore only broadly follow the ship's course. It was difficult to get close to the ice at greater depths, because the fish had to be kept clear of the under-water "noses". (We did hit the ice with the fish and cable, without

(b) <u>Direct soundings</u>

Data on the under-water shapes were also collected by direct soundings. Most soundings were done from a small boat, using a plumb-line. This was not easy in rough seas. Soundings were also done by plumb-line from the bow of the expedition vessel, K/V *Andenes*, by nudging the ice front, but this was also a problematic procedure. We sounded all sections where we used the side-scan sonar. Altogether, 47 profiles were sounded. Typical examples of soundings are given in Fig.4.

FIELD LOCALITIES

The programme was aimed particularly at obtaining data from localities that could be identified as having been exposed to the open sea for specific time periods. How this was achieved requires explanation.

Any ice face exposed to the open sea will develop undercutting at the water line. Our observations show that an undercut typically extends a few metres inwards, and suggest that it takes a couple of weeks to develop a 1 m cut under "normal" sea states. Thus the criterion "lack of undercutting" was used to identify those faces that had not been exposed to the open sea.

Two types of such faces were observed. The first type,

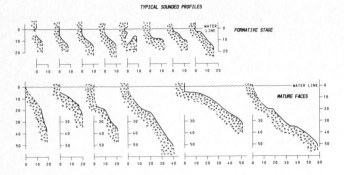

TYPICAL SOUNDED PROFILES

Fig.4. Examples of plumb-line soundings. The distance between each sounding was at 3–5 m, and the number of sounding points varies from 4–12 for the sections shown. The soundings were in all cases continued out to a vertical (or near-vertical) edge below which no bottom was encountered. All distances in meters.

noted especially in the early part of the season, was seen where the ice front had until recently been bounded by fast ice. Parts of the ice front would usually have a snow "foot" at the water line, i.e. snow adhering to the lower few metres of the ice front. (This snow would be the remnant of a snow bank formed on the fast ice, up against the ice front.) The under-water shape of this type could take a variety of forms.

The second type occurred when the ice front or iceberg had recently calved so that undercutting had not yet developed. Identification of this initial stage in the evolution of the ice face was particularly important to this study, e.g. iceberg Billie (B). This iceberg had clearly calved recently. It had well-developed undercutting on one side which extended round the corners to the two adjacent sides, but elsewhere showed no undercutting. Fig.3a and b are from two opposite sides of the iceberg. Horizontal profiling at different depths confirmed that the undercutting criterion distinguished between different developments. The sections without undercutting were smooth, while the mature side had a "rough" appearance.

Data were also required from sections that had been exposed to the sea for known periods. Studies were done of the fronts of Riiser-Larsenisen and Brunt Ice Shelf, where in both cases information on the calving history was available.

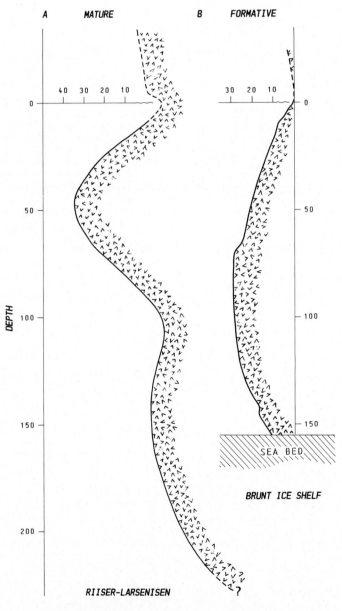

Fig.5. Rectified under-water profiles of (a) mature stage (Riiser-Larsenisen), and (b) formative stage (Brunt Ice Shelf). All distances in metres.

The most comprehensive studies were at Riiser-Larsenisen, where a detailed survey was done of a 4.8 km section from long. 16°07'W. to 16°16'W. Our surveys of Riiser-Larsenisen in 1977, 1979, and 1985 showed that this section of the ice shelf retreated due to major calving between 1977 and 1979. The ice front had thereafter advanced 450 m between 1979 and 1985. Our velocity measurements at the ice shelf indicate an outflow rate of around 100 m a⁻¹, suggesting no major calving during this period. Recognition of characteristic features in soundings both in 1979 and 1985 (see below) also indicated no major calving. Thus, this ice front could be taken as an approximately 7-year-old, "mature" face.

Nine profiles of the front of Riiser-Larsenisen were collected at controlled depths, spaced ≈40 m apart, during a 24 h period. Precision navigation was achieved by use of Motorola Miniranger transponders positioned on the ice shelf, together with the ship's own composite navigation systems. The distance from the ship to the ice front was recorded at frequent intervals. A second repeated profile was obtained 1 month later.

Part of this section was also covered by side-scan sonar studies during NARE 1978–79 (Klepsvik and Fossum 1980). These revealed, in map view, a 1 km section of distinctive waves, with amplitude of 10–15 m and wavelength of 30–100 m. The surveys repeated after 6 years showed that the waves were still present but the amplitudes had been reduced to 3–7 m.

The studies of Brunt Ice Shelf were done near Mobster Creek, the landing ramp for the British station Halley, and where station personnel informed us that various calvings had taken place over the recent few years. Thus, those sections were taken as examples of the young, formative stage, perhaps 1–3 years old.

OBSERVATIONS – LARGE FEATURES

Initial stage
(a) The *subaerial* face. As explained earlier, freshly fractured faces were recognized in the open sea by the absence of the effects of wave-action. The profile of such newly ruptured faces appears approximately vertical and smooth on a scale of metres.
(b) The *submarine* face. The initial under-water profile is essentially vertical, and appears smooth at our resolution of a few metres. Fig.3a shows the profile of a freshly formed face.

Formative (young) stage, 0.1–≈3 years
(a) The *subaerial* face. High ice fronts are generally over-hanging, often accentuated by snow-drift cornices. Crevasses are often seen in the ice shelf a few metres in from the edge. The undercut at the water line typically extends 2–3 m inwards, and 0.5–1 m up into the ice (firn).
(b) The *submarine* face. The upper part of the face is in the form of an approximately plane ramp sloping at high angles, usually >45°. The size of the ramp depends upon age. It extends from a few metres to a few tens of metres. The ramp starts about 5 m below the water line, and the maximum depth varies from 5–10 m to a few tens of metres. Below the ramp, the face is nearly vertical for a considerable depth, before sloping backwards below 100–150 m depth (Figs 4 and 5b).

Mature stage, several years old
(a) The *subaerial* face is like the formative stage described above.
(b) The *submarine* face has an under-water ramp observed at Riiser-Larsenisen up to 60 m from the ice front. This corresponds to an average melt rate of about 10 m a⁻¹, from the time of calving at this locality (higher if the melt rate at the nose is large). The slopes may be plane, concave, or convex, and are generally less steep than at the formative stage. The face has a prominent "nose" at around 50 m water depth, and below this the ice face slopes backwards at angles around 30°→45° (Figs 3b, 4 and 5a). Limited data from one locality suggest that the face is smoothed in the direction along the front, i.e. that the under-water back-melting is greatest on what were originally protruding cusps. This smoothing seems to increase with depth.

OBSERVATIONS – SMALL FEATURES
It is beyond the scope of this paper to discuss in detail the various smaller features that can be observed in the side-scan sonar. However, some aspects relevant to the discussion of processes should be noted.
(1) The mature face appears "rough", and includes isolated areas, especially at less than 100 m water depth, which do not reflect the sound waves (Fig.6). These have different shapes, vary from 2.5 m to 10 m in extent, and may be caused by protruding knobs or by depressions in the ice surface, in the latter case presumably reflecting where blocks of ice had "calved" in a submarine environment. An example of a sounded hollow is seen in Fig.4.

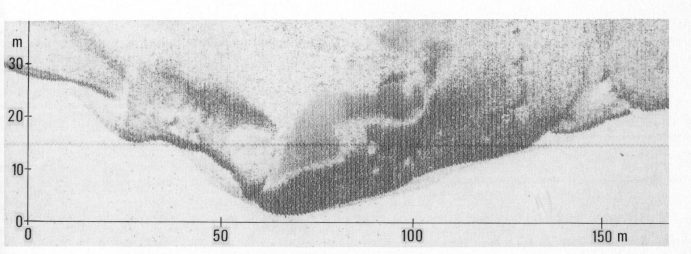

Fig.6. Sonograph looking up from 90 m water depth at a mature face. Note the difference in scale along the face and towards the face. The latter scale is converted from the return time of the sound waves, and later echoes are from higher on the ice face. The variation in strength of the return signal is related to varying angle of incidence and to the roughness of the face. Note step-like features and hollows or protruding knobs.

(2) Vertical grooves, with a regular spacing of a few metres, were seen during vertical profiling (Fig.3b). Such shapes would not be recorded by horizontal profiling, and it is not known whether they are common. However, these grooves were observed in partly successful under-water photography along Riiser-Larsenisen, and they have been seen in overturned icebergs (Fig.7). I believe these grooves result from upward-flowing sea-water, and hence demonstrate vertical circulation along the ice front.

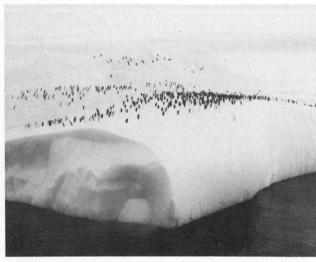

Fig.7. Submarine vertical grooves and wave-cut platform exposed in iceberg tilted on its side. Photograph: January 1981 in northern Weddell Sea. Scale of grooves can be judged from Adélie penguins; the grooves are similar in size to those shown in Fig.3b.

EVOLUTION OF THE ICE FRONT: PROCESSES AND MODEL

The above observations lead to a description of the evolution of the ice front with the following main components (see Fig.8).

(1) <u>Initial stage</u>: fracturing of the ice shelves takes place along smooth, curvi-linear segments with vertical faces.

THREE STAGES IN EVOLUTION OF ICE FRONT

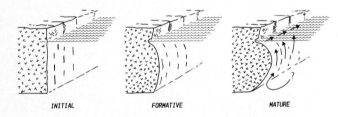

INITIAL   FORMATIVE   MATURE

Fig.8. The three stages in the evolution of a "typical" ice front.

(2) <u>Formative stage</u>: the freshly formed vertical face is eroded (a) by wave and swell action around the water line, (b) by small calvings from the undercut, overhanging subaerial face, and (c) by submarine melting. The melting increases with depth to ≈10 m a⁻¹ at 200 m, which is about twice the rate of erosion around the water line.

(3) <u>Mature stage</u>: erosion continues, with a maximum rate at around 50 m water depth, where the under-water "nose" has its maximum projection, observed extending up to 60 m.

Processes governing the formation of the undercut face

Our observations indicate that ice fronts with a freeboard >20 m were practically always overhanging. The main reason must be undercutting at the water line and related crevassing of the ice shelf.

Three processes can contribute to the undercutting:

(1) Higher local water velocities produced by waves and swell action cause higher effective heat-transfer rates between the ice and the near-surface waters, heated by short-wave radiation in the "summer". Martin and others (1978) showed that this was a very effective process, producing a 3 m wave cut in 1 day in ≈+2 °C water.
(2) Calved pieces of ice moved by the turbulent water cause mechanical erosion.
(3) Ice and firn is wetted by splashing, causing lower albedos and hence under appropriate conditions more absorbed solar radiation.

These are mainly "summer" processes. Most of the year the sea ice effectively prevents turbulent motion, and in the winter the surface waters are also at freezing temperatures.

Only the first of these three processes can contribute to the deeper erosion of the submarine ramp observed in the formative and mature stages. The heat transfer seems to be very effective to 5 m depth. Below this the efficiency falls rapidly with depth.

Processes governing back erosion at greater depths

The observations indicate varying degrees of melting with depth, with a minimum around 50–100 m depth, and a maximum annual melt of ≈10 m at 200 m depth. The heat transfer to the ice front depends upon the water temperature, and mixing coefficient and small-scale boundary-layer dynamics.

Field observations on the water temperatures show two phenomena:

(1) In the summer, the upper ≈70 m of the water in contact with Riiser-Larsenisen are heated to about 0.5 °C above the pressure melting-point (Foldvik and others 1985[a], Fig.15), with highest temperatures at the surface. Correspondingly, the above-freezing temperatures extend to 30–50 m depth along the Filchner–Ronne Ice Shelf (Foldvik and others 1985[b], Figs 14 and 15). In the winter, the upper waters are at pressure melting-point. Hence, this condition leads to back melting to such depths in summer.
(2) Foldvik and Kvinge (1974, 1977) observed that the melting potential of water masses in contact with the Filchner Ice Shelf increases with depth, because the pressure melting-point decreases with depth (Fig.9). Temperatures were about 0.2 °C above the pressure melting-point at 200 m depth. This condition should lead to melting which increases with depth from a minimum around 100 m, and takes place year-round.

We therefore see that the combination of these temperature observations explains qualitatively the observed shapes as resulting from annual melting rates with a minimum around 50–100 m depth, and with greater annual melting at 200 m depth than near the surface.

Melting of vertical ice walls in the ocean has been discussed by Gill (1973), Martin and Kauffman (1977), Gade (1979), Greisman (1979), Huppert (1980), and Russell-Head (1980). These give a range of melt-rates which bracket the rates presented here. The present observations give a melt-rate of 10 m a⁻¹ for a temperature difference (ΔT) above pressure melting-point of 0.2 °C at 200 m depth, and a rate of 20 m a⁻¹ for a ΔT of 0.5 °C, the latter derived from the observed rate of ≈5 m a⁻¹, and taking this to apply only for the summer period of 1/4 year. The differences between the various melt-rates proposed by the above authors are related to varying models for heat transfer to the ice face and are outside the theme of the present discussion.

The observed vertical ripples suggest vertically moving water and may confirm upwelling. Such processes, at various scales, have been proposed, e.g. by Neshyba (1977) and Josberger (1980), and would enhance melt-rates by their

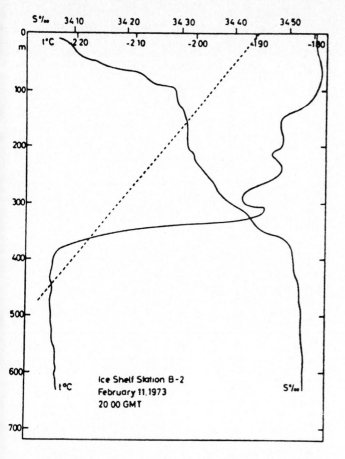

Fig.9. Temperature and salinity of water near the Filchner Ice Shelf (lat. 77°44'S., long. 41°44'W.). The dashed line shows the change in pressure melting-point with depth for water of the salinity at lower levels. (From Foldvik and Kvinge 1977.)

effect on the boundary-layer dynamics.

Observations from the closest comparable field situation are those of Holdsworth (1982). He calculated a lateral melt-rate of 6–10 m a^{-1} in −1.3°C water for the Erebus Glacier tongue, based on repeated mapping and calculated lateral creep spreading.

Many of the field observations presented here were anticipated by Robin (1979), who used the work of Foldvik and Kvinge (1974, 1977) in a theoretical discussion of melting/freezing under, and in front of, ice shelves. He recognized the potential for increased melting with depth, and that this process could take place year-round, and he suggested a melt-rate at depth on the order of 10 m a^{-1}.

Buoyancy effects

As noted, even at the mature stage, the ice shelves are not generally up-warped at the front. Calculations show that profiles such as Figs 3b and 5 are essentially in balanced buoyancy, because the effect of the "nose" is compensated by the ice removed at greater depths.

Indeed, down-sloping towards the front is more commonly observed, and this is indirect confirmation of back-melting at depth, as recognized by Robin (1979). Such bending from the effect of extensive melting of the lowest ice could also increase the depth of the ramp.

Buoyancy probably accounts for the rough appearance of the mature face, causing large and small pieces of ice to "calve" off along faults and inhomogeneities (Fig.6). The face seems roughest at shallow depths and such calving is presumably assisted by wave and swell action.

Under-water plastic deformation

A protruding under-water "nose" will experience deformation as a result of the buoyancy stresses. Taking as a first approximation that the ice will begin to deform

rapidly at a shear stress of 100 kPa (1 bar), we see that the shape of a "nose" extending 50 m as in Fig.3b or Fig.5a will be determined by the melt-rates, as these shapes will have maximum shear stresses less than 50 kPa at the line of the vertical wall. The shear stress increases with extension of the nose, and a 100 m nose could experience rapid plastic deformation. Thus plastic flow will limit the extension of the nose to approximately 100 m.

CONCLUDING REMARKS

The preceding results are based on field data which have been consistent, and are therefore used as the foundation for a model. However, some words of caution are needed. As mentioned earlier, the records are poorer for the deeper part of the ice front, and in particular for the mature case. In many cases, no records were obtained because the ice retreated and was out of range for the sonar. Thus, we know that the ice was sloping backwards, but not its exact shape. The records are also from a small part of the Antarctic ice shelves, and possibly the Riiser-Larsenisen locality is one where water enters beneath the ice shelf to circulate around the ice rise Kvitkuven. For this reason, the melting-rates at depth here for the mature case may be abnormally high, and it is likely that there is less melting of ice shelves further south. Further data are needed from different ice shelves, and in different oceanographic environments, to test whether these observations are typical and the model is universal.

ACKNOWLEDGEMENTS

Execution of this programme involved several of my colleagues at NARE 1984–85, especially Eystein Hansen who operated the side-scan sonar, and Arne Foldvik and Guttorm Jakobsen who sounded from the small boat. Thomas Martinsen did the laborious compilation of the individual sections into a more usable form. I extend my gratitude to all those who helped. This is publication No. 88 of the Norwegian Antarctic Research Expeditions (1984–85).

REFERENCES

Flemming B W 1976 Side-scan sonar: a practical guide. *International Hydrographic Review* 53: 65–92

Foldvik A, Kvinge T 1974 Conditional instability of sea water at the freezing point. *Deep-Sea Research* 21(3): 169–174

Foldvik A, Kvinge T 1977 Thermohaline convection in the vicinity of an ice shelf. *In* Dunbar M J (ed) *Polar oceans; proceedings of the Polar Oceans Conference ... Montreal ... 1974.* Calgary, Arctic Institute of North America: 247–255

Foldvik A, Gammelsrød T, Slotsvik N, Tørresen T 1985[a] Oceanographic conditions on the Weddell Sea shelf during the German Antarctic Expedition 1979/80. *Polar Research* 3(2): 209–226

Foldvik A, Gammelsrød T, Tørresen T 1985[b] Physical oceanography studies in the Weddell Sea during the Norwegian Antarctic Research Expedition 1978/79. *Polar Research* 3(2): 195–207

Gade H G 1979 Melting of ice in sea water: a primitive model with application to the Antarctic ice shelf and icebergs. *Journal of Physical Oceanography* 9(1): 189–198

Gill A E 1973 Circulation and bottom water production in the Weddell Sea. *Deep-Sea Research* 20(2): 111–140

Greisman P 1979 On the upwelling driven by the melt of ice shelves and tidewater glaciers. *Deep-Sea Research* 26A(9): 1051–1065

Holdsworth G 1982 Dynamics of Erebus Glacier Tongue. *Annals of Glaciology* 3: 131–137

Huppert H E 1980 The physical processes involved in the melting of icebergs. *Annals of Glaciology* 1: 97–101

Josberger E G 1980 The effect of bubbles released from a melting ice wall on the melt-driven convection in salt water. *Journal of Physical Oceanography* 10(3): 474–477

Klepsvik J O, Fossum B A 1980 Studies of icebergs, ice fronts and ice walls using side-scanning sonar. *Annals of Glaciology* 1: 31–36

Leenhardt O 1974 Side-scanning sonar – a theoretical study. *International Hydrographic Review* 51: 61–80

Martin S, Kauffman P 1977 An experimental and theoretical study of the turbulent and laminar convection generated under a horizontal ice sheet floating on warm salty water. *Journal of Physical Oceanography* 7(2): 272–283

Martin S, Josberger E, Kauffman P 1978 Appendix: Wave-induced heat transfer to an iceberg. *In* Husseiny A A (ed) *Iceberg utilization; proceedings of the First International Conference ... Ames, Iowa ... 1977.* New York etc, Pergamon Press: 260–264

Neshyba S 1977 Upwelling by icebergs. *Nature* 267(5611): 507–508

Robin G de Q 1979 Formation, flow, and disintegration of ice shelves. *Journal of Glaciology* 24(90): 259–271

Russell-Head D S 1980 The melting of free-drifting icebergs. *Annals of Glaciology* 1: 119–122

Annals of Glaciology 9 1987
© International Glaciological Society

MAPPING OF AMERY ICE SHELF, ANTARCTICA, SURFACE FEATURES BY SATELLITE ALTIMETRY

by

K.C. Partington, W. Cudlip, N.F. McIntyre and S. King-Hele

(Mullard Space Science Laboratory, University College London, Holmbury St. Mary, Dorking, Surrey RH5 6NT, U.K.)

ABSTRACT

Ice shelves are important regions to observe because they are likely to be sensitive indicators of climatic change. The satellite-borne radar altimetry is highly suited to ice-shelf monitoring; experience with Seasat, which flew in 1978, has demonstrated that a height-measurement precision of the order of 1 m can be obtained over ice surfaces (Brooks and others 1983).

We identify subtle changes in altimeter wave forms associated with crevassed zones and the grounding line. Normal retracking procedures are shown to be inadequate in detecting such changes, and so methods which provide sensitive indication of the presence of these features in the sampled areas are devised. By ranging to the first return in the echo, the grounding line is identified, and by differencing this measurement with the half-peak power range, a measure of surface roughness is obtained which can be used to detect crevassed zones.

Detection of crevassed shear zones allows delimitation of distinct zones of flow in the ice shelf which can be monitored by future altimeter missions. Monitoring of the grounding-line position can provide sensitive indication of mass-balance conditions over the grounded part of the drainage basin.

INTRODUCTION

Theoretical studies suggest that ice shelves play a crucial role in maintaining the stability of large regions of the Antarctic ice sheet (Hughes 1973; Thomas and others 1979). Their health, however, is sensitive to the temperature of the ocean and ice-sheet mass-balance conditions (Robin 1979). Early indications of climatic change may therefore be provided by changes in the morphology of the ice shelves. Amongst the surface features which are likely to be responsive indicators of climatic change are shear zones and grounding lines.

Shear zones delimit discrete zones of flow (flow bands) within the ice shelf (Crabtree and Doake 1986). Fluctuations in mass balance within a drainage basin are likely to result in movement of shear zones and changes in flow-band thickness and velocity. Changes in mass balance within a drainage basin may well occur, as suggested by the current stagnation of Ice Stream C in West Antarctica (Shabtaie and Bentley in press), and past surging of Fisher Glacier in the drainage basin of the Amery Ice Shelf, Antarctica (Wellman 1982).

Grounding lines mark the boundary between grounded and floating ice. Weertman (1974) showed that, for an idealized bedrock and perfectly plastic ice sheet, the surface slope across the grounding line must decrease in order to reduce the basal shear stress. The grounding line may therefore be observable on the surface as a distinct feature marked by a sharp change in surface gradient (e.g. Budd and others 1982). However, where the bed in the grounding zone has complex morphology, the transition from fully grounded to fully floating ice may take place over several kilometres (e.g. Stephenson and Doake 1982). As ice shelves are characterized by low surface gradients, the horizontal position of the grounding line is very sensitive to thickness changes in the ice shelf and ice sheet (Weertman 1974).

The detection of these features by remote-sensing techniques would enable their long-term monitoring for climate studies and glaciological applications. Satellite altimeters provide sufficiently precise measurements of surface elevation to detect many surface features. For example, Thomas and others (1983) have demonstrated an ability to monitor the ice-shelf margin using altimeter data. It is the intention of this paper to demonstrate that monitoring of the grounding line and crevassed zones is possible with satellite altimetry.

THE AMERY ICE SHELF

The study area chosen was the Amery Ice Shelf, since this is the largest Antarctic ice shelf lying within the coverage of Seasat. Further, it is a relatively narrow (*c.* 250 km) outlet for perhaps the fourth largest drainage basin in Antarctica and is therefore likely to be a sensitive indicator of mass-balance changes inland. The estimated area of the drainage basin is 1.63×10^6 km^2 ± 4.9×10^5 km^2 (Giovinetto 1964). The mass flux across the ice front has been estimated as 2.7×10^{16} g year^{-1} ± 35%, which discharges at a velocity of 1.2 km a^{-1} (Budd 1967; Allison 1979). Approximately one-third of this is contributed by Lambert Glacier, which flows in from the south and is itself fed by several major tributaries which converge about 200 km south of the grounding line. Wellman (1982) presented geomorphic evidence to suggest that one of these, Fisher Glacier, has surged in the past. Most estimates of mass balance for the basin indicate net accumulation, although recent study has shed doubt on this (McIntyre 1985).

Previous observations

The Amery Ice Shelf has been the subject of ground, aircraft and satellite survey. Early aerial exploration by Australian National Antarctic Research Expeditions (ANARE) provided information on ice-shelf surface features, including an approximate course for the grounding line (Mellor and McKinnon 1960). The course indicated in the vicinity of Lambert Glacier was oblique to the main direction of flow, and this has recently been confirmed with satellite observations by Swithinbank (in press). Later, ground survey, airborne radio echo-sounding, and drilling were carried out by ANARE to measure ice movement, elevation, accumulation, and thickness in a comprehensive series of studies of the ice shelf in the 1960s and 1970s. These measurements were taken along traverse lines following the approximate centre line of flow, and across the ice shelf near the ice front and near the grounding line (Budd and others 1967; Morgan and Budd 1975; Budd and others 1982). The data produced from these surveys allowed re-assessment of the dynamics of the ice shelf (Budd 1966) and enabled new estimates of the mass balance of the drainage basin to be made (e.g. Budd and others 1967; Allison 1979). Importantly for this study, the position of a grounding point in the vicinity of Lambert Glacier was determined (Budd and others 1982).

Seasat, which flew in 1978, provided the first coverage by a satellite-borne altimeter, allowing Brooks and others (1983) to produce a topographic map of the ice shelf with a contour interval of 5 m and an accuracy of the order of 1 m. From the map, they compared corrected altimeter elevations with the centre-line profile obtained by Australian ground survey (Budd and others 1982). The two profiles agree well in shape from near the ice front to near the grounding zone. However, the apparent position of the grounding point, as observed in the altimeter profile, was displaced approximately 40 km up-stream from the true position, indicating a decrease in the reliability of the altimeter measurements in this region (discussed later). Elsewhere, the presence of crevasses in elevation profiles showed up as zones of rapid elevation fluctuations of amplitude ±(4–5) m and wavelength the order of a few kilometres. In addition, the presence of two grounded regions within the ice shelf was indicated by the presence of topographic highs.

THE DATA

Fifty-seven passes of Seasat altimeter data across the Amery Ice Shelf were available for the present study, including complete coverage from a 17-day repeat orbit. To validate interpretation of altimeter wave-form data over the Amery Ice Shelf, this study uses digitally enhanced Landsat, AVHRR (Advanced, Very High Resolution Radiometer) imagery and ANARE ground-survey data. Imagery obtained for the ice shelf was digitally enhanced by applying an auto-Gaussian contrast stretch, which forces the frequency histogram of pixel values to take on a Gaussian form. In addition, edge enhancement was carried out to produce an apparent increase in spatial resolution. With a spatial resolution of 1.1 km, crevassed zones are clearly visible after enhancement, and strong flow features can be seen in the region of lower Lambert Glacier, both in Landsat and AVHRR.

Seasat altimeter-range measurements

The Seasat radar altimeter was designed for use over the open ocean, which can be assumed to be a planar, horizontal, and diffuse scatterer over the area sampled. For such surfaces, information on the surface dielectric and geometric properties within the pulse-limited footprint is contained in the leading edge of the echo wave form (McGoogan 1975). The instrument attempted to locate the centre of the digitizing window at the half-power point of the leading edge, and the range was estimated from the delay time associated with the centre of the sample. The on-board alignment of the digitizing window was carried out by prediction from previous estimates of range and range rate, and was modified by the generation of a height-error signal. As long as the centring was carried out successfully, the calculated range corresponded to the mean elevation within the pulse-limited footprint (Brown 1977). However, the instrument was not designed to track rough and rapidly varying ice-sheet surfaces (Brenner and others 1983; Partington and Rapley 1986).

Fig.1 shows four altimeter wave forms recorded over the Amery Ice Shelf, along with positions on each wave form associated with elevation calculated in four different ways, as discussed below. Wave form 1 is a typical signal received from a flatter part of the ice shelf. The on-board calculated elevation (method a) can be seen to align with the half-power position on the leading edge of the wave form, as a result of the low and slowly varying relief of the surface. The lengthened leading edge of wave form 2 suggests that it is from a rougher region of the ice shelf. The tracker again finds the half-power position because the surface elevation is varying slowly along track and the wave form similar in shape to rough ocean return. The complexity of wave forms 3 and 4 suggests that they are from geometrically more complex surfaces. Wave form 3 is double-ramped, suggesting the presence of two distinct reflecting regions in the sampled area. Wave form 4 is still more complex and difficult to interpret, with the leading edge hard to define. Areas contributing to the pulse-limited footprint in such cases may be spatially separated and variable in scattering properties and relief (Griffiths 1984).

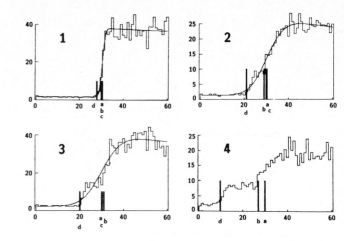

Fig.1. Four ice-shelf wave forms with position ranged to by the on-board tracker (a), half-peak power retracked elevation (b), least-squares fitting of a model ocean return (c), and first-return retracked elevation (d).

Wave forms 3 and 4 were given poor elevation estimates by the on-board tracker.

Retracking methods

As the on-board tracker often failed to align the centre of the digitizing window with the half-power point of the leading edge, it is normally necessary to re-align these so that a more suitable position on the wave form is selected. Three methods have been reported in the literature to date.

Brooks and others (1983) retracked to the half-peak value in the wave form (Fig.1, method b) to produce their topographic map of the Amery Ice Shelf. For ocean return, this produces a range to the half-power position of the leading edge. Wave form 1 is successfully retracked using the half-peak method. The low slope of the leading edge of wave form 2, however, results in increased sensitivity to noise. Wave form 3 shows that the retracked elevation obtained from double-ramped wave forms is very sensitive to the relative amplitudes of the two ramps. In this case, it is the second ramp which is ranged to. A gradual increase in the amplitude of the first ramp relative to the second along a sequence of double-ramped wave forms would result in a spurious step in the elevation profile. The profiles of elevation produced by Brooks and others (1983) across crevassed zones may well suffer from this problem and therefore may not represent the true surface.

Martin and others (1983) used a more sophisticated re-tracking technique (method c). The model ocean wave form assumed in design of the on-board tracking procedure (Brown 1977) was fitted to altimeter ice-sheet data by the method of least squares. They also used a "double-ramp" model to fit wave forms of type 3 (Fig.1), so that two range estimates could be made. Compared with the half-peak method, this technique results in reduced sensitivity to noise. In addition, the method provides several parameters which may be used to describe wave-form shape, including leading-edge width, which have not been used to extract information on surface character to date. A similar version of the single-ramp model has been implemented for least-squares fitting in this study, for comparative purposes.

Methods a to c all assume that the return is ocean-like, which is not necessarily appropriate for detection of subtle changes in wave-form shape associated with passage across surface features such as crevassed zones and the grounding line.

Martin and others (1983) attempted to overcome this problem by fitting a double-ramp model wave form to return from more complex surfaces. Where the wave form exhibits a form which implies return from two distinct reflecting surfaces, then this method works well. However, visual inspection of returns from the Amery Ice Shelf indicates that many wave forms fail to fall into the "ocean-like" or "double-ramp" categories.

Thomas and others (1983) used a different retracking method to enable them to map the ice margin. They showed that the location of the ice margin could be determined from the position at which the return signal from sea ice begins a parabolic migration away from the satellite. In order to demonstrate this, they retracked the altimeter data to the first return associated with the sea ice, in a signal which contained returns from both sea ice and ice shelf. The retracked elevation then referred to the closest part of that surface to the satellite, and was a spot measurement rather than a mean over an area.

First-return retracking

A new method of retracking adopted here (d in Fig.1) is specially designed to provide meaningful elevation measurements over rougher ice-shelf topography, where the other methods, which assume an ocean return, produce highly ambiguous measurements. The method ranges to the first-return signal in the wave form, so providing a spot elevation measurement associated with the closest surface to the satellite.

The method used to retrack data employs a threshold retracker set at a value of ten "counts" (undimensionalized units of signal amplitude). Seasat used an on-board automatic gain control to attenuate the signal and keep it within the dynamic range of the instrument (Townsend 1980). Counts are the signal-amplitude units after application of automatic gain control to the signal. Over the main body of the Amery Ice Shelf, the signal-amplitude value varied little, so noise was held at a fairly constant count level. This allowed the use of a signal-count value to detect the first return.

The first-return method results in a higher elevation value than other retracking methods, the difference being proportional to the width of the leading edge for an ocean return (Fig.1, wave forms 1 and 2). The method is particularly useful, however, for retracking of complex wave forms such as 3 and 4. Here, definition of the area of the surface contributing to the leading edge of the return is not required, so easing interpretation, though the closest surface to the satellite may not lie at nadir. Such cases may be flagged by the presence of parabolic sections in elevation profiles, suggesting that the retracking method is ranging to a single surface feature as the satellite approaches and recedes (Gundestrup and others 1986).

RESULTS

Fig.2 shows a map of the Amery Ice Shelf and lower Lambert Glacier. Five altimeter tracks pass within 5 km of the grounding point identified by Budd and others (1982), and marked on Fig.2 by a circle. Several tracks of altimeter data cross crevassed zones propagating down-stream from Gillock Island (A) and Charybdis Glacier/Jetty Peninsula (B), and these zones can be seen in digitally enhanced Landsat and AVHRR imagery.

Crevassed zones

Fig.3b shows a sequence of wave forms recorded during satellite passage across crevassed area B, which delimits flow originating from Charybdis and Lambert

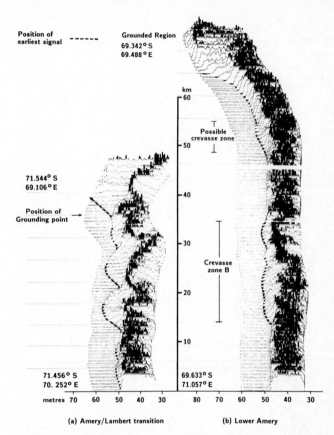

(a) Amery/Lambert transition   (b) Lower Amery

Fig.3. Two altimeter wave-form sequences shown crossing the Amery Ice Shelf from east to west (bottom to top of the plot). Each wave form is a histogram of received signal power (z-axis) against time/range (with non-absolute scale shown across base of plot). The wave forms overlap and are scaled to force the peak value of each wave form to be the same height. The high-amplitude parts of the wave forms are shaded to emphasize the passage of migrating features through the sequence.

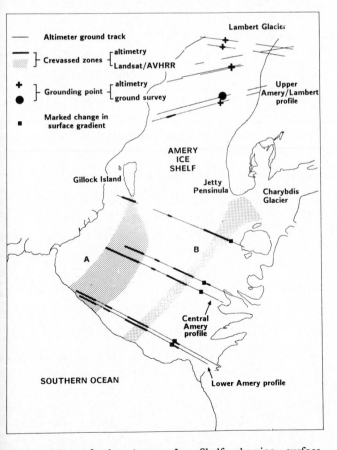

Fig.2. Map of the Amery Ice Shelf showing surface features positioned by satellite altimetry, together with features identified from imagery and ground survey (Budd and others 1982).

Glaciers. The crevassed area is clearly marked by an increase in the width of the leading edge of the wave forms, and by an increase in surface elevation indicated by the position of the first return. Over crevassed surfaces the precise nature of the relationship between surface roughness and leading edge is likely to be complex and therefore lead to difficulties in the interpretation of any elevation measurements which assume an ocean return (e.g. Brooks and others 1983).

The plot suggests that crevassed zones may be delimited

by measurement of leading-edge width. One such measurement is provided as a by-product from least-squares fitting of a model ocean return to the data (Martin and others 1983). However, the complex wave forms found in crevassed areas often defeat the fitting procedure, which either fails to converge iteratively or produces a poor fit. Another measure of leading-edge width is provided by significant wave height, which is automatically telemetered from the satellite. However, this measure is damped by a 0.8 s time constant, so tends to be insensitive. The most reliable method was found to be a difference between the first-return and the half-peak elevations. Fig.4 shows three transects across the Amery Ice Shelf, each containing the

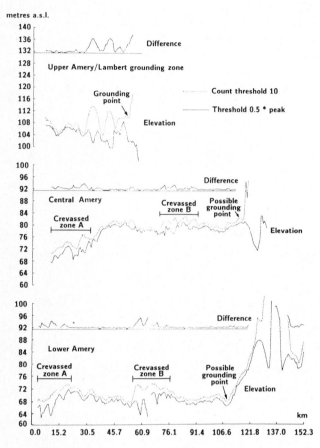

Fig.4. Three transects of altimeter-derived surface elevation across the Amery Ice Shelf. Each transect contains two profiles obtained from first-return and half-peak retracked elevations.

first-return and the half-peak elevations along with the difference between the two. Zones of crevassing are indicated by a large difference between the two values, with the contrast between crevassed and non-crevassed regions most clear in the lower regions of the ice shelf.

This latter technique is sufficiently sensitive to detect regions of possible crevassing on the ice shelf which are not visible in digitally enhanced AVHRR and Landsat data. An example can be seen in Fig.3b, at the position of a marked change in surface gradient. The leading edge of the wave form increases in width, suggesting that a zone of crevasses may be located at this position. The possible shear zone can be detected intermittently along the western side of the ice shelf north of Charybdis Glacier in other sequences of altimeter data. Flow lines visible in AVHRR imagery show that its location is consistent with that of the edge of a relatively stagnant region determined from velocity measurements in the extreme north-west of the ice shelf (Budd and others 1982). Other possible crevassed zones are indicated by bold sections of ground tracks in Fig.2. The criterion for delimitation of crevassed zones for this map was a difference of 1.5 m in the elevations as determined by the first-return and half-peak methods, on purely empirical grounds.

Flow bands are delimited by shear zones. In Fig.4, two flow bands are visible, separated by crevassed zone B. Repeated measurements of the elevations and widths of flow bands delimited by shear zones can be used to infer mass-balance conditions within sub-catchments.

Measured widths of the leading edge cannot, however, be used indiscriminately to detect crevassed zones. There are a number of realistic surface geometries and scattering properties which can combine to produce similar wave forms and not all of these are crevassed surfaces. For example, the transect across the Amery Ice Shelf/Lambert Glacier transition region (Fig.4) shows a large and fluctuating difference between the first-return and half-peak elevations. The first-return elevation can be seen to follow a parabolic trajectory in two places, which indicates that a single surface feature, rather than the nadir surface, is being ranged to (Gundestrup and others 1986). For this reason, initial detection of shear zones using altimetry requires validation from imagery, either from direct confirmation of the presence of crevasses or from the use of flow lines to indicate that a likely shear zone lies down-stream of a promontory or glacier confluence.

The grounding line
Fig.3 shows a sequence of wave forms over the grounding zone of Lambert Glacier. A grounding point (at station T4) identified by Budd and others (1967) is located approximately 4 km to the north of the closest approach of the ground track at the position indicated. The closest approach of the satellite to the grounding point is marked by a sharp change in surface gradient indicated by the position of the first return. In addition, the wave forms in the vicinity of the grounding point show an evolution similar to those generated by simulation over a change in slope (Rapley and others 1985).

Fig.5 shows profiles of on-board height and three retracking heights along a track in the vicinity of the grounding point at T4. The first-return retracking method

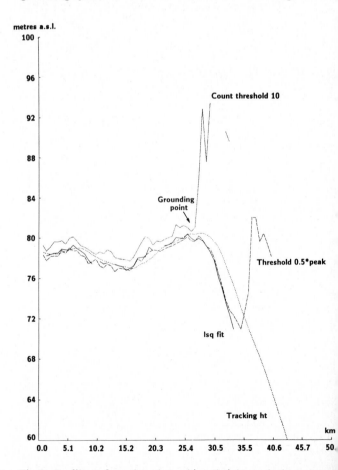

Fig.5. Profiles of on-board tracking height and three types of retracked elevation in the region of the grounding point (T4) reported by Budd and others (1982) for Lambert Glacier.

produces a profile of the nearest surface to the satellite as a series of spot measurements. With low surface slopes, this will be a good representation of the sub-satellite surface. The other profiles in Fig.5 show a decrease in surface elevation in the vicinity of the grounding line, which in each case is principally the result of assuming a model ocean return. As the return is focussed by the surface geometry associated with the change in slope, these methods continue to produce range measurements to the grounding zone after passing over the grounded ice. This demonstrates that the first-return method is the only reliable method for ranging across the grounding line.

Fig.4a shows a longer profile of the same track, with elevation associated with the first-return and half-peak methods. In addition to a grounding point, the first-return profile indicates the presence of two features sampled from off nadir, as discussed in the previous section. It is possible that these are grounded regions located to one side of the ground track, which would suggest that the grounding line is "wavy" in form.

Two methods are available for locating the position of the change in slope associated with the grounding line. The first utilizes the first-return method and applies a correction. The second employs the method of Thomas and others (1983) to locate the position at which return from the grounding zone begins a parabolic migration away from the satellite.

The grounded ice is detected prior to the satellite being positioned directly above it, and is detected by the first-return method as a "toe" moving out of the main ramp associated with the floating surface. Assuming the grounding line is a straight line separating two planar surfaces, the grounding line will be located by the first-return method a distance d before its true position on the satellite ground track where:

$$d = \frac{h \tan(\alpha/2)}{\sin\beta} \quad (1)$$

or

$$d = \frac{hm}{2n \sin^2\beta} \quad (2)$$

Here, h is the distance of the satellite from a floating surface, α is the slope of the grounded surface, β is the angle of incidence of the satellite with the grounding line, m is the elevation difference in metres between consecutive wave-form first returns, and n is the along-track distance between wave-form samples (662 m). For a horizontal floating ice surface, a grounded ice gradient of 0.29° (Table I and Fig.3a), a β value of 90° and an h value of 800 km,

TABLE I. GRADIENTS EITHER SIDE OF GROUNDING POINT (MEAN OVER 12 km)

	Floating ice	Grounded ice
Upper Amery Ice Shelf	6.1×10^{-3}°	2.9×10^{-1}°
Central Amery Ice Shelf	1.4×10^{-2}°	2.9×10^{-1}°
Lower Amery Ice Shelf	9.9×10^{-3}°	

d is 2025 m. The slopes associated with the floating ice in these areas will have a negligible effect on the magnitude of d. Where the satellite does not cross the grounding-line at right-angles, the estimated slope will be lower than it actually is, leading to an underestimate of d. In the example given above, the value of d will increase to 2338 m for an approach angle of 60° to the grounding line, and to 4049 m for an approach angle of 30°. In principle, it is possible to gain a first-order estimate of the angle of incidence of the altimeter from the first-order estimate of the grounding-line position and iterate towards

its true position where ground-track density is high enough.

In view of the limitations on accuracies achievable in positioning of the grounding point using this method alone, it is useful to employ an additional method analogous to that employed by Thomas and others (1983) in their mapping of the ice margin. They ranged to the earliest signal associated with the lower (sea-ice) surface on approach to the ice shelf. The point at which this began a parabolic trajectory away from the satellite marked the position of the ice margin. Although the grounding line in this case is a change in slope rather than a step, the same technique can be applied here to provide an additional measure of the grounding-point position. In the example given above (Fig.3a), the earliest signal associated with the grounding zone begins a parabolic trajectory away from the satellite at a distance of 2648 m beyond the uncorrected estimation of the grounding-line position obtained from the first-return method. The near-parabolic profile produced by two retracked elevations and on-board calculated elevation can be seen in Fig.5. All three methods continue to track to the grounding zone once the satellite has passed on to the grounded ice.

In the example above, the two techniques produce a discrepancy of 623 m, assuming the grounding line is at right-angles to the altimeter ground track. The accuracies achievable are limited by the along-track distance between wave-form samples and by the rough nature of the surface, which causes the assumptions behind the techniques to break down. The accuracies in positioning the beginning of the parabolic trajectory and positioning of the increase in surface gradient are each of the order of half the distance between wave-form samples along track (331 m). This places the discrepancy of 623 m within the maximum expected (662 m). The rough nature of the surface will cause additional errors in estimation of the beginning of the parabolic trajectory and in estimation of surface gradient, which may decrease accuracies substantially when the satellite crosses the grounding line obliquely. Along the track closest to T4, these methods produce a location for the grounding line approximately 2.8 km to the north-east, with an accuracy in location of the order of ±500 m.

ACKNOWLEDGEMENTS
The authors would like to acknowledge the help of R.H. Merson of the National Remote Sensing Centre in processing and preparation of AVHRR imagery.

REFERENCES
Allison I 1979 The mass budget of the Lambert Glacier drainage basin, Antarctica. *Journal of Glaciology* 22(87): 223-235
Brenner A C, Bindschadler R A, Thomas R H, Zwally H J 1983 Slope-induced errors in radar altimetry over continental ice sheets. *Journal of Geophysical Research* 88(C3): 1617-1623
Brooks R L, Williams R S Jr, Ferrigno J G, Krabil W B 1983 Amery Ice Shelf topography from satellite radar altimetry. *In* Oliver R L, James P R, Jago J B (eds) *Antarctic earth science.* Canberra, Australian Academy of Science: 441-445
Brown G S 1977 The average impulse response of a rough surface and its applications. *IEEE Transactions on Antennas and Propagation* AP 25(1): 67-74
Budd W F 1966 The dynamics of the Amery Ice Shelf. *Journal of Glaciology* 6(45): 335-358
Budd W F, Smith I L, Wishart E 1967 The Amery Ice Shelf. *In* Oura H (ed) *Physics of Snow and Ice: International Conference on Low Temperature Science ... 1966 ... Proceedings, Vol 1, Pt 1.* [Sapporo], Hokkaido University. Institute of Low Temperature Science: 447-467
Budd W F, Corry M J, Jacka T H 1982 Results from the Amery Ice Shelf Project. *Annals of Glaciology* 3: 36-41
Crabtree R D, Doake C S M 1986 Radio-echo investigations of Ronne Ice Shelf. *Annals of Glaciology* 8: 37-41

Giovinetto M B 1964 The drainage systems of Antarctica: accumulation. *In* Mellor M (*ed*) *Antarctic snow and ice studies*. Washington, DC, American Geophysical Union: 127-155 (Antarctic Research Series 2)

Griffiths H 1984 Special difficulties of retrieving surface elevation over continental ice. *In* Guyenne T D, Hunt J J (*eds*) *ERS-1 Radar Altimeter Data Products; proceedings of an ESA workshop held at Frascati, Italy, on 8-11 May, 1984*. Paris, European Space Agency: 61-65

Gundestrup N S, Bindschadler R A, Zwally H J 1986 Seasat range measurements verified on a 3-D ice sheet. *Annals of Glaciology* 8: 69-72

Hughes T J 1973 Is the West Antarctic ice sheet disintegrating? *Journal of Geophysical Research* 78(33): 7884-7910

McGoogan J T 1975 Satellite altimetry applications. *IEEE Transactions on Microwave Theory and Techniques* MTT 23 (12): 970-978

McIntyre N F 1985 A re-assessment of the mass balance of the Lambert Glacier drainage basin, Antarctica. *Journal of Glaciology* 31(107): 34-38

Martin T V, Zwally H J, Brenner A C, Bindschadler R A 1983 Analysis and retracking of continental ice sheet radar altimeter waveforms. *Journal of Geophysical Research* 88(C3): 1608-1616

Mellor M, McKinnon G 1960 The Amery Ice Shelf and its hinterland. *Polar Record* 10(64): 30-34

Morgan V I, Budd W F 1975 Radio-echo sounding of the Lambert Glacier basin. *Journal of Glaciology* 15(73): 103-111

Partington K C, Rapley C G 1986 Analysis and simulation of altimeter performance for the production of ice sheet topographic maps. *Annals of Glaciology* 8: 141-145

Rapley C G and 16 others 1985 *Applications and scientific uses of ERS-1 radar altimeter data; final report*. Noordwijk, European Space Agency

Robin G de Q 1975 Ice shelves and ice flow. *Nature* 253(5488): 168-172

Robin G de Q 1979 Formation, flow, and disintegration of ice shelves. *Journal of Glaciology* 24(90): 259-271

Shabtaie S, Bentley C R In press Mass balance of the Ross embayment ice streams, West Antarctica. *Journal of Geophysical Research*

Stephenson S N, Doake C S M 1982 Dynamic behaviour of Rutford Ice Stream. *Annals of Glaciology* 3: 295-299

Swithinbank C W M In press Do Antarctic ice streams surge? *Journal of Geophysical Research*

Thomas R H 1973 The creep of ice shelves: interpretation of observed behaviour. *Journal of Glaciology* 12(64): 55-70

Thomas R H, Sanderson T J O, Rose K E 1979 Effect of climatic warming on the West Antarctic ice sheet. *Nature* 277 (5695): 355-358

Thomas R H, Martin T V, Zwally H J 1983 Mapping the ice-sheet margins from radar altimetry data. *Annals of Glaciology* 4: 283-288

Townsend W F 1980 An initial assessment of the performance achieved by the Seasat-1 radar altimeter. *IEEE Journal of Oceanic Engineering* OE 5(2): 80-92

Weertman J 1974 Stability of the junction of an ice sheet and an ice shelf. *Journal of Glaciology* 13(67): 3-11

Wellman P 1982 Surging of Fisher Glacier, eastern Antarctica: evidence from geomorphology. *Journal of Glaciology* 28(98): 23-28

Zwally H J, Bindschadler R A, Brenner A C, Martin T V, Thomas R H 1983 Surface elevation contours of Greenland and Antarctic ice sheets. *Journal of Geophysical Research* 88(C3): 1589-1596

Annals of Glaciology 9   1987
© International Glaciological Society

MASS BALANCE OF SOUTH-EAST ALASKA AND NORTH-WEST BRITISH COLUMBIA GLACIERS FROM 1976 TO 1984: METHODS AND RESULTS

by

M.S. Pelto

(Department of Geological Sciences, Institute of Quaternary Sciences, University of Maine, Orono, ME 04469, U.S.A.)

ABSTRACT

The annual surface mass balance for 1983 and 1984 and the 10 year cumulative mass balances for 1975–85 were calculated for 60 south-east Alaskan and north-west British Columbia glaciers. At present, the mass balance is positive on nine, at equilibrium on nine, and negative on 42 glaciers. The ratio of glaciers with positive and equilibrium mass balance to glaciers with negative mass-balance has not changed significantly since 1946; however, the magnitude of negative balances has declined on 39 of the 42 glaciers.

The annual mass balance of south-east Alaska and north-west British Columbia glaciers cannot be measured on more than a few glaciers. This paper presents the methods and results for a mass-balance model using as input local weather records, Juneau Icefield field studies, and satellite imagery. The primary variable in mass balance from one glacier to another is the budget gradient. The budget gradient varies predictably according to three parameters: ocean proximity, surface slope, and valley width–valley height. The annual fluctuation of the budget gradient can be determined by examination of local weather records, determination of activity indexes, and delineation of the equilibrium-line gradient from the maritime to the continental part of each icefield. The latter two variables are determined using largely satellite imagery, keyed to topographic maps.

This procedure, where applicable, yielded mass-balance errors of ±0.16–0.22 m and 10 year cumulative mass-balance errors of ±0.08–0.15 m.

INTRODUCTION

The behaviour of glaciers in south-east Alaska is a function of surface mass balance and glacier dynamics. The surface mass balance and annual equilibrium-line altitude (ELAa) are controlled solely by climate and can be used as climatic indicators regardless of glacier type. Surge and tide-water glaciers, during parts of their surge or advance–retreat cycle, are insensitive to climate; their behaviour is dictated by either surge or calving dynamics. Climatic changes control the surface mass balance and therefore the behaviour of other types of glaciers. There is an immediate secondary and delayed primary response of south-east Alaskan glaciers to each climatic change. The terminus response to changing ablation conditions is immediate but secondary, since in this region the percentage change in ablation for any given climatic change is less than the change in accumulation. The primary response due to changing accumulation lags by an unknown period.

The fluctuation in climatic parameters which control glacier behaviour are recorded by the glacier. Examining a large number of glaciers within a restricted region and combining the data with existing weather records permits a better climatic identification than can otherwise be obtained. This is particularly true in south-east Alaska, where there are several thousand glaciers, but only seven weather stations, all at sea-level.

Before Alaska glaciers can be used to identify climate, the annual mass balance of a large number of glaciers must be known. Because the mass balance cannot be measured directly on more than two or three glaciers, a model had to be developed that would predict accurately the mass balance of large glaciers, utilizing local weather records, satellite imagery, and field studies on the Juneau Icefield.

The model was developed for use as a substitute for detailed annual field work. The accuracy of the equations was tested against the exisiting mass-balance record of Andrei, Berendon, Lemon, and Taku Glaciers.

The model is based primarily on field work conducted by the Juneau Icefield Research Program (JIRP). JIRP has conducted annual climatic and mass-balance studies since 1946. Under the direction of Maynard M. Miller and the Foundation for Glacier and Environmental Research, annual field measurements include determination of: ELAa, budget gradient, glacier movement, glacier-surface level, heat budget, melt-water transport, meteorologic conditions, terminus fluctuations, and glacier thickness along selected transects. Methods utilized in mass-balance studies have been

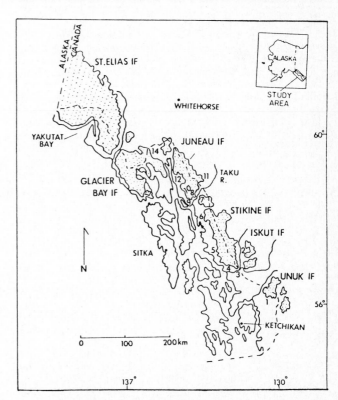

Fig.1.   South-east Alaska and north-west British Columbia; glacier-covered areas indicated by stippling, ice field is abbreviated IF, and selected glaciers are numbered, see Table III for numbering scheme.

discussed by LaChapelle (1954), Miller (1954, 1975), Nielsen (1957), and Hubley (1957). The majority of the work has been conducted on Taku Glacier, thereby allowing annual mass-balance calculations to be completed for Taku Glacier during the 1946–85 period (Pelto in press).

STUDY-AREA CLIMATIC CONDITIONS

A vast system of ice fields extends along the Coast Range system of south-east Alaska and adjacent Canada (Fig.1). At the Coast Range is an interaction zone between maritime and continental climatic conditions. South-eastern Alaska has the highest cyclone activity in North America due to its location east of the Aleutian low. The cyclonic frequency moderates temperatures and provides the maritime flanks of the Coast Range with 2000–4000 mm of annual precipitation. Precipitation decreases with distance from the ocean, with a sharp decline at the Coast Range crest. Table I gives the climatic conditions during the ablation season and accumulation season for three coastal weather stations of south-east Alaska. From these data, climatic conditions at specific elevations in the vicinity of each station can be estimated using data and techniques employed by Marcus (1964) and Adkins (1958), and radiosonde data from each station.

During the accumulation season the weather in south-east Alaska and north-west British Columbia is dictated by the relative strengths of the maritime Aleutian low and the continental Canadian polar high (Miller 1975). The boundary separating continental and maritime climatic conditions generally intersects each ice field. The mean boundary position is 3–25 km inland of the Coast Range divide. The ELAa is dramatically higher under continental climatic conditions. Hence, even slight shifts in the mean boundary position can be identified by observing changes in ELAa on glaciers of each ice field. The primary storm track and prevailing wind direction are from the south-south-east to east.

During the ablation season the Aleutian low is no longer present. The cyclonic activity decreases and the prevailing wind is from the south-south-east to east-south-east.

MASS-BALANCE CONTROLS

The surface mass balance of glaciers in south-east Alaska is controlled by three climatic parameters: accumulation-season cyclonic activity, ablation-season temperature, and summer cyclonic activity. Precipitation and winter balance increase as accumulation-season cyclonic activity increases. Cyclonic conditions are associated with mild temperatures in south-eastern Alaska, indicating a direct relationship between accumulation-season temperature and winter balance. This is also the case in southern Alaska (Mayo 1984).

Ablation-season temperature dictates ablation below the snow line. Above the snow line, summer cyclonic activity is as important as temperature, since ablation is highest during storm conditions, high winds, and precipitation in the form of rain (Hubley 1957). Ablation increases as summer cyclonic activity and ablation-season temperature increase.

In the case of the maritime glaciers of south-east Alaska, 65–70% of the annual mass-balance variance is accounted for solely by the accumulation cyclonic activity (Pelto in press). Ablation-season temperature and cyclonic activity explain 25–30% of the variance. Glaciers on the continental side of the ice fields have estimated variances of 55–60% for accumulation-season cyclonic activity and 30–40% for ablation-season conditions. Therefore, the primary control of annual mass-balance fluctuations is the changing intensity and duration of maritime conditions for each glacier.

On a geodynamic basis, the mass balance of south-east Alaskan glaciers is dictated by the calving rate, the percentage of a glacier's area in the zone of maximum accumulation (MAA), and the accumulation-area ratio (AAR: percentage of a glacier's area above the ELA). The maximum accumulation area in south-east Alaska and north-west British Columbia is the region where mean accumulation-season temperature is estimated from meteorologic and winter-balance measurements to be in the range of $-5°$ to $-13°$C. Changes in the regimen on non-surging glaciers occur only when AAR and MAA percentages cross the equilibrium threshold values (Mercer 1961). As determined in this study, the respective threshold values for AAR and MAA are: 67 and 50 for a non-calving or low-calving-rate glacier, 76 and 55 for a moderate-calving-rate glacier, and 84 and 62 for a high-calving-rate glacier. The mean calving rate for a low-calving-rate glacier is less than 200 m/a, 200–1100 m/a for a moderate-calving-rate glacier, and greater than 1100 m/a for a high-calving-rate glacier.

Of 120 glaciers in south-east Alaska and north-west British Columbia examined with these criteria, only three do not fall within the above guidelines, proving the usefulness of this method for mass-balance estimation. Changes in terminus configuration or fjord geometry alter the calving rate and cause pronounced shifts in the threshold AAR and MAA values for the glacier.

MASS-BALANCE CALCULATION METHOD

The mass balance of a glacier is the sum of the surface and calving flux. The surface flux is dependent on the budget gradient, ELAa, and glacier surface at each elevation. The variation of the mean budget gradient on large south-east Alaskan glaciers is small and each is a reasonably uniform curve (Mayo 1984). The mean budget gradient varies predictably from one glacier to another according to: ocean proximity, surface slope, and valley

TABLE I. CLIMATIC CONDITIONS OBSERVED FOR THREE COASTAL ALASKAN WEATHER STATIONS AND EXTRAPOLATED FOR DIFFERENT ELEVATION IN THEIR RESPECTIVE AREAS

			Ablation season		Accumulation season	
Location	Elevation	Period	Mean temperature	Mean precipitation	Mean temperature	Mean precipitation
	m		°C	m	°C	m
Annette	200	4/1–10/30	10	1.40	−2	1.20
	1000	5/1–10/1	7	1.05	−4	1.85
	2000	6/1–8/15	4	0.70	−10	2.60
Juneau	200	4/15–10/15	9	1.35	−3	1.25
	1000	5/15–10/1	7	0.95	−5	2.00
	2000	6/1–8/15	4	0.65	−11	2.70
Yakutat	200	4/15–10/15	8	1.50	−3	1.80
	1000	5/1–10/15	6	0.90	−6	2.10
	2000	6/1–8/15	3	0.50	−14	2.40

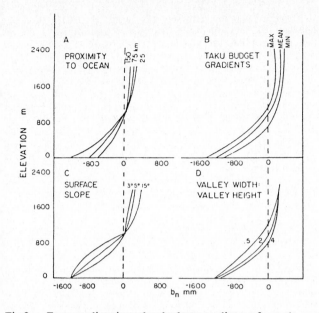

Fig.2. Factors dictating the budget gradient of south-east Alaskan glaciers. Proximity to ocean, surface slope, and valley width—valley height ratio. Also shown are the maximum and minimum boundary conditions for the annual budget gradient of Taku Glacier.

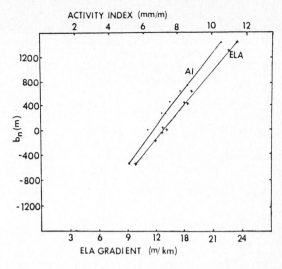

Fig.3. The effect of varying activity index (AI), and equilibrium-line gradient from the continental to the maritime side of the Juneau Icefield, on the mass balance of Taku Glacier.

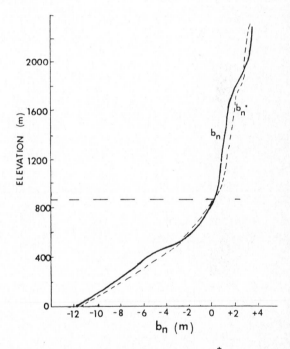

Fig.4. The predicted budget gradient (b_n^*) and measured budget gradient (b_n) for Taku Glacier in 1967.

width—valley height ratio (Fig.2). Ocean proximity is measured along the path of primary winter and summer storm movement. Each of the parameters fluctuates from the terminus to divide and one tributary to another on each glacier. The effect of each parameter was determined from comparison of the budget gradients for Andrei, Berendon, Columbia, Kaskawulsh, Lemon, Malaspina, Salmon, Taku, and Wolverine Glaciers (Adkins 1958; Meier and Post 1962; Marcus and Ragle 1970; Mayo 1984; Young unpublished). This method was then tested on Herbert, Llewellyn, Sawyer, Twin, and Vaughan Lewis Glaciers, proving accurate to within ±0.15 mm/m for any elevation.

The mean budget gradient is then known for each glacier; however, annual fluctuations are significant. Annual budget-gradient variations are due almost entirely to the changing intensity and duration of maritime conditions for each glacier. Corrections are based on annual observation of each glacier's activity index and the ELAa gradient from the maritime to the continental side of each ice field, and utilization of local weather records to determine accumulation-season cyclonic activity, ablation-season temperature, and summer cyclonic activity. The activity index is the budget gradient in the vicinity of the ELA (Meier and Post 1962), and can be measured by observing the snow-line rise with time. The snow line and ELAa are nearly coincident at the end of the ablation season in this region. The ELAa gradient is the rise in elevation of the ELAa with distance inland from the maritime to the continental side of the ice field. This distance is measured along the primary winter storm-path direction. As the activity index and ELAa gradient increase, so does mass balance (Fig.3). This method proved accurate in replicating the annual budget gradient of Taku Glacier without using any field data for the 1977–84 period (Fig.4). The values of activity – index and ELAa-gradient values, and their effect on mass balance, vary slightly from one ice field to another.

The surface area within each 160 m elevation was determined from U.S.G.S. topographic maps using a polar planimeter. The annual surface mass balance could then be determined using Equation (1):

$$b_n = \bar{b}_n \frac{A}{A_t} \qquad (1)$$

where A_t is the total surface area, A is the surface area within each 160 m elevation band, and b_n is the mean balance within each 160 m elevation band. The net annual mass balance (b_n), in meters of water equivalent, is the sum of the mass balance for each elevation band.

An equilibrium line-based equation similar to that of Braithwaite (1984) yields extremely similar results. It is actually more useful because it does not require a complete budget gradient. However, a more complete budget gradient does yield more accurate results

$$b_n = ([(AG)_1 + (AG)_2 + (AG)_3] \, [ELAo - ELAa])/ A_t. \qquad (2)$$

In Equation (2) G is the budget gradient, 1 is the ablation zone, 2 is the lower *névé*, 3 the upper *névé*, and ELAo the zero-budget equilibrium-line altitude. The ELAo is determined from the AAR and MAA mass-balance relationship for each glacier and checked by mass-balance calculation from the mean budget gradient and glacier surface area-elevation distribution.

The only variables are ELAa, determined from satellite imagery and keyed to topographic maps, and G determination, which has been described. The major error in this method is obtaining a satellite image at the end of the ablation season to determine the ELAa. There has been a usable image for the Juneau Icefield for nine of the last 11 years.

TABLE II. THE ANNUAL EQUILIBRIUM-LINE ALTITUDE (ELAa), ANNUAL MASS BALANCE (b_n: IN mm OF WATER EQUIVALENT), NUMBER OF DATA POINTS, AND PREDICTED MASS BALANCE (b_n*). 1985 DATA ARE PRELIMINARY AND WERE NOT USED FOR CALCULATION OF MEAN VALUES

Year	ELAa	b_n	Data points	b_n*
1946	980	−40	38	−90
1947	900	360	40	320
1948	879	510	43	510
1949	800	930	71	840
1950	1010	−180	74	−220
1951	1160	−340	79	−260
1952	950	160	93	110
1953	1010	−150	82	−280
1954	980	−70	57	−180
1955	780	970	34	900
1956	1000	−130	37	0
1957	1010	−40	35	−50
1958	930	210	63	290
1959	915	350	59	310
1960	950	160	51	200
1961	885	480	47	570
1962	900	390	71	340
1963	875	570	41	600
1964	750	1130	62	1260
1965	810	790	57	690
1966	965	80	48	170
1967	930	250	53	230
1968	885	460	46	520
1969	730	1170	44	1270
1970	825	760	37	700
1971	850	630	56	540
1972	880	420	35	470
1973	870	520	39	560
1974	850	580	44	640
1975	800	850	52	810
1976	850	660	38	590
1977	885	470	49	410
1978	915	310	47	370
1979	950	140	34	160
1980	870	540	38	480
1981	980	120	43	120
1982	950	150	91	250
1983	1085	−420	86	−570
1984	875	640	178	520
1985	600	1400	29	1350
Total	-	15 800	-	15 430
Mean	910	370	55	360

The above methods and Equation (2) were used to calculate the mass balance of Taku Glacier using a split-sample technique, in which no field data were utilized in model development or mass-balance calculations from odd-numbered years (Table II). Errors for one standard deviation on Taku Glacier are ±0.14 m annually and ±0.08 m for a 10-year period. The expected errors for other glaciers, where the equation proved applicable, are ±0.18–0.24 m annually and ±0.10–0.16 m for a 10-year period. Table III lists the ELA, AAR, MAA, 1976–84 mass balance and 1984 mass balance for 14 selected glaciers. Similar calculations were carried out for all 60 glaciers and the mean values are shown in Table III.

The mass-balance prediction method was not satisfactory for glaciers which have an area less than 20 km². It did prove accurate for Iskut, Juneau, Stikine, and Unuk Icefields, Brady Glacier, and the eastern part of Glacier Bay (Fig.1). These are all nunatak ice fields, with an interconnected plateau *névé* zone, drained by outlet glaciers. Accumulation via wind drifting and avalanching is insignificant on nunatak ice fields. There are no data to check this method in the case of valley ice fields, which consist of individual valley-glacier systems separated by high mountain ridges. It is unlikely that this method would be accurate for valley ice fields, such as Fairweather and St. Elias Icefields of south-east Alaska, where avalanching and wind drifting are significant sources of accumulation.

The calving flux is the product of calving velocity, ice thickness, and glacier width. There is no suitable method for determining the calving flux, although Brown and others (1982) demonstrated that water depth at the glacier front is the primary variable. In the current study only changes in the calving flux were estimated. These estimates were based on annual aerial photographs and satellite images. The most useful check of the calving flux is to calculate volume flux at a point near the terminus where glacier depth, width, and ice velocity are known. These data are obtainable for most south-east Alaskan tide-water glaciers. If a glacier is not retreating rapidly or being down-drawn, then the calving flux cannot exceed the winter balance or the calculated volume flux by a significant amount. Use of only summer glacier velocities leads to an overestimate of the calving flux, as was the case in the Brown and others (1982) study.

CLIMATIC TRENDS

The mass-balance record during the 1946–85 period for Iskut, Juneau, Stikine, and Unuk Icefields can be divided into four climatic intervals. The climate and mass-balance trends for this period are shown in Fig.5. During the 1946–62 period, ablation-season conditions were relatively

TABLE III. THE EQUILIBRIUM-LINE ALTITUDE, ACCUMULATION-AREA RATIO, MAXIMUM ACCUMULATION-AREA RATIO, AND MASS BALANCE OF SELECTED SOUTH-EAST ALASKAN AND NORTH-WEST BRITISH COLUMBIA GLACIERS; SEE FIG.1 FOR MAP LOCATION. THE MEAN VALUES ARE THOSE FOR ALL 60 GLACIERS OF THIS STUDY

Map no.	Glacier name	ELA	AAR	MAA	1976–84 b_n	1984 b_n
		m	%	%	mm	mm
1	Chickamin	1175	65	45	− 80	+ 30
2	Great	1200	63	29	− 90	−210
3	Porcupine	1400	59	38	− 60	−280
4	Le Conte	1125	77	61	− 40	+260
5	Baird	1175	66	55	− 90	+100
6	Sawyer	1125	68	48	− 10	−120
7	Wright	1125	64	53	− 70	−270
8	Taku	925	82	62	+230	+540
9	Norris	950	62	38	− 00	−180
10	Mendenhall	1050	65	52	− 80	+210
11	Llewellyn	1475	64	30	− 40	−290
12	Meade	1300	63	50	− 40	− 60
13	Brady	575	66	46	− 90	+140
14	Tsirku	1150	69	52	+100	+260
	Mean	1075	64	51	−340	+190

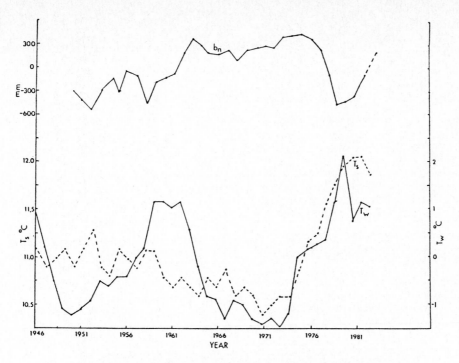

Fig.5.　5 year running means of accumulation-season temperature (T_w), ablation-season temperature (T_s), and mass balance of south-east Alaska and north-west British Columbia glaciers, in mm of water equivalent. From 1946 to 1971 the mass-balance record is based on 16 glaciers for which ELAa data were available. Between 1972 and 1985, the mass-balance curve is based on all 60 glaciers of this study.

constant. The single variable determining mass balance was accumulation-season cyclonic activity fluctuations. Winter temperature and cyclonic activity decreased from 1945 to 1953, causing moderately negative mass balances. Winter temperature and cyclonic activity then increased from 1954 to 1962, producing declining negative balances. From 1963 to 1975, low ablation-season temperatures and above-average winter precipitation caused slightly positive glacier balances. Lower winter temperature was due to a global temperature decline and not to decreasing cyclonic activity, hence winter precipitation did not decline. A rapid increase in annual temperature delineates the 1976–83 interval. Record ablation-season warmth offset slightly above-average winter precipitation, causing moderately negative mass balances. From 1984 through 1986, record accumulation-season warmth and precipitation, in addition to average ablation-season temperatures, has resulted in large positive mass balances. This climatic interval is especially noteworthy because of an eastward shift in the wind drift moats that had been stationary since 1946 on the Juneau Icefield.

CONCLUSIONS

It is evident that the surface mass balance of many glaciers in south-east Alaska can be identified using satellite imagery, local weather records, and Juneau Icefield glaciologic data. Modification of the budget-gradient estimation methods is necessary to obtain mass-balance records for glaciers of the Alaska, Chugach, Fairweather, St. Elias, and Wrangell Mountains. To predict the future behaviour of glaciers in south-east Alaska and north-west British Columbia requires an understanding of the dynamics which determine the lag time and magnitude of terminus response.

During the 1900–62 period no more than 12 of the 60 glaciers were near to equilibrium conditions. At present, 29 of the 60 glaciers are near equilibrium. The positive mass-balance pulse of the 1963–75 period has not reached the termini of most of the glaciers. This, in conjunction with the recent warm winter temperatures, causing record winter balances, could lead to a stabilization of the ice volume of

glaciers in south-east Alaska for the remainder of this century.

Braithwaite (1984) posed the question, "can the mass balance of a glacier be estimated from its equilibrium-line altitude?" The answer in south-east Alaska and north-west British Columbia is "no". If rephrased, can the mass balance of a glacier be estimated from its snow line and ELA at different times during the ablation season and from regional variations of the ELAa? The answer for south-east Alaska and north-west British Columbia is "yes".

REFERENCES

Adkins C J 1958 The summer climate in the accumulation area of the Salmon Glacier. *Journal of Glaciology* 3(23): 193–206

Braithwaite R J 1984 Can the mass balance of a glacier be estimated from its equilibrium-line altitude? *Journal of Glaciology* 30(106): 364–368

Brown C S, Meier M F, Post A 1982 Calving speed of Alaska tidewater glaciers with application to the Columbia Glacier, Alaska. *US Geological Survey. Professional Paper* 1258-C

Hubley R C 1957 An analysis of surface energy during the ablation season on Lemon Creek Glacier, Alaska. *Transactions of the American Geophysical Union* 38(1): 68–85

LaChapelle E R 1954 *Snow studies of the Juneau Ice Field.* New York, American Geographical Society (Juneau Ice Field Research Program Report 9)

Marcus M G 1964 *Climate-glacier studies in the Juneau Ice Field region, Alaska.* Chicago, University of Chicago. Department of Geography (Research Paper 88)

Marcus M G, Ragle R H 1970 Snow accumulation in the Ice Field Ranges, St. Elias Mountains, Yukon. *Arctic and Alpine Research* 2(4): 277–292

Mayo L R 1984 Glacier mass balance and runoff research in the U.S.A. *Geografiska Annaler* 66A(3): 215–227

Meier M F, Post A 1962 Recent variations in mass net budgets of glaciers in western North America. *International Association of Scientific Hydrology Publication* 58 (Colloque d'Obergurgl, 10-9 – 18-9 1962 – *Variations of the Regime of Existing Glaciers*): 63–77

Mercer J H 1961 The estimation of the regimen and former firn limit of a glacier. *Journal of Glaciology* 3(30): 1053–1062

Miller M M 1954 *Juneau Ice Field Research Program, summer field season 1950*. New York, American Geographical Society (Juneau Ice Field Research Program Report 7)

Miller M M 1975 *Mountain and glacier terrain study and related investigations in the Juneau Icefield region, Alaska-Canada. Final report*. Seattle, WA, Foundation for Glacier and Environmental Research. Pacific Science Center (Monograph Series)

Nielsen L E 1957 Preliminary study on the regimen and movement of the Taku Glacier, Alaska. *Bulletin of the Geological Society of America* 68(2): 171–180

Pelto M S In press Mass balance of the Taku Glacier, Alaska. *Arctic and Alpine Research*

Young G Unpublished Mass balance data from the Canadian National Hydrologic Research Institute. Ottawa, Ontario, Canadian National Hydrologic Research Institute

Annals of Glaciology 9   1987
© International Glaciological Society

POSSIBILITIES AND LIMITS OF SYNTHETIC APERTURE RADAR FOR SNOW AND GLACIER SURVEYING

by

H. Rott

(Institut für Meteorologie und Geophysik, Universität Innsbruck, Austria)

and

C. Mätzler

(Institut für angewandte Physik, Universität Bern, Bern, Switzerland)

ABSTRACT

The physical background for the interpretation of microwave measurements on snow and ice is summarized. The angular and spectral behaviour of backscattering is shown for various snow types based on measurements carried out in the Swiss Alps. The information content of SAR images in regard to snow and glacier applications is discussed, and examples are shown for Seasat SAR and airborne SAR images. The preliminary specifications are given for an optimum SAR system for snow and glacier monitoring. The main advantage of SAR is due to its weather independence; for special applications the SAR information on the physical state of snow and ice may be of interest. Future SAR sensors can become important components in a snow and glacier monitoring system, but in order to fulfil all tasks, high-resolution optical sensors and improved passive microwave sensors will also be required.

INTRODUCTION

Spaceborne synthetic aperture radar (SAR) systems are of considerable interest for snow and glacier monitoring because of the capability to penetrate clouds and because of the high spatial resolution. So far, only experimental SAR systems have been operating in space, but, from the year 1989 onward, various SAR sensors for earth observation are due to be launched. In order to learn about the possibilities of SAR for snow and glacier applications, detailed investigations have been carried out based on airborne and spaceborne SAR data and on radar image simulations. Ground-based scatterometer measurements and theoretical considerations provided the basic information on the backscattering properties of snow and ice. The investigations resulted in the preliminary definition of a spaceborne SAR system for snow and glacier monitoring and in an assessment of the SAR potential for various scientific and operational applications.

BASIC CONSIDERATIONS ON BACKSCATTERING OF SNOW AND ICE

To understand the microwave signatures of snow cover lying on the ground or on an ice surface, it is essential to know which layers are contributing to backscattering. In the case of a semi-transparent medium such as snow, the power attenuation of a propagating wave is described by the volume extinction coefficient κ_e or its reciprocal value, the penetration depth $\delta_p = 1/\kappa_e$. For dry snow δ_p is in the order of 10 m at 10 GHz and decreases to about 1 m at 40 GHz (Rott and others, 1985). The presence of liquid water strongly increases the dielectric losses. δ_p-values for wet snow with liquid water content of 2-4% by volume are typically of the order of one wavelength only.

The significance of wetness for the interaction of microwaves with the snow cover is evident in Fig.1, which shows the results of measurements carried out at Weissfluhjoch on 10 and 11 April 1980. The site was covered with 1.9 m of snow, the total water equivalent was 0.60 m; the snow temperatures were between $-1°$ and $-2°C$ except for the top 20 cm, where the temperature varied between $0°$ and $-18°C$. Melting took place during the noon and early afternoon hours, the maximum column height of liquid water was 0.17 mm on 10 April and $\geq$ 1 mm on 11 April when it was warmer and the sky was clear. However, the top millimeters of the snow cover remained frozen during the days because the air temperature was clearly below $0°C$. The active and passive microwave signatures are strongly influenced by the diurnal variation of wetness in the surface layer. Blackbody properties are approached when the surface layer is wet, resulting in high emission and low backscattering values. The differences in the minimum value of the backscattering coefficient γ between 10 and 11 April are related to differences in snow wetness. During the night penetration of the microwaves increases significantly and scattering in the snow pack is effective, for the longer wavelength (γ at 10.4 GHz) scattering at the ground/snow interface is important. γ is related to σ_0, the scattering cross section per unit surface area, by $\gamma = \sigma_0/\cos\theta$, where θ is the angle off nadir.

The radar return signal from glacier ice is dominated by surface scattering which is dependent on surface roughness in relation to the wavelength. The radar return from snow-covered targets may include contributions from scattering at the air/snow boundary and at internal inhomogeneties and boundaries. For wet snow, scattering at the surface and in the uppermost snow layer is effective.

Calculations of scattering from snow and ice in dependence of various physical parameters are of great importance for the analysis of radar data, because only limited measurement data are available. However, accurate modelling of the scattering behaviour of snow and ice surfaces is a difficult task because the characteristic dimensions of the surface roughness are similar in magnitude to the radar wavelengths. Roughness measurements of wet snow at an Alpine test site revealed rather gentle undulations with auto-correlation lengths of surface height of several centimeters, allowing application of the Kirchhoff method for backscatter calculations at X- and C-band wavelengths for the surface scatter contribution (Rott 1984[a]). For the reflectivity at horizontal polarization also the simple Fresnel formula provides good results (Mätzler and others 1984).

Due to the characteristic roughness features, the problem of surface scattering from glacier ice is of great interest, but so far thorough theoretical and experimental investigations are lacking. Measurements of small-scale surface roughness on an Alpine glacier revealed bimodal frequency distributions of surface height for melting glacier ice with cryoconite holes. In addition to small-scale roughness, surface undulations in the horizontal scale of

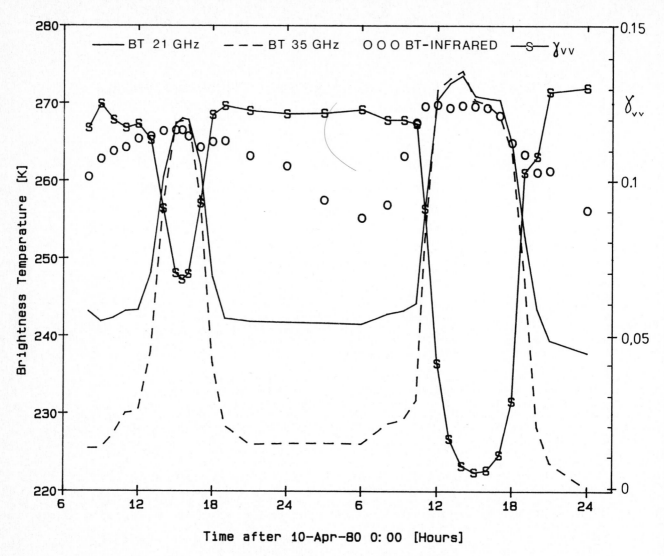

Fig.1. Temporal variations of brightness temperatures BT at 21 GHz and 35 GHz in vertical polarization, of infrared brightness temperature, and of the backscattering coefficient γ vv at 10.4 GHz, all at 40° off nadir, measured over snow cover on Weissfluhjoch, Switzerland, on 10 and 11 April 1980.

meters to tens of meters are typical. Generally the roughness of the ice surfaces is significantly higher than the roughness of snow surfaces, resulting in higher radar return (Rott 1984[a]). However, in relation to longer wavelengths (e.g. L-band), glacier ice surfaces sometimes appear smooth and show low backscattering. Surfaces of superimposed ice are frequently smooth; this may cause problems for discrimination against wet snow.

For a dry snowpack, the contribution of surface scattering is small; scattering at snow grains and at internal layers dominates. The wavelengths of the earth observation radars are significantly larger than the diameters of the individual snow grains. For calculating the volume scattering contribution, the Born approximation can be applied, in which the dielectric constant ϵ is considered as a function of position and is described by a stationary mean part $\bar{\epsilon}$ and a randomly fluctuating part e_f For closely packed scatterers, a modification to the Born approximation was derived, applying an exponential autocorrelation function for the amplitude of ϵ_f (Mätzler 1985). The calculations show that volume scattering increases with the frequency f according to f^4 for low frequencies, similar to Rayleigh scattering, but in the case of dense packing of the irregular ice grains the exponent is slightly reduced.

The main conclusions for the interpretation of radar images are summarized as follows. Due to high dielectric losses, wet snow cover is characterized by low return; only the uppermost layer contributes to backscattering. High return is only observed at angles of specular reflection. Dry snow is transparent at L- to X-band frequencies (0.4 to 11 GHz). If dry snow is lying on soil, the scattering contribution of the snow/soil interface dominates the radar signals. Therefore the detection of dry snow is hardly possible. Higher microwave frequencies (> 15 GHz) may be useful for mapping dry snow because of the increase in volume scattering, but presently no earth observation radars are planned at these frequencies.

If the thickness of a dry snow pack is in the order of many meters to tens of meters, as in dry accumulation zones of glaciers and ice sheets, a strong return signal can be observed, because the absorption is low and a thick layer contributes to backscattering. In this case not only the snow grains act as scatterers, but also internal interfaces due to density variations may be effective. Layers of thick, refrozen crusts can substantially increase the backscattering at X-band and higher frequencies (Reber unpublished). If a comparatively thin layer of dry snow is lying on glacier ice or on a wet snow pack, the return signal from these surfaces will only be slightly modified by the dry snow pack. This offers the possibility to detect the boundary between snow and ice areas on glaciers through a layer of fresh snow.

GROUND-BASED BACKSCATTERING MEASUREMENTS

Systematic measurements of microwave emission and backscattering from an Alpine snow pack have been carried out for several years at the Swiss test site, Weissfluhjoch

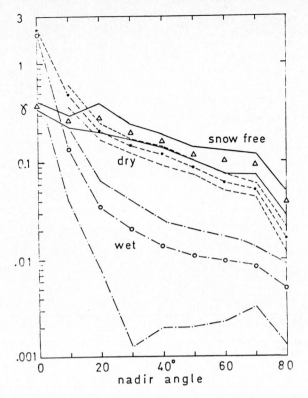

Fig.2. Backscattering coefficients γ at 10.4 GHz (mean values of hh and vv polarizations) versus nadir angle measured at the test site, Weissfluhjoch, for the dry snow, wet snow, and snow–free situations. The symbols represent the mean values, the lines the standard deviations resulting from different target conditions.

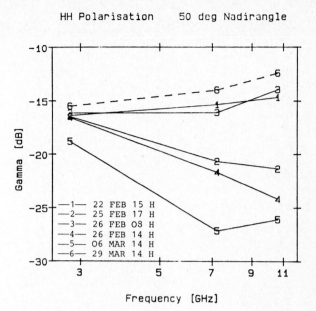

Fig.3. Backscattering coefficient γ versus frequency for various snow-cover situations measured in an Alpine valley between 22 February and 29 March 1985. (1) dry snow of 50 cm depth, (2) and (4) slightly wet snow, (3) refrozen after situation 2, (5) wet snow of 25 cm depth, and (6) the snow-free site (from Hüppi 1986).

(Mätzler and others 1982; Mätzler and Schanda 1984). Fig.2 summarizes the results of the backscatter measurements conducted with a noise scatterometer operating in the X-band (at 10.4 GHz). The mean values and standard deviations of the backscattering coefficient in parallel polarizations (hh and vv) are shown for incidence angles between 0 and 80 degrees off nadir for the snow-free test site, for dry snow, and for wet snow. At incidence angles $\geq 20°$ the backscattering coefficient of wet snow is on the average ten times smaller than for snow-free ground. When the ground is covered with dry snow, the differences from the snow-free site are small. Only at vertical incidence, where specular reflection at the snow surface and internal interfaces dominates, the backscattering coefficient is higher for all types of snow cover.

The highest relative variability of backscattering was observed for the wet snow cover. This is primarily an effect of variations in surface wetness and grain size and secondly due to surface roughness variations. Relative variability of γ for the snow-free site with a rough surface and for dry snow cover was smaller. For the dry snow measurements, a weak decrease in γ was observed with increasing snow–water equivalent (Mätzler 1986). Ulaby and Stiles (1980), on the other hand, measured a clear increase of γ at 9 GHz and $57°$ incidence angle, when dry snow was piled up over a smooth surface. This points out the importance of the surface below a dry snow pack for the intensity of backscattering. The frequency dependence of backscattering in the S- to X-band range for various snow conditions is illustrated in Fig.3. The measurements were taken in an Alpine valley in spring 1985 (Hüppi 1986). The dry snow situations (curves 1 and 3) show little differences in γ compared with the snow-free site, a mown meadow. Melting of a thin surface layer during the days 25 and 26 February results in a clear decrease of γ at frequencies >5 GHz. The completely wet snow pack gives the lowest signal; the contrast to the snow-free site is optimum in the X-band range.

When the results of the scatterometer measurements are applied to radar image analysis, additional factors have to

be taken into account. One problem of radar imagery is the limited radiometric accuracy because of speckle effects and difficulties of calibration. Other problems are related to the spatial variability of the physical properties of natural targets. This means that clear discrimination of two targets is only possible if an appreciable contrast in the mean backscattering intensities is observed. For wet snow cover, good contrast can be expected in the frequency range between about 5 to 15 GHz. At lower frequencies many natural surfaces appear smooth, similar to the snow cover (Rott and others 1985).

RADAR IMAGES OF SNOW AND GLACIERS

Available spaceborne SAR images of snow and glaciers are limited to L-band data, which have been acquired at 1.28 GHz by the Seasat SAR system and Shuttle Imaging Radars SIR-A and SIR-B. The look angle of the Seasat SAR antenna was 20 degrees off nadir, which is not suitable for application in areas with high relief because of geometric distortions and layover effects. However, in areas with gentle topography the Seasat SAR was found quite useful and a large number of good images of glaciers was acquired by Seasat during its 3 months of operation in 1978. SIR-A and SIR-B were short-term missions and data were taken only over a few glacier-covered areas. The antenna look angle of SIR-A ($47°$ off nadir) was adequate for imaging in mountainous regions, the look angle of SIR-B was selectable between $15°$ and $60°$. Airborne SAR experiments were carried out to investigate the SAR capabilities for mapping the seasonal snow cover.

Fig.4 shows a SAR image of the northern part of the Langjökull ice cap in Iceland which was acquired by Seasat on 24 August 1978. The nominal spatial resolution of the SAR was 25 m with 4 looks averaged, the swath width was 100 km. The altitude differences in the area are not very large; the glacier extends from about 800 m to 1300 m a.m.s.l. The central part of the firn plateau shows low radar return, corresponding to wet snow with a smooth surface. The intensity of the return is increasing at the lower parts of the firn area and reveals similar values to those of glacier ice. This behaviour can be explained by increasing roughness of the snow surfaces at lower altitudes. According to theory, the scattering contribution from the surface is dominant in the L-band at incidence angles below at least $40°$. Along the boundary of the snow area a zone of very low return appears. This corresponds to the signature of

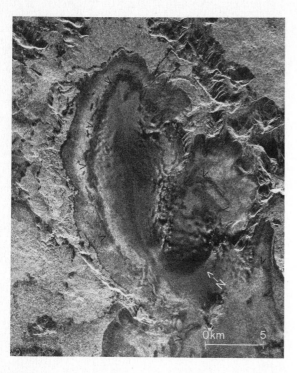

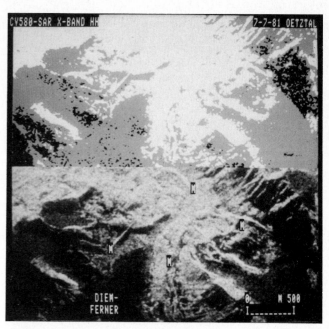

Fig.4. Part of Seasat SAR image of Iceland acquired on 24 August 1978, showing the northern part of the Langjøkull ice cap, processed at DFVLR, Oberpfaffenhofen, Germany. SAR look direction is from bottom to top. The black arrows point to the snowline on the glacier.

Fig.5. Evaluation of digital airborne SAR data, acquired on 7 July 1981 over the Austrian SAR-580 test site. Lower image: display of digital SAR data after 5 × 5 pixel low-pass filtering (radar illumination is from top to bottom). M = moraines from historic glacier advances. Upper image: map of snow extent by density slicing. White = snow-free; grey = snow; black = radar shadow.

very wet snow or slush which reveals high dielectric losses due to the water content. The interpretation of brightness differences in the image is further complicated by variation in backscattering due to topographic effects. This example shows clearly that textural information is essential for the discrimination of snow and ice areas. Similar characteristics of backscattering were observed in Seasat images of other glaciers in Iceland (Rott 1984[a]).

If we look at radar images from high mountain areas, effects of imaging geometry are dominant, the slope orientation relative to the antenna being of main importance for the strength of the return signal. SIR-A acquired an image swath over the Karakoram mountains in Pakistan, where the large glaciers cover altitude ranges of more than 3000 m. The main glacier streams can be easily identified in these images due to their characteristic dendritic shape. However, it is hardly possible to derive information on ice or snow cover on steep slopes or to delineate the glacier boundaries in the accumulation zone. An interesting feature on the glaciers is the high return signal of the dry firn zones at high altitudes, which results in strong contrast to the wet firn areas with low return (Rott 1984[b]).

Airborne SAR images of seasonal snow cover were acquired in the X-, C-, and L-bands at Alpine test sites in Switzerland (Mätzler and Schanda 1984) and Austria (Rott and Domik 1984) in June and July 1981 during the European SAR-580 experiment which was conducted by the European Space Agency and by the Joint Research Centre of the European Communites. Fig.5 shows an evaluation from the SAR data of the Austrian test site in the Ötztal mountains; the antenna incidence angles range from 60° to 65° off nadir across the swath shown. The lower part of the Figure shows a window of the X-band radar data after application of a 5 × 5 pixel low-pass filter. In this enlargement the speckle structure is still visible in spite of filtering of the original single-look data, which had a nominal resolution of 3 m × 3 m.

The snow cover was wet at the time of the SAR overflight, and the average liquid water content was 6% by volume. This resulted in low radar return in the X-band, which enables clear discrimination against the snow-free surfaces. If a limited area with little variation of incidence

angles is considered, the discrimination of snow-covered and snow-free areas is possible by application of simple density-slicing methods as shown in the upper part of the figure. In some cases ambiguities were found between radar shadow and snow, because the transmitter power of the radar system was low. For machine classification over larger mountain areas, the variations in signal strength due to the local incidence angle of the radar beam have to be taken into account. This requires more complicated techniques such as matching of real and simulated images (Rott and Domik 1984). Quantitative analysis of the airborne SAR data indicated that the contrast between wet snow cover and snow-free surface decreases from X-band to longer wavelengths; the same is true for the discrimination of snow and ice areas on glaciers (Rott 1984[a]) and was also found by ground-based scatterometer measurements (Hüppi 1986).

CONCLUSIONS

Concerning snow and glacier applications, the main advantage of SAR compared to imaging sensors in the visible and near infrared is the all-weather capability. For mapping wet snow cover, an X-band system may be considered optimum; a C-band system should also be useful. It is questionable if SAR can be applied for mapping a thin layer of dry snow. According to our present knowledge dry snow mapping would require a dual-frequency system with one frequency in the X-band and the second frequency $\geq$ 18 GHz. More research is needed on this topic to come to firm conclusions. Radar image simulations were carried out to learn about the requirements in imaging geometry. For snow mapping in mountain areas, antenna look angles between 40 and 50 degrees and spatial resolutions of 15 to 20 m are needed (Rott and others 1985). The spatial resolution has to be higher than for optical sensors, because spatial averaging is needed for the SAR data to reduce the speckle effects.

Considering the glacier applications, SAR will be of main interest for those tasks which require frequent coverage or regular repetition. Due to the imaging geometry, glacier mapping by SAR is problematic in areas with high relief. Generally, for glacier mapping high-resolution optical sensors such as Landsat-TM or SPOT are preferable. However, for monitoring fast-changing glaciers, a weather

independent system may be needed. Also over the polar ice sheets SAR systems are able to fulfil important complementary tasks such as mapping of ice shelf boundaries and of surface flow features.

Snow and bare ice areas on glaciers usually can be discriminated in SAR images, though it is often more difficult than in multispectral optical imagery. If the equilibrium line is within the snow area, the detection by SAR will be hardly possible, whereas differences in snow reflectivity often make it visible in optical imagery. On the other hand, microwave sensors offer some advantages due to the penetration of snow. Thus SAR may enable the detection of the boundary between ice and firn through a layer of dry snow in the autumn. In radar images the extent between the wet and dry zones in the accumulation areas can be detected. This is, for example, of interest for investigations of the Greenland ice sheet.

The European Space Agency (ESA) plans to launch a C-band SAR system on the remote sensing satellite ERS-1 in 1990. This system was designed for ocean and sea-ice applications. According to the imaging geometry it will be of interest for glacier applications only in areas with gentle relief. For the nineties, a SAR system for land application is under consideration by ESA; the requirements for snow and land-ice applications have been investigated in a detailed study (Rott and others 1985). The following SAR system parameters were specified for a snow and glacier monitoring system:

Frequency single	X-band	(for wet snow)
dual	X and $\geq$ 18 GHz	(for dry snow, prel.)
Polarizations	hh or vv	
Incidence angle	40 to 50 degrees	(off nadir)
Spatial resolution	15 to 20 m. 1 look	
Range of δ°	-25 dB to 0 dB	(for X-band hh, vv)
	-20 dB to +5 dB	(at 20-30 GHz)
Radiometric resolution	3 dB for 15 m × 15 m (at -25 dB)	
Radiometric accuracy	1 dB (mean)	

Because the launch of a spaceborne SAR system at frequencies $\geq$ 18 GHz is not under consideration for technical reasons, a supplementary sensor is required for dry snow mapping. Due to the capability for mapping dry snow and its water equivalent, a multispectral imaging microwave radiometer should be included in a snow monitoring system. The main drawback of passive microwave sensors is the limited spatial resolution, which is not better than about 20 km for present sensors, but in the future can be improved to a few kilometers with acceptable efforts. This should be adequate for large- to medium-scale hydrological applications. The main sensors for glacier applications, in particular for glacier mapping, are certainly high resolution optical sensors; SAR can fulfil important complementary tasks.

REFERENCES

Hüppi R 1986 S- to X-band signature measurements of snow. *In Proceedings of MIZEX-EARSel workshop: Microwave Signatures of Arctic Sea Ice under Summer Melt Conditions, University of Bern, March 1986*: 70–81

Mätzler C 1986 Can microwave signatures be used to retrieve the water equivalent of a dry snow pack? *In Spectral Signatures of Remote Sensing. Proceedings of 3rd International Colloquium ISPRS*: 277–284 (ESA SP-247)

Mätzler C Unpublished Interaction of microwaves with the natural snow cover. (Habilitation thesis, University of Bern, 1985)

Mätzler C, Schanda E 1984 Snow mapping with active microwave sensors. *International Journal of Remote Sensing* 5(2): 409–422

Mätzler C, Schanda E, Good W 1982 Towards the definition of optimum sensor specification for microwave remote sensing of snow. *IEEE Transactions on Geoscience and Remote Sensing* GE 20(1): 57–66

Mätzler C, Aebischer H, Schanda E 1984 Microwave dielectric properties of surface snow. *IEEE Journal of Ocean Engineering* OE 9(5): 366–371

Reber B Unpublished Volumenstreuung von Mikrowellen an gefrorenem Schnee. (Diploma thesis, University of Bern, 1986)

Rott H 1984[a] Synthetic aperture radar capabilities for snow and glacier monitoring. *Advances in Space Research* 4(11): 241–246

Rott H 1984[b] The analysis of backscattering properties from SAR data of mountain regions. *IEEE Journal of Ocean Engineering* OE 9(5): 347–355

Rott H, Domik G 1984 The SAR-580 experiment on snow and glaciers at the Austrian test site. Final report. *In European SAR-580 Campaign, JRC, Ispra, Italy.* Vol 2: 217–231

Rott H, Domik G, Mätzler C, Miller H 1985 *Study on use and characteristics of SAR for land snow and ice applications. Final report to ESA.* Innsbruck, Universität Innsbruck (Institut für Meteorologie und Geophysik. Mitteilungen 1)

Ulaby F T, Stiles W H 1980 The active and passive microwave response to snow parameters. 2. Water equivalent of dry snow. *Journal of Geophysical Research* 85(C2): 1045–1049

Annals of Glaciology 9   1987
© International Glaciological Society

LARGE-SCALE PATTERNS OF SNOW MELT ON ARCTIC SEA ICE MAPPED FROM METEOROLOGICAL SATELLITE IMAGERY

by

G. Scharfen, R.G.   Barry

(Co-operative Institute for Research in Environmental Sciences, University of Colorado, Boulder, CO 80309, U.S.A.)

and

D.A. Robinson, G. Kukla, M.C. Serreze

(Lamont-Doherty Geological Observatory of Columbia University, Palisades, NY 10964, U.S.A.)

ABSTRACT

The seasonal progression of snow melt on the Arctic pack ice is mapped from satellite shortwave imagery (0.4-1.1 micrometers) for four spring/summer seasons (1977, 1979, 1984 and 1985). This provides the first detailed information on the temporal change of the ice surface albedo in summer and of its year-to-year variability. The average surface albedo of the Arctic Basin for the years investigated falls from between 0.75 and 0.80 in early May to between 0.35 and 0.45 in late July and early August. In the central Arctic, where ice concentration remains high and ponding on the ice is limited, the July albedo ranges from 0.50 to 0.60. Overall, melt progresses poleward from the Kara and Barents Seas and from the southern Beaufort and Chukchi Seas, with the melt fronts meeting on the American side of the Pole. There are substantial year-to-year differences in the timing, duration and extent of the melt interval. The progression of melt in May and June of the earliest melt year (1977) was about 3 weeks ahead of the latest year (1979). By late July, the central Arctic was essentially snow free in 1977 and 1979, but more than 50% snow covered in 1984. Although limited in extent, our data base suggests relationships between snow melt and Arctic surface air temperatures in spring, spring cloudiness and the extent of late summer ice.

INTRODUCTION

The timing, duration and extent of snow melt on the pack ice has long been recognized as a critical variable influencing the summer climatic regime in the Arctic Basin, with potential impacts on other parts of the Northern Hemisphere (Fletcher 1966). The snow melt has implications for the long-term mass balance and stability of the sea ice and may serve as an indicator of CO_2-induced climatic change (Barry 1985).

Up to now, direct information on melt and the resultant changes in surface reflectivity (albedo) in the basin has been limited to measurements taken at drifting stations or on fast ice and to a few aircraft programs (e.g. Laktionov 1953, Untersteiner 1961; Langleben 1971; Bryazgin and Koptev 1970; Weaver and others 1976, Holt and Digby 1985). Others have used these data to estimate regionally averaged summer albedo (e.g. Larsson and Orvig 1962; Marshunova and Chernigovskiy 1966; Posey and Clapp 1964; Hummel and Reck 1979, Robock 1980; Kukla and Robinson 1980). Estimates of July surface albedo in the central Arctic range from 0.40 to 0.65.   More recently, microwave satellite data have been used to identify the earliest phase of snow melt onset (Anderson and others, 1985) and also to estimate sea ice concentration and, indirectly, Arctic surface albedo (Carsey 1985). In the latter study, the areal coverage of bare ice was deduced from the data and the proportion of melt ponds and leads was estimated from published reports. Parameterized albedos were then assigned to each surface type and weighted according to their coverage. Central Arctic albedo in mid-July 1974 was estimated to be 0.58 with values about 0.10 lower around the margins of the basin.

In the present study, we have used operational daily meteorological satellite imagery to map the changes of surface brightness and texture associated with four spring/ summer seasons in the Arctic Basin (Figure 1). Parameterized albedos have been assigned to different brightness classes by analyzing satellite data on an image processor.

DATA AND METHODS

Imagery from the Defense Meteorological Satellite Program (DMSP) near-polar orbiter served as the primary data source for assessing surface conditions. Shortwave (0.4-1.1 micrometers) images with resolutions of 0.6 km for direct read-out products and 2.7 km for orbital swath

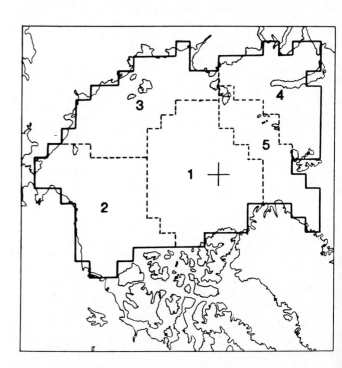

Fig.1.   Arctic Basin study zone (heavy line) divided into five regions (dashed lines): 1) Central Arctic, 2) Beaufort/Chukchi Seas, 3) East Siberian/Laptev Seas, 4) Kara/Barents Seas and 5) Northwest North Atlantic.

format images were used. Supplemental data included NOAA Very High Resolution Radiometer (VHRR) and Advanced VHRR 1.1 km resolution imagery.

Snow melt on the ice is recognized in the shortwave images by a characteristic decrease in brightness and a change in surface texture. As the snow dissipates, melt ponds form and bare ice is exposed (Barry 1983) (Figure 2). Landfast ice and first-year pack ice brighten somewhat when melt ponds drain (Zubov 1945; Lapp 1982; Holt and Digby 1985). Comparisons of DMSP imagery with synchronous 80 m resolution Landsat imagery confirm the recognition of surface features in the lower resolution products. The comparisons also show that, in most cases, brightness and textural changes due to melt processes on the ice may be successfully distinguished from variations in ice concentration. This distinction was most difficult where U.S. Navy/NOAA ice charts (Godin 1981) depict ice concentrations of less than 75%; less than 10% of the basin during most of the melt season (Figure 3).

Basin-wide maps of surface brightness and texture were constructed manually in three-day increments from May through August 1977, 1979, 1984, and 1985. Repetitive coverage and characteristic textures permitted the differentiation of moving clouds from the surface (Robinson and others 1985). Interactive image processor analyses of selected scenes showed good agreement with visual classification of surface brightness. Maps were digitized using the National Meteorological Center standard grid which divides the basin into 212 cells. At least one cloud-free scene per 3-day interval was available over more than 80% of the basin from May to mid-August. Missing cells were either: (1) assigned observed brightness values from an immediately preceding or subsequent chart, (2) considered to be open water, if shown as such on the Navy/NOAA weekly ice chart closest to the analyzed interval, or (3) handled as missing data (less than 10%, on average).

Four ice-surface classes are identified. The literature on ground and aerial observations, referred to earlier, shows that these classes represent: class (1) fresh snow cover over 95% of the ice; class (2) snow covers between 50-95% of the surface, with the remainder being bare or ponded ice (in spring this is considered the initial stage of active snow melt); class (3) the final stage of active snow melt, with between 10-50% of the ice surface snow covered and with numerous melt ponds, or, following pond drainage, predominantly bare ice, with snow patches and scattered ponds; and class (4) heavily ponded or flooded ice with less than about 10% snow cover, or exposed bare ice.

Local observations indicate that when class 3 is first identified the surface is in the final stage of active melt and remains so for approximately the next 2-3 weeks. At later dates, when class 3 is charted, the surface is predominantly drained bare ice (Kuznetsov and Timerev 1973; Hanson 1980). Classes 2 and 3 are illustrated for the Beaufort Sea in Figure 2. Class 4 is most commonly found in regions of fast ice. The absence of the numerous flaws and leads in the fast ice and its proximity to surface run-off from land results in more extensive flooding of this ice than over pack ice.

Large-scale surface albedo values adopted for the charted classes are: class (1) 0.80; class (2) 0.64; class (3) 0.49; and class (4) 0.29, with standard deviations between 0.08 (class 3) and 0.05 (class 4). These values were adjusted for average summer cloudiness (Robinson and others 1985) by ±0.05 for brightness classes 1-3 and ±0.02 for class 4, based on data from Buzuyev and others (1965), Kuznetsov and Timerev (1973) and Grenfell and Perovich (1984). Albedo is also decreased to account for the presence of open water within the pack (Cogley, 1979). For class 1, the correction ranges from -0.17 when the ice concentration is 75%, to -0.01 when the ice concentration is 99%. Ice concentrations were based on U.S. Navy/NOAA weekly ice charts.

Class albedos were calculated by measuring the brightness of selected clear-sky scenes on a digital image processor. Linear interpolation was made between homogeneous bright snow on multiyear ice and dark open-water. Clear-sky albedos of these targets were estimated from measured ground and aerial data (e.g. Hanson 1961, Nazintsev 1964, Langleben 1971, Bryazgin and Koptev 1970,

78N/176W   78.5N/144W

70N/168W   70N/148W

Fig.2. High–resolution (0.6 km) DMSP shortwave image of the Beaufort Sea on June 17, 1978 showing stages of snow melt (classes 2 & 3) on the sea ice.

Payne 1972, Grenfell and Maykut 1977, Pautzke and Hornof 1978, Cogley 1979). The snow-covered targets were assigned an albedo of 0.80 until late June, after which a value of 0.70 was used. The open–water albedo was taken as 0.12. Specular reflectance is minimal over these surfaces at relatively high solar zenith angles during the Arctic summer and at the satellite viewing time and angle (Taylor and Stowe 1984). This procedure has been used in other studies (e.g. Preuss and Geleyn 1980; Robinson and Kukla 1985) and found to be particularly well suited for use with the broad-band DMSP imagery (Shine and Henderson-Sellers 1984).

RESULTS

Our maps show that in all four years melt began in the Barents and Kara Seas and the southern Beaufort and Chukchi Seas, then progressed along the Arctic coast of Asia and towards the American side of the Pole. In 1977, active melt (classes 2 and 3) covered 50% of the basin by the end of May (Figure 3A), while over half of the basin was categorized as class 3 or 4 by the end of June. Melt began almost 3 weeks later in 1979 (Figure 3B), yet, as in 1977, more than half of the basin was classified as class 3 or 4 by the end of June. In 1984, melt began prior to May 1 in the southern Beaufort and Chukchi Seas (Figure 3C). However, it was not as extensive in May as it was in 1977. In 1984, areas with concentrated narrow leads in seas bordering the Asian continent resulted in a large-scale surface brightness equivalent to unbroken ice undergoing partial melt (class 2). This gives the appearance in Figure 3C of quite extensive basin snow melt early in May. However, an examination of high–resolution DMSP and AVHRR images indicates that while melt did begin quite early in the southern Beaufort and Chukchi Seas and portions of the Kara and Barents Seas in 1984, it did not begin over more than 10-15% of the basin until late May. Over half of the basin was classified as classes 3 and 4 by the end of June 1984. Melt began at about the same time in 1985 (Figure 3D) as in 1979, with about 25% of the basin categorized as class 2 at the end of May. Approximately 50% of the basin was classified as classes 3 and 4 by the end of June.

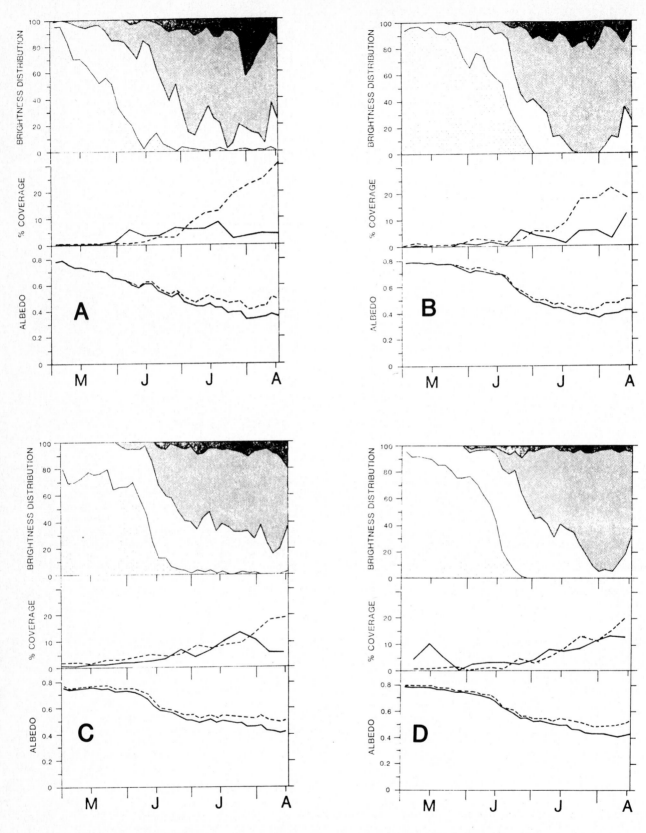

Fig.3. A. top: Progression of snow melt and subsequent ponding and drainage on Arctic sea ice from
May to mid-August 1977, as shown by the changing distribution of brightness classes (cf. text for
class descriptions). Classes shaded from light grey (class 1) to dark grey (class 4). Areas with less
than 12% ice concentration or open water omitted.

   middle: Percentage of the basin with open water or with less than a 12% (1/8)
concentration of ice (dashed line) and percentage of the basin with a 12%–75% (1–6/8) ice
concentration (solid), according to 1977 weekly Navy/NOAA ice charts.

   bottom: Basin-wide albedo (including sea ice and open water) from May–August 1977 (solid
line). Albedo of ice (sea water excluded) in areas with at least 75% ice concentration (dashed).

   B. Same as 3A, except for 1979.

   C. Same as 3A, except for 1984 and ice concentrations of 10% (1/10) and 80% (8/10) where
marked as 12% and 65% in A.

   D. Same as 3A, except for 1985 and ice concentration of 10% and 80%.

All sectors of the basin showed signs of melt by the beginning of July, although there were considerable year-to-year variations in melt intensity during this month. In 1977, a mid-month snowfall in the central Arctic temporarily brightened the surface and increased the albedo. Melt advanced steadily in July 1979, leaving the area essentially snowfree by the last week of the month. The central Arctic continued to have a considerable snow cover throughout July 1984, as was the case until the last week of the month in 1985.

Soon after the minimum albedo was reached in late July or early August of each year, fresh snow began once again to cover the central part of the basin. The newly snow-covered areas did not appear as bright on the imagery as full spring cover, probably due to the presence of open leads and undrained ponds. Out of the four years, snow cover was most extensive by mid-August 1985. Heavy cloudiness allowed only sporadic views of the basin surface in the latter half of all Augusts, when it appeared that fresh snow increasingly covered the basin.

Due to the earlier melt in 1977 and the combination of melt and lower ice concentration in 1984, basin albedo averaged 0.73 in May 1977 and 1984, compared with 0.77 in 1979 and 0.76 in 1985 (Table I). Basin albedo was highest in June 1979 (0.66), and lowest in June 1977 (0.58).

TABLE I. AVERAGE SURFACE ALBEDO OVER THE ARCTIC BASIN (FOR EACH MONTH IN THE FOUR STUDY YEARS)

	1977	1979	1984	1985
May	0.73	0.77	0.73	0.76
June	0.58	0.66	0.61	0.65
July	0.43	0.44	0.48	0.48
August*	0.36	0.40	0.42	0.42

* for the period August 1-17

By July, basin albedo dropped to 0.43 in 1977, 0.44 in 1979, and 0.48 in 1984 and 1985. The first half of August averaged 0.36 in 1977, and between 0.40 and 0.42 in the other years, the major difference in 1977 being the relatively large amount of open water in the basin (Figure 3). The late July and early August albedo of the ice surface (sea water excluded) in those parts of the basin where ice concentration exceeded 75% was approximately 0.45 in 1977 and 1979 and 0.50 in 1984 and 1985 (Figure 3).

A regional breakdown (Figure 1) of surface albedo shows that the early advance of extensive snow melt in 1977 compared with the other three years was most pronounced in the Central Arctic (Table II). Later in the summer the albedo differences between years in the coastal seas were a result of differences in ice extent. This is particularly evident in the East Siberian and Laptev Seas, where the early August albedo ranged between 0.25 (1977) and 0.47 (1984).

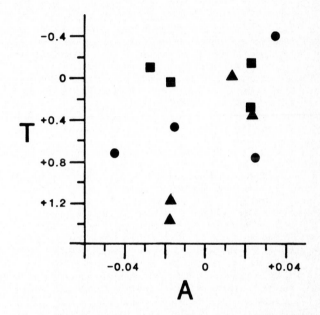

Fig.4. Comparison of monthly anomalies of surface albedo over the Arctic Basin and surface air temperature from 65-85°N for May (triangles), June (circles) and July (squares) of the four study years. Albedo anomalies from the average of the four years. Temperature anomalies based on 1951-1970 means (Jones 1985; Climate Monitor 1985a, 1985b).

TABLE II. AVERAGE SURFACE ALBEDO OVER REGIONS OF THE ARCTIC BASIN (FOR EACH MONTH IN THE FOUR STUDY YEARS)

		Central Arctic	Beaufort/ Chukchi	E. Siberian/ Laptev	Kara/ Barents	NW North Atlantic
May	1977	0.78	0.74	0.73	0.66	0.72
	1979	0.79	0.76	0.79	0.77	0.73
	1984	0.79	0.75	0.74	0.63	0.70
	1985	0.79	0.77	0.79	0.69	0.74
June	1977	0.66	0.58	0.55	0.48	0.60
	1979	0.76	0.62	0.67	0.58	0.62
	1984	0.69	0.59	0.63	0.51	0.59
	1985	0.73	0.65	0.62	0.54	0.62
July	1977	0.52	0.42	0.40	0.27	0.46
	1979	0.53	0.41	0.42	0.32	0.45
	1984	0.58	0.45	0.49	0.30	0.50
	1985	0.59	0.47	0.48	0.31	0.47
August*	1977	0.50	0.33	0.25	0.19	0.44
	1979	0.53	0.37	0.39	0.22	0.43
	1984	0.54	0.42	0.47	0.15	0.43
	1985	0.54	0.41	0.45	0.17	0.44

* for the period August 1-17

In the central Arctic, where ice concentration generally remains above 95% throughout the summer, the July albedo was 0.52 in 1977, 0.53 in 1979, 0.58 in 1984 and 0.59 in 1985. These satellite-derived albedos are within the range of published estimates (e.g. Hanson 1961, Larsson and Orvig 1962; Posey and Clapp 1964; Marshunova and Chernigovskiy 1966; Schutz and Gates 1972, Hummel and Reck 1979; Robock 1980; Kukla and Robinson 1980; Barry 1983; Carsey 1985), which vary from 0.40 to 0.65. The large range of the estimates reflects the high year-to-year variability in the extent of snow melt, degree of ponding and frequency of summer snowfalls. Differing approaches and limited data bases available to earlier researchers may also be a factor.

Towards the middle of August, when the extent of fresh snow cover in the central Arctic began to increase, albedo rose by 0.05 to 0.10 in all years; however, our satellite-derived albedo in the first half of August is lower than that reported earlier from drifting stations (Nazintsev 1964, Pautzke and Hornof 1978). This may be the result of our large-scale averaging, as opposed to the ground observations made mostly on thick multi-year ice surrounding drifting stations. Alternatively, the August snowfalls may have come relatively late or been less extensive in the four years of our study.

DISCUSSION

The geographic progression of seasonal melt identified in our maps was compared with earlier studies. The climatological summary of Marshunova and Chernigovskiy (1978) shows a concentric pattern of melt over the Arctic pack ice progressing towards the Pole by early July. From a study of sequential ESMR data for 1974, Campbell and others (1980) suggested that melt began along the Siberian coast in May and moved as a roughly linear front across the Pole, reaching the Canadian and Greenland coasts about a month later. Re-examination of their data by Crane and others (1982), indicated that the color coding of the microwave imagery created a partially spurious impression of a progressing melt front.

Climate models suggest that an early loss of snow cover may have an impact on ice extent later in the summer (Maykut and Untersteiner 1971; Semtner 1976). In 1977, when the snow melt and ponding came early, the subsequent late summer ice extent was considerably reduced compared to ice cover in the other years. This relationship was most evident in the Beaufort and Chukchi Seas and the East Siberian and Laptev Seas regions. Whether such a dependence was coincidental or not must remain speculative until more years are examined.

Cloud cover, mapped at approximately three-day intervals in the summers of 1977 and 1979, showed a late May–early June maximum in extent (averaging approximately 90%) and thickness over the basin, followed by a period of less extensive (approximately 75%) and somewhat thinner cover extending into early August (Robinson and others 1985). This suggests that early stages of surface melt may be related to the early season cloudiness, which appears to be associated with the poleward retreat of the Arctic front and advection of air from lower latitudes by synoptic disturbances (Barry and others 1986). Early melt may also be enhanced by the increase in infrared radiation at the surface due to the cloud cover. This results in an increase in surface net radiation over high-albedo Arctic surfaces (e.g. Ambach 1974).

The relationship between snow melt onset and large-scale Arctic surface air temperatures was examined using monthly temperature and albedo anomalies (Figure 4). Temperature anomalies are based on 1951-1970 means for 65-85°N (Jones 1985; Climate Monitor 1985[a] and 1985[b]). Albedo anomalies are calculated from the average of the four study years (cf. Table I). A relationship between positive anomalies of temperature and negative anomalies of basin-wide surface albedo is apparent in May and three of the Junes studied, but is not evident in July. This suggests that the temperature data are not representative of the inner Arctic Basin, but are related to melt in the coastal seas located near the reporting stations.

CONCLUSION

This work provides the first direct evidence of the fluctuations of snow cover and surface albedo across the entire Arctic Ocean in spring and summer. The timing of snow melt on the Arctic sea ice is shown to vary greatly from region to region as well as from year to year. This variation will have a significant impact on the heat and mass balance of the Arctic, since the surface albedo and the thermodynamics of the ice are strongly related to the presence of snow. Improved knowledge of the spring and summer surface albedo in the Arctic is important in climate models and may also help in recognizing the initial signs of any climatic changes induced by CO_2 and other trace gases.

ACKNOWLEDGEMENTS

We thank D. van Metre and S. Innes for assistance in the analysis. DMSP data are archived for NOAA/NESDIS at the University of Colorado, CIRES/National Snow and Ice Data Center, Campus Box 449, Boulder, CO 80309. This work was supported by NSF grant ATM 83-18676 and the Air Force Office of Scientific Research, Air Force Systems Command, USAF, under grant AFOSR 86-0053. The U.S. Government is authorized to reproduce and distribute reprints for Governmental purposes notwithstanding any copyright notation thereon. This is LDGO contribution 4026.

REFERENCES

Ambach W 1974 The influence of cloudiness on the net radiation balance of a snow surface with high albedo. *Journal of Glaciology* 13(67): 73–84

Barry R G 1983 Arctic Ocean ice and climate: perspective on a century of polar research. *Annals of the Association of American Geographers* 73: 485–501

Barry R G 1985 The cryosphere and climate change. *In* MacCracken M, Luther F (*eds*) *Detecting the climatic effects of increasing carbon dioxide.* Washington, DC, United States Department of Energy: 109–148 (DOE/ER–0235)

Barry R G, Crane R G, Newell J P, Schweiger A 1986 *Empirical and modeled synoptic cloud climatology of the Arctic Ocean.* Boulder, CO, University of Colorado. Cooperative Institute for Research in Environmental Sciences

Bryazgin N N, Koptev A P 1970 Spectral albedo of snow-ice cover. *Problems of the Arctic and the Antarctic* 31: 355–360

Buzuyev A Ya, Shesterikov N P, Timerev A A 1965 Al'bedo l'da v arkticheskikh moryakh po dannym nablyudeniy s samoleta [Albedo of ice in Arctic seas from air observation data]. *Problemy Arktiki i Antarktiki* 20: 49–54

Campbell W J, Ramseier R O, Zwally H J, Gloersen P 1980 Arctic sea-ice variations from time-lapse passive microwave imagery. *Boundary-Layer Meteorology* 18(1): 99–106

Carsey F D 1985 Summer Arctic sea ice character from satellite microwave data. *Journal of Geophysical Research* 90(C3): 5015–5034

Climate Monitor 1985[a] Update table for graphs. *Climate Monitor* 14: 42

Climate Monitor 1985[b] Update table for graphs. *Climate Monitor* 14: 82

Cogley J G 1979 The albedo of water as a function of latitude. *Monthly Weather Review* 107: 775–781

Crane R G, Barry R G, Zwally H J 1982 Analysis of atmosphere-sea ice interaction in the Arctic Basin using ESMR microwave data. *International Journal of Remote Sensing* 3: 259–276

Fletcher J O 1966 The Arctic heat budget and atmospheric circulation. *In* Fletcher J O (*ed*) *Proceedings of the Symposium on the Arctic Heat Budget and Atmospheric Circulation... 1966, Lake Arrowhead, California.* Santa Monica, CA, Rand Corporation: 23–43

Godin R H 1981 Sea ice charts of the Navy/NOAA Joint Ice Center. *Glaciological Data Report* GD-11: 71–77

Grenfell T C, Maykut G A 1977 The optical properties of ice and snow in the Arctic Basin. *Journal of Glaciology* 18(80): 445–463

Grenfell T C, Perovich D K 1984 Spectral albedos of sea ice and incident solar irradiance in the southern Beaufort Sea. *Journal of Geophysical Research* 89(C3): 3573–3580

Hanson A M 1980 The snow cover of sea ice during the Arctic Ice Dynamics Joint Experiment, 1975 to 1976. *Arctic and Alpine Research* 12(2): 215–226

Hanson K J 1961 The albedo of sea-ice and ice islands in the Arctic Ocean basin. *Arctic* 14(3): 188–196

Holt B, Digby S A 1985 Processes and imagery of first-year fast sea ice during the melt season. *Journal of Geophysical Research* 90(C3): 5045–5062

Hummel J R, Reck R A 1979 A global surface albedo model. *Journal of Applied Meteorology* 18: 239–253

Jones P D 1985 Arctic temperatures 1951–1984. *Climate Monitor* 14: 43–49

Kukla G, Robinson D 1980 Annual cycle of surface albedo. *Monthly Weather Review* 108(1): 56–68

Kuznetsov I M, Timerev A A 1973 The dependence of ice albedo changes on the ice cover state as determined by airborne observations. *Problems of the Arctic and the Antarctic* 40: 67–74

Laktionov A F (ed) 1953 *Rukovodstvo dlya nablyudeniyy nad l'dami arkticheskikh morey, rek i ozer na polyarnykh gidrometeorologicheskikh stantsiyakh* [*Handbook for observing the ice of Arctic seas, rivers and lakes at polar hydrometeorological stations*]. Leningrad, Izdatel'stvo Glavsevmorputi (Posobiya i Rukovodstva 31)

Langleben M P 1971 Albedo of melting sea ice in the southern Beaufort Sea. *Journal of Glaciology* 10(58): 101–104

Lapp D 1982 *A study of ice meltponds.* Toronto, Atmospheric Environment Service

Larsson P, Orvig S 1962 *Albedo of Arctic surfaces.* Montreal, McGill University (Publications in Meteorology 54)

Marshunova M S, Chernigovskiy N T 1966 Numerical characteristics of the radiation regime in the Soviet Arctic. *In* Fletcher J O (ed) *Proceedings of the Symposium on the Arctic Heat Budget and Atmospheric Circulation... 1966, Lake Arrowhead, California.* Santa Monica, CA, Rand Corporation: 279–297

Marshunova M S, Chernigovskiy N T 1978 *Radiation regime of the foreign Arctic.* New Delhi, Indian Scientific Documentation Centre for the Office of Polar Programmes; Washington, DC, National Science Foundation (Technical Translation 72-51034)

Maykut G A, Untersteiner N 1971 Some results from a time-dependent, thermodynamic model of sea ice. *Journal of Geophysical Research* 76(6): 1550–1575

Nazintsev Yu L 1964 Teplovoy balans poverkhnosti mnogoletnego ledyanogo pokrova v Tsentral'noy Arktike [Surface heat balance of the perennial ice sheet of the central Arctic]. *Trudy Arkticheskogo i Antarkticheskogo Nauchno-Issledovatel'skogo Instituta* 267: 110–126

Pautzke C G, Hornof G F 1978 Radiation regime during AIDJEX: a data report. *AIDJEX Bulletin* 39: 165–185

Payne R E 1972 Albedo of the sea surface. *Journal of the Atmospheric Sciences* 29: 959–970

Posey J W, Clapp P F 1964 Global distribution of normal surface albedo. *Geofisica Internacional* 4: 33–48

Preuss H J, Geleyn J F 1980 Surface albedos derived from satellite data and their impact on forecast models. *Archives for Meteorology, Geophysics and Bioclimatology* Ser A 29: 345–356

Robinson D A, Kukla G 1985 Maximum surface albedo of seasonally snow-covered lands in the northern hemisphere. *Journal of Climate and Applied Meteorology* 24: 402–411

Robinson D A, Kukla G J, Serreze M C 1985 Arctic cloud cover during the summers of 1977–1979. Palisades, NY, Lamont-Doherty Geological Observatory (Technical Report L-DGO-85-5)

Robock A 1980 The seasonal cycle of snow cover, sea ice and surface albedo. *Monthly Weather Review* 108(3): 267–285

Schutz C, Gates W L 1972 *Global climatic data for surface, 800 mb, 400 mb: July.* Santa Monica, CA, Rand Corporation

Semtner A J Jr 1976 A model for the thermodynamic growth of sea ice in numerical investigations of climate. *Journal of Physical Oceanography* 6(3): 379–389

Shine K P, Henderson-Sellers A 1984 Cryosphere-cloud interactions near the snow/ice limit: sensitivity testing of model parameterizations. *In* Barry R G, Shine K P, Henderson-Sellers A *Cryosphere-cloud interactions near the snow/ice limit.* Boulder, CO, University of Colorado. Cooperative Institute for Research in Environmental Sciences; Liverpool, University of Liverpool. Department of Geography: 77–236

Taylor V R, Stowe L L 1984 Reflectance characteristics of uniform earth and cloud surfaces derived from NIMBUS-7 ERB. *Journal of Geophysical Research* 89(D4): 4987–4996

Untersteiner N 1961 On the mass and heat budget of Arctic sea ice. *Archives for Meteorology, Geophysics and Bioclimatology* Ser A 12: 151–182

Weaver R L, Barry R G, Jacobs J D 1976 Fast ice studies in western Davis Strait. *In POAC 75: the Third International Conference on Port and Ocean Engineering under Arctic Conditions, Fairbanks, Alaska, 1975. Proceedings Vol 1*: 455–466

Zubov N N 1945 *L'dy Arktiki* [*Arctic ice*]. Moscow, Izdatel'stvo Glavsevmorputi [English translation: Washington, DC, US Navy Oceanographic Office]

Annals of Glaciology 9 1987
© International Glaciological Society

A DIGITAL RADAR SYSTEM FOR ECHO STUDIES ON ICE SHEETS

by

Donald G. Schultz, Lee A. Powell and Charles R. Bentley

(Geophysical and Polar Research Center, University of Wisconsin–Madison, Madison, WI 53706, U.S.A.)

ABSTRACT

A digital radar system comprising multiple microprocessors, for use with 50 MHz radar units modified from the Scott Polar Research Institute Mark IV design, is described. The major features of the system include coherent integration of radar traces, storage of data in raw digitized form without demodulation, real-time play-back of digitized information, and high system performance resulting in good spatial sampling with integration even in airborne operations. Unfocused synthetic beam shaping also results from the integration of echoes, thus reducing clutter or incoherent scattering from the sides of the beam pattern along the profiling track.

Examples of data collected during the austral summer of 1985–86 in the Antarctic on ice stream B, in both ground and airborne programs, illustrate both the flexibility in data presentation and features present in the records.

INTRODUCTION

With the advent of commercially available high-speed digitizers, it has become possible to sample relatively wide-band radar signals. Conventional recording techniques in radar-sounding programs usually employ an optical recording process (e.g. Crabtree and Doake 1986; Shabtaie and Bentley 1987). The resultant form of the recording is frozen into an image that contains limited amplitude and no phase information, and is difficult to use for quantitative analyses. In contrast, an 8 bit digitizer has a dynamic range of greater than 40 dB, all of which is preserved in a recording from which the form of data presentation can be selected.

For the quantitative analysis of radar information, recording the data digitally obviates the necessity of digitizing the data from analog records, which often compromises the quantity and quality of the results. The primary focus of radar studies has been the mapping of ice thicknesses and surface elevations, which can be facilitated by computer automation. Programs that select arrival times for echoes, with or without user intervention, can build data bases from which maps may be drawn directly (Bamber 1987). Other types of analyses, such as generating frequency distributions of echo amplitudes for the determination of small-scale roughness characteristics (Neal 1982), are relatively easy to perform after the echo amplitudes have been determined from digital records. For example, the back-scatter images made by Musil and Doake (1987) could have been created directly from processing digital data that retain the phase information.

Beyond advantages related to recording, the computer can be used to enhance the quality of the radar signal itself. For example, complex systems have been constructed to steer or shape the beam of antenna arrays. The digital radar system in this application preserves phase information, and adds successive radar traces together coherently to increase both the signal-to-noise ratio and the dynamic range of the recorded signal. Coherent integration (stacking) of traces, as distinct from integration after demodulation (e.g. Dowdeswell and others 1986), permits echoes with amplitudes smaller than the mean noise level to emerge from the noise. From a fast-moving vehicle such as an aircraft, the integration process has the additional advantage of significantly narrowing the effective antenna aperture along the direction of motion, thus reducing the clutter from surface scattering that interferes with primary echoes.

The radar transmitter and receiver units used in the study were similar to the Scott Polar Research Institute Mark IV design described by Evans and Smith (1969). The radar signal is amplitude-modulated at a carrier frequency of 50 MHz; the signal band width is 10 MHz. The pulse-repetition frequency is software–controlled and arbitrarily variable, being tied to the acquisition cycle for a single trace. Buffered outputs were provided for both pre-detected and post-detected signals. The post-detected signal was used for monitoring, the pre-detected one for digitizing. The digital circuitry that triggered the transmitter and recorded the receiver output was shielded to prevent RF contamination of the received signal from emitted digital noise. Hardware components in the digital system were chosen for functionality in polar operations (operation in a heated enclosure is still necessary) and rapid data-acquisition capabilities. The software was designed primarily for high-speed data recording, ease of use in controlling the system, and versatility.

SYSTEM HARDWARE DESCRIPTION

The digital radar-system hardware consists of a "Camac sub-system", a "Multibus sub-system", and a bit-slice processor that passes information between them. The Camac dataway and radar instrumentation modules, which reside on the dataway, together constitute the Camac sub-system (Fig.1). The Multibus sub-system comprises computer peripherals and devices connected to "Multibus" (a trademark of the Intel Corporation), a general purpose, multi-processor, microcomputer bus.

The Camac dataway, originally developed by the ESONE Committee of European Laboratories, is a standardized "data highway", by means of which an assemblage of instrument modules can communicate with each other and with external computers and computer components (Costrell 1971). There are four primary Camac modules which conform to IEEE Bus Standard 583-1975. One module is a four-channel arbitrary wave-form generator (AWG) that is used for data play-back. The other modules are involved in the data-acquisition process. The suppressed-carrier, amplitude-modulated radar signal passes through a wide-band pre-scaling module to an 8 bit analog-to-digital converter (ADC). A clock module provides an external sample clock for the ADC. The sample clock consists of two suites of interdigitated pulses, each produced at a 25 MHz rate. One suite is delayed in time by 270° of the center frequency to produce sets of samples that are approximately in quadrature. A transmitter trigger that is synchronized with the sample clock is derived from the master oscillator in the clock module. Auxiliary modules provide local memory for the ADC and the AWG. All of the modules except the pre-scaler are controlled by Camac function commands issued by the host processor via the bit-slice processor.

The host processor (Intel 8085 microprocessor), and other peripherals connected in the Multibus sub-system, together constitute a general purpose microcomputer. Many Multibus components were, however, specialized for data

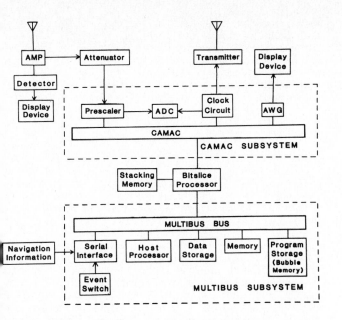

Fig.1. Hardware schematic of the digital radar system. The digital system consists of a Camac sub-system for bi-directional translation of signals from the analog and digital realms, a Multibus sub-system that constitutes a general purpose microcomputer, and a bit-slice processor for Camac control and communication between sub-systems. Scott Polar Research Institute Mark IV radar units were adapted for coherent integration of echoes.

acquisition in the field. Data were archived on two drives of a Bernoulli box (cf. Linton 1982). A 10 Megabyte floppy disk resides within a rigid cartridge that protects it from chemical and mechanical contamination. Because it floats on an air cushion during recording, the disk is additionally accorded a high degree of shock immunity. The acquisition code was stored in bubble memory, which is non-volatile and also provides a high degree of protection from the environment. For recovery from catastrophic failures like a loss of power, a small amount of battery-backed random-access memory was used to save and restore system-status information, thus providing a means of rapidly repairing a corrupted data structure on disk. A serial port was utilized to allow the operator to record external events (e.g. passage by reference points on survey lines), thus providing a data link between the radar data-stream and positioning information. An auxiliary serial port is available for the input of navigational data.

A micro-programmable bit-slice processor (Am 2910) allows fast and complex management of Camac operations. A bus-master interface also gives the bit-slice processor direct access to the memory of the host processor and to other devices in the Multibus sub-system. Bits within a 48 bit microcode instruction word directly control the hardware connected to the processor. The micro-coding and other architectural features of the bit slice lend themselves to parallel processing and high-speed control of hardware operations, which are essential in phase-sensitive recording if summing is performed from a moving vehicle. The bit-slice processor stores accumulated data in a memory bank (8 kilobytes by 24 bits) referred to as the "stacking" memory.

SOFTWARE DESCRIPTION

The host processor executes instructions in FORTH, a computer language that operates within the CP/M programming environment. The user configures and controls the acquisition process through the host processor. The user is first presented with a set of configuration menus from which the desired data-acquisition parameters, such as the sample clock source for the ADC, the number of samples in a data buffer, and the sweep rate of the AWG, are selected. The menu selections provide the flexibility required for generally recording transient wave forms, such as other

types of radar signals. During processing, the user can monitor the data flow on a terminal screen and, whether or not the data are being recorded, monitor the digitally acquired data on an oscilloscope or oscillograph. In addition, he has a switch that is used to start data recording and mark events.

After a configuration is selected, several steps are performed before data recording commences. The recording status from the previous recording session is checked and if a failure is detected the data structure on disk is repaired if necessary. Then the disk directory is read and a new record header containing configurational details is written. At this stage, the system is ready for data acquisition.

The system architecture lends itself to processing tasks concurrently and therefore at high speed during data acquisition. A start-up process precedes the main acquisition loop. In this process, the first radar time series, which is the normalized sum of several received radar traces, is acquired. In preparation for writing on to disk, the data are transferred from stacking memory to Multibus memory. An analog replay of the data is also initiated on the AWG. At the completion of these procedures, the activities of both processors in the data-acquisition process (Fig.2) are repeated until the user intervenes.

SYSTEM ACTIVITIES

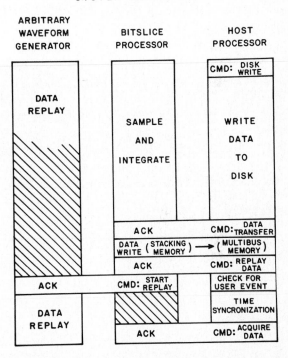

Fig.2. Block diagram of system activities during a data acquisition cycle. Intercommunication between processors and between the bit-slice processor and the AWG are indicated by command issue (CMD) and command acknowledge (ACK) pairs. Time increases toward the bottom of the figure. For maximum speed, the three major time-consuming tasks of data replay, sampling and integration, and writing data to disk are performed concurrently.

There are three major time-consuming tasks performed during the main acquisition loop. First, the data are written to the archiving disk, a process that is carried out entirely within the Multibus system. Transferring typically 1–2 kilobytes of data from the buffer to the disk requires approximately 0.4 s. Secondly, the sampling and integrating process, performed by the bit-slice processor in conjunction with the ADC, requires an amount of time that is linearly dependent on the number of traces to be superimposed. Each trace typically required 0.3 ms. To acquire a single trace, the ADC, and consequently a trigger pulse to the radar transmitter, are activated by the bit-slice processor. An additional pulse, synchronized to the transmitter pulse, is

also sent to the ADC to start the sampling. After a trace has been digitized, the ADC interrupts the bit-slice processor, which in turn retrieves and, in parallel, adds the data to the stacking memory. The sampling procedure repeats until the number of desired traces (monitored by a trace counter) has been integrated.

The third major task is analog play-back of the integrated data on the AWG. The quadrature sample pairs are combined and the result is normalized to 10 bits before it is sent to the AWG by the bit-slice processor. The samples are combined, or in effect the trace is demodulated two samples at a time, by taking the larger sample of a quadrature pair and adding it to half the smaller sample. (This approximates the square root of the sum of the squares within an error of 12%.) After transfer of the entire demodulated trace, the AWG will replay the data independently, at a sweep rate set previously during initialization. Of course, the original digitized data remain in storage after play-back.

At the start of the acquisition cycle, the AWG is presenting the demodulated form of the latest set of acquired traces, a process initiated at the end of the previous acquisition cycle. At the same time, the bit-slice processor is acquiring the next suite of integrated traces, and the host processor is writing the latest set of data on to the disk. With the three main processes running concurrently, the highest data-flow rate will be provided if sampling-and-integration takes the same length of time as

writing the data on to the disk. In our system, summing 128 traces maximizes the data-flow rate. (For airborne profiling at 120 knots, that rate results in a suite of samples being integrated over, and recorded, every 20 m.) In our work, however, we commonly used more stacking to improve the signal-to-noise ratio and narrow the effective antenna beam.

RESULTS

The digital radar system was used both for ground profiling at Upstream B camp on ice stream B and in an airborne program on the Siple Coast employing a DeHavilland Twin Otter. At this time, we have only begun to examine the vast amount of data collected. Two examples of airborne profiles across the ice stream are shown in fig.2 of Bentley and others (1987). Both cross-sections are intensity-modulated analog records, each constructed from approximately 2000 digital records. Each digital record is the sum of the radar returns of 256 individual radar pulses. In both profiles, internal layers show clearly over the slow-moving ice between ice streams B and C as one approaches the margin of ice stream B. Numerous bumps exist on the bed that diminish in size in cross-sections farther down-stream. Clutter associated with surface crevassing sharply delineates the marginal zones because of the beam-narrowing effect of the integration. The clutter reduction by spatial averaging also has made the bottom

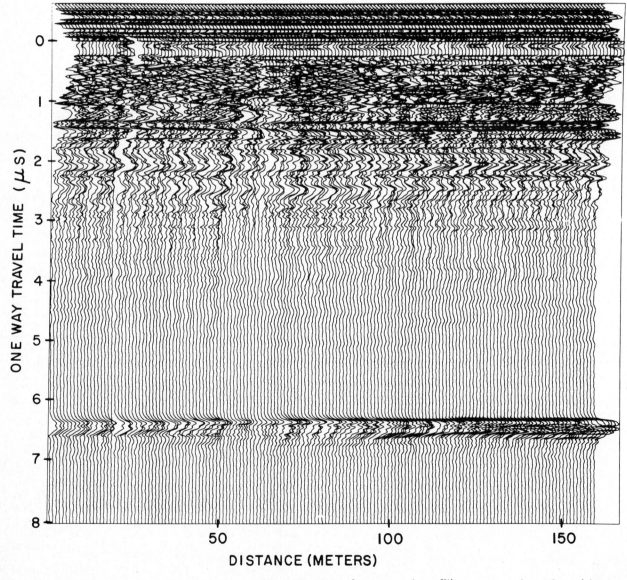

Fig.3. Displays of successive integrated (stacked) traces from ground profiling; survey along the axial direction of ice stream B.

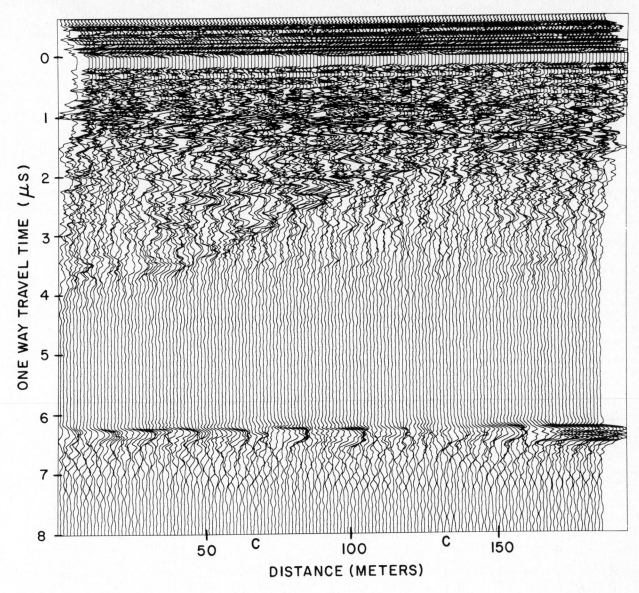

Fig.4. Displays of successive integrated (stacked) traces from ground profiling; survey transverse to the axis of ice stream B.

show more clearly, and occasionally even internal layers in the ice stream can be discerned.

Parts of two recorded ground profiles are shown in Figs 3 and 4. Each figure shows a sequence of integrated traces reconstructed from quadrature samples. Each integrated trace is the result of adding 1024 individual radar traces.

Along the flow direction (Fig.3), internal layers that are laterally continuous and parallel to the surface are clearly evident in the upper two-thirds of the section. In the lower third of the section, the internal layers are less clearly continuous. A train of echoes can be observed at or near the ice-stream bed. The first basal echo is the strongest, and the temporal invariance of this echo indicates a bed that is fairly smooth in the longitudinal direction. Modulation of the echo-train amplitude resulting from surface crevassing is apparent at 11 m and 53–60 m along the profile. A second echo follows the first, but the amplitude of this echo varies more than the primary echo. The fading pattern changes slowly over the 160 m survey distance. A closer examination of this section and accompanying data will be required to identify the sources for these subsidiary echoes at the bed.

A section transverse to the ice-stream axis (Fig.4) shows internal layers similar to those in the longitudinal section for one-way travel times less than 1.5 μs. Internal layers may exist but, if so, they are obscured by interference from structures closer to the surface of the ice

stream. The basal echo train is also more complicated on the transverse section. The echo train from the bed varies over a distance of 10–20 m. Two or more echoes from undulating layers that are marginally resolved may be present throughout the entire section. Approximately 0.1 μs before the main bottom echo is a small-amplitude echo that is spatially discontinuous. This layer also exists above two indentations or channels in the bottom 5-meter-wide (marked "C" in Fig.3). Interpretation of these features is in progress.

ACKNOWLEDGEMENTS

The development of the radar system required the talents of several individuals. In particular, we thank R B Abernathy, W L Unger, B Karsh, R Eastwood, and N Lord. We should also like to thank those who assisted in the acquisition of the radar data: R B Abernathy, S N Stephenson, J Firestone, S T Rooney, R Flanders, and S Anandakrishnan. Financial supprt was provided by the National Science Foundation under grant DPP-8412404. This is contribution No. 464 of the Geophysical and Polar Research Center, University of Wisconsin–Madison.

REFERENCES

Bamber J L 1987 Internal reflecting horizons in Spitsbergen glaciers. *Annals of Glaciology* 9: 5–10

Bentley C R *and 6 others* 1987 Remote sensing of the Ross ice streams and adjacent Ross Ice Shelf, Antarctica. *Annals of Glaciology* 9:

Costrell L 1971 Camac instrumentation system: introduction and general description. *IEEE Transactions on Nuclear Science* NS-18 (2): 3–8

Crabtree R D, Doake C S M 1986 Radio-echo investigations of Ronne Ice Shelf. *Annals of Glaciology* 8: 37–41

Dowdeswell J A, Drewry D J, Cooper A P R, Gorman M R, Liestøl O, Orheim O 1986 Digital mapping of the Nordaustlandet ice caps from airborne geophysical investigations. *Annals of Glaciology* 8: 51–58

Evans S, Smith B M E 1969 A radio echo equipment for depth sounding in polar ice sheets. *Journal of Scientific Instruments (Journal of Physics E)* Ser 2, 2: 131–136

Linton R J 1982 Flexible disk drive closes in on hard-disk densities. *Electronics* 55(8): 117–120

Musil G J, Doake C S M 1987 Imaging subglacial topography by a synthetic aperture radar technique. *Annals of Glaciology* 9:

Neal C S 1982 Radio echo determination of basal roughness characteristics on the Ross Ice Shelf. *Annals of Glaciology* 3: 216–221

Shabtaie S, Bentley C R 1987 West Antarctic ice streams draining into the Ross Ice Shelf: configuration and mass balance. *Journal of Geophysical Research* 92(B2): 1311–1336

Annals of Glaciology 9 1987
© International Glaciological Society

FRACTURES IN ARCTIC WINTER PACK ICE

(NORTH WATER, NORTHERN BAFFIN BAY)

by

K. Steffen

(Department of Geography, Swiss Federal Institute of Technology, Zürich, Switzerland)

ABSTRACT

Profiles of the ice cover in the North Water area were obtained in the winter of 1980/81 by using low-level infrared thermometry. The flight measurements were carried out from December to March. The statistical analysis of the sea ice surface temperature was carried out to yield distributions, frequencies and widths of fractures. Ice-free as well as ice-covered fractures with a maximum ice thickness of 0.4 m were analysed. Typical fracture frequencies were 0.25 per km for Lancaster Sound and 0.14 per km for Baffin Bay and the North Water area, with 90% of fractures being less than 0.6 km wide. From December to March, the fractures occupied 8.8% of the ice cover in the North Water area, 8.7% along the Baffin Bay profile and 10% in the Lancaster Sound. In the North Water area the distance (y) between fractures for different fracture widths (x) is an exponential function of the form y=Aexp(ax) (A,a are constants), for fractures between 50 and 800 m wide. In the North Water area during winter, fractures of all widths occur 5 times more frequently than in M'Clure Strait and about 7 times more frequently than in southern Beaufort Sea. The heat loss in Lancaster Sound at the ice–air interface was found to be 40 to 100% larger due to the fractures compared to a fast ice situation in the same winter.

INTRODUCTION

The occurrence of fractures in pack ice is of major geophysical interest. In sea ice mechanics, the knowledge of, for example, fracture distribution, is important, since each fracture is a potential pressure ridge, and the distribution of fractures is a measure of the capacity of the ice cover to sustain deformation through convergence and shear. A direct application of fracture statistics deals with the openings in Arctic and Antarctic pack ice which are a natural environment for sea mammals during winter and spring. A drastic change in the number of fractures between the years can be fatal for large sea mammals (e.g. whales). In climatology, information on fracture occurrence is vital for energy exchange calculations at the ice–air interface. At this boundary layer the heat flux for ice-free fractures and those covered with young ice is about ten times larger than over first year ice. Other applications are of more practical use — fractures may for instance serve as surfacing sites for submarines, or as landing sites for aircraft, now widely used in support of ice research.

So far, little is known about the number of fractures, their width and their distribution in Arctic winter pack ice. Fracture patterns from infrared satellite imagery (Ackley and Hibler 1974) have been available for some time, but not until recently was infrared satellite imagery suitable for the detection of fractures less than 1 km wide. The best suited instrument platforms for high resolution fracture analysis in winter are submarines or low-flying aircraft. Measurements obtained from low-flying aircraft will be discussed in more detail.

The study was carried out as part of the North Water

Project (Müller and others 1973). North Water, well-known for its young ice and open water areas during winter, caused by the existence of polynyas (Steffen 1985[a]), is situated between Greenland and the Canadian islands, Ellesmere and Devon Island (Fig.1). The objective of the present paper is (1) to report on the number of fractures, as well as their width and distribution in winter pack ice of the North Water region, and (2) to give some estimates of the energy loss through fractures during the winter months.

METHODS AND MEASUREMENTS

In winter 1980/81, six low level remote sensing flights were carried out over the North Water area. The sea surface temperature (SST) was measured from a flight altitude of 300 m with an infrared sensor (Barnes PRT-5) along profiles in Lancaster Sound and the northern Baffin Bay (Fig.1). The radiation temperature measured with the PRT-5 intrument in the spectral range 9.5 to 11.5 μm was corrected by taking into account the emissivity of water (ϵ=0.991), snow (ϵ=0.997) and ice (ϵ=0.987); the absorption and emission by atmospheric water vapour as well as the multiple reflection between the cloud base and the ground. The temperature dependence of the emissivity for dry snow was neglected. Young ice surfaces remain wet due to the high brine content in the thin ice. Consequently, brightness temperature measurements of sea ice up to 0.25 m thick were corrected by using the emissivity of water. A detailed account of this correction procedure is presented in Steffen (1985[b]). After having applied the corrections, the accuracy of the PRT-5 temperature was estimated at ±0.15°C at a flight altitude of 300 m.

Fractures in the pack ice could be detected due to the SST-difference between open water and ice surfaces. However, only fractures wider than 50 m could be classified because of the ground resolution of the instrument. Fractures covered with young ice (<0.4 m thick) still have a large conductive heat flux at the ice–air interface (Fig.2). Therefore, not only ice-free fractures are of importance for energy flux calculations over pack ice regions. The conductive heat flux, the only energy source over Arctic winter pack ice, amounts to 840 W m^{-2} in ice-free fractures and to approximately 125 W m^{-2} for fractures covered with young ice (Maykut 1978: January situation, air temperature -30°C). The conductive heat flux decreases to about 20 W m^{-2} for ice 2 m thick.

Both sensible and latent heat fluxes are dependent on surface temperatures, which in turn are influenced by the ice thickness. From the SST-measurements, the ice thickness can be calculated by the parametrisation of the energy balance equation (Steffen 1986). The following ice types could be classified: dark nilas, light nilas, grey ice and grey-white ice. Therefore, the occurrence frequency and distribution of ice-free and ice-covered fractures (maximum ice thickness: 0.4 m) could be calculated from the SST-measurements. The analysis was carried out for six flight dates from December to March for the following areas: Lancaster Sound (Fig.1, point 5 to 7), Baffin Bay

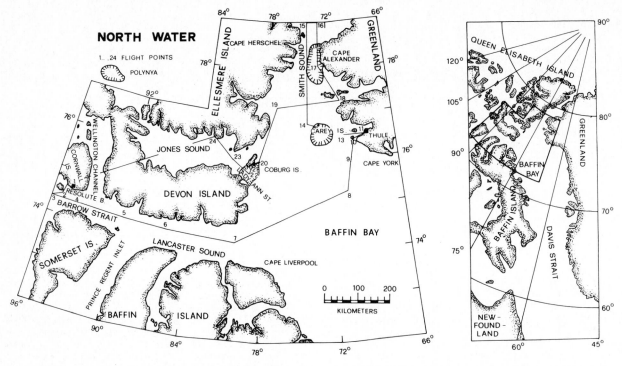

Fig.1. Map of North Water region and profile of remote sensing. Note position and average extension of the polynyas in Smith Sound, around the Carey Islands and in Lady Ann Strait during winter 1980/81.

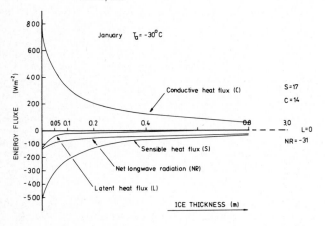

Fig.2. Energy fluxes and ice thicknesses at the ice–air interface in January for an air temperature of −30°C (after Maykut 1978).

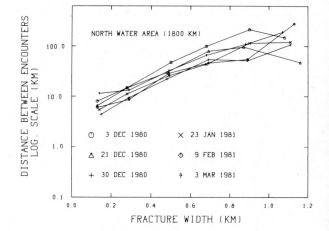

Fig.3. Relation between encounters with fractures and fracture widths in the North Water area during winter 1980/81.

(point 7 to 9), and North Water area (point 13 to 23). The polynyas in Smith Sound, Lady Ann Strait and around the Carey Islands, all located between the flight points 13 to 23 (Fig.1), were excluded from the fracture analysis. The fractures were classified according to their width: (1) 50 − 200 m, (2) 200 − 400 m; (3) 400 − 600 m; (4) 600 − 800 m, (5) 800 − 1000 m, (6) 1000 − 1200 m. Fractures in class (1) are called small fractures, those in class (2) medium fractures and those in the classes (3) to (6) large fractures, according to the ice classification (WMO 1970).

RESULTS AND DISCUSSIONS

For pack ice in the North Water area, the distances between fractures for different fracture widths are plotted in Fig.3. The analysed profile, which excluded polynyas in Smiffh Sound and around Carey Islands, has a total length of 1800 km. Only fractures between 50 m and 1.2 km in width were analysed; those below 50 m could not be detected due to the instrument resolution; fractures wider than 1.2 km were rare. For fractures of 50 to 200 m wide, the distance between encounters varies from 5 to 11 km. There was no systematic change observed in fracture distribution during winter. The relation between distance of

encounters between fractures (y) and fracture widths (x) was an exponential of the form $y=A\exp(ax)$ (a=1.786, A=3.899) for fracture widths of up to 0.8 km as a mean of six flights in winter 1980/81. Fractures between 1 and 1.2 km in width were found with spacings varying from 50 to 280 km. For a statistically relevant statement on fractures wider than 800 m, the profile length should be much longer. There is no reason to expect an exponential law or any other simple analytical form to be valid for all fracture width classes. Rothrock and Thorndike (1984) have shown that the floe size distribution (mean caliper diameter and number of floes per unit area) can be approximated by a power law for certain Arctic pack ice regions. Further, they showed other possible distribution laws which are of some interest for this work as floe size and fractures are somehow related to each other. However, for a detailed discussion more data on fracture-width frequency from different regions and years are needed. Fig.4 depicts the situation for Lancaster Sound, Baffin Bay and the North Water profile as a mean for the six flights between December and March. Lancaster Sound is 300 km long, borders with Barrow Strait in the west and leads into Baffin Bay in the east. As the narrow channel is only

60 km wide, pack ice movement is limited by its coastal boundaries. Pack ice in Baffin Bay by contrast has more freedom to move around due to the polynyas in Smith Sound and around the Carey Islands. However, shortest spacings between fractures were found in Lancaster Sound. Fractures along Baffin Bay and the North Water profile had an almost identical spacing for fracture widths of up to 0.5 km.

Typical fracture frequencies were 0.25 per km for Lancaster Sound and 0.14 km for the Baffin Bay and the North Water area. For the three areas in discussion about 55% of all fractures were between 50 m and 200 m wide, about 25% between 0.2 and 0.4 km, and 10% between 0.4 and 0.6 km (Fig.5). The occurrence frequency of fractures 0.6 to 1.2 km wide was less than 10%.

During the AIDJEX pilot study in April 1972 high resolution aerial photographs were taken along profiles of 600 km in length in the Beaufort Sea. The visible interpretation of the photographs showed that on the average every 4 km a fracture occurred which was less than 200 m wide (Hall 1980). This fracture-width frequency measured in the Beaufort Sea in April is similar to the one found for the North Water region during winter 1980/81. However, the AIDJEX study includes fractures less than 50 m in width and therefore the two data sets can not be compared directly. Further, it is not clear which fracture ice types (age of fracture) were included in the AIDJEX study.

The fracture distribution along Lancaster Sound, Baffin Bay and the North Water profile is comparable to the results published for M'Clure Strait (Wadhams and Horne 1980) and the Beaufort Sea (McLaren and others 1983). M'Clure Strait is located at the west end of the North West Passage, whereas Lancaster Sound is at its eastern end. M'Clure Strait leads into the Beaufort Sea, which is part of the Arctic Ocean. The fracture analysis from the western part of the Canadian Arctic was obtained from submarine vessels by using a narrow-beam upward-looking sonar. For the sonar measurements a fracture was defined as a continuous sequence of depth points in which no point exceeds 1 m in draft (McLaren and others 1984). The results of Wadhams and Horne (1980) and McLaren and others (1984) include fractures with an ice cover of up to 1 m compared to our analysis, which includes a maximum ice thickness of 0.4 m.

Fig.4 shows the difference between the fracture distribution in winter and summer in M'Clure Strait (including the neighbouring Beaufort Sea shelf) as well as the winter situation for the southern Beaufort Sea, the North Water area, Lancaster Sound and the northern Baffin Bay. In M'Clure Strait, fractures of all widths occur almost ten times more frequently in summer than in winter. Compared to the North Water area, the fracture distribution in M'Clure Strait is about 4 to 5 times less frequent in

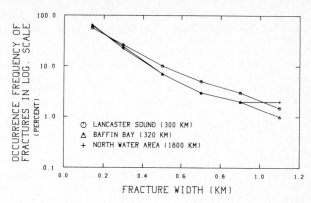

Fig.5. Occurrence frequency of fractures for fracture width between 50 m and 1.2 km in Lancaster Sound, northern Baffin Bay and the North Water area during winter 1980/81 (mean of six measurements).

winter. It is surprising that in M'Clure Strait there are only twice as many fractures in summer as in the North Water in winter. For the southern Beaufort Sea, the fracture distribution in spring is even 7 times smaller compared to the North Water in winter. This can be explained by the influence the polynyas (Smith Sound, Carey Islands, Lady Ann Strait) have on the pack ice cover in northern Baffin Bay and Lancaster Sound. The polynyas are ice-free or covered by young ice throughout winter (Steffen 1985[b]; 1986), which has a direct influence on the surrounding pack ice regions.

The energy loss caused by open water and the thin ice covering the fractures was calculated for the Lancaster Sound (profile points 5 to 7, Fig.1) and on the basis of the heat flux rates for different ice types (Maykut 1978) and the fracture analysis along the flight profile. From December to March, the calculated heat loss on the ice–air interface due to fractures is 40 to 100% larger than for a fast-ice situation. The mean heat loss over the fractures in Lancaster Sound was 240 W m^{-2} for the six flight dates in winter 1980/81.

SUMMARY AND CONCLUSIONS

The distribution of fracture frequency and widths has been found to fit an exponential function y=Aexp(ax), with y as distance between fractures of the same width in km, x as fracture width in km and A,a empirical constants (a=1.786, A=3.899) for 0.05<x<0.8. This function is valid for the North Water area and northern Baffin Bay. For Lancaster Sound, the constants are: a=2.036 and A=2.065. This parametrisation of the fracture distribution might be useful for energy flux modelling in pack ice regions which are influenced by the existence of recurring polynyas. The analysis of the fracture distribution showed that during winter (December to March), fractures occupied 8.8% of the surface in the North Water area, 8.7% along the Baffin Bay profile and 10% in Lancaster Sound. In the North Water area in winter, fractures of all widths occur 5 times more frequently than in M'Clure Strait and about 7 times more frequently than in southern Beaufort Sea in winter. The fracture occurrence of the North Water region in winter is almost the same as in M'Clure Strait in summer.

ACKNOWLEDGEMENTS
Flight time and the logistic support in the field were generously provided by the Polar Continental Shelf Project of the Department of Energy, Mines and Resources of Canada. Financial support for the project came from the Government of Canada (Contract No. OSU76-00151), the US National Science Foundation (Contract No. DPP-7826132), the Swiss National Science Foundation (Grant No. 2.8807.-0.77) and Petro Canada Ltd., Calgary, Canada.

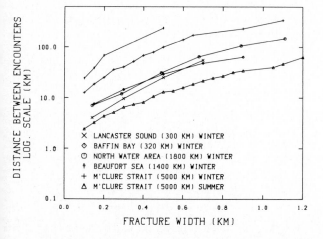

Fig.4. Relation between encounters with fractures and fracture widths in North Water area (point 13-23, Fig.1), Lancaster Sound (point 5-7), northern Baffin Bay (point 7-9), M'Clure Strait (after McLaren and others 1984) and Beaufort Sea (after Wadhams and Horne 1980)

REFERENCES
Ackley S F, Hibler W D III 1974 Measurements of Arctic ocean ice deformation and fracture patterns from satellite imagery. *AIDJEX Bulletin* 26: 33-47

Hall R T 1980 AIDJEX modeling group studies involving remote sensing data. *International Association of Hydrological Sciences Publication* 124 (Symposium of Seattle 1977 — *Sea Ice Processes and Models*): 151–162

McLaren A S, Wadhams P, Weintraub R 1984 The sea ice topography of M'Clure Strait in winter and summer of 1960 from submarine profiles. *Arctic* 37(2): 110-120

Maykut G A 1978 Energy exchange over young sea ice in the central Arctic. *Journal of Geophysical Research* 83(C7): 3646-3658

Müller F, Ohmura A, Braithwaite R J 1973 Das North Water Projekt (kanadisch-grönländische Hocharktis). *Geographica Helvetica* 28(2): 111-117

Rothrock D A, Thorndike A S 1984 Measuring the sea ice floe size distribution. *Journal of Geophysical Research* 89(C4): 6477–6486

Steffen K 1985[a] Surface temperature and sea ice of an Arctic polynya: North Water in winter; Canadian and Greenlandic high Arctic. *Zürcher Geographische Schriften* 19

Steffen K 1985[b] Warm water cells in the North Water, northern Baffin Bay during winter. *Journal of Geophysical Research* 90(C5): 9129-9136

Steffen K 1986 Ice conditions of an Arctic polynya: North Water in winter. *Journal of Glaciology* 32(112): 383–390

Wadhams P, Horne R J 1980 An analysis of ice profiles obtained by submarine sonar in the Beaufort Sea. *Journal of Glaciology* 25(93): 401-424

WMO 1970 *Sea ice nomenclature*. Geneva, Secretariat of the World Meteorological Organization (WMO/OMM/BMO 259 TP 145)

Annals of Glaciology 9 1987
© International Glaciological Society

USE OF REMOTE-SENSING DATA IN MODELLING RUN-OFF FROM THE GREENLAND ICE SHEET

by

H.H. Thomsen and R.J. Braithwaite

(Grønlands Geologiske Undersøgelse, Øster Voldgade 10, DK 1350, København K, Denmark)

ABSTRACT

Run-off modelling is needed in Greenland to extend the short series of measurements. However, the delineation of hydrological basins on the Greenland ice sheet is difficult because of the lack of information about surface and subglacial drainage patterns. Low Sun-angle Landsat data have been used for mapping local surface features which has led to an improvement in basin delineations and thereby run-off simulations. Work is now in progress to map subglacial topography by electromagnetic reflection (EMR) from a helicopter. This information will be used for calculating hydraulic potentials within the basin and to assess the possibilities of future changes in drainage-basin delineation.

INTRODUCTION

The possibilities of developing hydroelectric power in Greenland have been investigated since the oil crisis of the 1970s. The locations of possible hydropower basins, denoted by four-letter codes, are shown in Fig.1. Although run-off measurements have been started in all these basins, the measured series are still too short to give a reliable basis for planning. This can be overcome by using run-off – climate models to calculate past run-off from climatological data and thereby to extend the measured series (Braithwaite 1984). This involves the calculation of specific run-off which is then integrated over the whole area of the drainage basin to give the run-off volume. The purpose of the present paper is to point out that there can be problems in determining areas of hydrological basins in Greenland, and that remote-sensing data are useful for solving these problems.

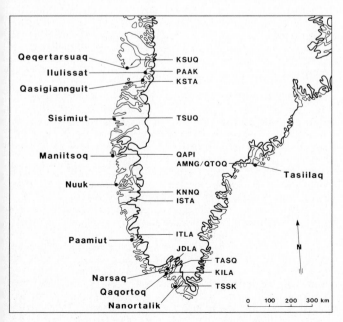

Fig.1. Basins in Greenland where hydro-electric power might be developed.

Most of the basins shown in Fig.1 are conventional hydrological basins bounded by impervious bedrock and there are no serious problems in delineating their boundaries using available topographic maps. However, four of the basins, i.e. PAAK, KSTA, ISTA, and ITLA, contain sectors of the inland ice sheet which makes it difficult to delineate their boundaries because drainage occurs on, within, and under the ice. The surface topography of the ice sheet is poorly mapped, making it difficult to determine the surface-drainage pattern, while the subglacial topography is generally unknown so that it is impossible to determine the englacial and subglacial drainages. The existence of large errors in drainage-basin areas can be illustrated by the results of run-off calculations for the PAAK and KSTA basins which were made at the start of the present project using surface contours from the available maps, i.e. the 1 : 1 000 000 ICAO and 1 : 250 000 Danish Geodetic Institute map series. The calculated run-off for the two basins over-estimated the measured run-off by 121 and 81%, respectively (Braithwaite and Thomsen 1984[a], [b]). As described below, these results were very much improved subsequently by the use of remote-sensing information.

DRAINAGE CONDITIONS AND THE NEED FOR DATA

Melting on the Greenland ice sheet generally starts in May and lasts until August or September, with greatest run-off in July to August. The melt water decreases with elevation, mainly due to the decrease of temperature. The first phase of melting at any elevation involves the removal of winter snow cover (if any) followed by the melting of glacier ice. Run-off from melting ice is almost instantaneous while run-off from melting snow may be hindered, or even prevented, in various ways. For example, water from a melting snow surface may be refrozen within or under the snow cover, i.e. as ice lenses or as superimposed ice, so that it will not contribute to run-off. There are also large areas of the ice sheet with relatively gentle slopes where melt water can be stored for shorter or longer times in slush fields and snow-dammed lakes. These tend to lie around the transient snow line and move to progressively higher elevations with time. Drainage areas on the ice sheet therefore increase with time and have their maximum extent at the end of the summer.

After the winter snow cover has been melted away, melt water drains through a system of rivers on the ice surface which generally follows hollows on the often strongly undulating surface. In some cases, rivers run in canyons up to 6 m deep which cut through local ridges on the surface and can even lead the water against the trend of surface topography. Rivers sometimes run on the surface directly to the ice margin but more often the melt water disappears into the ice through moulins or crevasses several kilometers from the ice margin. The surface drainage therefore depends on the general surface topography, local surface topography reflecting subglacial topography, and structural features such as shear bands, healed crevasses, and ridges indicating the ice-flow pattern.

The melt water drainage within and beneath glaciers is very difficult to study because of the inaccessibility of this

environment. Björnsson (1982) described a theory for sub-glacial drainage which shows that water flows down the gradient of hydraulic potential which depends upon both surface and subglacial topography, so that marked ridges and valleys under the ice also influence the drainage pattern.

Detailed maps of surface and subglacial topography are therefore needed for a description of the overall drainage pattern and a delineation of drainage areas.

DRAINAGE-BASIN DELINEATION BY REMOTE SENSING

Detailed mapping of surface and bottom topography over large areas of the inland ice sheet would be a time-consuming and expensive process beyond the limits of the present project. However, remote sensing offers a quick and cheap way to extract some of the information needed for improved run-off simulations. For example, Landsat MSS data recorded under conditions of low Sun angle have been used in glacier-covered areas to map surface features which reflect the subglacial topography (Krimmel and Meier 1975; Thorarinsson and others 1973).

For the present study, digital Landsat MSS data recorded late in the melt season were chosen. The image processing was done at the DK.IDIMS facility at the Electromagnetics Institute, Technical University of Denmark. The data analysis included geometrical correction and enhancement to increase the contrast ratios in the images so that individual surface features can be more clearly differentiated (Thomsen 1983[a], [b]). The data were plotted on an Applicon ink plotter to give Landsat-image maps which were used for interpretation and later plotting of maps of surface features (Thomsen 1986). These maps show surface features related to ice and melt-water drainage as well as' the local topography which reflects the subglacial topography.

So far, maps of surface features have been made for the PAAK, KSTA, and ISTA basins (Thomsen 1983[b]). In all three cases they led to a radical change in the basin delineation. This is illustrated by Fig.2, which shows the delineation of the KSTA basin before (top figure) and after (bottom figure) taking account of information from the surface feature map. The large-scale contour lines in the two cases are identical.

The above example shows that the hydrological drainage area is smaller than first thought. However, there is still an element of subjectivity in inferring the direction of drainage and hydrological boundaries, i.e. signatures (7) and (8) in Fig.2, from the surface feature information, i.e. signatures (3) to (6). Two different estimates of drainage area, corresponding to maximum and minimum estimates, were therefore made and used as input data for further run-off simulations.

RUN-OFF SIMULATION

Simulations of run-off are made by calculating specific run-off and then integrating it over the presumed area of the drainage basin to give the run-off volume. The specific run-off at various elevations is calculated from temperature and precipitation data which are extrapolated from long-term climate stations on the coast of Greenland: Ilulissat for the PAAK and KSTA basins, Nuuk for the ISTA basin, and Paamiut for the ITLA basin. The calculations are made separately for glacier-covered and glacier-free conditions by the MB1 (mass-balance) and SM1 (snow-melt) models of Braithwaite (1984). Results from the two models are then passed into a third model which calculates the total run-off from the basin after the elevation distributions of glacier-covered and glacier-free areas within the basin have been specified.

The details of how to calculate specific run-off from climate data using the MB1 and SM1 models are beyond the scope of the present paper as they have been briefly discussed by Braithwaite (1984) and in Braithwaite and Thomsen (1984[a], [b]), and will be fully described elsewhere. The point of the present discussion is that the calculated run-off volume is sensitive to assumptions about the distribution and total amount of ice cover in basins

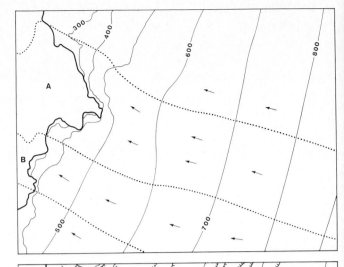

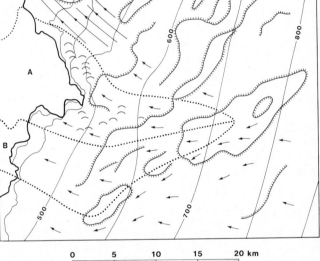

Fig.2. Sketch map showing the marginal part of the Greenland ice sheet draining to the ice-free areas A and B. The upper map shows basin delineation based only on a large-scale surface topography. The lower map shows drainage-basin delineation also taking into account inferred subglacial topography. (1) Ice margin, (2) contour line, (3) flow lines, (4) marked positive relief, (5) marked change in slope, (6) undulating terrain, (7) inferred direction of melt-water drainage, (8) inferred hydrological boundary.

including sectors of the ice sheet. This is illustrated by Fig.3, which shows the relation between calculated run-off and assumed area of the PAAK drainage basin. There are two curves corresponding to different estimates of the degree-day factor used to calculate ablation from temperature (Braithwaite and Olesen 1985).

The curves in Fig.3 are constructed by making run-off simulations for different versions of the PAAK drainage basin; the original version based only upon large-scale contours and the maximum and minimum estimates based upon the remote-sensing information. The climate data used in this case are the averages for the period 1980–84 for which the run-off from the basin has been measured. The measured annual run-off for the 5 years is on average 280 x 10^6 m³ which, according to Fig.3, corresponds to a basin area of between 380 and 200 km². These hydrologically based estimates are in rough agreement with the maximum and minimum estimates of the basin area using remote-sensing data, i.e. 395 and 265 km², and are much less than the original delineation with an area of 788 km².

The remaining uncertainties in the size of drainage

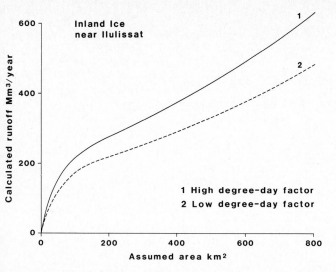

Fig.3. Relation between calculated annual run-off and assumed drainage area for the PAAK basin, Ilulissat, West Greenland.

basins on the ice sheet can be overcome by calibrating the run-off model with the few years of measured run-off data and then using the calibrated model to reconstruct past variations in run-off (Braithwaite and Thomsen 1984[a], [b]). However, this reconstruction assumes a static relation between climate and run-off which may not be true. For example, if the present retreat of the ice margin continues, the ablation area of the basin will get smaller and there may also be a diversion of the drainage pattern.

WORK IN PROGRESS

Work is now in progress to improve the delineation of drainage basins on the ice sheet and to assess the possibilities of future changes. For example, previous work involved the inference of bottom topography from low Sun-angle Landsat images, while the present programme includes the direct measurement of bottom topography by electromagnetic reflection (EMR) techniques. Despite some initial problems (Thomsen and Madsen 1985), detailed profiles of ice thickness have now been measured in the PAAK basin by Thorning and others (1986). The measurements were made with a 300 MHz radar (developed by the Electromagnetics Institute, Technical University of Denmark) mounted in a small helicopter.

The EMR results are still being analysed but it is planned to make a map of subglacial topography which, combined with an improved map of surface topography by Thomsen (1986), will be used for calculation of hydraulic potentials following the method of Björnsson (1982). This should give a more objective basis for drainage-basin delineation than the method used at present. It may also be possible to extrapolate the present thinning of the ice margin into the future to see whether the drainage-basin delineation will change.

ACKNOWLEDGEMENTS

This paper is published by permission of the Director, the Geological Survey of Greenland. The work is a contribution to investigations of hydro-electric power in Greenland and was partly funded by the European Economic Community (EEC) through the European Regional Development Fund. The digital image processing was done at the DK.IDIMS facility, Electromagnetics Institute, Technical University of Denmark. The Electromagnetics Institute also loaned a 300 MHz radar set which was operated by Leif Thorning and Egon Hansen of the Geological Survey of Greenland.

REFERENCES

Björnsson H 1982 Drainage basins on Vatnajökull mapped by radio echo soundings. *Nordic Hydrology* 13(4): 213–232

Braithwaite R J 1984 Hydrological modelling in Greenland in connection with hydropower. *Grønlands Geologiske Undersøgelse. Rapport* 120: 90–94

Braithwaite R J, Olesen O B 1985 Ice ablation in West Greenland in relation to air temperature and global radiation. *Zeitschrift für Gletscherkunde und Glazialgeologie* 20, 1984: 155–168

Braithwaite R J, Thomsen H H 1984[a] Runoff conditions at Kuussuup Tasia, Christianshåb, estimated by modelling. *Grønlands Geologiske Undersøgelse. Gletscher-hydrologiske Meddelelser* 84/2

Braithwaite R J, Thomsen H H 1984[b] Runoff conditions at Paakitsup Akuliarusersua, Jakobshavn, estimated by modelling. *Grønlands Geologiske Undersøgelse. Gletscher-hydrologiske Meddelelser* 84/3

Krimmel R M, Meier M F 1975 Glacier applications of ERTS images. *Journal of Glaciology* 15(73): 391–402

Thomsen H H 1983[a] Glaciological applications of Landsat images in connection with hydropower investigations in West Greenland. Proceedings of the EARSeL/ESA Symposium on Remote Sensing Applications for Environmental Studies. *ESA Special Publication* 188: 133–136

Thomsen H H 1983[b] Satellitdata – et redskab til studier af indlandsisens randzone i forbindelse med vandkraftundersøgelser. *Grønlands Geologiske Undersøgelse. Gletscher-hydrologiske Meddelelser* 83/8

Thomsen H H 1986 Photogrammetric and satellite mapping of the margin of the inland ice, West Greenland. *Annals of Glaciology* 8: 164–167

Thomsen H H, Madsen P S 1985 Radio ekko målinger af indlandsisens randzone i Disko Bugt området 1984. *Grønlands Geologiske Undersøgelse. Gletscher-hydrologiske Meddelelser* 85/1

Thorarinsson S, Sæmundsson K, Williams R S Jr 1973 ERTS-1 image of Vatnajökull: analysis of glaciological, structural, and volcanic features. *Jökull* 23: 7–17

Thorning L, Thomsen H H, Hansen E 1986 Geophysical investigations over the margin of the inland ice at Påkitsoq. *Grønlands Geologiske Undersøgelse. Rapport* 130: 114–121

Annals of Glaciology 9 1987
© International Glaciological Society

IMPULSE RADAR SOUNDING OF FOSSIL ICE WITHIN THE KURANOSUKE PERENNIAL SNOW PATCH, CENTRAL JAPAN

by

K. Yamamoto

(Water Research Institute, Nagoya University, Furo-cho, Chikusa-ku, Nagoya 464, Japan)

and

M. Yoshida

(Hakusan Industry Co. Ltd., Musashidai 1 - 18, Fuchu-shi, Tokyo 183, Japan)

ABSTRACT

The impulse radar was found to be useful in surveying the internal structures of wet snow patches in mountain regions. Radar profiles revealed that the thickness of the perennial ice was 30 m, possibly the thickest in Japan. An unconformity widely extended nearly parallel to the surface at the depth of 2 to 9 m and divided the ice into two parts.

1. INTRODUCTION

The Kuranosuke snow patch is located in the Mt. Tateyama (3015 m) region in the Northern Japanese Alps, where the snow accumulation in winter is about 15 m and the snow melt in summer is nearly the same. It has been known that moulin-like vertical holes appear once every few years after the firn has melted. Yoshida and others (1983) investigated these holes in detail and considered that there had been crevasses or cracks in the snow patch in the past.

In the present work, the internal structure and the bottom topography of this snow patch were investigated to understand the mechanism of its formation.

2. DIRECT OBSERVATION OF THE PERENNIAL ICE STRUCTURE

M. Yoshida entered into a vertical hole (Y in Fig.1) and observed the stratification and the grain size of the

perennial ice in October 1979. The vertical cross-section of the hole and the stratification are illustrated in Figure 2(a). The hole was 19 m deep. The dip of the layer boundary changed from 10° near the surface to 30° near the bottom. The thickness of the layers became narrower with increasing depth. The grain diameters were measured and found to be 10–40 mm near the bottom.

The same observations were carried out at two holes, J and L in Fig.1, in September 1980. The surface of the perennial ice at that time was 1 m lower than that in October 1979. It was seen at both holes that the dip of the boundary layer abruptly changed across the gravel layer at the depth of 4 m.

Some core samples were taken in August 1980 to examine the stratification and the grain size profile. The results for one of them are illustrated in Fig.2(b). In this figure, granular snow covering the perennial ice is not shown. The level of the ice surface was approximately the same as that in October 1979. The stratification was quite different between both sides of the gravel layer at the depth of 5 m. Furthermore, the grain diameter of the upper ice was nearly uniform and approximately 1.5 mm, while that of the lower ice was widely scattered and there were large grains up to 30 mm. These suggest this gravel layer is an unconformity. This was also identified in the other core samples.

It is reasonable to consider that the gravel layers at a

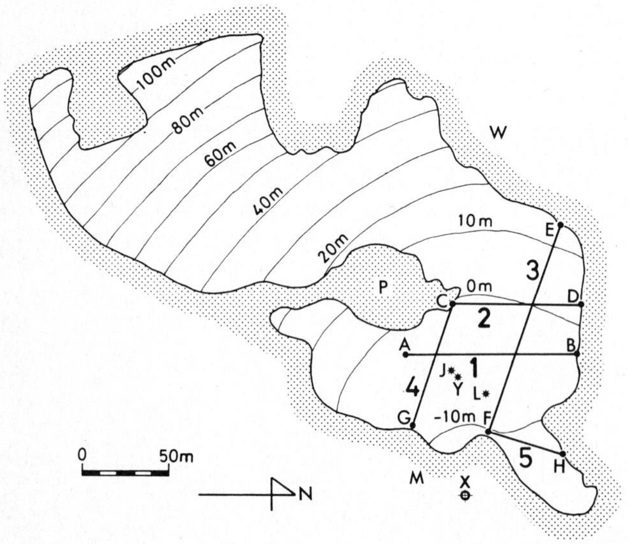

Fig.1. Topographic map of Kuranosuke snow patch. Symbols: 1–5: measuring lines; A–H: end points of lines; J, L and Y: vertical holes; M: end moraine; P: pro-talus rampart; W: cirque wall.

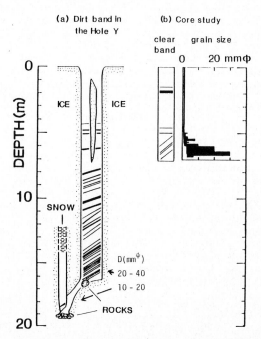

Fig.2. Stratification and grain-size profile.

depth of 4 m, observed in the holes J and L, are identical with the unconformity found in the core samples. In the case of the hole Y, it was impossible to confirm from the observational results whether the dirt layers at the depth of 5 or 7 m were the unconformity. The radar profiles described later suggest the layer at depth of 7 m to be the unconformity.

The grain size of the ice was very different on opposite sides of the unconformity. According to Tusima's (1978) experimental formula on the recrystallization rate of ice grains immersed in water, the formation age of the lower ice is at least 1000 years if it has recrystallized without non-hydrostatic stress. It is interesting to know the kind of climatic and stress conditions under which the ice has been formed.

3. RADAR SOUNDING
3.1. Procedure
Radar echo-sounding for the bottom topography and the internal structure of the perennial ice was carried out in September 1983. The topographic map of the snow patch at that time is shown in Fig.1. No vertical hole appeared in that year. The level of the snow patch's surface was close to that in October 1979. The soundings were made along five lines, 1-5, as shown in Fig.1.

The impulse radar used in the present investigations was a model KSD-2A developed by KODEN Electronics Co. Ltd. The main feature of the equipment is summarized in Table I. The equipment consists of eight blocks: two antenna units (for 140 and 400 MHz), a control unit, an oscilloscope for A-scope, an image processor unit, a coloured CRT unit for B-scope, a cassette data recorder and an engine generator.

The 140 MHz antenna was set on a snow boat and moved along measuring lines at as constant speed as possible. The control unit and the data recorder were carried by an investigator. The monitoring units were unconnected during traverse because of easy movement. Echo data recorded on cassette magnetic memory tapes were reproduced on A- and B-scopes and analyzed after returning to the laboratory. Digital image processing techniques were also used.

3.2. Dielectric constants of snow and ice
The dielectric constant of wet snow depends on its density and water content and the frequency of the electromagnetic wave used for measurements. The constant,

however, was not measured. Then, the permittivity of snow was taken to be 4.0, which was derived from our permittivity measurements using the same equipment on various types of snow cover (Yamamoto and others 1986). According to the theoretical formula of Yamamoto and others (1984), this value is realized in snow having the wet density of 0.53 Mg/m^3 and the water content of 15 weight%.

The dielectric constant of ice also depends on its density and water content and the frequency. It is, however, shown from both a theoretical consideration and an impulse radar measurement at the ice body formed within a lava cave called "Fuji fuketu", that 3.2 (the permittivity of pure ice at very high frequency) can be used for this purpose (Yamamoto and others 1986). Under these approximations, we suspect that the error of the estimated depth of interfaces is within 10%.

3.3 Observation results
A radar profile along line 1 is shown in Fig.3. In the figure, the horizontal line (ss) at the travel time of 0 ns shows the surface of the snow patch. The snow/ice interface (si) is observed at depths from 1 to 3 m. The strong intensity from the unconformity (u) described above is seen at the depths from 4 to 8 m. It is reasonable to consider that the gravel deposited on the unconformity gives rise to strong reflection. The bottom of the ice body is bounded by the foot of a pro-talus rampart at distances from 0 to 35 m and that of a cirque wall from 70 to 100 m. The slope of the cirque wall is so steep that its echo is very weak. The strong echo from bedrock is observed at distances between 25 and 90 m. It becomes deeper under the pro-talus rampart and the cirque wall, since the permittivity of the detritus is higher than that of ice. In this figure, the maximum depth of the snow patch is approximately 30 m at a distance of 70 m. It is possibly the thickest perennial ice in Japan. No echo shows that fossil ice is buried under the pro-talus rampart and/or cirque wall.

In Fig.3, at least three interfaces (a, b, and c) are distinguished in the ice body under the unconformity. Simple geometrical analyses show that they are not ghost echoes from the pro-talus rampart, the cirque wall and the end moraine. It is reasonable to consider that these interfaces are gravel layers, since such strong echoes cannot occur due to other layer boundaries. In the cross-section of Fig.3, the unconformity and gravel layers form a south-facing slope.

TABLE I. THE MAIN FEATURE OF A KODEN IMPULSE RADAR, MODEL KSD-2A (FACILITIES NOT USED IN THE PRESENT INVESTIGATIONS ARE NOT DESCRIBED)

Antenna system and type	monostatic bow-tie type
Transmission pulse	
Pulse shape	monocycle sinusoidal
Pulse width	6 ns
Centre frequency	140 MHz
Peak power	50 W
Pulse repetition rate	50 kHz
Receiving system	
RF gain	40 dB
LF gain	40 dB with the sensitivity time control
Display	
A-scope	
B-scope (colour CRT)	
Image processing	smoothing, space filter, correlation, Rosenfeld method, difference method
Data record	
Medium	cassette magnetic memory tape
Available input	echo, synchronous, displacement, voice signals

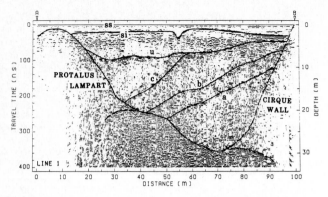

Fig.3. Radar profile of line 1. Symbols are explained in the text.

Radar profiles of lines 2-5 were also analyzed. The details have already been described elsewhere (Yamamoto and others 1986). Figs 4 and 5 show the estimated sections of the internal structure and bottom topography of the snow patch under lines 2 and 3, respectively. The unconformity is raised 5 m near the distance of 20 m of line 2 (Fig.4). Three gravel layers (d, e and f) were distinguished from the radar profile of line 2.

Line 3 was set approximately parallel to the fall line. The snow/ice interface, the unconformity and five gravel layers were identified as shown in Fig.5. The depths of these interfaces in Fig.5 coincide with those in Figs 3 and 4 at the points of intersections between lines 1 and 3, and between lines 2 and 3. Figs 3, 4 and 5 show that the unconformity is widely extended, nearly parallel to the snow patch surface and at depths from 2 to 9 m. The radar profiles of lines 4 and 5 showed that the unconformity did not exist under these lines.

4. DISCUSSION AND CONCLUSIONS

The impulse radar was found to be useful in surveying the internal structure of wet snow patches.

The stratification of vertical hole Y, and the radar profile at the distance of 35 m along line 1 agree well with each other. The unconformity and the gravel layers in the radar profile coincide in depth with thick dirt layers.

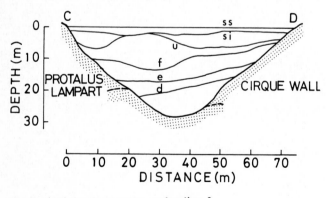

Fig.4. Estimated structure under line 2.

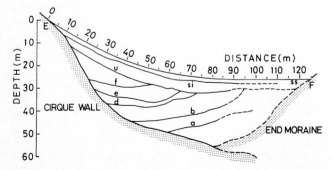

Fig.5. Estimated structure under line 3.

The unconformity bent under line 2. It is reasonably explained that the bend was caused from enhancement and reduction of snowmelt due to deposition of detritus on the surface of the snow patch (Yamamoto and others 1986). This process plays an important role in the layered structure formation of the snow patch.

The radar stratifications and the grain-size profile show that there exist two ice bodies of different age, separated by an unconformity. It is possible to infer that the lower ice is old glacier ice, although there is no active glacier at present in Japan. The ice might have been prevented from surface melting by the heavy snow accumulation in the winter and the thick debris cover. Furthermore, bottom melting might have been small due to the existence of permafrost, which is conceivable, as this location is close to the lower limit for discontinuous permafrost. Core samples will be taken to determine the formation age of the bottom ice to confirm the above.

ACKNOWLEDGEMENT

The authors wish to thank Chikyu-Kogaku-Kenkyujo Ltd for help with transport of equipment. This work is partly supported by a Grant-in-Aid for Scientific Research 61540301 from the Ministry of Education, Science and Culture and a grant from the National Institute of Polar Research. The digital image data was processed by the use of the system in the Computation Centre of Nagoya University.

REFERENCES

Tusima K 1978 Grain coarsening of ice particles immersed in pure water. *Seppyo* 40(4): 155-165 [In Japanese with English abstract]

Yamamoto K, Kusu S, Higuchi K 1984 Dielectric properties of wet snow in the microwave region. *Seppyo* 46(1): 1-9 [In Japanese with English abstract]

Yamamoto K, Iida H, Takahara H, Yoshida M, Hasegawa H 1986 Impulse radar sounding in Kuranosuke snow patch, central Japan. *Seppyo* 48(1): 1-9 [In Japanese with English abstract]

Yoshida M, Fushimi H, Ikegami K, Takenaka S, Takahara H, Fujii Y 1983 Distribution and shape of the vertical holes formed in a perennial ice body of the Kuranosuke snow patch, central Japan. *Seppyo* 45(1): 25-32 [In Japanese with English abstract]

Annals of Glaciology 9 1987
© International Glaciological Society

BOTTOM TOPOGRAPHY AND INTERNAL LAYERS IN EAST DRONNING MAUD LAND, EAST ANTARCTICA, FROM 179 MHz RADIO ECHO-SOUNDING

by

Minoru Yoshida*

(National Institute of Polar Research, Tokyo 173, Japan)

with

Kazunobu Yamashita

(Fifth Region Maritime Safety Headquarters, Maritime Safety Agency, Kobe 650, Japan)

and

Shinji Mae

(Department of Applied Physics, Faculty of Engineering, Hokkaido University, Sapporo 060, Japan)

ABSTRACT

Extensive echo-sounding was carried out in east Dronning Maud Land during the 1984 field season. A 179 MHz radar with separate transmitting and receiving antennae was used and the echoes were recorded by a digital system to detect minute reflections. The results gave cross-sections of the ice sheet along traverse routes from lat. 69°S. to 75°S. Detailed observations on the ground at Mizuho station showed that there was elliptical polarization in the internally reflected echoes when two antennae, kept in parallel with each other, were rotated horizontally. The internal echoes were most clearly distinguished when the antenna azimuth was oriented perpendicular to the flow line of the ice sheet. The internal echoes with a high reflection coefficient were detected at depths of 500–700 m and 1000–1500 m at Mizuho station. Since a distinct internal echo at a depth of 500 m coincides with a 5 cm thick volcanic ash-laden ice layer found in the 700 m ice core taken near the observation site, these echoes may correspond to the acidic ice layers formed by past volcanic events in east Dronning Maud Land.

INTRODUCTION

The Japanese Antarctic Research Expedition (JARE) has carried out extensive glaciological work in the inland area of east Dronning Maud Land since 1982. During the last two seasons, radio echo-sounding was conducted at the lower altitudes (Nakawo and others 1984; Nishio and others 1986). They used 60 MHz radar for the ground survey and 179 MHz radar for the airborne survey.

The authors carried out the ground survey by 179 MHz radar at the higher altitudes during the 1984 field season. A-scope data were recorded every 10–20 km along the new traverse route (Fig.1). The object of the observations was to measure the ice thickness along the route and the positions of the internal echoes.

RADAR SYSTEM

The radar transmits a 1.5 kw (peak) pulse in a repetition frequency of 1 kHz. The receiver has a sensitivity

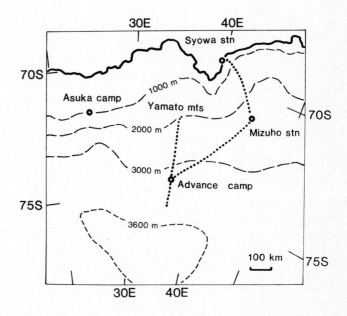

Fig.1. Inland traverse route during 4 October–5 December 1984.

of –104 dBm, whose receiving frequency is 179 MHz ± 2.5 kHz. Six-element Yagi antennae were used as the transmitting and receiving aerials which had linear polarizations. These antennae were set up on the snow, maintaining parallelism to each other at a distance of 3 m during the observations.

Wada and others (1982) reported that this radar with a three-element Yagi antenna failed to detect echoes from the bed deeper than 1500 m when the radar was used for airborne surveys. In order to detect minor echoes, we have analyzed the A-scope data by a digital recording system. After having been digitized at 50 ns intervals, 256–512 received echoes were stacked in a memory to eliminate random noises. The A-scope echo shown on the CRT was also recorded as a photograph which was exposed for 20–30 s. This photographic film also represents the optically averaged echo from 20 000–30 000 received signals.

* Present address: Hakusan Industry Co. Ltd, Tokyo 183, Japan.

RESULTS AND DISCUSSION

The radar system successfully detected all of the echoes from the bed along the route. Fig.2 shows a cross-section of the ice sheet along the traverse route obtained both by the echo-sounding and the surveyed altitudes of the ice

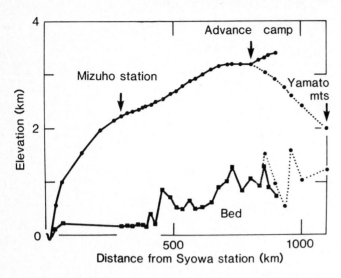

Fig.2. Cross-section of the ice sheet along the traverse route. Surface elevations of the ice sheet are based on the data of Fujii and others (1986).

surface. The ice thickness was determined by the echo time from the bed, assuming that the radio-wave velocity in the ice is 169 m/μs. The bed elevations in latitudes lower than 72°S. are 100–200 m a.s.l. In the higher-latitude area, the bed elevations increase to 500–1000 m and the bottom topography is extremely rough. The thickest ice along the route was 2650 m at lat. 75°S. The traverse party of JARE-26 extended this route to lat. 78°S. in 1985 and reported 3500–4000 m thick ice there (personal communication from K. Kamiyama).

The observations on the internal echoes were carried out at Mizuho station (lat. 70°42'S., long. 44°20'E., 2230 m a.s.l.). Although polarization in the ice sheet, observed when the transmitting antenna was rotated around the fixed receiving antenna R, has been reported in a number of papers (e.g. Hargreaves 1978, Woodruff and Doake 1979), few observations on polarization when two aerials are kept parallel to one another have been reported (e.g. Bentley 1975). Furthermore, these reports are based on analyses of the bottom-reflected signals; it seems that there have been virtually no reports on polarization resulting from internal echoes. Since the bottom-reflected echoes are affected not

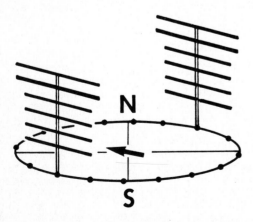

Fig.3. Setting of the antennae. The ice-flow direction is shown at the center. This direction is equal to that of the principal axis of the extending strain of the ice sheet (Naruse and Shimizu 1978).

only by the birefringence of the ice but also by the morphology of the glacier bed (Berry 1975), it is necessary to analyze the internal echoes in order to determine the birefringence of the ice body itself. We observed the A-scope data for the internal echoes, when the receiving antenna was rotated around the transmitting antenna, and kept parallel to one another at a distance of 3 m. Measurements were repeated at 22.5° intervals (Fig.3). In order to evaluate the relationship between the polarization and the strain of the ice sheet, the antennae were installed on the snow at the center of a strain-grid which had been set up for observation of the strain-rate of the ice sheet.

Fig.4 shows the noise-reduced A-scope data at the same travel-time (depth) scale which clearly indicates

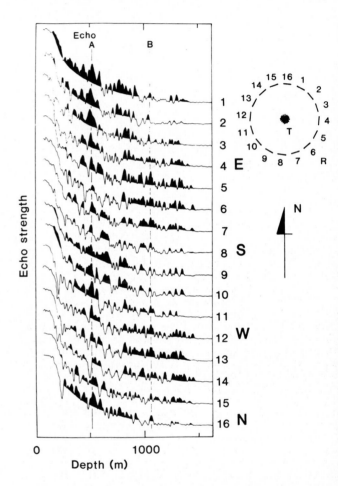

Fig.4. Internal echoes at 16 different antenna azimuths observed by A-scope data. Echo strengths above the average level of the 16 data sets obtained by A-scope are coloured dark. R and T represent the two parallel antennae, receiving and transmitting aerials, respectively. The observation was carried out in May 1984.

polarization of a radio wave in the ice sheet. The internal echoes from a depth of 1000–1500 m are strongest when the receiving antenna R is situated at east-south-east from the transmitting antenna T (No. 4, 5, 6 in the illustration) or west-north-west (No. 12, 13, 14).

In order to analyze the echo strength with respect to the antenna azimuth, two internal echoes at depths of 500 m (echo A) and 1050 m (echo B) were chosen. Fig.5 shows the results. Both echoes A and B show an elliptical pattern, which confirms the birefringence of the ice sheet. The difference between the maximum and minimum echo strengths is a gain of 7–10 dB, which is a significant value for the receiver sensitivity.

The A-scope records in Figs.4 and 5 indicate that the internal layers could be most clearly distinguished when the azimuth was perpendicular to the flow line. The internal echo from a depth of 1050 m (echo B) became strongest

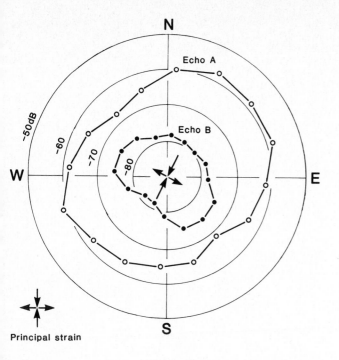

Principal strain

Fig.5. Echo strength as a function of the direction of the two parallel antennae. The positions of echo A and echo B are shown in Fig.3. The directions of the principal strain axes of the ice sheet observed at the same time are shown at the center of the circle graph.

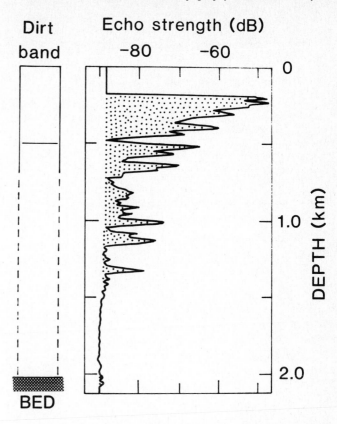

Fig.6. Internal echo and the depth of the volcanic ash layer at Mizuho station. The antennae azimuths were perpendicular to the direction of ice flow. Echo sounding was carried out in December 1984.

when the antenna azimuth faced 030° or 210° (R was situated at 120° or 300° (from T). This antenna azimuth is very close to that of the principal compressive strain of the ice sheet, which is perpendicular to the flow line at the observation site. This coincidence of the azimuth may suggest that the birefringence was caused by the ice flow. Narita and others (in press) have examined the 700 m deep ice core from Mizuho station and have reported that the c-axes of the ice crystals below 400 m depth become aligned along the vertical plane. It is plausible that the strain history of the ice has caused this orientation fabric, which may determine the birefringence of a polycrystalline ice sheet as Hargreaves (1978) has argued in a theoretical study.

Another possible cause of this polarization could be explained by the depositional pattern of the internal layers. Internal layers, such as volcanic ash layers, probably have prolonged patch-patterns due to the strain deformation of the ice sheet. The reflection coefficients of these irregular patches would depend on the azimuths of the antennae.

Millar (1981) has analyzed the internal echoes in the Greenland ice sheet and he concluded that the internal echo layering was formed by layers of ice of changed composition, which are probably composed of acidic ice formed after volcanic eruptions. Since we have found a 5 cm thick volcanic ash-laden ice layer at Mizuho station (Narita and others in press), the A-scope record when the antenna azimuth was perpendicular to the flow line has been compared with the position of the ash-laden ice layer. The result (Fig.6) represents a strong echo up to 20 dB peak at the depth of the volcanic ash-laden ice layer. This coincidence suggests that the echo strength/depth profile corresponds to acidic layers caused by past volcanic activity in east Dronning Maud Land.

Although dating of the ash-laden ice layer at 500 m depth has not yet been completed, from the annual accumulation rate of 7 g/year at Mizuho station (Watanabe and others 1978) the age can be estimated as $\sim 8 \times 10^3$ year B.P. Other internal layers with high reflection coefficients seem to be located at 1000–1500 m depth. The ages of these ice layers can be estimated as between 20 and 40×10^3 year B.P. These ages of the internal layers seem to coincide with those of volcanic ash bands found in the ice core at Byrd Station (Gow and Williamson 1971); this coincidence

may suggest the extensive fall-out of acidic aerosols on the Antarctic ice sheet. Further echo-sounding of internal layers will undoubtedly reveal the areal distribution of these volcanic ash layers during the last glaciation.

ACKNOWLEDGEMENT
The radio echo-sounding data were collected during the JARE-25 programme in 1983–85. The authors would like to thank the members of the wintering party, led by Professor T Hirasawa, for their enthusiastic field support. We also wish to express our thanks to Dr F Nishio, Dr H Ohmae, Dr K C Jezek and Dr R S Williams for their helpful advice.

REFERENCES
Bentley C R 1975 Advances in geophysical exploration of ice sheets and glaciers. *Journal of Glaciology* 15(73): 113–135
Berry M V 1975 Theory of radio echoes from glacier beds. *Journal of Glaciology* 15(73): 65–74
Fujii Y, Kawada K, Yoshida M, Matsumoto S 1986 Glaciological research program in east Queen Maud Land, East Antarctica. Part 4, 1984. *JARE Data Reports* 116 (Glaciology 13)
Gow A J, Williamson T 1971 Volcanic ash in the Antarctic ice sheet and its possible climatic implications. *Earth and Planetary Science Letters* 13(1): 210–218
Hargreaves N D 1978 The radio-frequency birefringence of polar ice. *Journal of Glaciology* 21(85): 301–313
Millar D H M 1981 Radio-echo layering in polar ice sheets and past volcanic activity. *Nature* 292(5822): 441–443
Nakawo M, Narita H, Isobe T 1984 Glaciological research program in east Queen Maud Land, East Antarctica. Part 2, 1983. *JARE Data Reports* 96 (Glaciology 11)
Narita H, Fujii Y, Nakawo M In press Morphological observations of structure and fabric patterns of 700 m deep ice core obtained at Mizuho Station, Antarctica. *Memoirs of National Institute of Polar Research*. Special Issue

Naruse R, Shimizu H 1978 Flow line of the ice sheet over Mizuho Plateau. *Memoirs of National Institute of Polar Research*. Special Issue 7: 227–234

Nishio F, Ohmae H, Ishikawa M 1986 Glaciological research program in east Queen Maud Land, East Antarctica. Part 3, 1982. *JARE Data Reports* 110 (Glaciology 12)

Wada M, Yamanouchi T, Mae S 1982 Radio echo-sounding of Shirase Glacier and the Yamato Mountains area. *Annals of Glaciology* 3: 312–315

Watanabe O, Kato K, Satow K, Okuhira F 1978 Stratigraphic analyses of firn and ice at Mizuho Station. *Memoirs of National Institute of Polar Research*. Special Issue 10: 25–47

Woodruff A H W, Doake C S M 1979 Depolarization of radio waves can distinguish between floating and grounded ice sheets. *Journal of Glaciology* 23(89): 223–232

Annals of Glaciology 9 1987
© International Glaciological Society

SATELLITE SNOW-COVER MONITORING IN THE QILIAN MOUNTAINS
AND AN ANALYSIS FOR CHARACTERISTICS OF STREAM SNOW-MELT
RUN-OFF IN THE HEXI REGION, GANSU, CHINA

by

Zeng Qunzhi, Zhang Shunying, Chen Xianzhang and Wang Jian

(Lanzhou Institute of Glaciology and Geocryology, Academia Sinica, Lanzhou, Gansu,
People's Republic of China)

ABSTRACT

The images of NOAA/TIROS-N APT, AVHRR, and a few Landsat MSS obtained from 1980 to 1985 are analysed in this paper. It is found that the snow-cover distribution in Qilian Mountains is above 3700 m a.s.l. during winter to spring every year. There are two concentrations of snow cover. One is on Mount Leng Longling in the upper reaches of the Shiyang River and the other is located between Hala Lake and Mount Danghe Nanshan.

Based on preliminary investigations, it is known that the surface water resource in the Hexi region is 68.8×10^8 m^3, of which about 24.8% is from glaciers and melting, and the snow-melt run-off is 7.63×10^8 m^3, equal to 62.6% of the total amount of spring run-off.

The average value of Cv for spring run-off in the Shiyang River, Heihe River, and Shule River is 0.32 and the Cv value of snow-melt run-off in spring is 0.41, about three times as much as that of the annual run-off in the Hexi region. A prediction model of spring snow-melt run-off at the Ying Louxia Hydrometric station in the Heihe River area can be constructed by using hydrometeorological data and snow-cover percentage for the Heihe River basin obtained from NOAA/TIROS-N APT, and AVHRR images. The prediction models (2) and (3) have been tested by the Water Resources Management Office of the Heihe River basin in the Zhangye and Flood Prevention Office of Gansu Province. The prediction accuracy is suitable for demands.

INTRODUCTION

The Qilian Mountains are located at the north-east edge of the Qinhai-Xizang Plateau (lat. 36°30'–39°30'N., long. 93°30'–103°E.), between the dry Hexi region and the Alashan Plateau in the north and the Qaidam basin and the Yellow River valley in the south, 850 km in length, 200–300 km in width and about 200 000 km^2 in area. About 30% of the whole mountain area is above 4000 m a.s.l.. Because of the high altitude and low air temperature, there are many glaciers in the Qilian Mountains. According to the statistics, there are 2859 glaciers with a total area of 1972.5 km^2. The altitude of the snow line (north slope) rises westward from 4409 m in the east at Mount Leng Longling to 5075 m in the north-west Mount Qaidam. Also, the altitudes of the glacier termini rise from 4296 m in the east up to 5021 m in the north-west. The total amount of glacier-melt run-off per year in the Qilian Mountains is 11.56 $\times$ 10^8 m^3, of which 9.46 $\times$ 10^8 m^3 flows into the Hexi region, equal to 13.1% of the outlet surface run-off from the Qilian Mountains. In addition, there is a lot of seasonal snow cover in winter and spring in the Qilian Mountains providing snow-melt water; for example, the amount of snow-melt run-off in spring is as much as 7.64 $\times$ 10^8 m^3 in the Shiyang River, Heihe River, and Shule River areas, about 11.7% of total run-off. Based on preliminary investigations it is known that the surface-water resource in the Hexi region is 68.8 $\times$ 10^8 m^3, of which

about 24.8% is from glacier and snow melting. The uneven distribution of water resources in space and time is the main problem in water supply, especially in the spring (April–June), and agricultural growth in the Hexi region is affected by a lack of water in the spring.

Because most of meteorological and hydrometric stations in the Qilian Mountains are distributed along the valleys at a low altitude, the data cannot be representative of information on precipitation and snow-cover in the high mountains. So, it is necessary to carry out snow-cover dynamic monitoring on a large scale and snow-melt run-off forecasting in that region using satellite remote-sensing technology. In this paper, the authors intend to discuss the characteristics of variation in the spring run-off in the Hexi region and their relation to snow cover in the Qilian Mountains, taking the Heihe River as an example, by using remote-sensing techniques.

1. DATA, METHODS AND ACCURACY OF ESTIMATIONS

NOAA/TIROS-N APT, AVHRR, and a few Landsat MSS cloud-free images in 1980–85 were used as the basic data. By optical interpretation techniques, the extent of snow cover in the Qilian Mountains was determined. Snow cover in satellite images was transferred on to a topographic map (1 : 150 000), using the Zoom Transfer Scope (ZTS); then the snow-cover area can be obtained by using a planimeter. Another method is as follows: snow cover in the satellite image was transferred on to a 1 : 100 000 scale topographic map, together with the elevation of the average seasonal snow line. The snow-cover area can be derived from an area–elevation curve for the Heihe River basin.

Bingou River, in the upper reaches of Heihe River was selected as a test basin for examining the dependability of snow-cover data derived from the methods mentioned above and for hydro-meteorological observations throughout the year, snow-cover surveys in the mountains and spectrum-reflectance measurement (0.38–1.20 μm) from different objects on the subaerial surface. The elevation of the seasonal snow-cover line in the Bingou River area on 3 November 1983 was 3900 m and the area of snow cover above this altitude was measured as 53.27 km^2 from the topographic map (1 : 50 000), while it was 58.5 km^2 in area from interpretation of NOAA/TIROS-N AVHRR images at 16.00 h on the same day. The value measured from images was 5.23 km^2 or 9.8% greater than that by observation. This method is economical and efficient, although it is less accurate. To measure the snow-cover area in a basin using NOAA/TIROS-N AVHRR digital data (CCTs), first of all, geometric correction of the AVHRR images must be carried out. Then, statistics of the snow-cover percentage can be obtained. Using this method, the position error is no more than 0.5 pixel. The NOAA/TIROS-N AVHRR digital data (CCTs) cannot be obtained within 24 h, although the accuracy is high, so it is not suitable for operational use.

2. DISTRIBUTION OF SNOW COVER IN THE QILIAN MOUNTAINS

Seasonal snow cover in the Qilian Mountains is formed in early September and ends in early June the following year. September to October and April to June the following year are the main supply periods of seasonal snow cover. Under the control of the Mongolian high pressure in winter, the Qilian Mountains are rather dry and cold, and with less precipitation. By interpreting Landsat MSS images on 7 and 8 October 1975, it is known that the altitude of the seasonal snow line in early winter in the eastern Qilian Mountains, including Heihe River and Shiyang River, is about 3800 m. Field observation shows that there is only fragmentary snow cover on shadow slopes and forest regions at an altitude of 3800 m in the upper reaches of the Heihe River in November. Between 3800 and 3900 m there is a discontinuous snow-cover zone; above 3900 m, there is a continuous (stable) snow-cover area, where the depth of snow cover reaches 5–20 cm. Based on an analysis of field-measurement data from 1983 to 1985 in the Bingou River test basin, within a certain range of altitude, snow depth increases with altitude, as follows:

$$D = -52.8 + 154 \times 10^{-2}*H + 3\sin(35.7 - \beta) \qquad (1)$$

where: D is snow depth (in cm), H is altitude (in m), β is slope (in degrees).

The analysis of NOAA/TIROS-N, APT, and AVHRR images in recent years indicates that the snow-cover area in the Qilian Mountains decreases from the east to the west and there are two centres of snow cover: one is in Mount Leng Longling and the other is located between Hala Lake and Mount Danghe Nashan.

The beginning of the snow-melt season in the eastern part of the Qilian Mountains is on 5 April, according to data from the weather station there. When the mean daily air temperature is stable above 0°C, both the snow cover below an altitude of 3700 m and the river ice begin melting. In May, the 0°C isotherm of mean daily air temperature reaches an altitude of 4053 m and the snow cover melts rapidly. From April to early June is a period for both snow-cover accumulation and melting intensity.

All the seasonal snow cover melts in summer except in the area of the firn basin.

3. CHARACTERISTICS OF SPRING RUN-OFF IN THE MAIN RIVERS IN HEXI REGION

According to an analysis based on observations at 12 hydrometric stations on ten main rivers, the characteristics of spring run-off in the Hexi region can be described as follows:

(1) The amount of spring run-off in the Hexi region (excluding the Sugan Lake basin) is 12.18×10^8 m^3.

(2) The snow-melt run-off is 7.63×10^8 m^3, equal to 62.6% of the total amount of spring run-off. The values of snow-melt run-off in Shiyang River, Heihe River, and Shule River are 2.46×10^8 m^3, respectively. In addition, the proportion of snow-melt run-off to the amount of spring run-off decreases westward; for example, it is 72.8% for the Shiyang River basin in the east and 42.8% for the Shule River in the west.

(3) The major annual change in precipitation in the Qilian Mountains during winter and spring causes major annual variation in spring run-off. The average value of Cv for spring run-off in Shiyang River, Heihe River, and Shule River is 0.32, and the Cv value for snow-melt run-off in spring is 0.41, about three times as much as that of annual run-off in the Hexi region.

(4) Based on an analysis of square deviations and power spectra, it can be shown that there are several fluctuation periods such as 3, 8, and 15 years in Shiyang River, Heihe River, and Shule River, corresponding to variation in periods (years 3, 5, and 8) of precipitation from October to June in the Qilian Mountains. But the variations in spring run-off in the eastern part are not synchronous with those in the western part.

4. PREDICTION OF SNOW-MELT RUN-OFF

The spring run-off in the Hexi region is fed by the seasonal snow-melt water in the mountain areas, and secondly by ground water. The variation in spring run-off is determined by winter–spring snow-cover and air-temperature conditions in the Qilian Mountains (Fig.1). A prediction model of spring snow-melt run-off at the Ying Louxia Hydrometric Station (Heihe River) can be constructed by using hydro-meteorological data and snow-cover percentages for the Heihe River basin obtained from NOAA/TIROS-N APT, and AVHRR images.

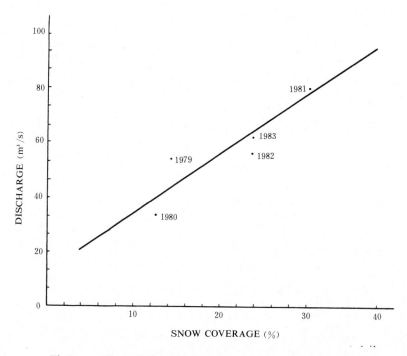

Fig.1. The relationship between snow cover in October and daily mean discharge from April to June at Yingluoxia hydrometric station.

(1) Daily model

$$Q_{n+1} = \exp(1.57 + 0.08\, T_n + 0.0165\, S_n + 0.01\, P_n) + $$
$$ + (Q_{nm} - Q_{nc})^{0.7} \qquad (2)$$

where

n	is	date series;
T_n	is	mean daily air temperature (°C) at the Qilian Weather Station on n day;
P_n	is	daily precipitation (mm) at the Qilian Weather Station on n day;
S_n	is	snow-cover percentage in the Heihe River basin on n day
Q_{n+1}	is	mean daily discharge predicted on $n+1$ day;
Q_{nm}	is	field-measured mean daily discharge on n day (m³/s);
Q_{nc}	is	calculated discharge on n day (m³/s).

(2) Prediction model for 2 days

$$Q_{n+2} = Q_o + [a_o + a_1 T_n + a_2 S_n + a_3(P_n + P_{n-1} + $$
$$ + P_{n-2})]^* \quad F(T_n) + Q_n[1 - \exp(-\beta t)] \qquad (3)$$

where

Q_0	is	the mean daily base flow;
a_0, a_1, a_2, a_3		are coefficients;
$S_n, T_n, P_n, P_{n-1}, P_{n-2}$		are the same as mentioned above;
$F(T_n)$	is	the function of mean daily temperature;
β	is	the depletion coefficient of water

$$(\beta = \frac{1}{230}\, Q_n^{0.79});$$

t is the term of prediction (2 d).

Equations (2) and (3) have been tested by the Water Resources Management Office of the Heihe River basin in the Zhangye and Flood Prevention Office of Gansu Province (Fig.2). The prediction accuracy is suitable for demands.

For an appraisal of the accuracy of Equations (2) and (3), a volume-error equation and a measure of model efficiency R^2 have been used, as suggested by J.E. Nash and J.V. Sutcliffe (1970).

$$D_v = \frac{V_c - V_m}{V_m} \qquad (4)$$

where

D_v	is	volume error (%);
V_m	is	measured amount of run-off;
V_c	is	calculated amount of run-off;

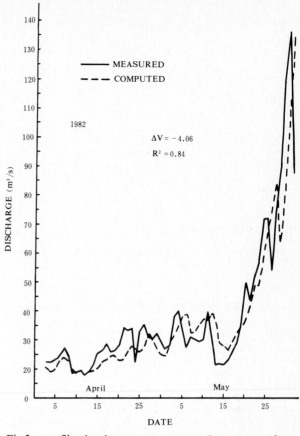

Fig.2. Simulated versus measured stream flow at Yingluoxia hydrometric station from April to May 1982.

$$R^2 = 1 - \frac{\sum\limits_{i=1}^{n}(Q_m - Q_c)^2}{\sum\limits_{i=1}^{n}(Q_M - Q)^2} \qquad (5)$$

where

Q_m	is	measured discharge (m³/s);
Q_c	is	calculated discharge (m³/s);
Q	is	mean discharge measured during the prediction interval (m³/s).

Table I lists the values of D_v and R^2 for Equations (2) and (3). For comparison, D_v and R^2 from run-off simulations in several other basins are given in Table II. In this

TABLE I. ACCURACY OF SNOW-MELT RUN-OFF FORECASTS FROM EQUATIONS (2) AND (3)

Basin (simulation period)	Volumetric difference D_v (%)	Model efficiency R^2	Area of basin km²	Elevation range m a.s.l.
Babao River, China (1980–84)	0.4–15.0	0.81–0.88	2452	2590–4700
Heihe River Zamashik station, China (1980–81)	1.7	0.85	4589	2635–4894
Heihe River-Ying Louxia station, China (1980–85)	(−7.0)–5.0	0.73–0.84	10009	1674–4894

TABLE II. ACCURACY OF SNOW-MELT RUN-OFF FORECASTS IN VARIOUS COUNTRIES*

Basin (simulation period)	Volumetric difference Dv (%)	Model efficiency R^2	Area of basin km^2	Elevation range m a.s.l.
Modry Dul, Czechoslovakia	1.7	0.95	1.32	1000–1554
Dischma Alps, Switzerland	−0.65	0.83	43.3	1668–3146
Dinwoody, Rocky Mountains, USA (1974, 1976)	3.3	0.85	223	1981–4202
Bull Lake, Rocky Mountains, USA	4.8	0.82	484	1970–4185

* From A. Rango and J. Martinec (1981).

case, however, the daily flow for the rivers was computed from the snow-covered areas, temperatures, and precipitation, without corrections for the measured discharge. Therefore, the values of the accuracy criteria are not directly known.

CONCLUSION

It has been known that mountain snow cover in various climatic regions and at altitudes in the Qilian Mountains has different altitudes of the seasonal snow line and snow-melt conditions, due to variations in precipitation and the heat-exchange processes. Analyses presented here show that the glacier and snow-melt water in the Qilian Mountains is a very important part of the surface-water resource in the Hexi region and the development of agriculture in the Hexi region is strongly influenced by the snow-melt run-off from the Qilian Mountains in the spring.

This work suggests methods which may be used to forecast snow-melt run-off in any basin fed from the Qilian Mountains.

ACKNOWLEDGEMENTS

The authors gratefully acknowledge assistance from the Meteorological Bureau of Gansu Province in supplying NOAA/TIROS-N APT, and AVHRR images, and from Qiu Guoqing and Mrs Yang Zhenniang for their valuable advice.

REFERENCES
Nash J E, Sutcliffe J V 1970 River flow forecasting through conceptional models. Part 1. A discussion of principles. *Journal of Hydrology* 10(3): 282–300
Rango A, Martinec J 1981 Accuracy of snowmelt runoff simulation. *Nordic Hydrology* 12(4–5): 265–274

Annals of Glaciology 9 1987
© International Glaciological Society

ANTARCTIC ICE-SHELF BOUNDARIES AND ELEVATIONS FROM SATELLITE RADAR ALTIMETRY

by

H.J. Zwally

(Laboratory for Oceans, NASA Goddard Space Flight Center, Greenbelt, MD 20771, U.S.A.)

S.N. Stephenson

(Science Applications Research, NASA Goddard Space Flight Center, Greenbelt, MD 20771, U.S.A.)

R.A. Bindschadler

(Laboratory for Oceans, NASA Goddard Space Flight Center, Greenbelt, MD 20771, U.S.A.)

and

R.H. Thomas

(Royal Aircraft Establishment, Farnborough, U.K.)

ABSTRACT

As part of a systematic analysis of Seasat radar altimetry data to obtain Antarctic ice fronts and ice-shelf elevations north of lat. 72°S., Fimbulisen (between long. 12°W. and 08°E.) and the Amery Ice Shelf (around long. 72°E.) are mapped. Interactive computer analysis is used to examine and correct the altimetry range measurements and derive the ice-front positions. Surface elevations and ice-front positions from radar altimetry are compared with ice fronts, ice rises, crevasse zones, and grounding lines identified in Landsat imagery. By comparison of the visible features in imagery and the computer-contoured elevations from radar altimetry, the radar-elevation mapping on some ice rises is confirmed, but some spurious contours are also identified. During the interval between the 1974 Landsat imagery and the 1978 radar altimetry, the central part of the Amery Ice Shelf front advanced 1.5 ± 0.6 km/a, which is in agreement with the ice-velocity measurements of 1.1 ± 0.1 km/a (Budd and others 1982), suggesting negligible calving in the central part of the ice shelf. The undulating surface and small mean slope from the grounding line to about lat. 70°S. suggest a zone of partial grounding similar to Rutford Ice Stream. On Fimbulisen, some previously unmapped ice rises are identified. The ridge of the Jutulstraumen ice tongue is shown to be about 20 m above the surrounding ice and laterally expanding as it flows northward to the ice front. Icebergs within the sea ice and a zone of shore-fast ice are also identified with the same technique used to map the ice-shelf front.

INTRODUCTION

Since 1975, satellite radar altimeters designed for measurement of ocean-surface topography have been used for measuring ice-sheet topography and for deducing other features of the ice configuration and reflective properties of the surface (e.g. Brooks and others 1978, Thomas and others 1983, Zwally and others 1983[a], Bindschadler 1984, Partington and others 1987). The use of satellite radar-altimeter data for glaciological studies has required the analysis of each radar wave form (corresponding to a range measurement every 1/10 s) to obtain the surface elevation at 670 m intervals (Martin and others 1983). Also, the effects of regional surface slopes and surface undulations has had to be considered, because the altimeter tends to measure the range to the closest surface location within about a 10 km

radius of the sub-satellite point (e.g. Brenner and others 1983, Gundestrup and others 1986).

Due to limitations in the width of the range window of the altimeter and its ability to change the position of the window as the range from the altimeter to the surface changes, abrupt changes in elevation at ice fronts, and even the changes in range due to surface undulations, often cause the altimeter to lose track of the range between measurements (e.g. Martin and others 1983). After loss of track, several seconds or more of data were not obtained while the altimeter searched through a much larger range window looking for a reflected signal from the surface. Consequently, substantial gaps occurred in the data record in regions of large undulations or at abrupt changes in elevation.

Nevertheless, Thomas and others (1983) showed that the margin of large ice sheets can be mapped using the characteristics of the Seasat altimeter range measurements just before the altimeter lost track in the vicinity of an abrupt change in elevation. They demonstrated that the altimeter continued to measure the slant range back to the sea ice in front of the sea-ice/ice-shelf boundary while passing from sea ice to ice shelf. A short section of the front, sometimes 10 km long, could be mapped at each crossing of a boundary, usually followed by a data gap of 10–20 km. After loss of track, the altimeter range search would find the reflected signal from the higher elevation of the ice shelf, and measurement of surface elevation would resume. A similar situation occurred in reverse during crossings from ice shelf to sea ice. The method has now been applied to about 400 orbits of the Seasat data set in the Antarctic, and in this paper some of the results for the Amery Ice Shelf and Fimbulisen are described.

METHOD

To process the large volume of data, an analysis procedure was developed on an HP9845C interactive computer. The altimeter wave-form shapes and intensities and the indicated heights, which had been previously range-corrected by automated computer re-tracking (Martin and others 1983), were interactively reviewed. The automatic computer re-tracking algorithm had also been modified to re-track the sharply peaked wave forms characteristic of the specular reflections from sea ice. The program allowed the operator to select a new re-track point, when needed, to

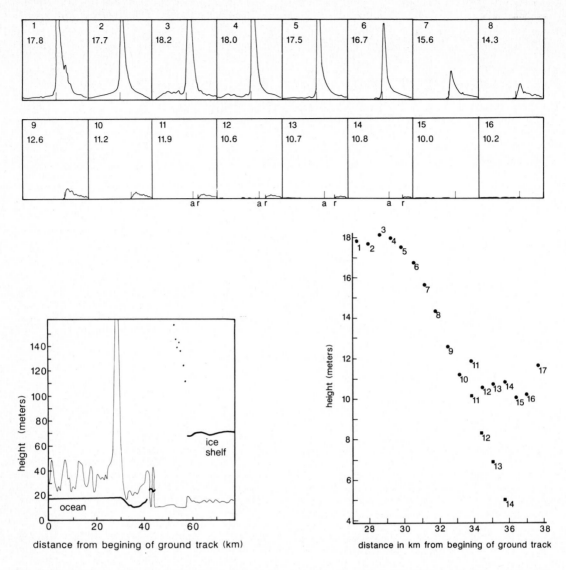

Fig.1. (a) Sequence of radar-altimeter wave forms at an ice-front crossing, from sea ice to ice shelf, selected by finding the apparent (false) lowering in the sea surface shown in Fig.1b. The top left-hand number in the sequence is also used in Fig.1c. The lower numbers indicate the level of the sea-ice surface with respect to a reference ellipsoid. The thin vertical tick mark labeled "a" shows the computer-calculated half-peak power position of the ramp. The thick tick mark labeled "r" shows the interactively selected position for the weaker returns. (b) Indicated surface heights (dark line) with respect to an ellipsoid and the relative intensity (light line) of the reflected radar signal before the ramp of the altimeter wave forms. (c) Uncorrected indicated heights at points corresponding to wave forms in (a) and corrected indicated heights for points 11 to 14.

determine the correct range (indicated height). In particular, interactive re-tracking is often needed for the weaker returns from the sea-ice/ice-sheet boundary, particularly at the larger distances from the boundary. The interactive program also calculates and plots the envelope of possible reflectors on a surface map for each boundary crossing and selects a set of reflecting points, which are reviewed by the operator.

A typical set of wave forms while traversing an ice-sheet margin is shown in Fig.1a. The segment of the record to be examined is first identified and selected by plotting the indicated heights along a longer section of the orbit, as shown in Fig.1b. The intensity of the wave form forward of the main wave-form ramp is also plotted to assist in the identification of a boundary crossing, which is characterized by an increase in intensity caused by early reflection from the part of the altimeter wave front intersecting the ice-shelf surface forward of the altimeter (Thomas and others 1983). Here, several typical features of a boundary crossing are evident. For the first 30 km from the beginning of this segment, there is sea ice at sea-level. After 55 km, there is ice shelf, 50 m above the level of the

sea ice. At the 30 km point, an apparent (false) lowering of the sea surface begins (shown in greater detail in Fig.1c), due to the increasing slant range to the reflecting sea-ice area near the ice front. After the drop-off in elevation, the wave-form signal becomes lost in the noise, and the altimeter range lock is lost. Loss of lock initiates a signal search, and range lock is regained on the ice shelf at the 55 km point.

Referring to Fig.1a and c, it can be seen that the on-board tracker and the re-tracking algorithm find it increasingly difficult to identify a wave-form ramp (sharp rise in reflected radar signal strength), as the slant range increases and the intensity decreases. It is also obvious that wave forms 11 through 16 were not properly re-tracked by the automatic algorithm, as indicated by the automatic re-track point (a) in Fig.1a. For the purpose of mapping the sea-ice/ice-shelf boundary, the chosen re-tracking point on the wave-form signal is where the rising ramp of the return signal begins to emerge from the noise. Wave forms 11 to 14 have adequate signal above the noise and are interactively re-tracked to provide four more range measurements to the sea ice, thereby extending the observed

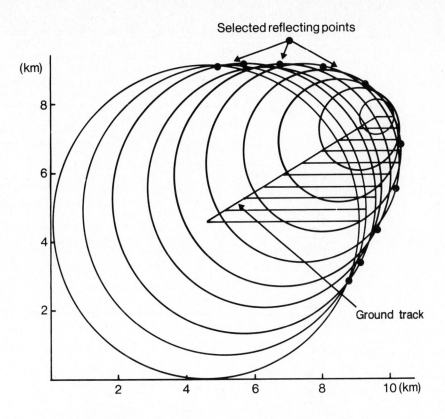

Fig.2. Sequence of circles representing the intersection of the slant range, derived from the corrected indicated heights for points 5 to 14 of Fig.1, with the Earth's surface. Dots between the intersections of the circles are chosen to represent the reflecting points, on the sea ice near the ice front, that correspond to the re-tracked radar signal. Series of dots form the arms of a "V" with left–right ambiguity resolved by plotting a series of "V"s from nearby crossings.

segment about 2 km. Subsequently, no useful returns were obtained between 36 and 58 km, at which point the profiling of the ice-shelf surface begins. It should also be noted that just prior to regaining range lock around the 58 km point, some erroneous elevation measurements (>100 m) are included in the computer re-tracked data set. Most of such erroneous elevations have been eliminated from the data set used for gridding and contouring elevations by visual editing of plots of the wave forms and the re-tracking parameters.

Using the sequence of corrected indicated heights, which represent the sequence of slant ranges back to the boundary crossing, a set of circles is then drawn representing the envelope of possible reflecting points at the boundary, as shown in Fig.2. The circles represent the intersection on the Earth's surface of a cone defined by the slant range and the height of the satellite above the surface. The sequence of heights is obtained by extrapolating the geoid inland from the measurements over sea ice. The mid-points between intersections of the circles are chosen as the average reflecting points at the boundary. A left–right ambiguity occurs at each crossing, with the mirror images of the two possible boundaries forming two arms of a "V". In many cases, the left–right ambiguity is resolved when several nearby crossing points are plotted and two arms of partially overlapping "V"s line up with each other. Where an orbit is isolated from its neighbors, the general direction of the front as indicated by the neighboring "V"s can help resolve the ambiguity, because the arms of the "V" represent steps in a known direction and usually one of the arms fits the general picture better than the other.

During some boundary crossings, the gain of the receiver is significantly less than in the case shown here and the elevation drop-off is not clearly defined. However, the power of the return decreases sharply at the boundary,

which corresponds to the beginning of the drop-off, as shown by wave forms 5, 6, and 7 in Fig.1a. A similar sharp decrease in the peak power is observed in the low-gain cases, and where this occurs a boundary crossing point is also mapped.

Thomas and others (1983) also described the reverse situation, when the satellite path traverses from ice sheet or ice shelf to sea ice. They showed that a similar drop-off of the indicated surface elevation occurs. However, a difficulty is encountered in this case, because the ice-shelf surface is not as flat as the sea-ice surface and, consequently, the changes in elevations over ice-shelf undulations sometimes reduce the drop-off or resemble the change in indicated height typically seen at such boundaries. Therefore, it is often difficult to judge where the decrease in elevation due to undulations stops and the drop-off caused by slant-range measurement backward to the ice-shelf front begins. Also, because the ice-shelf surface near the front tends to be tilted seaward, a marked drop-off in the signal strength from the ice-shelf reflection usually does not occur after the spacecraft passes seaward over the boundary. However, since the sea ice is lower in elevation than the ice shelf (i.e. greater range from the satellite) and the altimeter tracking circuit is adjusting to the increasing slant range backward to the ice shelf, the altimeter sometimes picks up the sea-ice return without loss of track, in contrast to the usual loss of track during a sea-ice to ice-shelf transition. Ice-front positions are mapped for these ice-shelf to sea-ice transitions but the location is usually not as well defined.

AMERY ICE SHELF

The result of mapping the Amery Ice Shelf front by Seasat radar-altimeter data obtained during July–October 1978 is shown in Fig.3. Seventeen orbital paths cross the

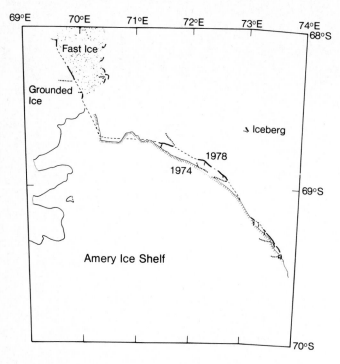

Fig.3. Amery Ice Shelf front in 1978 derived from radar altimetry compared with ice front sketched from Landsat imagery taken in 1974. Also shown is an area of fast sea ice and the seaward edge of the grounded ice sheet north-west of the ice shelf.

front of the main part of Amery Ice Shelf and one orbit (No. 545) intersects the edge three times giving 30 km of ice front. Both arms of the "V"s are shown, with the preferred arm indicated by the heavier line. Overall, about 40% of the 170 km front is mapped. Thomas and others (1983) estimated the errors of this method to be in the range of ±0.1–1.0 km, depending on the distance of the satellite from the ice front when the slant range is obtained. Although the actual accuracy of the radar-mapped boundary is difficult to establish, it is roughly estimated to be ±1 km.

Fig.3 also shows the ice front from Landsat imagery obtained in 1974, 4 years before the Seasat data. The images are the same as those used by Brooks and others (1983) in their illustration with surface contours of the Amery Ice Shelf, also from Seasat radar altimetry. The superimposed geographic grid from their illustration is used here to match the Landsat image with the altimeter-derived edge. Brooks and others (1983) assigned latitudes and longitudes to their mosaic by transferring coordinates of nunatak features from the 1 : 1 000 000 topographic map of the Amery Ice Shelf produced by the Australian Division of National Mapping in 1971. The accuracy of the Landsat-derived ice-front position is roughly estimated to be about ±2 km.

Along the eastern part of the ice shelf, several of the radar-derived ice-front positions and the Landsat ice front are in good agreement. For the central part of the ice shelf, there is a 6–7 km difference in the ice-front positions obtained by the two methods at an interval of 4 years. Allowing for the uncertainties in each of the lines and the 4 year period between the observations, the difference in positions indicates an advance of the ice front at a rate of 1.5 ± 0.6 km/a. This rate of advance is in agreement with the ice-shelf velocity measurement of 1.1 ± 0.1 km/a by Budd and others (1982), suggesting little or no calving of the central part of the ice front during this period.

Segments of the seaward margin of the grounded ice sheet north-west of the ice shelf are also mapped, demonstrating the application of the radar margin-mapping technique for grounded coastal margins as well as floating ice fronts. A single edge point located in the ocean at long. 73°W., about 40 km north of the ice front, is interpreted as an iceberg.

A number of edge features are also located in the ocean in the north-western corner of Fig.3. Here, several edge points line up, and they are interpreted as the edge of fast sea ice with a thick snow cover. The reflecting surface is about 2 m higher inside this boundary, the altimeter wave form is characteristic of reflection from a firn surface rather than the peaked specular return characteristic of sea ice, and the surface appears to be about 1–2 m above the sea ice to the east. Also, the edge does not vary in position noticeably during the period of the altimeter measurements, as it would if it were a polynya with possibly non-specularly reflecting sea-ice types. Furthermore, the area is in a location of highly consolidated sea ice at this time of year (Zwally and others 1983[b]), and the US Navy–NOAA maps indicated fast ice there the previous summer.

In addition to the new estimate of the ice-shelf front, automated contouring of the altimeter data has been carried out following the method described by Zwally and others (1983[a]). To summarize the procedure, the data have been re-tracked using the method described by Martin and others (1983), followed by visual editing of the re-tracked wave forms to eliminate additional invalid data. Radial adjustment of the orbits (r.m.s. minimization of height difference at orbital intersections over the ocean) and correction for the slope-induced error due to off-nadir reflections (Brenner and others 1983) have not yet been carried out on the Antarctic data set. The slope-induced error is small on the ice shelf due to the small surface slopes, but radial orbit errors up to several meters remain in some of the data. The data are then used to produce a surface on a regular grid at 5 km spacing, and then contoured at 2 m intervals over the ice shelf and 20 m intervals over the ice sheet. The gridding procedure also eliminates data that are more than 3 sigma from the surface fitted to the data in the vicinity of a grid point to obtain the value at the grid point. The elevations shown are referenced to sea-level by subtracting the GEM-10B (Goddard Earth Model version 10B) geoid from the ellipsoid-referenced altimeter elevations shown in Fig.1b.

The ice-shelf contours shown in Fig.4 basically reproduce the main features shown by Brooks and others (1983), although data from 40 additional orbits are used. Over the ice sheet surrounding the ice shelf, altimeter tracking was frequently lost, resulting in a sparse data set at the change in slope inside the grounding line and in the region to the west of the ice shelf. Since no orbit adjustment has been made, linear features in the direction of the satellite tracks (predominately east–west in these regions) with several meters relief need to be viewed with caution (cf. next section).

One significant difference between the map of Brooks and others (1983) and Fig.4 is the delineation of a region of the ice shelf with an undulating surface and small mean slope for about 200 km below the grounding line (from about lat. 72° to 70°S.). If the ice shelf were freely floating, such a pattern would not be expected. The characteristics of this region suggest a zone of patchy grounding of the ice shelf, similar to that observed on enhanced Landsat imagery of Rutford Ice Stream (Doake and others in press).

Another new feature in the contour map is a small topographic high point located at lat. 69°15'S., long. 71°30'E. about 40 km from the ice front. Four orbits define this feature, and it is interesting to speculate whether it is an ice rumple forming on a shoal which is helping to stabilize the ice front. However, the surface expression is small and the existence of this feature needs to be examined after the radial-orbit adjustments have been applied to the data.

FIMBULISEN

A similar analysis has been carried out for the altimeter data over Fimbulisen. Fig.5 shows the sub-satellite ground tracks in this area, with the gaps indicating loss of track. The data were computer re-tracked, but again, no radial-orbit adjustment has been applied. The contour plot is given in Fig.6. Residual radial errors in some of the orbits produce false linear features extending, for example, over the ocean from around lat. 69°S., long. 0°E. towards

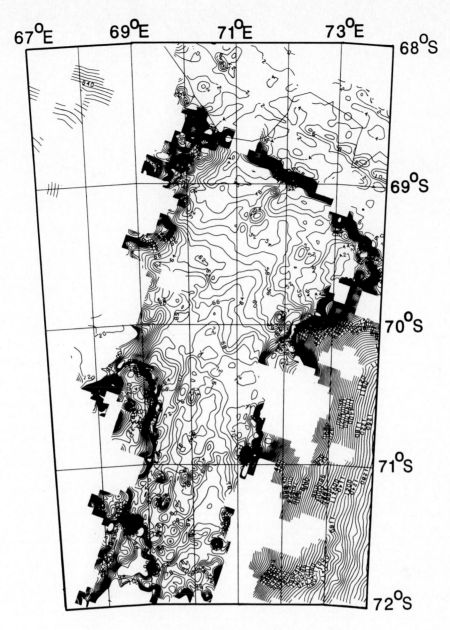

Fig.4. Surface elevations of the Amery Ice Shelf and adjacent grounded ice from Seasat radar altimetry, showing undulating surface with small mean slope extending from vicinity of grounding line near lat. 72°S. to around 70°S.

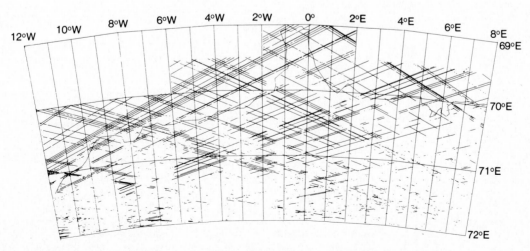

Fig.5. Seasat ground tracks in the vicinity of Fimbulisen, showing loss of track in places of large surface slope and after abrupt elevation change over the ice-shelf front. Solid line is continental boundary from 1965 AGS map prior to the 1967 calving of Trolltunga at long. 0°W.

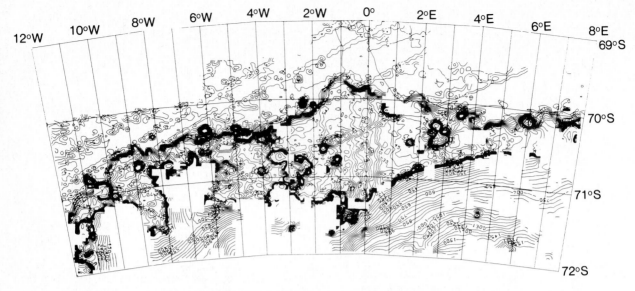

Fig.6. Surface elevations of Fimbulisen and adjacent grounded ice from Seasat radar altimetry showing Jutulstraumen ice tongue, ice rises, and other features discussed in the text.

the south-west. The error in these particular orbits appears to be about 5 m, which is significantly larger than the approximately 1 m r.m.s. precision of the orbit calculations. The continuation of these linear orbit-error features is also seen on the ice shelf, but they are typically obscured by the larger relief observed on the ice shelf. Another error source is the data collected whenever a large change in elevation occurs and the altimeter is unable to maintain track of successive wave forms. Fig.1b illustrated this situation for a sea-ice/ice-shelf crossing. Some of these errors may remain in the data set as a result of incomplete visual editing of the re-tracked data or 3 sigma editing in the gridding procedure. One topographic high resulting from this second error source is located at lat. 70°50'S., long. 01°10'W.

Other features seen on Fig.6 are the ice front, several large ice rises, and the Jutulstraumen ice tongue (long. 0°W.). The Jutulstraumen ice tongue is a ridge, 20 m above the surrounding ice shelf, which extends almost to the ice front. The enhanced lateral divergence expected for the ridge, compared to the surrounding thinner ice of the shelf,

is observed in the gradual spreading of the contours as the ice tongue flows northward. Similar strong flow features are not observed on any of the other ice-shelf units, which is consistent with previous knowledge of this area.

A sketch map is made of features visible on unenhanced Landsat imagery (Fig.7). The Landsat imagery confirms the presence of large areas of grounded ice in the ice shelf, as indicated by the contours in Fig.6. The eastern half of the ice shelf is characterized by a series of small ice rises near the ice front. Most of these ice rises were originally mapped by Orheim (1978) using Landsat imagery and included in his contribution to the SPRI Antarctic Folio (Drewry 1983). The western part of Fimbulisen is divided by peninsulas of grounded ice, the boundaries of which are reasonably well represented by the contours in Fig.6, but the topography over the ice rises varies rapidly causing the altimeter to lose track. Also visible on the Landsat imagery is a zone of patchy blue ice south of the ice-shelf grounding line, which is represented by a line in the sketch map.

Examination of the altimeter wave forms indicates that

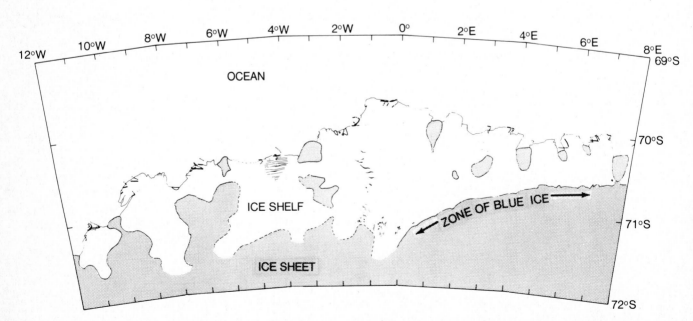

Fig.7. Fimbulisen ice front in 1978 derived from radar altimetry compared with ice front sketched from Landsat imagery taken in 1974. Also shown are ice rises, crevasse zones, and a line representing the northern edge of a zone of blue ice visible on the Landsat imagery inside the grounding line.

valid elevations are obtained over at least some of the ice rises, and good elevation profiles can be obtained from the altimeter data set. However, the contours over the ice rises, as shown in Fig.6, are not good representations of the ice elevations, particularly those near the ice edge, because the gridding process is not designed to handle the sudden changes in surface height. Also, one example of a spurious elevation feature in the contour map appears around lat. 70°40'S., long. 01°10'W., in the vicinity of a crevasse zone by the Jutulstraumen ice tongue. Crevasse zones were first detected in the altimetry data on the Amery Ice Shelf by Partington and others (1987). In general, the altimeter continues to track strong reflectors after it has passed overhead, such as the rift in the ice shelf at lat. 70°40'S., long. 09°30'W. and the grounding lines of ice rises, for example, at lat. 70°18'S., long. 02°40'W.

CONCLUSIONS

The slant-range method of mapping sea-ice/ice-shelf boundaries has been applied to a systematic mapping of the Antarctic ice-shelf fronts north of lat. 72°S. The method also applies to mapping at least some of the ice-sheet/sea-ice boundaries. At each satellite crossing of a boundary, a segment of the coastline up to about 10 km is mapped. Therefore, positions of parts of the seaward margin along the entire coastline north of lat. 72°S. can be derived from Seasat radar-altimetry data. However, the density of Seasat coverage is limited by the short 3 month lifetime of the satellite, and positions are not obtained at every crossing. The method is also sensitive to icebergs embedded in the sea ice, to fissures near the ice front, and to the edge of fast sea ice. Consequently, the boundary of an ice shelf or the ice sheet is not always clearly determined.

About 40% of the Amery Ice Shelf front is mapped and about 10% of Fimbulisen. The accuracy in absolute position is estimated to be ±1 km. Comparison with features in Landsat imagery is especially useful for interpolating between radar-derived boundaries, for helping to resolve the left–right ambiguity of isolated radar-derived boundaries, and for distinguishing between ice fronts and other features such as fissures, icebergs, and fast ice. Therefore, information from the Landsat images should be incorporated into the mapping process. For Fimbulisen and the Amery Ice Shelf, there are some geodetic ground-control points, but for other areas the altimeter data can be used to provide control for Landsat or similar high-resolution imagery, provided the data are acquired at approximately the same period of time.

Comparison of the radar-derived ice front of the Amery Ice Shelf with the front on earlier Landsat imagery indicates an advance of the central part of the ice shelf with little or no calving between 1974 and 1978. Application of this method to the radar-altimetry data expected from subsequent satellite missions will be especially useful for monitoring the margin of the Antarctic ice shelves and seaward margin of the ice sheet.

The inner 200 km of the Amery Ice Shelf are shown to be an area with small mean slope and undulations, suggesting it is a zone of partial grounding similar to part of Rutford Ice Stream. On Fimbulisen, the ridge of a major ice tongue is mapped and several previously unmapped ice rises are evident in the radar data and the Landsat imagery. The radar altimeter provides profiles across some ice rises but, in other areas of crevasse zones or marked changes in surface elevations, special care must be taken to eliminate invalid elevations from the altimeter data set.

ACKNOWLEDGMENTS

We especially thank Dr R S Williams for his kind assistance in helping us select the Landsat images from the files of the US Geological Survey and the SCAR photography library. Computer programming and data-analysis support has been provided by Anita Brenner, Judy Major, Yagyensh Pati, and Tom Martin of EG&G Washington Analytical Services. Support for this work has been provided by NASA's Oceanic Processes Program and NASA's Climate Program.

REFERENCES

Bindschadler R A 1984 Jakobshavns glacier drainage basin: a balance assessment. *Journal of Geophysical Research* 89(C2): 2066–2072

Brenner A C, Bindschadler R A, Thomas R H, Zwally H J 1983 Slope-induced errors in radar altimetry over continental ice sheets. *Journal of Geophysical Research* 88(C3): 1617–1623

Brooks R L, Campbell W J, Ramseier R O, Stanley H R, Zwally H J 1978 Ice sheet topography by satellite altimetry. *Nature* 274(5671): 539–543

Brooks R L, Williams R S, Ferrigno J G, Krabill W B 1983 Amery Ice Shelf topography from satellite radar altimetry. *In* Oliver R L, James P R, Jago J B (*eds*) *Antarctic earth science.* Cambridge, Cambridge University Press: 441–445

Budd W F, Corry M J, Jacka T H 1982 Results from the Amery Ice Shelf Project. *Annals of Glaciology* 3: 36–41

Doake C S M, Frolich R M, Mantripp D R, Smith A M, Vaughan D G In press Rutford Ice Stream, Antarctica. *Journal of Geophysical Research*

Drewry D J 1983 *Antarctica: glaciological and geophysical folio.* Cambridge, University of Cambridge. Scott Polar Research Institute

Gundestrup N S, Bindschadler R A, Zwally H J 1986 Seasat range measurements verified on a 3-D ice sheet. *Annals of Glaciology* 8: 69–72

Martin T V, Zwally H J, Brenner A C, Bindschadler R A 1983 Analysis and retracking of continental ice sheet radar altimeter waveforms. *Journal of Geophysical Research* 88(C3): 1608–1616

Orheim O 1978 Glaciological studies by Landsat imagery of perimeter of Dronning Maud Land, Antarctica. *Norsk Polarinstitutt. Skrifter* 169: 69–80

Partington K C, Cudlip W, McIntyre N F, King-Hele S 1987 Mapping of Amery Ice Shelf, Antarctica, surface features by satellite altimetry. *Annals of Glaciology* 9: 183–188

ice-sheet margins from radar altimetry data. *Annals of Glaciology* 4: 283–288

Thomas R H, Martin T V, Zwally H J 1983 Mapping ice-sheet margins from radar altimetry data. *Annals of Glaciology* 4: 283–288

Zwally H J, Comiso J C, Parkinson C L, Campbell W J, Carsey F D, Gloersen P 1983[a] *Antarctic sea ice, 1973–1976: satellite passive-microwave observations.* Washington, DC, National Aeronautics and Space Administration (NASA Special Publication SP-459)

Zwally H J, Bindschadler R A, Brenner A C, Martin T V, Thomas R H 1983[b] Surface elevation contours of Greenland and Antarctic ice sheets. *Journal of Geophysical Research* 88(C3): 1589–1596

Annals of Glaciology 9 1987
© International Glaciological Society

A STUDY OF THE FLUCTUATIONS IN SEA-ICE EXTENT OF THE BERING AND OKHOTSK SEAS WITH PASSIVE MICROWAVE SATELLITE OBSERVATIONS

(Abstract)

by

D.J. Cavalieri and C.L. Parkinson

(Laboratory for Oceans, NASA Goddard Space Flight Center, Greenbelt, MD 20771, U.S.A.)

ABSTRACT

The seasonal sea-ice cover of the combined Bering and Okhotsk Seas at the time of maximum ice extent is almost 2×10^6 km^2 and exceeds that of any other seasonal sea-ice zone in the Northern Hemisphere. Although both seas are relatively shallow bodies of water overlying continental shelf regions, there are important geographical differences. The Sea of Okhotsk is almost totally enclosed, being bounded to the north and west by Siberia and Sakhalin Island, and to the east by Kamchatka Peninsula. In contrast, the Bering Sea is the third-largest semi-enclosed sea in the world, with a surface area of 2.3×10^6 km^2, and is bounded to the west by Kamchatka Peninsula, to the east by the Alaskan coast, and to the south by the Aleutian Islands arc.

While the relationship between the regional oceanography and meteorology and the sea-ice covers of both the Bering Sea and Sea of Okhotsk have been studied individually, relatively little attention has been given to the occasional out-of-phase relationship between the fluctuations in the sea-ice extent of these two large seas. In this study, we present 3 day averaged sea-ice extent data obtained from the Nimbus-5 Electrically Scanning Microwave Radiometer (ESMR-5) for the four winters for which ESMR-5 data were available, 1973 through 1976, and document those periods for which there is an out-of-phase relationship in the fluctuations of the ice cover between the Bering Sea and the Sea of Okhotsk. Further, mean sea-level pressure data are also analyzed and compared with the time series of sea-ice extent data to provide a basis for determining possible associations between the episodes of out-of-phase fluctuations and atmospheric circulation patterns.

Previous work by Campbell and others (1981) using sea-ice concentrations also derived from ESMR-5 data noted this out-of-phase relationship between the two ice packs in 1973 and 1976. The authors commented that the out-of-phase relationship is "... surprising as these are adjacent seas, and one would assume that they had similar meteorologic environments". We argue here that the out-of-phase relationship is consistent with large-scale atmospheric circulation patterns, since the two seas span a range of longitude of about 60°, corresponding to a half wavelength of a zonal wave-number 3, and hence are quite susceptible to changes in the amplitude and phase of large-scale atmospheric waves.

REFERENCE

Campbell W J, Ramseier R O, Zwally H J, Gloersen P 1981 Structure and variability of Bering and Okhotsk sea-ice cover by satellite microwave imagery. *In* Halboury M T (ed) *Energy Resources of the Pacific Region.* American Association of Petroleum Geologists. Tulsa, OK: 343–354

AIRBORNE UHF RADAR MEASUREMENTS OF CALDERA GEOMETRY AND VOLCANIC HISTORY, MOUNT WRANGELL, ALASKA, U.S.A.

(Abstract)

by

G.K.C. Clarke and G.M. Cross

(Department of Geophysics and Astronomy, University of British Columbia, Vancouver, BC V6T 1W5, Canada)

and

C.S. Benson

(Geophysical Institute, University of Alaska, Fairbanks, AK 99701, U.S.A.)

ABSTRACT

The ice-filled caldera of Mount Wrangell, Alaska, provides an unusual opportunity to examine the interaction between a glacier and an active volcano. The caldera acts as a giant calorimeter, preserving a rough balance between snow precipitation at the glacier surface and bottom melting. In April 1982 we sounded the glacier using an airborne 840 MHz pulsed radar (Narod B B, Clarke G K C 1983 UHF radar system for airborne surveys of ice thickness. *Canadian Journal of Earth Sciences* 20(7): 1073–1086). The data were recorded on magnetic tape, then computer processed and plotted as depth, rather than time, sections.

In addition to mapping ice thickness, we detected extensive internal layers presumed to have been deposited during past eruptions of Mount Wrangell. The challenge of interpreting these internal reflectors inspired us to develop a unified interpretation model that incorporates both glaciological measurements and phenomenological equations for

firn and ice. The interpretation model yields the variation with surface-normal depth z of temperature T, heat flux q, ice pressure p, density ρ, surface-normal component of flow velocity w, down-slope component of flow velocity u, depositional age t_a, two-way wave travel time τ_2, and two-way propagation loss P_2. The following system of linear differential equations is integrated using the Runge–Kutta method:

$$\frac{dT}{dz} = -\frac{q}{K} \tag{1}$$

$$\frac{dq}{dz} = -pcw \frac{dT}{dz} + 2B_0 \exp(-Q/RT) (Fp \tan \alpha)^{(n+1)} + \frac{pw}{3\rho} \frac{d\rho}{dz} \tag{2}$$

$$\frac{dp}{dz} = pg \cos \alpha \tag{3}$$

$$\frac{d\rho}{dz} = m_1\rho^2 \left[\frac{\rho_I - \rho}{\rho_I}\right] \quad p \leqslant p^* \tag{4}$$

$$= m_2\rho^2 \left[\frac{\rho_I - \rho}{\rho_I}\right] \quad p > p^*$$

$$\frac{dw}{dz} = -\frac{w}{\rho} \frac{d\rho}{dz} - \Delta_0 \tag{5}$$

$$\frac{du}{dz} = -2B_0 \exp(-Q/RT) (Fp \tan \alpha)^n \tag{6}$$

$$\frac{dt_a}{dz} = \frac{1}{w} \tag{7}$$

$$\frac{d\tau_2}{dz} = \frac{2\sqrt{\epsilon}}{v_0} \tag{8}$$

$$\frac{dP_2}{dz} = 2D \tag{9}$$

together with physical property relations such as the following

$$c(T) = 2115.343 + 7.7929 (T - T_0) \tag{10}$$

$$K(\rho,T) = 2\rho K_I(T)/(3\rho_I - \rho) \tag{11}$$

$$K_I(T) = 2.1725 - 3.403 \times 10^{-3}(T - T_0) + {} + 9.085 \times 10^{-5} (T - T_0)^2 \tag{12}$$

$$\epsilon(\rho) = (1 + 8.5 \times 10^{-4}\rho)^2 \tag{13}$$

$$D(T) = D_0 \exp(-E/RT). \tag{14}$$

The boundary conditions on the integration variables are $T(0) = T_S, T(h) = T_M, p(0) = 0$, $\rho(0) = \rho_S$, $w(0) = w_S$, $u(h) = u_B$, $t_a(0) = 0$, $\tau_2(0) = 0$, and $P_2(0) = 0$, where T_S is surface temperature, T_M is bottom temperature, ρ_S is surface density, w_S is surface-normal velocity measured at the glacier surface, and u_B is sliding velocity. The remaining variables are thermal conductivity K, specific heat capacity c, flow-law constant B_0, creep activation energy Q, universal gas constant R, shape factor F, surface slope α, flow-law exponent n, gravity acceleration g, empirical constants m_1, m_2, and p^* from Benson's firn-densification law (Benson C S 1962 Stratigraphic studies in the snow and firn of the Greenland ice sheet. *SIPRE Research Report* 70), ice density ρ_I, surface value of two-dimensional flow divergence $\Delta_0 = (\partial u/\partial x + \partial v/\partial y)s$, relative dielectric permittivity ϵ, free-space electromagnetic wave velocity v_0, propagation loss rate $D, T_0 = 273.16$ K, thermal conductivity of ice K_I, loss-law constant D_0, and activation energy for propagation loss E. The units of temperature and density in the above equations are respectively K and kg m^{-3}.

Preliminary modelling has yielded an age-depth relationship for internal reflecting layers that roughly matches the known eruption record for Mount Wrangell and extends the volcanic history by more than 200 years. High signal absorption rates restrict interpretation of the caldera geometry. Reflections from the bed are not consistently detectable beyond a depth of approximately 350 m, suggesting a loss rate of 7–8 dB per 100 m of path length. The anomalous losses are attributed to abnormally high concentrations of ionic impurities and scattering sources.

INTERPOLATION TECHNIQUES FOR THE DISPLAY OF REMOTELY SENSED GLACIOLOGICAL DATA

(Abstract)

by

A.P.R. Cooper

(Scott Polar Research Institute, University of Cambridge, Cambridge CB2 1ER, U.K.)

ABSTRACT
The use of many display techniques for remotely sensed glaciological data requires the reduction of the data to a regularly spaced rectangular grid of values. Most remotely sensed data are not immediately suitable for display, because the area of interest is covered by more than one set of data on mutually incompatible grids (e.g. Landsat, AVHRR), or because the data are available as profiles along widely spaced ground tracks (e.g. radio echo-sounding, satellite altimetry). In addition, data may be sparsely and randomly scattered (e.g. surface elevations from TWERLE balloons).

A variety of techniques is available to reduce data to a specified grid system. These include spatial averaging, interpolation from nearest neighbours, and surface-fitting techniques, notably polynomial fitting and bi-cubic splines. All of these are useful under differing circumstances.

237

ACOUSTICAL REFLECTION AND SCATTERING FROM THE UNDERSIDE OF LABORATORY-GROWN SEA ICE: MEASUREMENTS AND PREDICTIONS

(Abstract)

by

K.C. Jezek

(Dartmouth College, Hanover, NH 03755, and NASA, Washington, DC 20546, U.S.A.)

with

T.K. Stanton

(University of Wisconsin, Madison, WI 53706, U.S.A.)

and

A.J. Gow

(U.S. Army Cold Regions Research and Engineering Laboratory, Hanover, NH 03755, U.S.A.)

ABSTRACT

We have studied acoustical reflection and scattering properties of the underside of laboratory-grown sea ice. Our purpose was to determine the morphologic characteristics of undeformed sea ice that control acoustic scattering and reflection. So our experiments included both detailed studies of the structure of the ice as well as the application of a variety of acoustic methods.

Ice sheets were grown in an outdoor pond (about 6 m by 13 m by 1.5 m) and exhibited features characteristic of undeformed, cold sea ice: an upper granular zone; a columnar zone of crystals with cross-sectional areas of about 1 cm^2; vertical sheets of brine pockets; a bulk salinity of 9.1$^{\circ}/_{\circ\circ}$ distributed over 9 cm of ice; a dendritic interface at the ice/water boundary with dendrites about 0.5 mm across at the time of our measurements. Echo-amplitude fluctuations of normal-incidence sonar pings (100 kHz to 800 kHz) were measured as the sonars moved horizontally under the ice and accumulated into echo-amplitude histograms. (Data from a deteriorating ice sheet as well as data on lake ice were also collected.) We fitted the Rice probability density function (PDF) to the data and combined the resultant statistical parameter with Eckart acoustic scattering theory. The reflection coefficients calculated using this method ranged from 0.06 to 0.12, depending on environmental conditions. RMS roughness calculated using data from new sea ice was estimated to be about 0.3 mm. Because our ice thin sections show the ice to be porous and permeable at the interface with dendrites 0.5 mm thick, we suspect that the dendrites control the scattering as described by the echo-amplitude histograms. Further, we attribute the low reflection coefficients to the dendritic structure which may act as an impedance-matching zone into the columnar section of the sea ice.

Transmission measurements were performed by positioning a transducer located at the ice/air interface directly over the transducer located in the water. The total attenuation through 18 cm of ice ranged from 12 dB at 50 kHz to 70 dB at 420 kHz (signal levels were measured relative to the same path in water).

A BACK-PORTABLE MICROPROCESSOR-BASED IMPULSE RADAR SYSTEM

(Abstract)

by

F.H.M. Jones, B.B. Narod and G.K.C. Clarke

(Department of Geophysics and Astronomy, University of British Columbia, Vancouver, British Columbia V6T 1W5, Canada)

ABSTRACT

We have developed and tested a portable impulse radar for ground-based sounding of glaciers. Noteworthy characteristics of the instrument are its portability, low power consumption, digital data storage, and the ability to be operated either manually or automatically under program control. Current system specifications include a band width of 46 MHz; a sampling interval of 10.76 ns; depth precision of 0.9 m; 1024 samples per record; amplitude resolution of 8 bits; minimum recordable signal at the receive antenna equal to 0.26 mV; an operating center frequency of 8.5 MHz and an antenna-damping coefficient of 300 ohms.

The transmitter uses paired SCRs and a 12 V to 800 V converter to impress a 1200 V step on to a resistively damped dipole antenna. This pulse is triggered from the receiving system via a fibre optics cable so that each record can include the complete surface-path wavelet. The receiver unit combines a wide-band amplifier (with variable front-end attenuation) with a microprocessor-controlled data-acquisition system of our own design. The result of each sounding can be replayed as an "A-scope" display on a small, low-cost oscilloscope and stored on or retrieved from digital cassettes. In the unattended mode, records are collected at programmable intervals. The system weighs about 7.5 kg and uses dry cells or rechargeable batteries for power.

Examples were presented in a poster session of sounding profiles taken in July 1986 on Trapridge Glacier, Yukon Territory in Canada, along lines coinciding with an extensive drilling program. Although not yet fully analysed, we feel that some of the results may represent the effects of crevasses, internal features such as morainal material, varying bed features, and changes in subglacial and englacial hydrology.

STUDYING THE PROCESS OF DISSIPATION OF ENERGY OF MOTION IN HIGHLY FRACTURED GLACIERS USING REMOTE-SENSING TECHNIQUES

(Abstract)

by

A.B. Kazanskiy

(Institute of Geography, U.S.S.R. Academy of Sciences, Staromonetny 29, Moscow 109017, U.S.S.R.)

ABSTRACT

New aspects of the physics of ice movement in glaciers are presented. An analysis of the equation for the energy of a moving ice mass shows that the energy which is dissipated into heat (internal friction in a glacier) occurs not only because of average ice-velocity profiles, as it is commonly considered when simulating glaciers by continuum laminar flow. It also turns out that a significant part of the internal friction is associated with the dissipation of motion energy on internal macroheterogeneities in a glacier: at places where fractures appear and, in the case of a block structure, at points of contact between individual blocks, on ice dams, ice bridges, etc. This mechanism seems to be dominant during rapid ice flow in highly fractured glaciers.

The process was previously examined on the surging Medvezhiy Glacier in the central Pamir Mountains. On Medvezhiy Glacier the above phenomenon was particularly distinctive because of anomalously high ice-movement velocities during a calm period; that is, between consecutive surges (ice velocity amounted to 3–5 m d^{-1}) and because of the block structure of the glacier against the background of its fracturing. Simple measurements of "instant" velocities of a large number of bench marks on the glacier surface, located across the middle flow course of the glacier, established that velocities at all points are of a fluctuating character, which outwardly resemble velocity fluctuations seen in turbulent liquid flow. It turns out that deviations of the "instant" velocity from the average velocity can amount to more than 50% of the latter. Using data obtained from periodic recordings of the "instant" velocity for fixed pairs of bench marks, which were located at different distances (so that it was possible to evaluate the difference in instant velocities), a structural function was constructed. This structural function meets the well-known, so-called Kholmogorov two-thirds law for local isotropic turbulence. This indicates that, as with turbulence, the energy of middle movement is transferred along a cascade of ice conglomerates of decreasing size down to the smallest, limited only by the dimensions of fundamental blocks. At the points of contact at dams and junctions, motion energy dissipates into heat. It is postulated that this phenomenon may have a much broader application; it may also be characteristic of the movement of any fractured mass of solid material.

These findings may be used to formulate a theory which describes the mechanism of auto-oscillations in a surging glacier. According to this theory, the pseudo-turbulent character of internal friction during a quiescent period (between surges) is one of the factors which prevents the appearance of a surge as long as possible, thus ensuring the growth of a new active glacier after each ice catastrophe.

These studies also conclude research on the general concept of auto-oscillations of mountain glaciers − the so-called surging glaciers, originally formulated in the course of research on the cause of a surge on Medvezhiy Glacier in 1963. This concept was subsequently adopted and is now widely used in glaciology.

CALCULATION OF MASS BALANCE OF GLACIERS BY REMOTE-SENSING IMAGERY USING SIMILARITY OF ACCUMULATION AND ABLATION ISOLINE PATTERNS

(Abstract)

A.N. Krenke and V.M. Menshutin

(Institute of Geography, U.S.S.R. Academy of Sciences, Staromonetny 29, Moscow 109017, U.S.S.R.)

ABSTRACT

A number of maps of component isolines of the Maruch Glacier's (West Caucasus) mass balance compiled on the basis of field observations during 11 years were analyzed.

The total balance of the glacier's mass seems to be closely associated with the relationship between accumulation and ablation areas by years, while the total values of net accumulation and ablation are fully determined by the surface-area values of accumulation and ablation regions. The layer of residual snow in the nourishment area by the end of the ablation season and the layer of many-year ice melting in the discharge area changes from year to year against the points fixed in space, but their mean specific value according to the above glaciological areas remains constant from year to year.

Using the results and the remote-sensing images, in which the many-year ice and the snow surplus are well distinguished, it is possible to estimate the areas of the ablation and accumulation regions for the studied glacier and then the mass balance, net accumulation, and, naturally, the run-off from the melting many-year ice.

Estimation of the distribution within the glacier of not only "net" but also total values of accumulation and ablation is possible, if this distribution is similar from year to year, and if total values of accumulation and ablation are known at certain points. The multitude of points, for which total accumulation and ablation may be determined by remote-sensing images, is found on the equilibrium line or on the transient snow line of the glacier at any date during the ablation season. In the first instance, ablation and accumulation, the latter being here equal to the former by definition, may be calculated by the mean summer temperature of the air, extrapolated for the corresponding altitude against the nearest meteorological station. Secondly, ablation is calculated by the sum of positive temperatures of the air and by the temperature coefficient of snow melting, extrapolated for the altitude of the transient snow line.

In order to check the hypothesis of the similarity of isoline systems for accumulation and ablation, all the final values for each year and measured at different points have been normalized according to their mean value for each year.

The normalized maps turned out to be similar to each other. Using the regular net of points, we calculated the variation coefficient of normalized values. They were smaller by 0.20 for the whole area of the maps of normalized ablation and for the greater part of the map area of normalized accumulation. A map of mean normalized values was compiled for 11 years. The isoline "unit" on it coincided with the many-year firn line. With the help of this topological map, one can compile maps for distribution of ablation and accumulation on a glacier for any year or moment, for which there are data on the location of the nourishment line or snow line according to a remote-sensing image and for which air temperature at the altitude of these lines may be estimated.

In conclusion, the degree of similarity of the maps of component isolines for the glacier mass balance between different glaciers of the same morphological type was analyzed. For this, we used normalization of not only the characteristics of the glaciers' mass balances but also morphological characteristics (altitudinal change, width, etc.). The results point to similarity of distribution of accumulation and ablation among glaciers. This will allow us to extrapolate the principal features of the isoline maps and the described methods of calculation from the studied glaciers to unexplored ones.

SATELLITE-ALTIMETER MEASUREMENTS OF SURFACE HEIGHT IN
SEA-ICE AREAS
(Abstract)

by

S.W. Laxon and C.G. Rapley

(Mullard Space Science Laboratory, University College London, Dorking, Surrey RH5 6NT, U.K.)

ABSTRACT

The Seasat radar altimeter was designed to operate over the open ocean and encountered problems over sea ice. In particular, the on-board measurements of surface height were noisy and unreliable. As a consequence, published mean sea-surface and geoid maps based on the Seasat on-board height estimates either omit sea-ice-covered areas or include suspect data. We have identified and investigated the problems encountered by the Seasat altimeter over sea ice and have developed a technique for extracting accurate surface-height values from the sea-ice echo wave-form data. The retracking method is based upon fitting real wave forms to a library of model returns.

CHARACTERISTICS OF ALTIMETRY SIGNATURES OVER SEA ICE
(Abstract)

by

N.F. McIntyre and S.W. Laxon

(Mullard Space Science Laboratory, University College London, Dorking, Surrey RH5 6NT, U.K.)

ABSTRACT

We report characteristics of Seasat altimetry signatures recorded over Antarctic sea ice. Up to four discrete zones can at times be seen in characteristic sequences in the Weddell and Ross Seas, and elsewhere. They are substantially larger than those reported in the Arctic, covering up to 2500 km at the time of maximum ice extent in 1978. Transitions between them can be abrupt, with marked changes occurring in less than a few kilometres. Some zones were found to persist through the 3 month satellite lifetime; others exhibited intermittent variations. Repeat data coverage has enabled temporal as well as spatial patterns to be investigated.

Interpretation of the geophysical cause of the patterns observed has been limited by available data. Some comparisons may be made with surface measurements of nadir back-scatter on first- and multi-year floes but these account for only a small proportion of the altimetry returns studied. Correlations with the NOAA Navy Ice Charts show significant disparities in the determination of the ice edge which may relate to the sensitivity of the altimeter to the presence of fresh ice or ice in very small quantities. Similar signatures can be found next to small coastal leads at the continental margin, an area known to be important for the growth of new ice.

ANALYSIS OF SIMULATED RADAR ALTIMETRIC WAVE FORMS FOR POLAR ICE AND SNOW SURFACES

(Abstract)

by

Eva Novotny[*]

(Scott Polar Research Institute, University of Cambridge, Cambridge CB2 1ER, U.K.)

ABSTRACT

The radar altimeter of the satellite Seasat has proved that ice and snow surfaces in the polar regions can return meaningful signals if the terrain is not excessively rugged or sloping. Because the use of the leading edge of the wave forms for height determination entails inherent uncertainties and, at best, provides only a single datum per wave form, the *entire* wave forms should be studied. Excesses or deficiencies in amplitude at various ranges within a single wave form, and the changes that occur in successive wave forms, can be analysed to yield information on the geometric and scattering properties of features observed by the altimeter.

Results from computer simulations are presented, showing how (1) a margin of sea ice (sinusoidal in the model) can be mapped, (2) the boundaries of two isolated ice floes can be outlined, (3) sea-ice concentrations can be derived within annuli about the nadir of an individual footprint, and (4) for land ice, the elevations of topographic features, together with the general slope of the ground, can be determined if an imaging instrument that operates simultaneously with the altimeter provides the outlines of these features. In examples (1) and (2), it is assumed that the ice is contiguous wherever that is possible, to permit the analytical reconstruction of the ice margin or individual ice floes in the presence of the inevitable ambiguity in the position of any feature with respect to the two sides of the satellite track. Example (4) requires that the altimeter record correctly records the strongest signals returned by ice-packs. This condition is not fulfilled by any existing radar altimeter, but it may be achieved in the next generation of these instruments. In additional examples of information from entire wave forms, the effects of crevasses and sastrugi in reducing or re-distributing the energy of the returned signals are also illustrated.

Full details of these analyses and results will be published at a later date.

ICEBERGS IN THE SOUTHERN OCEAN

(Abstract)

by

Olav Orheim

(Norsk Polarinstitutt, P.O. Box 158, 1330 Oslo Lufthavn, Norway)

ABSTRACT

Relatively little data on the distribution of Antarctic icebergs were available prior to 1980. The published literature included size data of about 5000 icebergs, and position data of 12 000 icebergs. There were indications that the size data were biased in favour of larger icebergs.

A programme of systematic iceberg observations was therefore initiated by Norsk Polarinstitutt in 1981 through the SCAR Working Group on Glaciology. This programme is based on standard "blue" forms distributed to all ships going to Antarctica. The icebergs are recorded every 6 h and in five length groups: 10–50, 50–200, 200–500, and 500–1000 m, and those over 1000 m are described individually.

The amount of data has increased greatly from the start in 1981–82. The position of 70 000 icebergs, including 50 000 that had been size classified, were on file at Norsk Polarinstitutt by December 1985, and the data set is growing rapidly. Most ships travelling to and from Antarctica now participate in collection of the data. (Fig.1 shows the locations of the icebergs sighted.)

The size distribution of the classified icebergs observed under this programme up to December 1985 is given in Table I:

TABLE I.

Size class m	Total number	%	"Standard size" m	Total volume 10^9 m^3	%
10–50	17 788	34.9	30 × 25 × 20	0.3	<0.1
50–200	17 187	33.7	120 × 100 × 80	17	1
200–500	10 437	20.5	320 × 280 × 200	187	8
500–1000	4152	8.2	750 × 600 × 250	467	20
>1000	1400	2.7	meas. individ.	1379	70
	50 954	100.0		2050	100.0

The "standard size" (length, width, and thickness) is based on our observations from three Antarctic expeditions which carried out dedicated iceberg studies. Many icebergs are of course not right-angled parallelepipedal in shape, but this is a good approximation for most of the larger icebergs.

The data are based both on visual sightings and on radar observations. Duplicate observations from a ship moving at slow or zero speed are as far as possible eliminated, both during observation, and by critical appraisal before the data are filed. The data editing also includes evaluation of data quality, especially in connection with radar observations, and comparison of positions and dimensions of the large icebergs in order to reduce to a minimum repeated observations from different vessels of

---

* Present address: Department of Applied Mathematics and Theoretical Physics, Silver Street, Cambridge, U.K.

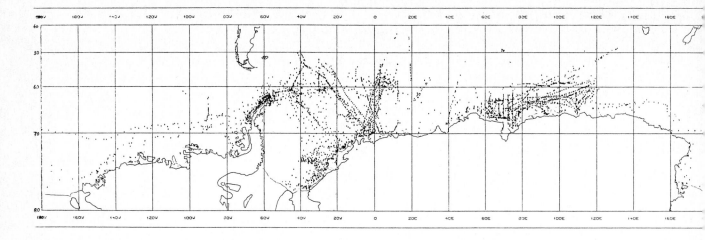

Location of iceberg observations under the programme initiated in 1981. Main ship tracks are clearly reflected. The average observation represents 14 icebergs.

icebergs >1000 m. These account for most of the iceberg mass (see Table I).

Consideration of iceberg-distribution patterns and the observed area of the Southern Ocean, and of duplicate observations, indicates more than 300 000 icebergs south of the Antarctic Convergence, with a total ice mass of about 10^{16} kg. Consideration of mean residence times indicates an annual iceberg production from the continent of $2-3 \times 10^{15}$ kg, which is considerably higher than most other recent estimates. This also suggests that the Antarctic ice sheet is in balance.

The data indicate large regional differences in iceberg sizes, the most noticeable being between the two sides of the Antarctic Peninsula, and between the Amery Ice Shelf/ Prydz Bay area and the remainder of East Antarctica.

These differences are probably mainly related to different calving sites.

About one-third of the observed icebergs are over the continental shelf of Antarctica. The total under-water area of these icebergs is two orders of magnitude less than the under-water area of the Antarctic ice shelves. The annual total iceberg melting and its effect on the water masses over the continental shelf has been calculated from ocean-water temperature variations at 200 m depth and estimated melt rates. This turns out to be an order of magnitude less than the annual effect of melting sea ice. The iceberg data considered here are probably under-represented with respect to the smallest sizes, and they do not include icebergs that have become <10 m. Inclusion of these ice bodies would increase the total melt.

EFFECT OF SURFACE ROUGHNESS ON REMOTE SENSING OF SNOW ALBEDO

(Abstract)

by

Stephen G. Warren, Thomas C. Grenfell and Peter C. Mullen

(Department of Atmospheric Sciences, AK-40, University of Washington, Seattle, WA 98195, U.S.A.)

ABSTRACT

Narrow field-of-view sensors on satellites monitoring solar radiation measure the reflected radiance in a particular direction. For climatic studies of the Earth's radiation budget, the albedo is needed, which is the integral of the upward radiance over all angles divided by the downward irradiance. In order to infer the albedo from a radiance measurement at only one angle, it is necessary to know *a priori* the distribution of reflected radiation with angle, i.e. the bi-directional reflectance-distribution function (BRDF). The BRDF is a function of four angles: solar zenith and azimuth, and satellite zenith and azimuth. For areal or temporal averages on many natural surfaces, only three angles are needed to describe the function, because only the difference between the two azimuths is important, not their individual values. This assumption was made when developing empirical BRDFs from Nimbus-satellite data for use in the Earth Radiation Budget Experiment (ERBE). However, in large areas of the polar regions, all four angles are needed, because the sastrugi are oriented parallel to a prevailing wind direction. The BRDF shows a forward peak when the solar beam is along the direction of the sastrugi, and an enhanced backward peak when it is perpendicular. Averaging over all solar azimuths (relative to the sastrugi azimuth) causes back-scattering to be averaged together with forward-scattering. The conclusion of the ERBE analysis, that snow is the most nearly isotropic of all Earth surfaces, is therefore at least partly a spurious result of this averaging.

Measurements of the BRDF were carried out from a 23 m tower at the South Pole during January and February at 900 nm wavelength for varying azimuths between the Sun and the sastrugi fabric. The wavelength was selected near the midpoint of the solar-energy spectrum but where scattered sky radiation is negligible. Measurements were made with 10° field of view at 15° intervals in viewing zenith and azimuth angles throughout the day, at intervals of 1 h (15° of solar azimuth). For BRDF normalized such that its angular average is unity, the principal features of the results include a forward-scattering peak with a value of about five together with a side- and back-scattering lobe of 1.1 to 1.3. Variations in solar azimuth produced a skewness in BRDF which was approximately consistent with enhanced scattering at the specular angle with respect to the

solar azimuth and the orientation of the principal fabric of the sastrugi pattern. The angularly averaged pattern was remarkably similar to the results of Taylor and Stowe even though their values were integrated over wavelength and were made through the atmosphere. Our studies thus suggest that, for mid- to late summer, the Taylor and Stowe results require only small corrections for sastrugi effects. This is not, however, expected to be true from sunrise through late November.

Spectral albedos showed values at visible wavelengths of 0.97 to 0.99 which agree very well with the model calculations of Wiscombe and Warren using our observed mean snow grain-sizes. Albedos for wavelengths above 1400 nm were higher than model predictions, indicating that the depth dependence of grain-size must be included in the analysis.

This research was supported by National Science Foundation grant DPP-83-16220.

Annals of Glaciology 9 1987
© International Glaciological Society

THE TIMING OF INITIAL SPRING MELT IN THE ARCTIC FROM NIMBUS-7 SMMR DATA

(Abstract)

by

Mark R. Anderson

(California Space Institute, A-021. Scripps Institution of Oceanography, La Jolla, CA 92093, U.S.A.)

ABSTRACT

The ablation of sea ice is an important feature in the global climate system. During the melt season in the Arctic, rapid changes occur in sea-ice surface conditions and areal extent of ice. These changes alter the albedo and vary the energy budgets. Understanding the spatial and temporal variations of melt is critical in the polar regions. This study investigates the spring onset of melt in the seasonal sea-ice zone of the Arctic Basin through the use of a melt signature derived by Anderson and others from the Nimbus-7 Scanning Multichannel Microwave Radiometer (SMMR) data. The signature is recognized in the "gradient ratio" of the 18 and 37 GHz vertical brightness temperatures used to distinguish multi-year ice. A spuriously high fraction of multi-year ice appears rapidly during the initial melt of sea ice, when the snow-pack on the ice surface has started to melt. The brightness-temperature changes are a result of either enlarged snow crystals or incipient puddles forming at the snow/ice interface.

The timing of these melt events varies geographically and with time. Within the Arctic Basin, the melt signatures are observed first in the Chukchi and Kara/Barents Seas. As the melt progresses, the location of the melt signature moves westward from the Chukchi Sea and eastward from the Kara/Barents Seas to the Laptev Sea region. The timing of the melt signal also varies with year. For example, the melt signature occurred first in the Chukchi Sea in 1979, while in 1980 the signature was first observed in the Kara Sea.

There are also differences in the timing of melt for specific geographic locations between years. The melt signature varied almost 25 days in the Chukchi Sea region between 1979 and 1980. The other areas had changes in the 7–10 day range.

The occurrence of these melt signatures can be used as an indicator of climate variability in the seasonal sea-ice zones of the Arctic. The timing of the microwave melt signature has also been examined in relation to melt observed on short-wave imagery. The melt events derived from the SMMR data are also related to the large-scale climate conditions.

MICROWAVE SNOW-WATER EQUIVALENT MAPPING OF THE UPPER COLORADO RIVER BASIN, U.S.A.

(Abstract)

by

W.J. Campbell and E.G. Josberger

(University of Puget Sound, Tacoma, WA 98416, U.S.A.)

P. Gloersen

(Laboratory for Oceans, NASA Goddard Space Flight Center, Greenbelt, MD 20771, U.S.A.)

and

A.T.C. Chang

(Laboratory for Terrestrial Physics, NASA Goddard Space Flight Center, Greenbelt, MD 20771, U.S.A.)

ABSTRACT

During spring 1984, a joint agency research effort was made to explore the use of satellite passive microwave techniques to measure snow–water equivalents in the upper Colorado River basin. This study involved the near real-time acquisition of microwave radiances from the Scanning Multichannel Microwave Radiometer (SMMR) aboard the Nimbus-7 satellite, coupled with quasi-simultaneous surface measurements of snow-pack depth and profiles of temperature, density, and crystal size within the basin. A key idea in this study was to compare, for the same space and time-scales, the SMMR synoptic physics data taken in the basin. Such a snow-measurement program was logistically difficult, but two field teams took detailed snow-pit measurements at 18 sites in Colorado, Utah, and Wyoming during the last 2 weeks of March, when the snow-pack is normally at its maximum extent and depth. These observations were coupled with snow–water-equivalent measurements from Soil Conservation Service SNOTEL sites. Microwave-gradient ratio, Gr (Gr is the difference of the vertically polarized radiances at 8 mm and 17 mm divided

by the sum), maps of the basin were derived in a near real-time mode every 6 days from SMMR observations. The sequential Gr maps showed anomalously low values in the Wyoming snow-pack when compared to the other states. This near real-time information then directed the field teams to Wyoming to carry out an extensive survey, which showed that these values were due to the presence of depth hoar; the average crystal sizes were more than twice as large as in the other areas. SMMR can be used to monitor the spatial distribution and temporal evolution of crystal size in snow-packs. Also, scatter diagrams of snow—water-equivalents from the combined snow-pit and SNOTEL observations versus Gr from the Wyoming part, and the Colorado and Utah part, of the basin can be used to estimate snow—water equivalents for various parts of the basin.

COMPARISON OF LANDSAT MULTISPECTRAL SCANNER AND THEMATIC MAPPER RADIOMETRIC AND SPATIAL CHARACTERISTICS OVER GLACIERS

(Abstract)

by

J.A. Dowdeswell*

(Scott Polar Research Institute, University of Cambridge, Cambridge CB2 1ER, U.K.)

ABSTRACT

For more than 10 years, images obtained from the four Landsat Multispectral Scanner (MSS) bands have provided important data for mapping and glaciological studies in the inaccessible polar regions. During this period, the specifications of the MSS have remained little altered, to allow data comparability. More recently, satellites 4 and 5 of the Landsat series have been equipped additionally with Thematic Mapper (TM) sensors. The TM has 7 bands in the visible, near infra-red, mid infra-red, and thermal infra-red, together with a larger dynamic range and improved spatial resolution relative to the MSS. The aim of this paper is to compare MSS and TM computer-compatible tapes (CCTs) from a glacierized area in order to demonstrate the advantages of using TM data in glaciological applications.

The digital MSS and TM scenes compared were imaged simultaneously from Landsat 5 on 5 May 1984 over the north-west part of Spitsbergen, Svalbard (path 218, row 3). This location was selected because of the range of glaciological features present: numerous valley glaciers, the ice field of Holtedahlfonna, fast ice, and ice floes. Partially cloud-covered imagery was preferred, to allow comparison of the two sensors in terms of their ability to distinguish between clouds and snow. The time of year is also advantageous, in that Sun elevation (27°) is high enough for detector saturation to occur in MSS band 2 (Dowdeswell and McIntyre 1986). Surface-elevation data from airborne radio echo-sounding, and other ancilliary glaciological information, are also available for this part of Svalbard.

Differences in the dynamic range and the wavelengths over which TM and MSS data are collected have two main implications for glaciological studies. First, snow and snow-covered ice masses can be distinguished easily from cloud cover in TM band 5 (1.57 to 1.78 μm). Snow appears dark whereas clouds are light at this wavelength. For example, thin clouds over part of Oscar II Land in Spitsbergen became apparent. In many MSS scenes of the Antarctic, the cloud-free ice-sheet surface has been misidentified as cloud-covered during quality-control analysis. Secondly, the wider dynamic range of the TM sensors means that saturation occurs less frequently over snow than was the case with MSS imagery. Digital analysis of MSS and TM scene radiance over Spitsbergen demonstrates this fact and implies that ice-surface topographic information will only rarely be degraded in TM imagery, although TM band 1 (0.45 to 0.52 μm) is most often saturated.

The nominal spatial resolution of TM sensors is 30 m, except for the thermal infra-red band. This is a significant improvement over the 79 m by 56 m resolution of the MSS. A major advantage of this is that ice margins and ice-surface features can be more precisely identified. More accurate glacier maps can be made, and smaller variations in termini positions of outlet glaciers can be monitored. Ice-surface features, such as crevasses, are more likely to be recorded on TM imagery, and examples are shown from Spitsbergen glaciers. The identification of such features is of major importance in studies of ice-surface velocities from Landsat imagery. For sea-ice applications, the ability to identify smaller floes is also important; for example, in the analysis of floe-size distributions.

The only significant drawbacks to the use of Landsat TM data in glaciological studies are the expense, particularly in the more useful digital format, and the small amount of coverage yet available for the polar regions.

---

*   Present address: Department of Geography, University College of Wales, Aberystwyth, Penglais, Aberystwyth, Dyfed SY23 3DC, U.K.

REFERENCE

Dowdeswell J A, McIntyre N F 1986 The saturation of Landsat MSS detectors over large ice masses. *International Journal of Remote Sensing* 7: 151—164

RADAR AND SEISMIC ICE-THICKNESS MEASUREMENTS COMPARED ON SUB-POLAR GLACIERS IN SVALBARD

(Abstract)

by

D.J. Drewry

(Scott Polar Research Institute, University of Cambridge, Cambridge CB2 1ER, U.K.)

ABSTRACT

A comparison has been made of 46 radar-determined ice thicknesses and those resulting from seismic sounding on Bakaninbreen, Skobreen, and Paulabreen in central Spitsbergen. Significant differences were recorded between the two techniques, with 50% of the comparisons exceeding 15 m. Systematic differences between the three glaciers were also observed: on Paulabreen the seismic ice depths are consistently deeper than those determined by radio echo-sounding, whilst the opposite is true on Skobreen.

Instrumental errors from the radar (SPRI 60 MHz unit) and seismic equipment (ABEM Terraloc) are considered small or insignificant. Factors affecting the respective propagation velocities may be responsible for differences in mean thickness particularly in the case of seismic waves, although the changes are obtained from the first returns. One hypothesis to explain the differences on Paulabreen, and to a lesser degree on Bakaninbreen, is that these glaciers are underlain by a seismic low-velocity layer due to the presence of moraine or till. Unfortunately, equipment problems in the field prevented the digital logging of the seismic data and the analogue records are not of sufficient quality for detailed analyses to reveal the possible presence of a till horizon and its seismic velocities. However, observations at the snout of Paulabreen show considerable thicknesses of basal till. With a "P"-wave velocity in such a layer close to or less than that of ice acoustic returns would possibly come from the till—bedrock interface, whereas radar returns would be from the region of the ice—till boundary.

For the seismic ice depths that are shallower than the radar soundings on Skobreen an alternative explanation is required. The valley occupied by the glacier is considerably narrower than in the case of the other two glaciers. According to one detailed radio echo-sounding cross-profile, the line of the combined seismic and radar sounding was displaced to one side of the centre and deepest part of the glacier. This would result in early seismic returns from the nearest facets of the valley side rather than the subjacent bed. The radio waves, however, undergo a focussing effect in the ice, giving rise to a considerably smaller footprint. The difference in slant length between the general area of the bed viewed by the radar and that returning seismic energy is approximately +15—20 m at the location of the cross-profile. This value is of the order of the differences between the two systems and could therefore account for the observed disparity here and at the other locations.

ACOUSTIC EMISSION METHODS APPLIED TO AVALANCHE-FORMATION STUDIES

(Abstract)

by

V.P. Epifanov

(U.S.S.R. Academy of Sciences, Institute of Problems of Mechanics, Moscow, U.S.S.R.)

and

V.P. Kuz'menko

(U.S.S.R. Goskomgidromet, Central Asian Regional Research Institute, Tashkent, U.S.S.R.).)

ABSTRACT

The relationship between the intensity of snow acoustic emission impulses and snow-cover stability is revealed by measuring the physical and mechanical properties of the snow cover in the starting zones of avalanches. This relationship is fundamental to the remote identification of an avalanche-hazard period.

In order to estimate the mechanical properties of a snow layer, a method of applying a rigid penetrometer equipped with a piezoelectric accelerometer is used. The viscosity coefficients of snow under destruction and the specific energy of destruction are determined. The overall effect of the different elements of destruction is assessed using both structural investigations and acoustic methods (acoustic emissions).

THE SEPARATION OF SEA-ICE TYPES IN RADAR IMAGERY

(Abstract)

Benjamin Holt and F.D. Carsey

(Jet Propulsion Laboratory, California Institute of Technology, Pasadena, CA 91109, U.S.A.)

ABSTRACT

The ability to distinquish the several major types of sea ice with active radar instruments has been well studied in recent years. The separation of sea-ice types by radar results principally from variations in radar back-scatter due to characteristic differences of these ice types in surface morphology and brine content. When sea ice is viewed with an active radar at angles greater than about $20°$ from nadir, undeformed ice reflects radar waves and results in a low return, while ridges, hummocks, and small-scale surface features scatter the radar waves and produce a high return. The presence of salt increases the dielectric constant of ice; penetration by radar into the ice is then negligible, and the return is essentially determined by surface morphology. The absence of salt reduces the dielectric properties of ice; radar waves can then penetrate the ice to some depth and are scattered by air bubbles and brine-drainage channels (called volume scattering), thereby enhancing the return even for roughened surfaces. All these properties vary significantly with radar frequency and polarization as well as seasonally. For example, higher radar frequencies respond to smaller-scale surface features, while lower radar frequencies penetrate further into the ice with resulting volume scattering.

The high-resolution imagery from synthetic aperture radars (SAR), mounted on aircraft, shuttle, or satellite platforms, is very effective for many sea-ice studies, including the separation of ice types. An aircraft-mounted X-band (9 GHz) SAR, for example, can discriminate smooth first-year ice, rough first-year ice, multi-year ice, and open water by the intensity (tone) of the radar returns and floe geometry. The preferred SARs to date for satellites and shuttle platforms have been L-band (1–2 GHz) systems. SAR imagery of sea ice was extensively acquired by Seasat in 1978 over the Beaufort Sea, with limited quantities obtained by the Shuttle Imaging Radar (SIR-B) over the Weddell Sea in 1984. While L-band SAR can discriminate rough and smooth ice along with roughened open water based on image intensity and floe geometry, the returns from thick first-year ice and multi-year ice are not clearly distinguishable. The fact that there is volume scattering from multi-year ice suggests that there may be textural or spatial frequency variations that could be used to separate these two major ice types in radar imagery. In order to investigate the separation of sea-ice types in the large amount of L-band SAR imagery available, image-analysis techniques including filtering and classification programs have been utilized, pointing towards an automatic classification algorithm for use in future SAR sea-ice data sets, especially from space.

An important characteristic of all SAR imagery is the presence of image speckle, a coherent form of noise caused by the random variability of scatterers across even a uniform surface. Most SAR processors reduce this effect by averaging multiple independent samples but this is done at the cost of reducing resolution. Speckle reduction can also be accomplished by filtering. Several filters have been tested including median, box, and adaptive edge filters. Each filter has different characteristics in terms of smoothing speckle and in the response to sharp gradients or edges, such as ridge or lead openings, as well as computational requirements. Optimization of each filter's parameters has been determined by the quality of classification of each ice type.

The classification programs that have been tested are based on tone and texture image characteristics. The programs are supervised; that is, a small training area for each class is pre-selected for statistical analysis. From these statistics, the remainder of the imagery is subjected to the particular classification algorithm. The tone program separates classes based on the mean, standard deviation, and number of standard deviations of each class, and includes a Bayesian maximum-likelihood classifier for ambiguous elements. The texture program determines the statistical homogeneity of each class and the optimal segmentation of each small area into the various classes.

CARTOGRAPHIC REMOTE-SENSING MONITORING OF GLACIOLOGICAL SYSTEMS (EXAMPLE, MOUNT EL'BRUS, U.S.S.R.)

(Abstract)

by

Yu. F. Knizhnikov, V.I. Kravtsova and I.A. Labutina

(U.S.S.R.)

ABSTRACT

Remote-sensing methods in monitoring the glacierization of Mount El'brus are used to produce base and dynamic maps, and to obtain quantitative information (dynamic indices) about the rate, intensity, and variations of the process. The monitoring system is divided, according to scope and territory covered, into small-scale for total glacierization and the periglacial zone, medium-scale for separate glaciers, and large-scale (detailed) for part of the glaciers or sectors of the adjoining slopes. The approximate relationship of even scales is 1 : 4.

Small-scale monitoring remote-sensing systems are important for making maps showing the complex characteristics of the glaciological system. A series of maps was produced including geographical, those of high-altitude zones, slope and exposure angles, geological, glaciomorphological, climatic (temperature, precipitation, and winds), distribution of direct solar radiation, hydrological (source of streams), seats of avalanches, and landslides. All these data serve as a cartographical basis in monitoring the glacierization of Mount El'brus. They are compiled from remotely sensed and Earth-based data.

Current monitoring on a small scale includes observations of the conditions which determine the existence of the glacial system – this includes data on winter snowfall and

the period of snow cover. These observations were obtained from meteorological and resource satellites, and from scanner data of medium and high resolution. Also important are observations of changes in the outline of glaciers, times of snowfall and character of the distribution of snow, and its redistribution due to avalanches and snowstorms. High-resolution space photographs, small-scale aerial photographs, and aerovisual observations provide the data for these observations. It has been determined that the area of the glaciers of Mount El'brus has been reduced by 1% in the last 25 years, i.e. the rate of its deglacierization dropped sharply as compared to preceding decades.

The role of quantitative information gains importance in the medium-scale level of monitoring. Topographical maps of separate glaciers compiled from aerial photographs or data from ground stereo-photogrammetric surveys constitute the base maps at this level. The main methods used in monitoring were large-scale surveys from aircraft, perspective surveys from helicopters, and phototheodolite surveys. Multi-date surveys of the glaciers provide data about the changes in their outlines and height, the character of their relief, their moraines, the amount of snow accumulation and ablation in separate years, the surface rates of ice flow and their fluctuations. The techniques by which quantitative information is obtained about changes in the glaciers are derived from processing the data of multi-date surveys. The organization and techniques of phototheodolite surveys have been improved. A theory evolved for determining the surface-ice movement by stereo-photogrammetric means and the technique for it has

also improved; algorithms and programs for machine processing of the data of multi-date surveys (ground and from aircraft) have been produced.

At this level of monitoring, it has been found that the retreat rate of most glaciers has slowed down and several glaciers are now in equilibrium. Several glaciers became active at the beginning of the 1970s and 1980s; this was accompanied by an increase in their height and forward movement. For example, activation of Kyukyurtlyu Glacier has been recorded (higher surface and increasing flow rate) which has caused the glacier to move forward 100 m. Surveys at an interval of 2 years recorded the beginning of the process of retreat of this glacier.

Detailed monitoring is used to detect the mechanism of the dynamic processes and to study it on local representative sectors. On a glacier it may take the form of annual surveys of its tongue, which makes it possible to observe the processes of formation of moraines and glacio-fluvial relief. Studies may also be made of the mechanism of the movement of avalanches and landslides, deducing their quantitative characteristics and appraising the results of avalanches and landslides. Multi-date surveys of sectors of the slopes provide information about processes in the periglacial zone. At this level, regularly repeated ground stereo-photogrammetric surveys are the main means of observation.

Glaciological remote-sensing monitoring provides a wealth of data for theoretical development in the field of glaciology. It makes it possible to forecast and produce warnings about hazardous processes and phenomena.

METHODS OF CALCULATION AND REMOTE-SENSING MEASUREMENTS FOR THE SPATIAL DISTRIBUTION OF GLACIER ANNUAL MASS BALANCES

(Abstract)

by

V.G. Konovalov

(U.S.S.R. Goskomgidromet, Central Asian Regional Research Institute, Tashkent, U.S.S.R.)

ABSTRACT

The areal distribution of glacier annual mass balance $b(z)$ is an important characteristic of the existence of glacierization and its evolution. At present the measured value of annual mass balance at different elevations is only available for a limited number of mountain glaciers of the globe, because of the great amount of labour required for such measurements.

The analysis of long-term mass-balance measurements made at Abramova Glacier, Limmerngletscher, White Glacier, Hintereisferner, and Peyto Glacier has revealed that for each year the spatial distribution of annual mass balance is well described by quadratic equations. The main variable in these equations is altitude (z). The various parameters of these formulae are estimated by the author for mean weighted height of the ablation and accumulation areas, and for the glaciers as a whole. It is found that the parameters of annual mass balance for each glacier can be calculated from formulae which include combinations of the following variables: annual balance at one of the three weighted altitudes, maximum annual snow-line elevation, annual and seasonal amounts of precipitation, and air temperatures at nearby meteorological stations.

Therefore, in order to calculate the distribution of annual mass balance as a function of absolute altitude, it is sufficient to obtain a value for mass balance measured only at a single point on a glacier, and common meteorological observational data. A comparison of actual and calculated values of mass balance has shown good agreement between them.

Considering the successful use of aerial remote-sensing for the measurement of snow depth in mountains by means of special stakes, it is satisfactory to accept this method for the assessment of annual mass balance at the mean weighted altitude of the ablation zone. It is possible to use aerial photo-surveys or stereophotogrammetry to resolve this problem. Then annual mass balance for the whole area of a glacier is calculated by using data from one point together with data from a nearby meteorological station.

AN EXPERIMENT ON IMAGE TRANSMISSION TO AN ICEBREAKER

(Abstract)

by

O. Korhonen

(Asiakkaankatu 3A, PL 33, 00937 Helsinki, Finland)

ABSTRACT

Finland has such a climate that all its harbours in the Baltic Sea are frozen every winter. Ice may sometimes be more than 1 m thick, as it was in the Bay of Bothnia, the northernmost basin of the Baltic Sea, during the winter of 1985.

The Finnish Institute of Marine Research has used the imagery of the Tiros-N series of satellites successfully in sea-ice mapping for some years. In the Institute daily ice charts have been prepared and sent by facsimile to icebreakers in the Baltic Sea. In 1985, an experiment was conducted to transmit the same imagery to an icebreaker operating in the Bay of Bothnia. Existing telecommunication networks were used. The image data transmitted by NOAA-6 and NOAA-9 satellites were received at Tromsø Telemetry Station in Norway and then transmitted to Espoo in Finland. The data processing consisted of geometric correction, edge enhancement, and drawing the coastline with location symbols. The most interesting area was extracted and transmitted by NMT mobile telephone to the icebreaker.

The almost real-time image transmission turned out to be useful from the point of view of icebreaker operation. The images could be used to identify cracks and narrow leads in the ice. Such very detailed information cannot be included in routine ice charts. The icebreaker can use this information for giving instructions to other ships to find easier routes. This reduces the need for icebreaker assistance. The images can also help the icebreaker to avoid wide heavily ridged areas.

This experiment was at low cost and can be technically developed further. It showed that this kind of assistance for icebreakers is economically profitable for winter navigation.

SPECTRAL RESPONSE PATTERNS OF SNOW AND ICE SURFACES FOR THE LANDSAT MULTISPECTRAL SCANNER

(Abstract)

by

M. Kristensen

(Norwegian Meteorological Institute, 0314 Oslo 3, Norway)

and

N.F. McIntyre

(Mullard Space Science Laboratory, University College London, Dorking, Surrey RH5 6NT, U.K.)

ABSTRACT

The high-resolution imagery recorded by systems such as the multi-spectral scanners (MSSs) of the Landsat satellites has revolutionized the study of all types of surface in the polar regions. Visible and near-infra-red imagery has found a wide range of glaciological uses. There is, however, a lack of comparability within and between MSS data which may be a contributary factor to some current problems in interpretation of remotely sensed glaciological data.

With the expected continuity of MSS coverage for the forseeable future, it is highly desirable to extend use of the data beyond the basic mapping and feature identification which has made it such a valuable resource. One of the most obvious developments is to investigate characteristics of the reflecting surfaces and to achieve absolute identification of snow and ice surfaces. Although conversion of digital MSS grey tones to radiances enables direct comparison with other sources, automatic identification requires detailed and extensive knowledge of the spectral and reflecting characteristics of surfaces which are to be monitored. This is often best achieved through ground-based observation.

In order to provide a base line against which corrected radiances from Landsat MSS data can be compared, a spectrally gated photometer has been used to measure albedo at MSS wave bands in a wide range of conditions. The surfaces monitored in several parts of Norway include sea ice, lake ice, snow, firn and glacier ice, permafrost, and reference surfaces. A range of supporting measurements (including grain-size, surface irregularity, density, level, and free-water content) allows accurate characterization of each surface. This enables identification of spectral-response patterns for each surface category and hence the classification of their reflectances as recorded by the MSS. Examples are given of the application of such classifications to imagery of the polar regions.

METHODS FOR SNOW MAPPING FROM METEOROLOGICAL SATELLITE IMAGERY OF THE QILIAN MOUNTAINS, NORTH-WEST CHINA

(Abstract)

by

Z.K. Liu

(Digital Image Processing Laboratory, Department of Radio Electronics, University of Science and Technology of China, Hefei, China)

and

S.Y. Chang

(Lanzhou Institute of Glaciology and Geocryology, Academia Sinica, Lanzhou, China)

ABSTRACT

A snow survey in north-west China has been carried out using meteorological satellite NOAA-9 imagery. A two-step geometric correction method is presented. The first step involves a large-scale geometric correction, and the second step concentrates on a small area of interest, using the control-point correction method. A high precision of geometric correction can be achieved by using this method. The snow-covered area of the Qilian Mountains basin is calculated.

MULTI-SPECTRAL LANDSAT IMAGE MAPS OF ANTARCTICA

(Abstract)

by

B.K. Lucchitta, K. Edwards, E.M. Eliason and J. Bowell

(U.S. Geological Survey, Flagstaff, AZ 86001, U.S.A.)

ABSTRACT

The US Geological Survey is conducting a program to prepare digitally enhanced Multispectral Scanner (MSS) Landsat images of Antarctica. The goal is to furnish accurate planimetric, false-colour composite image maps in Lambert conformal conic projection for the following purposes: (1) to locate and delineate blue-ice areas for the collection of meteorites; (2) to produce special purpose maps showing selected features; (3) to provide synoptic views that aid in the detection and interpretation of glaciological features associated with the inland ice sheet, outlet glaciers, ice streams, and ice shelves; (4) to monitor changes in coastline and glacial features; (5) to enable the superposition and correlation of different types of digital cartographic data; and (6) to furnish spectral and (or) structural information in areas of limited bedrock outcrop to aid in regional geologic interpretation. Only the first four of these objectives are addressed here.

About 170 Landsat computer-compatible tapes covering Victoria Land, the coastline of West Antarctica, the Antarctic Peninsula, and other selected areas were assembled into false-color, multi-spectral, digital composites of band 4 (0.5 to 0.6 μm, green), band 5 (0.6 to 0.7 μm, red) and band 7 (0.8 to 1.1 μm, near infra-red). The tapes were subjected to routine image-processing procedures, such as noise removal and radiometric and geometric corrections. Further processing included haze removal and enhancement by linear stretching of individual MSS bands based on inspection of gray-value (digital-number) histograms. Saturation of snow-covered scenes in bands 4, 5, and 6 is a severe problem in Landsat MSS images of Antarctica and makes many images unsuitable for multi-spectral work. We have developed special techniques to restore the saturated snow-and-ice information in these bands to overcome this problem.

The Landsat image maps have different formats, depending on their planned applications. An example of a planimetric image map is the one of the McMurdo Sound area; it is based on excellent ground control and processed at full spatial resolution. It comprises five complete and three partial 1 : 250 000 scale topographic quadrangles. One of these is the Convoy Range quadrangle which includes the Allan Hills meteorite-collection site. Blue-ice areas show exceptionally well on this quadrangle, and new information on blue-ice locations and delineations was obtained from it.

Thematic maps prepared for the Byrd Glacier area selectively show only rock or ice areas, thus depicting the location of desired features. Synoptic-view maps and mosaics provide information on flow lines associated with ice streams, the location of ice rises, ice rumples, and other possible grounded areas, and the location and extent of buried mountain ranges. Image maps which cover the same area at different times do show changes: a 10-year interval between Landsat images obtained in the Byrd Glacier area shows that crevasses had opened and rifts had drifted within the floating part of the glacier; measurements of the changed positions have yielded average velocities for glacier and ice-shelf movements in that area. The ease with which the dynamics of the coastline can be monitored on Landsat images is particularly useful, because such changes have implications for variations in world climate.

REMOTE SENSING OF FAST ICE IN LÜTZOWHOLMBUKTA, EAST ANTARCTICA, USING SATELLITE NOAA-7, 8 AND AIRCRAFT

(Abstract)

by

S. Mae

(Department of Applied Physics, Faculty of Engineering, Hokkaido University, Sapporo 060, Japan)

and

T. Yamanouchi and Y. Fujii

(National Institute of Polar Research, Tokyo 173, Japan)

ABSTRACT

Lützowholmbukta, East Antarctica, is covered by fast ice except during a short period in April and May, but occasionally the ice cover breaks up and floats out of the bay.

The fast ice was observed every day using NOAA-7, 8 infra-red imagery. The satellite signal was received at Syowa Station, located on Ongul Island. In addition, aerial photographs and video pictures were taken using aircraft every fortnight.

In 1983, before the break-up of the fast ice in April, a distinctive increase of the infra-red radiance (NOAA AVHRR) was observed, even though the aerial observation showed no change. The increase of the radiance was estimated to be $5°C$. It was in this area, where the higher infra-red signal was observed, where the fast ice broke up. In December, in the central part of the northern area of the fast ice, the hummock-ice zone formed in a triangular shape. Before the hummock-ice zone floated out, observation of the infra-red radiance showed that the temperature of the ice had decreased by $3°C$.

ESTIMATION OF SNOW-MELT RUN-OFF DURING PRE-MONSOON MONTHS IN BEAS SUB-BASIN USING SATELLITE IMAGERY

(Abstract)

by

K.P. Sharma and P.K. Garg

(Civil Engineering Department, University of Roorkee, Roorkee, India)

ABSTRACT

The increasing demand for water, coupled with the construction of multi-purpose reservoirs to control and regulate snow-melt run-off, requires accurate stream-flow forecast. For making an accurate prediction of spring run-off, information on the amount of snow accumulation in winter is necessary; this may be achieved through remote-sensing techniques in any inaccessible region.

This paper outlines the snow-melt run-off study carried out in a part of Beas basin, India, using Landsat imagery for the years 1973, 1975, 1976, and 1977. The Beas basin lies between long. $76°56'$ to $77°52'$E. and lat. $31°30'$ to $32°25'$N., covering an area about 4900 km^2, of which 1400 km^2 is permanently covered by snow. The gradual melting of snow accumulated over the catchment area during the winter months is responsible for the perennial character of the Beas River.

Photohydrological investigation of the part of the Beas basin up-stream of Barji was carried out and a study was made for the estimation of the snow-melt run-off during the pre-monsoon period in the sub-basin up-stream of Manali. For this purpose, the sub-basin has been divided into permanent and temporary snow-covered zones. The degree-day method and the melt due to rainfall on snow have been used to estimate snow-melt run-off. The routing of snow-melt, after accounting for losses as well as the run-off from the excess rainfall from the permanent and temporary snow-covered areas, has also been done taking the recession coefficient K as 0.90, and the excess rain from the non-snow-covered areas has been assumed to contribute directly to the run-off for that day. Run-off coefficients of 0.595 for rainfall on the snow-covered areas and 0.278 for rainfall on the non-snow-covered areas have been determined.

Reference can be made to similar work in India and Pakistan to establish the relationship between the snow cover and the cumulative discharges for the months of March, April, and May of the years 1973, 1975, 1976, and 1977, and an exponential trend was observed with the help of Landsat imagery. Furthermore, the snow-covered areas as determined from bands 5 and 7 of the Landsat imagery, for the same day, showed a linear trend.

The analysis of the results shows that remote-sensing data used in conjunction with conventional methods are likely to improve the accuracy of the snow-melt forecasts in remote areas like the Himalayan catchments.

APPLICATION OF STATIONARY GEODIMETER TO STUDY THE VELOCITY OF THE KOZEL'SKIY GLACIER MOVEMENT, KAMCHATKA, U.S.S.R.

(Abstract)

by

V.N. Vinogradov

(Institute of Volcanology, Petropavlovsk-Kamchatskiy, 683006, U.S.S.R.)

ABSTRACT

Kozel'skiy glacier is located in lat. 53.1°N., 50 km from the Pacific Ocean, and has a southern exposure. It flows from the saddle between Avachinskiy and Kozel'skiy volcanoes. Its highest point is at an altitude of 1850 m a.s.l. The glacier tongue descends to an altitude of 900 m a.s.l. At present, this is the best-studied glacier in Kamchatka.

In May and June 1981, for the purpose of organizing routine monitoring of the movement of Kozel'skiy glacier, systematic geodimeter measurements were made at the "Mishennaya" Observatory. A model 8 geodimeter (AGA, Sweden) was used for measuring straight-line distances. The source of radiation was a He–Ne laser beam with a power of 5 MW and a wavelength of 0.6328 m. In Kamchatka this geodimeter can measure distances of 60 km and in some cases 90 km. If a special (diurnal) procedure of measurements is used, one can attain an accuracy of 1×10^{-6} of the length of the line D. Sixteen prism-angle reflectors were installed on the glacier.

The entire firn area of the glacier could be seen well from the top of Mount Mishennaya. Two profiles consisting of six points on the glacier surface were constructed. A bench mark on the ancient lava flow of Kozel'skiy volcano was taken as static in elevation and was used to control the accuracy of the measurements. The points on the glacier were established by 2 m poles with a diameter of 1.5 cm. Reflectors were installed on tripods and were centred with the help of an optical plumb bob. The points for measurement of the movement velocity were located at a distance of more than 26 km from the instrument. Studies were carried out during periods of stationary images (without inversions), generally early in the morning and late in the evening.

Two-sided trigonometrical levelling was carried out simultaneously. Zenith distances were measured four times by a Teo-010A theodolite. Meteorological conditions were determined at 1.5–2 h near the reflectors and at the observatory during the whole cycle of measurements. The processing of the results of distance measurements was done according to the procedure published in *Lobachev*. The elevation of points over the observatory was determined from the measured zenith distances.

The error in distance measurements was ± 0.04 m and in elevation ± 0.8 m (an increase in accuracy up to 0.2 m is possible). The displacement of points of velocity changed from 0.3 to 2.37 m over the whole period of observation. This corresponds to a glacier velocity from 1 to 6 m/d. Thus, a cycle of geodimeter measurements may be carried out once a week, providing an accuracy of distance measurements of ± 4 cm. Routine measurements of the velocities of glacier movement, using geodimeters, may be organized subject to reliable long-term attachment of reflectors to the surface of the glacier.

Experience of the studies made indicated good opportunities for application of a stationary geodimeter to the investigation of glacier movement. The data obtained agree with previous measurements made by the usual geodetic methods.

INTER-RELATIONS BETWEEN THE ARCTIC SEA ICE AND THE GENERAL CIRCULATION OF THE ATMOSPHERE

(Abstract)

by

G. Wendler, M. Jeffries and Y. Nagashima

(Geophysical Institute, University of Alaska-Fairbanks, Fairbanks, AK 99775-0800, U.S.A.)

ABSTRACT

Satellite imagery has substantially improved the quality of sea-ice observation over the last decades. Therefore, for a 25-year period, a statistical study based on the monthly Arctic sea-ice data and the monthly mean 700 mbar maps of the Northern Hemisphere was carried out to establish the relationships between sea-ice conditions and the general circulation of the atmosphere. It was found that sea-ice conditions have two opposing effects on the zonal circulation intensity, depending on the season. Heavier than normal ice in winter causes stronger than normal zonal circulation in the subsequent months, whereas heavier than normal ice in the summer–fall causes weaker zonal circulation in the subsequent months. Analyzing the two sectors, the Atlantic and Pacific ones separately, a negative correlation was found, which means a heavy ice year in the Atlantic Ocean is normally associated with a light one in the Pacific Ocean and vice versa.

COMPARISON OF SURFACE CHARACTERISTICS OF THE ANTARCTIC ICE SHEET WITH SATELLITE OBSERVATIONS

(Abstract)

by

N. Young and I. Goodwin

(Antarctic Division, Department of Science, Kingston, Tasmania 7150, Australia)

ABSTRACT

Ground surveys of the ice sheet in Wilkes Land, Antarctica, have been made on oversnow traverses operating out of Casey. Data collected include surface elevation, accumulation rate, snow temperature, and physical characteristics of the snow cover. By the nature of the surveys, the data are mostly restricted to line profiles. In some regions, aerial surveys of surface topography have been made over a grid network.

Satellite imagery and remote sensing are two means of extrapolating the results from measurements along lines to an areal presentation. They are also the only source of data over large areas of the continent. Landsat images in the visible and near infra-red wavelengths clearly depict many of the large- and small-scale features of the surface. The intensity of the reflected radiation varies with the aspect and magnitude of the surface slope to reveal the surface topography. The multi-channel nature of the Landsat data is exploited to distinguish between different surface types through their different spectral signatures, e.g. bare ice, glaze, snow, etc. Additional information on surface type can be gained at a coarser scale from other satellite-borne sensors such as ESMR, SMMR, etc. Textural enhancement of the Landsat images reveals the surface micro-relief.

Features in the enhanced images are compared to ground-truth data from the traverse surveys to produce a classification of surface types across the images and to determine the magnitude of the surface topography and micro-relief observed. The images can then be used to monitor changes over time.

Annals of Glaciology 9   1987
© International Glaciological Society

SUMMARY REMARKS

by

Richard S. Williams Jr
(US Geological Survey, 927 National Center, Reston, VA 22092,
U.S.A.)

The First Symposium on Remote Sensing in Glaciology was held in Cambridge, England, on 16–20 September 1974, and the proceedings were published in Volume 15 (No. 73) of the Journal of Glaciology in 1975.   Most of the 31 papers and 20 abstracts described glaciological research using radio echo-sounding systems, traditional geophysical techniques, and side-looking airborne radar (SLAR). A few of the papers and abstracts were devoted to analysis of satellite image data for surveys of snow cover, sea ice, and glaciers. The first Landsat spacecraft (ERTS-1) had been launched on 23 July 1972, so early results of research with these data were also included.

The just-concluded Second Symposium on Remote Sensing in Glaciology, was held in Cambridge on 8–9 and 11–12 September 1986. On 10 September 1986, 10 invited lectures on the development of glaciological studies in the past 50 years were given during the special Fiftieth Anniversary Celebrations of the International Glaciological Society. Presentations at the Second Symposium on Remote Sensing in Glaciology covered topics similar to those represented 12 years earlier at the first symposium but had a greater emphasis on analysis of satellite data. Eight authors who presented a paper or abstract at the 1974 symposium were also involved with the 1986 symposium. The proceedings of the Second Symposium on Remote Sensing in Glaciology, to be published in Vol 9 of the Annals of Glaciology in 1987 (this volume) will include the 45 plenary-session and 10 poster-session papers and abstracts and the 14 abstracts accepted by the papers committee but not presented.

Progress has been made since the first symposium. Research described at the second symposium relies more on computers to process digital image data and other types of data, such as data from radio echo-sounding and seismic surveys. Solutions have been found to some of the complex equations used to model glaciological phenomena. Data in digital format are preferred to facilitate analysis, and digital processing of field data is now common. There is increased miniaturization of instruments, so that they can be used in the field in addition to the laboratory. There has also been increased research on the physical and spectral properties of snow, firn, and ice and the correlation of remotely sensed data with such properties. Cost and availability of satellite data remain serious impediments to their more widespread use on a regional or global basis by glaciologists, however.

A selection of 10 topics of papers and abstracts given in the 9 plenary sessions and 1 poster session of the second symposium provides a good overview of the type of remote-sensing research underway in glaciology:

1. Increased use of computers, especially the trend towards all-digital data and miniaturization of instrumentation, permitting the use of complex field instruments, such as "a back-portable microprocessor-based impulse radar system".

2. Mapping of glacier facies on Landsat multispectral-scanner (MSS) and thematic-mapper (TM) images of ice caps and ice sheets.

3. Preparation of Landsat MSS and TM image and image-mosaic maps of ice sheets and ice caps from enhanced digital image data.

4. Use of Landsat MSS images in conjunction with field studies of glacier hydrology in West Greenland.

5. Measurements of the surface topography of ice shelves by satellite radar altimetry.

6. Radio echo-sounding and seismic sounding of the till layer under "Ice Stream B".

7. Determination of physical properties of snow and ice from Landsat digital data.

8. Use of sequential satellite images and computer graphics technology to determine the dynamics of arctic sea ice.

9. Mapping glacier facies on synthetic-aperture-radar (SAR) data and mapping the surface morphology of the ice sheet in West Greenland from satellite radar images.

10. Use of sequential Nimbus 5 electrically scanning microwave radiometer (ESMR) images of polar sea ice to determine seasonal and inter-annual change in concentration and areal extent.

Although given in the special fiftieth anniversary symposium on 10 September, as one of 10 lectures on the development of glaciological research during the past half century, the outstanding lecture by H. Jay Zwally on "Technology in the Advancement of Glaciology" deserves special citation because of his superb review of the development of glaciological remote sensing during this period. The 10 lectures will be published in a special issue of the Journal of Glaciology in 1987.

Approximately 14 years from now, the Third Symposium of Remote Sensing in Glaciology will be held in Cambridge, on 11–15 September 2000, and the proceedings will be published in Vol 30 of the Annals of Glaciology. The expected themes of the third symposium will be global in their orientation, including results of research carried out under various large-scale international programs, such as Earth System Science, Global Geoscience, Global Change, and the International Geosphere–Biosphere Program (IGBP), and may even include results of astroglaciological research, such as satellite glaciological studies of the Martian polar ice caps from data acquired by the Mars Observer spacecraft.

The types of remotely sensed data and instrumentation expected to be available to glaciologists for research reported on at the third symposium include: higher resolution satellite images with at least 10 m instantaneous field-of-view (IFOV); inclusion of additional parts and (or) narrow bands of the electromagnetic spectrum; global stereo-scopic image coverage; more field verification of remotely sensed data; global coverage with satellite SAR systems, radar altimetry, and scatterometry; global coverage with laser altimetry; data from full operation of the 18 satellites in the Global Positioning System (GPS), which will permit high positional accuracy in three dimensions of data acquired from ground, aircraft, and satellite platforms; routine use of optical disk technology for local storage and retrieval of vast quantities of image data; ready availability and accessibility of powerful microcomputers comparable in storage capacity and computational speed with present-day main-frame computers; and proliferation of national satellite programs.

In closing, I thank all the speakers and the session chairman for a most stimulating and productive symposium. I also want to give special acknowledgement to my colleagues on the papers committee, who selected the papers and abstracts presented and served as scientific editors for the scientific reviews of each paper for Vol 9, and to my good friend the House Editor for Vol 9 of the Annals of Glaciology, Eric L. Richardson. His superb business and editorial skills have markedly lessened my burden as Chief Scientific Editor for Vol 9. Special thanks go to the headquarters staff of the International Glaciological Society, to the many local volunteers who helped out in so many ways, and to the warm hospitality, which more than offset the frosty environment of the auditorium, offered by them and by the staff of the Scott Polar Research Institute, Cambridge University Chemical Laboratory, King's College, and the citizenry of Cambridge. It was a most memorable week, and I thank you all very, very much. I look forward to seeing you all at our next symposium here in Cambridge in September 2000.

INDEX OF AUTHORS